# Physics of Space Storms

## From the Solar Surface to the Earth

Hannu E. J. Koskinen

# Physics of Space Storms

## From the Solar Surface to the Earth

Springer

Published in association with
Praxis Publishing
Chichester, UK

PRAXIS

Professor Hannu E. J. Koskinen
University of Helsinki and Finnish Meteorological Institute
Helsinki
Finland

SPRINGER–PRAXIS BOOKS IN ENVIRONMENTAL SCIENCES
SUBJECT *ADVISORY EDITOR*: John Mason, M.B.E., B.Sc., M.Sc., Ph.D.

ISBN 978-3-642-00310-3 e-ISBN 978-3-642-00319-6
DOI 10.1007/978-3-642-00319-6
Springer Heidelberg Dordrecht London New York

Library of Congress Control Number: 2010934386



Cover design: Jim Wilkie
Project copy editor: Mike Shardlow
Author-generated LaTex, processed by EDV-Beratung Herweg, Germany

# Contents

# Preface

*Space weather* can be defined as a subtopic of solar–terrestrial physics, which deals with the spatially and temporally variable conditions in the Sun, solar wind, magnetosphere, and ionosphere that may disturb or damage technological systems in space and on the ground and endanger human health. *Space storms* are the strongest and most harmful appearances of space weather.

During the 1990s space weather grew to a prominent, if not the dominant, sector within solar–terrestrial physics. Also a significant fraction of basic space plasma physics research became motivated by its potential to contribute to useful space weather applications including more accurate forecasts. A key reason for the evolution of space weather activities is the growing understanding that a great number of systems in space, human beings included, and on the ground are vulnerable to severe space weather conditions. In fact, due to miniaturization and increasing complexity many technological systems are becoming more sensitive to the radiation environment than before. At the same time modern society is getting increasingly dependent on space infrastructure. In future the human presence in space, including space tourism, is expected to become more prominent. Some day we most likely will return to the Moon and, perhaps, initiate manned missions to Mars. On the ground the effects of space storms, such as saturation of transformers in electric power transmission networks or perturbations in telecommunication and global positioning systems, may be easier to handle, but this requires that the underlying physics be understood much better than today.

The developers of space weather services have done their best to follow the needs, sometimes real, sometimes imagined, of potential users of space weather applications. There is growing activity to produce tools for modeling and forecasting space weather conditions based on a limited set of observations, for specification of environmental conditions during storms, and for after-the-fact analysis of anomalous behavior of technological systems and hazards caused by severe space weather. Unfortunately, this activity is often based on insufficient knowledge of the underlying physical systems, sometimes even at the cost of basic research aiming at increasing this knowledge. This development is not always healthy in the long-term perspective. Furthermore, it is not enough just to solve the acute problems: the knowledge being gained today also needs to be maintained tomorrow.

While a large number of research articles and review papers on space storms have been published over the last several years, there is no comprehensive systematic textbook approach to the relevant physics of the entire chain of phenomena from the surface of the Sun to the Earth. The goal of the present monograph is to fill this gap. The text is aimed at doctoral students and post-doctoral researchers in space physics who are familiar with elementary plasma physics and possess a good command of classical physics. The topics reach from the storms in the solar atmosphere through the solar wind, magnetosphere, and ionosphere to the production of the storm-related geoelectric field on the ground. In the selection of material, preference has as much as possible been given to analytical and quantitative presentation over handwaving, while keeping the volume of the book reasonable.

Of course, several good plasma physics textbooks are available, which are useful in the education of space physicists, e.g., the rewritten classic of Boyd and Sanderson [2003], the little more challenging Sturrock [1994], or the recent volumes written by Gurnett and Bhattacharjee [2004] and Bellan [2006]. However, these books are written for very wide audiences from laboratory and fusion communities to space plasma physicists. Consequently, many important issues in the physics of tenuous space plasmas have had to be dealt with in a brief and cursory manner. For astrophysicists interested in the most abundant form of conventional matter in the universe the book by Kulsrud [2005] is strongly recommended, although quite demanding reading. There are also several textbooks with a clear focus on fundamental space plasma physics [e.g., Baumjohann and Treumann, 1996; Treumann and Baumjohann, 1996; Parks, 2003], but their approach too is more general than the thematically focused topic of the present volume. The multi-authored textbook edited by Kivelson and Russell [1995] covers large parts of the physical environment of this book. However, it does not go very deeply into the plasma physics and suffers to some extent from the different styles of the individually written chapters.

The rapid growth of space weather activities has led to a large number of compilation works of highly variable quality. An inherent problem of multi-authored collections is that each article is relatively short but at the same time written in a complete article style from introduction to conclusions and often with individual reference lists. Thus the books easily become thick but none of the articles can penetrate the basic physical principles. Some of the most useful collections in the present context are those edited by Crooker et al [1997], Tsurutani et al [1997], Daglis [2001], Song et al [2001], Scherer et al [2005], Baker et al [2007], Bothmer and Daglis [2007], and Lilensten et al [2008]. These books contain many excellent articles and provide students with a large body of study material with up-to-date observational data. However, these volumes rather complement than compete with this self-contained monograph.

This book can be interpreted to consist of three parts. The long Chapter 1 forms the first part. It contains a phenomenological introduction to the scene, from the Sun to the Earth, where space weather plays are performed. A reader familiar with basic physics of the Sun, solar wind, magnetosphere and ionosphere can jump over this chapter and only return to it when there is a need to check definitions or concepts introduced there.

The second part of the book consists of several chapters on fundamental space plasma physics. While this part is written in a self-consistent way, it is aimed at readers who already have been exposed to basic plasma physics. Chapter 2 briefly introduces the fundamental concepts and tools of plasma physics inherited from both electrodynamics and statistical physics. Chapter 3 reviews the classical guiding center approach to single particle motion and adiabatic invariants, including motion in the dipole field, near a current sheet, and in a time-dependent electric field.

Common problems to all plasma physics texts are in what order the microscopic and macroscopic pictures should be introduced and at what stage the waves and instabilities be discussed. The strategy in the present volume is to start with the wave concepts in the cold plasma approximation in Chapter 4. The chapter includes a discussion of radio wave propagation in the ionosphere as an example of dealing with wave propagation in inhomogeneous media in the WKB approximation, which is a powerful theoretical tool in problems where the wavelength is short as compared to the gradient scale lengths of the background parameters. Chapter 5 is a standard discussion of the Vlasov theory starting from Landau's solution and extending to the wave modes in uniformly magnetized plasma. Only after these is magnetohydrodynamics (MHD) treated in Chapter 6. Here more emphasis is placed on the field-aligned currents (i.e., force-free fields) than in many other plasma physics texts because they are of such great importance in the solar atmosphere, solar wind, and magnetosphere and in magnetosphere–ionosphere coupling. The chapter is concluded with a brief peek beyond the MHD approximation, including a quasi-neutral hybrid approach and the introduction of kinetic Alfvén waves.

Space plasma instabilities are the topic of Chapter 7. In whatever way you approach this complex, you end up being incomplete if you wish to keep the discussion within reasonable limits and focused. Here the approach is to introduce the basic ideas, such as the free-energy sources and stability criteria, behind several of the most important instabilities studied in the context of space storms, but most of the long and tedious derivations of the equations have been omitted. The reader interested in the details is recommended to consult more advanced textbooks in plasma theory and relevant research articles. Another choice motivated by the theme of this book is to discuss the magnetic reconnection and the tearing modes separately from other instabilities in a dedicated Chapter 8. Whatever the microphysical mechanisms associated with reconnection are, the understanding of its basic characteristics is an essential part of literacy in space physics, regardless of whether one is interested in solar flares, coronal mass ejections, solar wind interaction with the magnetosphere, or the substorms therein. Unlike other textbooks, the concept of dynamo is introduced in this chapter because the annihilation and generation of magnetic flux can be seen as two faces of related physical processes.

The primary goal of this book is to bridge the gap between the fundamental plasma physics and modern research on space storms. This is the challenge of the third part of the book. As in modern concertos, transition from the second to the third movement is not necessarily well-defined. In some sense Chapter 8 already opens the third part as here the treatise begins to focus more on the key issues in space storm research. Chapter 9, in turn, discusses the mechanisms giving rise to radiation that we see coming from the solar atmosphere at the time of solar storms as well as the scattering of radio waves from electrons and plasma fluctuations in the ionosphere. In Chapter 10 the adiabatic invariants introduced in Chapter 3 are used in formulating the kinetic equations for studies of plasma transport and acceleration in the inner magnetosphere.

Fluid turbulence remains one of the toughest problems in classical physics and turbulence in collisionless magnetized plasmas is an even harder problem. Particularly interesting environments, where turbulence is critical, are the interplanetary and planetary shocks with the associated sheath regions. Shocks and shock acceleration are discussed in Chapter 11.

Finally the treatise returns to the more phenomenological treatment of space storms in various parts of the solar–terrestrial system. Chapter 12 deals with the storms on the Sun and their propagation into the solar wind. In Chapter 13 magnetospheric storms and substorms and their drivers are investigated. As storm phenomena in the inner magnetosphere are of particular practical interest, they are discussed separately in Chapter 14. At the end of the journey some effects of space storms on the atmosphere and the current induction on the ground during rapid ionospheric disturbances are briefly discussed in Chapter 15.

The great variety of phenomena from the Sun to the Earth and the vast amount of different theoretical and modeling approaches to explain them make some hard choices necessary, in particular, the choice between a Sun–centered and an Earth–centered approach. The solar atmosphere, in particular the corona, is a much more stormy place than the Earth's environment. The Sun is also the driver of practically all space storm phenomena in the solar–terrestrial system. These facts would suggest adoption of the Sun–centered view on space storms. On the other hand, we live on the Earth and here we have to learn to handle the consequences of space storms. Thus the present choice is Earth-centered but more emphasis is put on the entire space storm sequence than in traditional textbooks on magnetospheric physics. There is a recent very comprehensive textbook on the physics of solar corona by Aschwanden [2004]. Actually just browsing through that volume, containing citations of about 2500 scientific articles, illustrates how difficult it is to compile a concise text on that end of the space storm chain. The first decade of the 21st century also forms a "golden age" of solar physics when several multi-wavelength spacecraft are producing an enormous amount of new empirical information on the active Sun. To digest all this will certainly take some time.

Another choice taken here is not to deal with space weather effects or practical modeling approaches. Concerning these we point the interested reader to the recent volumes by Bothmer and Daglis [2007] and Lilensten et al [2008] and references therein. In fact, the present book and those by Aschwanden [2004] and Bothmer and Daglis [2007] are strongly complementary to each other. They have quite different approaches but are dealing with closely related issues.

As one of the goals of this book is to provide material for advanced students, exercise problems of varying difficulty have been embedded within the text. They are grouped into three categories: Problems labeled *Train your brain* are mostly straightforward, often boring, derivations of expressions that are useful for students learning to master the basic material of the book. The label *Feed your brain* refers to problems or tasks that add to the reader's knowledge beyond the actual text and can also be useful for testing the reader's understanding of the material. Problems identified as *Challenge your brain* are a little harder (at least to the author), dealing also with unsolved or controversial issues. Creative solutions to some of these may be worth publishing in peer-reviewed journals.

A textbook discussing basic physics necessarily borrows material from earlier sources. The author was introduced to plasma physics through the classic texts by Boyd and Sanderson [1969], Krall and Trivelpiece [1973], and Schmidt [1979], which certainly can be recognized in the presentation of the fundamental plasma issues. When discussing "generally known" (or believed to be known) topics, in particular in Chapter 1, references to the scientific literature have been used sparsely. However, a number of references to some of the truly classic reports have been included. New generations of scientists every now and then tend to forget the original works with the risk of independent reinvention of the wheel. For students it is sometimes useful to recall that there was intelligent life even before they were born. In this respect the internet has actually made life much easier. We do no more need to have physical access to the best equipped libraries to read many of the classic reports in the scientific literature. Unfortunately, books like this are harder, or more expensive, to access electronically.

## Acknowledgements

A large part of the material of this book comes from notes for space plasma physics, solar physics, and space weather lectures that I have been giving over the years to both master's and doctoral students, mostly at the University of Helsinki but also at several summer schools and other special occasions. I realized that there was a need for a book along the approach that I have taken, when I was leading a nation-wide space weather consortium in space research programme Antares of the Academy of Finland in 2001–2004. However, it was not until the academic year 2008–2009 that I was able to invest enough time in the project as the result of an appropriation for a senior scientist from the Academy of Finland, which facilitated a full year of sabbatical leave. I spent the autumn 2008 at the Laboratory for Atmospheric and Space Physics of the University of Colorado, Boulder, and the spring 2009 at the International Space Science Institute (ISSI) in Bern, Switzerland. I wish to express my sincere thanks to the directors, Dan Baker and Roger-Maurice Bonnet, and their staffs for the hospitality and support I received. Boulder provided an excellent academic environment for writing the main part of the text, whereas ISSI was the exactly right place for the hard work of editing and organizing the material.

Several people have influenced my thinking of space physics. Of my former mentors I wish to express my gratitude to Rolf Boström and Risto Pellinen, the latter of whom introduced me to the field of magnetospheric physics and supported me in many ways until his retirement. I am heavily indebted to the space physics community of the Finnish Meteorological Institute and the Department of Physics of the University of Helsinki. These two institutes and their close collaboration form a unique space research environment whose role in this exercise cannot be overestimated. In the context of the present book I wish most gratefully to thank Tuija Pulkkinen for excellent collaboration in research on storms and substorms over more than 20 years. Another person deserving special acknowledgment is Rami Vainio whose contributions to our space physics curriculum, in particular on the Sun and space plasma shocks, have been invaluable in writing the corresponding chapters of this book. Of my past and present local collaborators who have, explicity or implicitly, contributed to the book I wish to thank Olaf Amm, Natalia Ganushkina, Heli Hietala, Pekka Janhunen, Riku Järvinen, Esa Kallio, Kirsti Kauristie, Emilia Kilpua (née Huttunen), Tiera Laitinen, Jakke Mäkela, Anssi Mälkki, Heikki Nevanlinna, Minna Palmroth, Risto Pirjola, Antti Pulkkinen, Ilkka Sillanpää, Eija Tanskanen, Petri Toivanen, and Ari Viljanen. For technical help with scanning and editing of several figures I am grateful to Artturi Pulkkinen.

It is practically impossible to acknowledge all the colleagues whose ideas have somehow migrated into the the text. Both over the years and in the context of the present project I have had particularly useful collaboration and discussions with Mats André, Vassilis Angelopoulos, Dan Baker, Stas Barabash, Wolfgang Baumjohann, Joachim Birn, Joe Borovsky, Pontus Brandt, Jörg Büchner, Tom Chang, Eric Donovan, Lars Eliasson, Scot Elkington, Karl-Heinz Glassmeier, Georg Gustafsson, Gerhard Haerendel, Walter Heikkila, Bengt Holback, Gunnar Holmgren, Richard Horne, Mary Hudson, Bengt Hultqvist, Mike Kelley, Paul Kintner, Jim LaBelle, Xinlin Li, Mike Lockwood, Ramon Lopez, Bill Lotko, Tony Lui, Rickard Lundin, Larry Lyons, Bob Lysak, Göran Marklund, Bob McPherron, Tuomo Nygrén, Terry Onsager, Hermann Opgenoorth, Götz Paschmann, Tom Potemra, Geoff Reeves, Gordon Rostoker, Alain Roux, Ingrid Sandahl, Rainer Schwenn, Victor Sergeev, Jim Slavin, Yan Song, Rudi Treumann, and Don Williams.

Finally I wish to acknowledge the most helpful support provided by PRAXIS Publishing Ltd., in particular for reviewing by John Mason, copy-editing by Mike Shardlow, cover design by Jim Wilkie, and LaTeX help from Frank Herweg. I am extremely grateful to the publisher Clive Horwood for his enthusiasm and support during the process. I am particularly indebted to his patience when problems with my schedule after returning from the sabbatical led to a long delay with the delivery of the files for copy-editing.

Helsinki, September 2010 *Hannu E. J. Koskinen*

# Units and Notation

SI units are used throughout the book. As a common exception energy and temperature are often expressed in electronvolts (eV), but in equations involving the temperature the Boltzmann constant $k_B$ is written explicitly, in which case the temperature is given in kelvins (K). Furthermore, physical distance measures, such as the radius of the Sun ($R_\odot$), the radius of the Earth ($R_E$), or the astronomical unit ($AU$), are in frequent use. Also, when dealing with densities of a few particles per $cm^3$, or magnetic fields of a few nT, it is preferable to use these as units in order to avoid unnecessary use of powers of ten.

A person working within theoretical plasma physics or solar physics must also master the Gaussian cgs unit system, as much of the literature in these fields is still written in these units. Transformation from grams to kilograms, from centimeters to meters, or ergs to joules is trivial, but in formulas involving electrodynamic quantities the different unit systems are a nuisance. This sometimes leads to erroneous calculations, not only by factors of 10, but examples of errors by a factor of 3 or $4\pi$ are not too difficult to find in the literature, peer-reviewed articles included.

Macroscopic quantities in the three-dimensional configuration space are denoted by capital letters, e.g., electric current $\mathbf{J}$, fluid velocity $\mathbf{V}$, pressure $P$, etc., vectors in boldface and scalars in italics. The lowercase $\mathbf{v}$ is reserved to denote particle velocity as a function of time and the velocity coordinates in the phase space, e.g., in expressions as $f(\mathbf{r},\mathbf{v},t)$, whereas the lowercase $\mathbf{p}$ denotes the particle momentum $\mathbf{p}(t)$. In order to avoid conflict electric potential is denoted by $\varphi$, whereas $\phi$ is an angular variable. Similarly volume is denoted by $\mathcal{V}$ in order not to mix up it in some expressions with speed $V$. The volume differential in integral expressions is denoted by either $d^3r$ or $d\mathcal{V}$.

In an ideal world a textbook should have a unique system of symbols. However, this is not a practical goal for a book that combines material from several different disciplines of physics, all with their own and by no means common or unique notations. Thus the most usual conventions are followed in the book, accepting that some symbols become heavily overloaded. One of them is $\mu$, that in this book may denote the magnetic permeability of a medium, the magnetic moment of a charged particle, or the cosine of the pitch angle. $J$ can denote the second adiabatic invariant, the absolute value of electric current $|\mathbf{J}|$, and omni-directional particle flux. $\gamma$ in turn appears as the polytropic index, as the Lorentz factor and in some instances as the wave growth rate, $n$ as the particle density, the index of refraction and in vector form the unit normal vector, $\sigma$ as electrical conductivity and the collision cross-section, etc. However, none of these ambiguities should lead to misunderstanding. After all, physicists are expected see the forest for the trees.

# 1. Stormy Tour from the Sun to the Earth

In addition to light and other wavelengths of electromagnetic radiation the Sun affects our environment through complicated plasma physical processes. The study of these interactions is known as solar–terrestrial physics. Already long before the space era there were indications that solar activity and geomagnetic perturbations must somehow be connected. A remarkable event was the large flare on the Sun observed, independently, by Carrington [1859] and Hodgson [1859] on September 1, 1859, after which a major magnetic storm commenced only 17 hours later. Today we understand that the storm was caused by a magnetic cloud associated with a coronal mass ejection (CME) that reached the Earth exceptionally quickly. The storm was very strong, evidently much stronger than any event recorded during the present era of space weather sensitive equipment in space and on the ground.

During the early 20th century the Sun was found to possess a highly variable magnetic field and the violent solar eruptions were found to somehow be related to strong magnetic variations observed on the Earth. But it was not until the dawn of spaceflight that the highly variable but continuously blowing solar wind was shown to be the agent that carries the perturbations from the Sun to the Earth. The variations in the solar wind shake the magnetic environment of the Earth, the magnetosphere. If the perturbations are strong enough, we call them "storms". We borrow terminology from atmospheric sciences and call the short-term variations in the solar–terrestrial system "space weather" and the longer-term behavior "space climate". In this book the term "space storm" is not limited to storms in the magnetosphere but includes stormy weather on the Sun, in the solar wind, and in the Earth's magnetosphere and ionosphere. Space storms at other planets form an interesting and intriguing complex of physics issues, the discussion of which, however, is beyond the scope of the present treatise.

## 1.1 Source of Space Storms: the Sun

Space weather and space climate are controlled by the temporal variability of the Sun in different time scales from minutes to millennia. In fact, when looking at the Sun with the

H.E.J. Koskinen, *Physics of Space Storms: From the Solar Surface to the Earth*,
Springer Praxis Books, DOI 10.1007/978-3-642-00319-6_1,

present observational tools, its surface and atmosphere are seen to be very stormy and noisy environments. In this section we review some of the basic properties of our active Sun. A modern introduction to the Sun itself is Stix [2002] and a wealth of material about the corona and its activity can be found in the comprehensive volume by Aschwanden [2004].

### 1.1.1 The Sun as a star

The physical picture of the Sun started to develop in the dawn of modern physical sciences when Galileo, one of the first developers and users of the telescope, observed sunspots on the solar disk. He showed in 1613 that they are structures on the surface of the Sun and not small planets as Schreiner had argued a few years earlier. After this promising start progress in solar physics remained slow. In 1802 Hyde discovered that solar spectrum contained several absorption lines, which were later cataloged by Fraunhofer. In 1844 Schwabe showed that the sunspot activity varies in an 11-year cycle and in 1859 Carrington and Hodgson observed a solar flare in white light. The second most common element in the universe was identified as late as 1868 in the solar spectrum by Lockyer and was later named helium.

Most of our present understanding of the Sun did not exist before the 20th century. Among the first major advances were Hale's measurements of intense magnetic fields in the sunspots in 1908, showing that whatever generated the solar activity, it was closely related to highly variable magnetism. An important enigma remained, however. In 1862 Sir William Thomson (later Lord Kelvin) had demonstrated that the largest imaginable energy source for solar radiation, the gravitational binding energy of the Sun, would not, at the present solar luminosity, be sufficient for more than 20 million years, which already at that time was considered far too short a history for the solar system. The solution to this problem required the development of quantum mechanics and finding of the nuclear forces. In 1938 Bethe and Critchfield described the dominant proton–proton reaction chain that powers the Sun. In this process 600 million tons of hydrogen is transformed to 596 million tons of helium, and the remaining 4 million tons is released as radiation.

After the revelation of nuclear fusion in the Sun an intensive puzzle work of fitting solar models to the increasing amount of detailed observation started with the goal of describing both the present structure and the past evolution of the Sun. From the mid-1970s the observations of solar oscillations and their interpretation, known as *helioseismology*, have become most important tools for reaching a very accurate description of the interior of the Sun.

Today we know that the Sun is a typical cool magnetic star. Its mass ($m_\odot$) is $1.99 \times 10^{30}$ kg (330 000 times more massive than the Earth) and radius ($R_\odot$) 696 000 km (109 times the Earth's radius, $R_E$). The present Sun irradiates with a luminosity of $3.84 \times 10^{26}$ W with an effective black body temperature of 5778 K. The Sun was formed about $4.55 \times 10^9$ years ago when an interstellar gas cloud with a mass of the order of $10^4\ m_\odot$ collapsed due to some interstellar gravitational perturbation, probably a shock wave, and further disintegrated, leading to the formation of the solar system. The collapse was not spherically symmetrical due to the presence of angular momentum and magnetic flux of the cloud.

While most of the angular momentum and magnetic flux were carried away by matter not ending up in the solar system, rotation and magnetic field are still today essential elements of the Sun and the solar system.

An intriguing obstacle on the road toward an acceptable solar model was the *solar neutrino problem*. Ever since the first neutrino experiments by Davis and Bahcall in the Homestake gold mine in 1967, observations based on different detection techniques indicated that the Sun would produce only 30–50% of the neutrino flux that the standard solar model predicts to arise from the fusion process in the core. Attempts to solve this problem, e.g., by adjusting the temperature of the central core, lowering the relative abundance of heavy elements, assuming a rapidly rotating core, or assuming a strong magnetic field in the core, all led to contradictions elsewhere in the solar models.

Meanwhile developments in neutrino physics started to point toward another solution based on the properties of the neutrinos themselves. Finally, strong evidence in favor of the nuclear physics explanation was obtained at the beginning of the 21st century with a Cherenkov experiment within a large water tank with a heavy water ($D_2O$) core at the Sudbury Neutrino Observatory [Ahmed and SNO Collaboration, 2004]. In that experiment it is possible to observe both the electron neutrinos, which are produced by the fusion, and the $\mu$ and $\tau$ neutrinos, to which a considerable fraction of the electron neutrinos are transformed through *neutrino oscillations* during the propagation from the Sun to the Earth

Figure 1.1 illustrates the main regions of the Sun (for a detailed discussion of the solar model, see Stix [2002]). The energy production takes place in the *core* within a radius of $0.25 R_{\odot}$ from the center of the Sun where temperature is $1.57 \times 10^7$ K and pressure $2.34 \times 10^{16}$ Pa. From the core energy propagates outward through a very slow process of

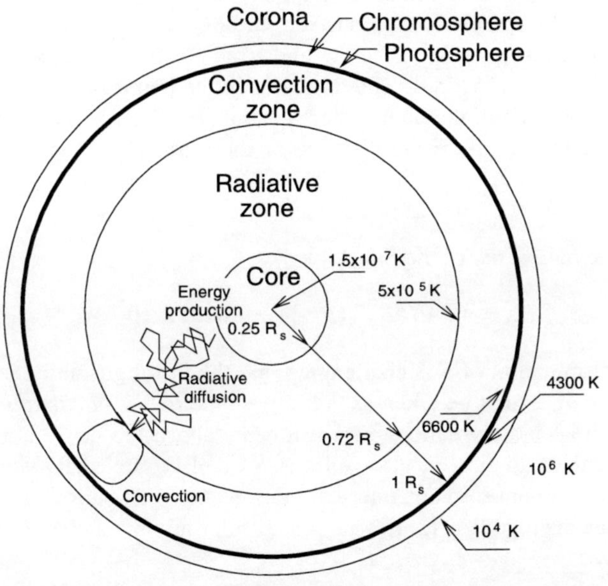


**Fig. 1.1** The structure of the Sun. (Figure by courtesy of R. Vainio.)

radiative diffusion during which the photons are absorbed and re-emitted by the dense solar matter over and over again. The energy propagation time of the distance of 2 light seconds is of the order of 170 000 years. Due to collisions and absorption–emission processes in this *radiative zone* the photons are redshifted toward the visible wavelengths.

At the distance of about $0.72R_\odot$ the solar gas becomes opaque to the photons and the energy transport toward the surface takes the form of turbulent convection, which is much faster than the radiative transfer. The plasma motion in this *convection zone* is extremely complex and of specific relevance to the topic of the present text, as the ever-changing magnetic field of the Sun is created within this zone, according to the present understanding close to its bottom. The radiation does not stop completely at the base of the convection zone. About $0.05\,R_\odot$ into the convection zone the convective energy flux exceeds the radiative flux and within the last $0.1\,R_\odot$ below the surface practically all energy transport is convective.

While the radiation zone is stably stratified, the convection zone is unstable: gas parcels move up, dissolve, and cool down, and the cool gas returns back along narrow lanes between the upward-moving gas parcels. The whole convection zone is continuously mixed, which makes it chemically homogeneous. This does not make the mean molecular mass constant because close to the surface the degree of ionization drops rapidly. However, within most of the convection zone the mean molecular mass is about 0.61.[1]

Finally the convection reaches the solar surface and introduces a granular structure on it. The intergranular lanes are about 100 K cooler than the regions of upward motion. Granules appear in various sizes, diameters ranging from about 1000 km up to a few times $10^4$ km, the latter being called supergranules. The smallest granules represent small convection cells close to the surface, whereas the larger granules are related to larger convection cells reaching deeper into the convection zone.

Above the convection zone a thin surface, the *photosphere*, absorbs practically all energy carried by convection from below and irradiates it as (almost) a thermal black body at the temperature of 5778 K. The thickness of the photosphere is only 500 km. The temperature at the bottom of the photosphere is about 6600 K and at its top 4300 K.

The total irradiance at the mean distance of the Earth (1 *AU*) is known as the *solar constant*

$$S = 1367 \pm 3\,\mathrm{W\,m^{-2}}\,. \tag{1.1}$$

It is related to the *luminosity* of the Sun $L_\odot$ by

$$L_\odot = 4\pi AU^2 S = (3.844 \pm 0.010) \times 10^{26}\,\mathrm{W}\,. \tag{1.2}$$

Accurate determination of $S$ is challenging and the last digits and uncertainties in the expressions above must not be taken as definitive. The *total solar irradiance* (TSI) must be observed with accurately calibrated instruments above the dense atmosphere, which absorbs most of the radiation in ultraviolet (UV) and infrared (IR) wavelengths. Early in the 21st century a consensus of inter-calibrations between various space observations was reached of an average $S \approx 1366\,\mathrm{W\,m^{-2}}$ near solar minima and $S \approx 1367\,\mathrm{W\,m^{-2}}$ near

---

[1] In a plasma free electrons are counted as particles. Thus the mean molecular mass of electron–proton plasma is 0.5.

solar maxima. However, observations with the Total Irradiance Monitor (TIM) onboard the *The Solar Radiation and Climate Experiment (SORCE)* satellite launched in 2003 indicate that the actual TSI would be some 4–5 $\mathrm{W\,m^{-2}}$ smaller than previously thought [Kopp et al, 2005]. By the time of writing this book the reason for this discrepancy had not been clarified.

For space storms the exact total irradiance is not as important as its relative variations. In particular, near solar maxima the irradiance varies by several $\mathrm{W\,m^{-2}}$ depending on the sunspot activity (Sect. 1.1.5).

The luminosity can be given in terms of the *effective temperature* defined by

$$L_\odot = 4\pi R_\odot^2\, \sigma T_{eff}^4\,, \tag{1.3}$$

where $\sigma = 5.6704 \times 10^{-8}\,\mathrm{W\,m^{-2}\,K^{-4}}$ is the *Stefan–Boltzmann constant.* The effective temperature of the Sun is $T_{eff} = 5778 \pm 3\,\mathrm{K}$. The photospheric gas has this temperature at the optical depth $\tau \approx 2/3$, which can be taken as the definition of the solar surface (for the definition of $\tau$, see, e.g., Stix [2002]).

"Solar constant" is actually one of many historical misnomers that we will encounter in this book. The Sun is a variable star in both short and long time scales. Fortunately for us, the variations are about a factor of three weaker than is typical for many other Sun-like stars. In the longest time perspective the luminosity of the newly-born Sun was about 72% of its present value. After some 2 billion years from now the Sun will have become so bright that the Earth will turn too dry for the present type of life. The slow rise of solar luminosity is due to the increase of the core temperature when more and more hydrogen is fused to helium.

In space weather and space climate time scales, $S$ varies by a factor of

- $10^{-6}$ over minutes
- $2 \times 10^{-3}$ (0.2%) over several days
- $10^{-3}$ over a solar cycle (the number is quite uncertain because the solar cycles are different)

The physical reasons and apparent periodicities for these variations are not fully understood.

### 1.1.2 Solar spectrum

The solar spectrum from $\gamma$-rays to metric radio waves is given in Fig. 1.2. Most of the solar energy is irradiated in the visible and near-infrared parts of the spectrum with peak irradiance in yellow light around 450–500 nm. The red end of the spectrum is an almost continuous black-body spectrum with some strong absorption lines, e.g., H$\alpha$ at 656.3 nm (not visible in the scale of Fig. 1.2). At the blue end there are more absorption lines.

About 44% of the electromagnetic energy is emitted at infrared wavelengths $\lambda > 0.8\,\mu\mathrm{m}$. This part of the spectrum is approximately thermal and can be represented by the *Rayleigh–Jeans law*

$$S(\lambda) \simeq 2ck_BT\,\lambda^{-4}\,(R_\odot/AU)^2\,. \tag{1.4}$$

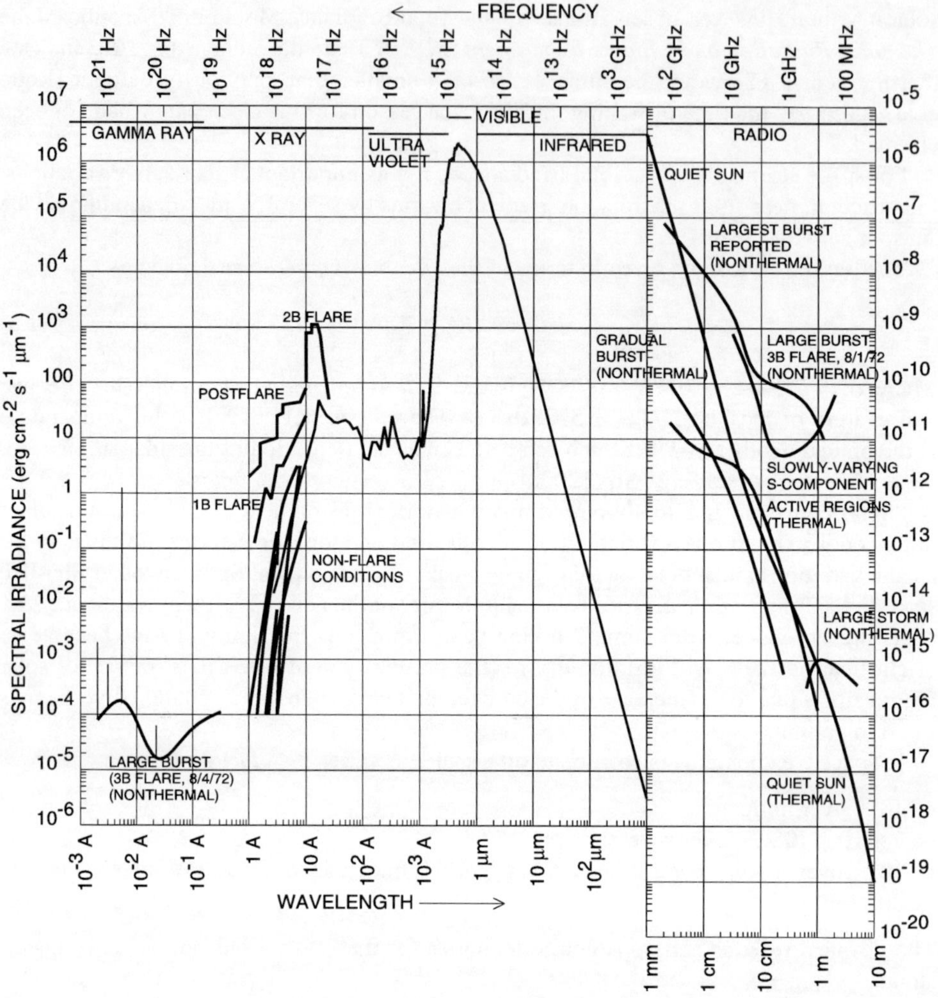


**Fig. 1.2** Solar spectrum from $\gamma$-rays to radio waves. The radio wave part of the spectrum is shifted up in irradiance by 12 orders of magnitude. The irradiance is given in cgs units and ångström (1 Å = 0.1 nm) is used below one 1 μm, which is common practice in solar physics. (From Aschwanden [2004].)

The infrared spectrum is absorbed mostly by water vapor in the Earth's atmosphere.

At radio wavelengths (> 1 mm) the spectrum is commonly presented as a function of frequency (recall the conversion: $\lambda(\mathrm{m}) = 300/f(\mathrm{MHz})$; e.g., 1 mm $\leftrightarrow$ 300 GHz). The Sun is strongly variable at these wavelengths because the radio emissions originate from nonthermal plasma processes in the chromosphere and corona (discussed in Sect. 1.1.3). As indicated in Fig. 1.2, the radio emissions during strong solar storms can exceed the quiet levels by several orders of magnitude. Note that there is an ankle in the slope of the quiet-Sun spectrum at around 10 cm indicating higher temperatures ($\sim 10^6$ K) than the main black body radiation. This is a signature of the chromosphere and corona being much hotter than the visible Sun.

In the ultraviolet side of the spectrum absorption lines are dominant down to 210 nm. At shorter wavelengths the intensity is reduced to correspond to the temperature of 4700 K. This reduction is due to absorption by the ionization of Al I. (Recall the notation: Al I represents non-ionized aluminum, Al II is the same as $Al^+$, Al III is $Al^{2+}$, etc.) Below 150 nm emission lines start to dominate the spectrum. The strongest is the hydrogen Lyman $\alpha$ line centered at 121.57 nm. Its average irradiance, 6 mW m$^{-2}$, is as strong as all other emissions below 150 nm together and the line is also clearly visible in Fig. 1.2 .

At shorter wavelengths the spectrum becomes highly variable, illustrating a nonuniform distribution of the emission sources in the solar atmosphere. The nonuniformity is both spatial and temporal. The wavelength band below 120 nm is called *extreme ultraviolet* (EUV). These emissions come both from neutral atoms and from ions up to very high ionization levels, e.g. Fe XVI ($Fe^{15+}$) in the solar corona. This facilitates the observations of the wide range of temperatures from 8000 K to $4 \times 10^6$ K, from the chromosphere to the corona.

Solar flares increase the EUV and soft X-ray (0.1–10 nm) spectra quite considerably. Also hard X-rays and $\gamma$-rays are emitted in these processes, as will be discussed in Chap. 12.

### 1.1.3 Solar atmosphere

That there is an atmosphere above the photosphere is evident already visually. The irradiance decreases from the center of the disk to the limb by an order of magnitude due to the absorption of the atmospheric gas, which is known as *limb darkening*. The temperature continues to decrease in the photosphere reaching its minimum at an altitude of about 500 km. Thereafter, the temperature starts to rise again in the *chromosphere*. The chromosphere has got its name from the colorful flash seen just at the beginning and at the end of a total solar eclipse. The most prominent color is the red H$\alpha$-line at 656.3 nm. Traditionally the chromosphere was thought to be a layer of thickness of about 2000 km, but as illustrated in Fig. 1.3 the present view to the structure of the solar atmosphere is much more complicated and dynamic than the old picture of a gravitationally stratified atmosphere.

At the upper end of the chromosphere the temperature begins to rise more rapidly. The chromosphere is sometimes defined to end at the temperature of 25 000 K. Above the chromosphere there is a thin *transition region* to coronal temperatures of the order of $10^6$ K. The *corona* is a key region of many aspects of space storms to which we will return in Sect. 1.1.6.

The steep temperature increase from the chromosphere to the corona remains one of the major insufficiently understood topics in solar physics. As illustrated in Fig. 1.3 the chromospheric and coronal plasmas partly overlap, flowing up and down with complicated dynamic magnetic field structures involving waves, shocks, magnetic reconnection, etc., which will be discussed in later chapters of this book. At the same time when this dynamism complicates the picture, it also indicates that there free energy is available for the heating. In fact, a steep temperature gradient in a gravitationally stratified atmosphere

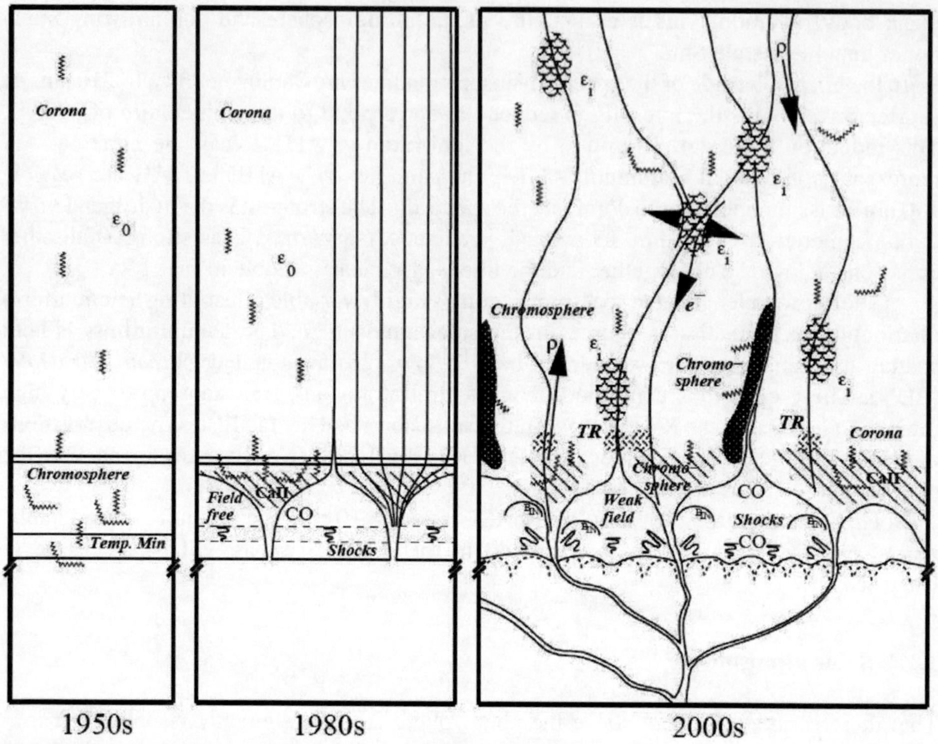


**Fig. 1.3** Evolution of the concepts the solar atmosphere from gravitationally stratified layers in the 1950s to a highly inhomogeneous mixing of the photosphere, chromosphere, and corona at the beginning of the 21st century. (From Schrijver [2001].)

might be much more difficult to explain than a spatially and temporally variable environment.

### 1.1.4 Rotation of the Sun

That the Sun rotates was discovered soon after the advent of telescope in about 1610. Around 1630 it became clear that the rotation is not rigid, but the equatorial surface rotates faster than the high-latitude regions. The origin of this *differential rotation* is not yet fully understood. It is related to the transport of angular momentum inside the Sun and it also plays a central role in the generation of the solar magnetic field. Differential rotation appears to be a general property of self-gravitating large gaseous bodies and is also observed in the giant planets of the solar system.

The rotation axis of the Sun is given by two angles: the *inclination* $i$ between the ecliptic plane and the equatorial plane, and the *angle of the ascending node* $\alpha$ of the Sun's equator, i.e., the angle in the ecliptic plane between the direction of the vernal equinox and the direction where the solar equator cuts the ecliptic from below. The Earth's precession shifts the equinox direction by 0.0196°, i.e., 50″, per year, and thus $\alpha$ increases by the same rate. Consequently, the *epoch* must be given when coordinates related to the equinox are used. Carrington determined these angles in 1863 as $i = 7.25^\circ$ and $\alpha(1850) = 73.67^\circ$. The latter is still valid but the Greenwich sunspot data from the period 1874–1976 imply $i = 7.12^\circ \pm 0.05^\circ$.

We denote the *heliographic latitude* by $\psi$, thus the polar angle (co-latitude) is $\theta = \pi/2 - \psi$. There is no physically unique way to define the longitude on the differentially rotating surface. For this purpose Carrington introduced a notation that is still in use. He divided time into intervals of 27.2753 days. These intervals are called *Carrington rotations*. Carrington rotation 1 was defined to have commenced on 9 November 1853. In one year of 365 days there are 13.38 Carrington rotations and thus the present rotation numbers are well over 2000. At the commencement of a new rotation longitude $\phi = 0$ is attached to the center of the solar disk. Note that the Carrington rotations are related to the motion of the Earth around the Sun, i.e., the "same place" at the solar equator is toward the Earth after one Carrington rotation. This is known as the *synodic period*. The "true" rotation period with respect to the stars is the *sidereal period* of about 25 days.

Carrington determined the surface rotation rate from sunspot data as a function of the heliographic latitude in (sidereal) degrees per day

$$\Omega(\psi) = 14.25 - 2.75 \sin^{7/4} \psi . \tag{1.5}$$

The power 7/4 is a bit awkward. A more modern approach is to expand the rotation rate as

$$\Omega(\psi) = A + B \sin^2 \psi + C \sin^4 \psi + \ldots \tag{1.6}$$

and in most studies only coefficients $A$ and $B$ are determined. Here $A$ is the equatorial rotation rate.

In addition to sunspot data, Doppler shifts, edges of coronal holes and surface magnetograms are used in studies of the rotation rate. The different methods yield slightly different results and there is some variability within the individual methods as well. Furthermore, different sunspot cycles are different. For example, Pulkkinen and Tuominen [1998] used the sunspot data from cycles 10–22 (years 1853–1996) and found that the coefficients varied in the ranges $A = (14.38, 14.85)$ and $B = (-3.19, -2.51)$.

It is interesting to note that the larger the structure used to determine the rotation, the more uniform rotation is found. The extreme are observations of large coronal holes, which sometimes show very little differential rotation at all. During the last decades helioseismology has revolutionized the studies of differential rotation. Now it is possible to empirically determine the rotation also inside the Sun, as illustrated in Fig. 1.4, which has been derived from the observations of solar oscillations using the MDI instrument onboard the *SOHO* spacecraft.

A rotating non-rigid body is not fully spherical. Even the Earth is elastic and has an *oblateness* $f = (r_{eq} - r_{pol})/r_{eq} \approx 1/300$. The fast-rotating gas giant planets Jupiter and Saturn are much more oblate, $f_J = 0.065$ and $f_S = 0.098$, which can be perceived already in rather low-resolution pictures. But how oblate is the slowly rotating Sun, whose exact diameter is difficult to measure?

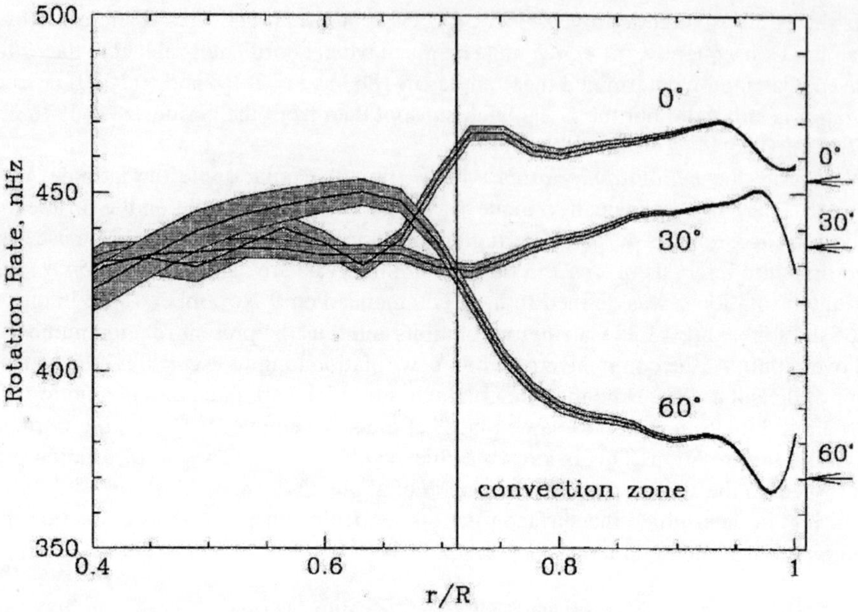


**Fig. 1.4** The internal rotation rate of the Sun. The radial profiles are calculated for three different latitudes. The grey regions indicate the estimated error in the inversion procedure. (From Kosovichev et al [1997].)

Neglecting the differential rotation and expanding the external gravitational field up to the quadrupole term (the first non-zero correction)

$$\Phi_{ext} = -\frac{Gm_\odot}{r}\left[1 - J_2\left(\frac{R_\odot}{r}\right)^2 P_2(\theta)\right] \tag{1.7}$$

the oblateness expressed as $\Delta r/R_\odot$ is

$$\frac{\Delta r}{R_\odot} = \frac{1}{2}\frac{\Omega^2 R_\odot}{g_\odot} + \frac{3}{2}J_2\,, \tag{1.8}$$

where $\Omega$ is the angular velocity of the solar surface, $J_2$ the quadrupole moment and $P_2(\theta)$ the second Legendre polynomial. Using the Carrington rotation rate, the first term in (1.8) is about $10^{-5}$.

In the past the Sun has rotated faster than today. The specific angular momentum (i.e., the angular momentum per unit mass) of the cloud collapsing to form the Sun was much larger than the angular momentum of the present solar system. Much of this was lost in a very early phase of the solar evolution. We know that the so-called *T Tauri stars*, which are in the early phase of their evolution, rotate much faster than the Sun. Their surface velocities are about 15 km s$^{-1}$ compared to 2 km s$^{-1}$ of the present Sun.

According to pre-main-sequence stellar models, the Sun was fully convective before the hydrogen burning started. The convection was turbulent and the rapid exchange of momentum between parcels of gas evened out the gradients in the angular velocity. The total angular momentum $J_0$ has been estimated to have been $8 \times 10^{42}$ kg m$^2$ s$^{-1}$, whereas it presently is $1.7 \times 10^{41}$ kg m$^2$ s$^{-1}$.

Matter leaving the Sun carries angular momentum, but the material loss since the time of large $J_0$ has been negligible. The magnetic field, however, is a very efficient lever arm for a torque. As we will discuss in the context of the solar wind (Sect. 1.2.2), the magnetic field forces the escaping material to rotate with the Sun out to the so-called *Alfvén radius* $r_A \approx 12 R_\odot$. Thus the angular momentum density increases up to $r_A$, and it is this angular momentum that is conserved in the escaping flow beyond $r_A$. The rate of angular momentum loss is

$$\frac{dJ}{dt} = \Omega\, r_A^2 \frac{dm}{dt} . \tag{1.9}$$

How much such *magnetic braking* really has taken place in the history is difficult to estimate because we do not know the history of the magnetic field on which $r_A$ depends. The magnetic field is generated by the solar dynamo (Sect. 8.3.2), which depends on $\Omega$ and in particular on its gradient. As long as the Sun was fully convective the slowing down affected the whole Sun. When the radiative core developed, the motion of the outer convective zone was disconnected from the interior. The convective part continued to lose angular momentum by magnetic braking, but what happened to the core? Because the central core contracted further, the first guess would be that its rotation rate should have increased.

However, the recent results of helioseismology (e.g., Fig. 1.4) do not support the idea of a fast-rotating core. The central core may rotate somewhat faster than the radiative zone but something seems to have slowed down the rotation also in the inner parts of the Sun. A strong inward gradient $d\Omega/dr$ would mean strong shear flows. These could drive instabilities, which, in turn, could transport the excess angular momentum, resulting in smoother $d\Omega/dr$. It has also been speculated that there could be an internal magnetic field in the core. Indeed, already a relatively weak magnetic field would be sufficient to slow down the core.

### 1.1.5 Sunspots and solar magnetism

The magnetic field of the Sun is very complicated both in time and in space. The existence of solar magnetic fields was first found in *sunspots* by Hale in 1908. Although we can today measure much weaker magnetic fields on the Sun, the sunspots have retained a central role in studies of solar magnetism. The theory of magnetic field generation is a difficult topic of plasma physics, and after a century of intensive study we still lack a fully satisfactory physical description of the generation and evolution of the solar magnetic field.

A sunspot corresponds to an intense magnetic flux tube emerging from the convection zone to the photosphere. Large spots can have diameters of about 20 000 km. The center of the spot is called the *umbra* whose temperature is about 4100 K, and the largest observed magnetic fields are about 0.3 T. The strong magnetic field is the cause of the low temperature and thus the relative darkness of the spot because it inhibits the hot plasma of reaching the surface. Around the spot there may be a *penumbra* that consists of dark and bright filaments. Young spots do not have penumbrae and in about 50% of the cases the spot evolution stops before a penumbra has developed.

The magnetic field is measured by observing the *Zeeman splitting* of atomic spectral lines. Because the Zeeman effect is weak, the observations have traditionally been limited to determination of the line-of-sight component of the magnetic field. However, the state-of-the-art spectropolarimetric observations with the Japanese *Hinode* satellite have contributed important advances in observations of the horizontal magnetic field in the photosphere [Lites et al, 2008]. This progress is important toward better understanding of the role of the magnetic fields in the heating of the chromosphere and corona.

The cyclic appearance of the sunspots, with a quasi-period of about 11 years was found by Schwabe in 1844. When a new cycle begins, spots start to appear at mid-latitudes (around 30–40°) on both hemispheres. The life-time of individual spots is relatively short, from days to weeks, but with time more and more new spots appear. The new spots are located closer and closer to the equator, resulting in the famous *butterfly diagram* (Fig. 1.5). After the maximum occurrence the sunspot number starts to decrease to the solar minimum of practically no sunspots at all.

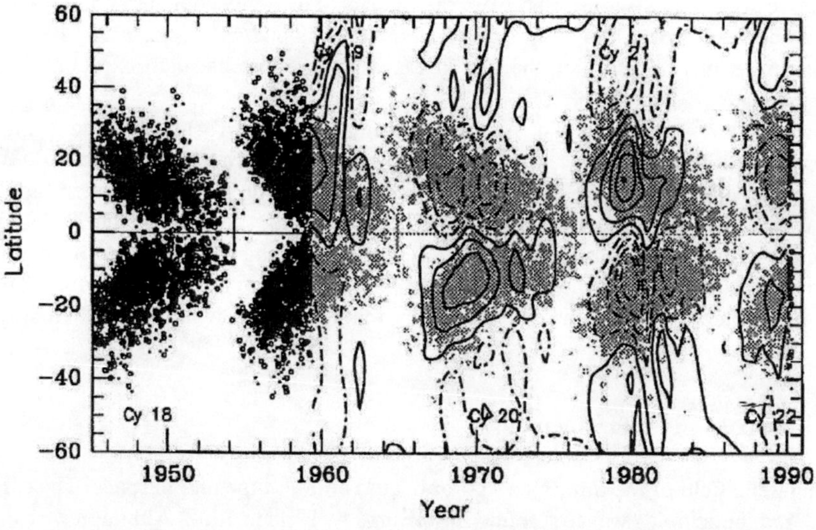


**Fig. 1.5** The butterfly diagram of sunspot appearance. The contours are $\pm 20\,\mu T$, $\pm 60\,\mu T$, $\pm 100\,\mu T, \ldots$, solid lines indicate positive polarity, dashed lines negative. (From Schlichenmaier and Stix [1995].)

The sunspots usually appear in pairs or in larger groups. The magnetic flux emerging from one spot returns to another. In 1923 Hale was able to confirm the *polarity rules* of sunspots that he had formulated with his colleagues in 1919:

- The magnetic orientation of leader and follower spots in bipolar groups remains the same in each hemisphere over the whole 11-year cycle.

- The bipolar groups in the two hemispheres have opposite magnetic orientation.
- The magnetic orientation of bipolar groups reverses from one cycle to the next.

As it takes two sunspot cycles to return to the same orientation, the length of the magnetic cycle of the Sun is 22 years. This is known as the *Hale cycle*, whereas the 11-year sunspot cycle is called the *Schwabe cycle*.

The mean magnetic field inside the Sun can be described as a sum of *toroidal* and *poloidal* components. The systematic behavior of bipolar sunspot groups can be understood in terms of a subsurface toroidal magnetic field. "Toroidal" means in this context that the field lines form closed loops around the solar rotation axis. Locally this field may be driven to the surface by convection and magnetic buoyancy, which forms a bipolar sunspot pair. The total flux of the toroidal field is of the order of $10^{15}$ Wb. If we assume that it is distributed within the latitudinal range of the sunspots and throughout the convection zone, the mean toroidal field is $B_t \approx 0.02\,\mathrm{T}$. It is possible that most of the flux is concentrated in a thin overshooting layer at the bottom of the convective zone, where the turbulent convective motion partially penetrates into the stable radiation zone. In that region the mean field can be of the order of 1 T.

The field lines of the poloidal field are in the meridional planes in the same way as the field lines of the familiar magnetic dipole. The differential rotation drags the poloidal field lines to enhance the toroidal component. This takes place during the rising solar activity. In order to establish the cyclic behavior there must be another process to return toward a more poloidal configuration with reversed polarity during the decaying activity (see Sect. 8.3.2).

Daily sunspot observations were started in 1749 at the Zürich Observatory. With later addition of observations from other observatories continuous sunspot data are available from 1849. The intensity of sunspot activity is usually given by the *relative sunspot number* $R$ introduced by Wolf in 1848

$$R = k(10g + f)\,, \tag{1.10}$$

where $g$ is the number of spot groups and $f$ is the total number of spots (an isolated spot is calculated also as a group). The coefficient $k$ is determined individually for each observatory to take into account the instrument properties and local seeing conditions. $R$ is approximately proportional to the area of the Sun that is covered by the spots. Thus it is a rough measure of the total absolute magnitude of the magnetic flux penetrating the visible hemisphere within the sunspots.

The sunspot cycles are enumerated so that cycle 1 began in 1756. Figure 1.6 shows the entire Zürich sunspot number time series from 1750 to the end of cycle 23 in 2008. While the solar cycle is remarkably repetitive, it also shows great variability which cannot be properly predicted yet. Both the intensity and the shape of the peaks in the sunspot time series are different from one cycle to another. Also the length of the cycles varies up to a few years. Most of the text of this book was written during a peculiarly long and deep solar minimum after cycle 23, the recovery from which did not start until early 2010.

The strongest recorded maximum took place in 1957 (cycle 19). During the last century there was an increasing trend of the peak sunspot numbers with the exception of cycle 20. However, the peak of cycle 23 in 2000 was weaker than the previous two. It may be a sign of the so-called *Gleissberg cycle* of about 80 years superposed on the 22-year Hale cycle. In that case the coming maxima would be smaller than the recent ones. The

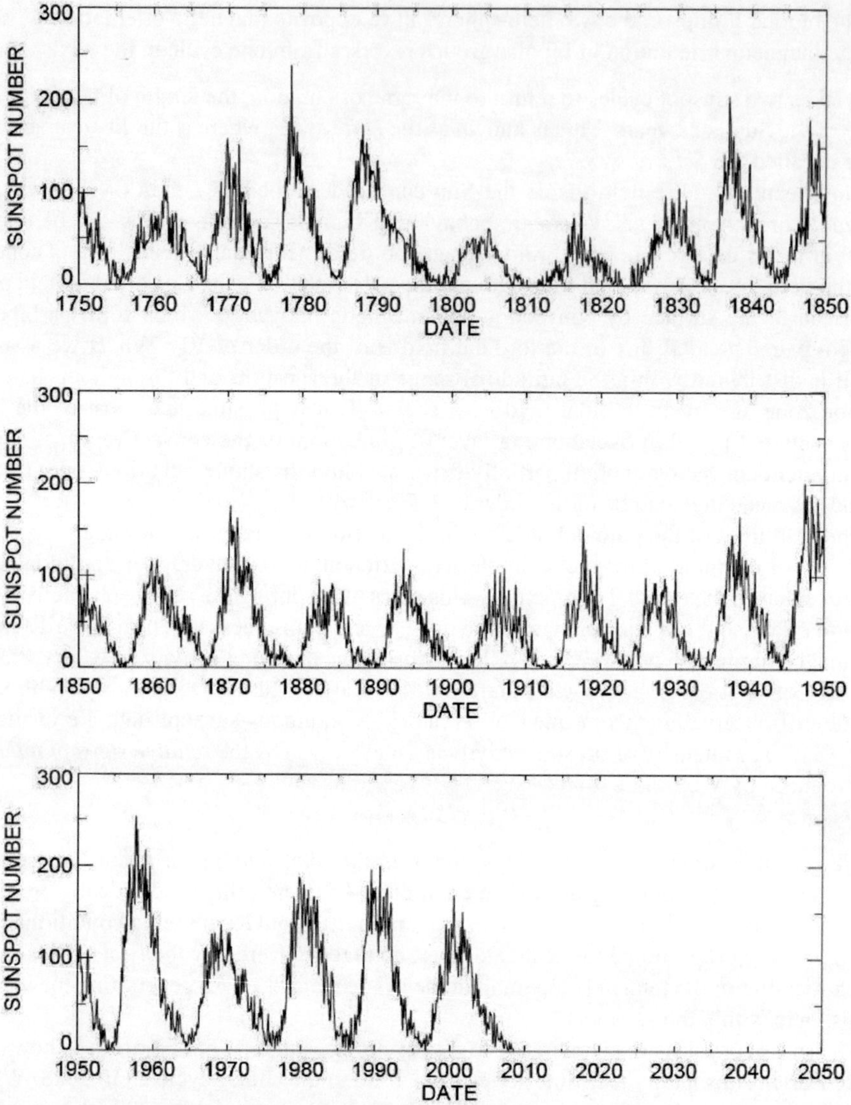


**Fig. 1.6** The Zürich sunspot number time series. At the time of writing this book, the official record was available to late 2009. The recovery from the last minimum was very slow and did not start until 2010. For updated information see, e.g., `http://sidc.oma.be`

Gleissberg cycle superposed with the about 200-year *de Vries cycle* is consistent with long-term minima in the 17th century (the *Maunder minimum*), around the year 1800 (the *Dalton minimum*) and around the year 1900 (the *Modern minimum*) (Fig. 1.7).

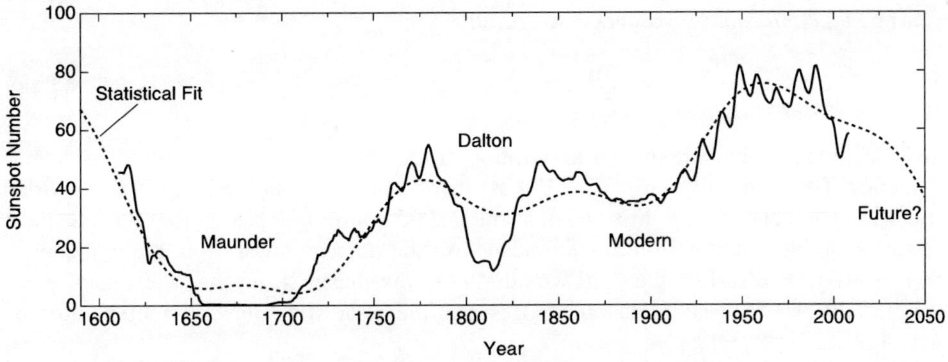


**Fig. 1.7** Long-term sunspot number variation after the 11-year cycle has been filtered away (solid line) and the superposition of the Gleissberg and de Vries cycles (dotted line). (Figure by courtesy of H. Nevanlinna.)

The most remarkable feature in the long time series is that the solar activity seems to have been almost nil during the Maunder minimum. This is not an artifact of poor observations; there simply were almost no spots on the Sun. This coincided with the so-called little ice age when the climate in Europe was exceptionally cool. This may have been a consequence of the fact that solar activity is related to the brightness of the Sun, luminosity being a factor of about $10^{-3}$ higher at the sunspot maximum than at the minimum. However, the effects of the solar activity on the terrestrial climate, if any, are not really understood.

What is the origin of the magnetic field of the Sun? In principle it could be a remnant of the magnetic field in the interstellar cloud that once collapsed to form the Sun. If the cloud's weak field, less than 1 nT, were compressed with the matter without any losses, the resulting flux density would be huge, some $10^6$ T. Much of this was lost in the early evolution of the Sun, but considering the fact that the Ohmic diffusion time $\tau_\eta$ for the Sun is of the order of $10^{10}$ years, the mere existence of the field does not require its continuous generation. The case is different for the planets, e.g., for the Earth $\tau_\eta \approx 10^4$ years, thus the Earth must possess a *dynamo* of some type. Otherwise the only magnetism would be remanence in magnetic materials in the ground, as appears to be the case in Mars.

Not even the 22-year magnetic cycle of the Sun is a fully convincing signature of an active solar dynamo. It might be a sign of oscillatory behavior of a slowly decaying fossil field. However, the detailed features of the differential rotation and its association to the migration of the sunspots are considered as the strongest evidence of the dynamo. The present Sun, the Earth, and other magnetized planets are able to manifold the pre-existing flux through a dynamo process. In the Sun this takes place in the convection zone, most likely close to its bottom. The excess magnetic energy is expelled away with the solar wind. The energy sources for the magnetic field generation are the rotation and the heat produced in the core.

The *induction equation* of magnetohydrodynamics (Chap. 6)

$$\frac{\partial \mathbf{B}}{\partial t} = \nabla \times (\mathbf{V} \times \mathbf{B}) + \eta \nabla^2 \mathbf{B} \tag{1.11}$$

gives a simple phenomenological description of the basic idea of magnetohydrodynamic dynamos. The convective term $\nabla \times (\mathbf{V} \times \mathbf{B})$ involves the plasma motion, which provides free energy to generate new flux, whereas the diffusive term $\eta \nabla^2 \mathbf{B}$ describes how the field is decaying. Note that both terms are needed in the description of a plasma dynamo. If there were just diffusion, the field would simply disappear. If, on the other hand, there were no diffusivity at all, (1.11) would describe the ideal MHD flow without creation of new flux.

The problem of *dynamo theory* is to find solutions for the induction equation where the convection and diffusion together result in creation of new magnetic flux, or more exactly, manifolding of the existing flux. This is somewhat analogous to a traditional bicycle dynamo. If you just have the dynamo rotating, not connected anywhere, the only effect would be weak friction that would make the driving a little harder. But if you connect the dynamo through a load, e.g., a lamp, a current flows in the cable and gives rise to a magnetic field according to Ampère's law. The energy to create the new flux is not drained from the magnetic energy of the magnet but from the mechanical work you are doing to keep the magnet rotating. This way we have natural roles for $\mathbf{V}$, the rotation, and for $\eta$, the dissipation, in the cable and the lamp. Both are needed!

This analogy should not be taken too literally. Technological dynamos are multiply-connected systems where the load is external to the dynamo itself. In MHD plasma there are no cables nor circuits. The new flux is directly superposed on the pre-existing field in the same simply-connected volume of fluid whose motion creates the flux and the flux is also dissipated in the same volume.

An important property of cosmic dynamos is *self-excitation*. In a bicycle the seed magnetic field is provided by a permanent magnet. We can imagine setting up a self-exciting dynamo by winding the wire connected to the load around the system so that it creates a magnetic field that is in the same direction as the seed field. Thereafter we remove the original magnet and the seed field is now provided by the field generated by the dynamo itself. This is not a perpetuum mobile, as the energy source for the magnetic field generation is the motion that has to be strong enough to balance the dissipation. We will discuss the dynamo processes in the Sun and in the magnetosphere in more detail in Chap. 8.

### 1.1.6 Coronal activity

The beauty and the dynamism of the corona is impossible to capture in the pages of a textbook. The reader is strongly recommended to visit the web pages of various solar spacecraft, in particular *SOHO*, *TRACE*, *STEREO*, and *SDO*. The two-spacecraft mission *STEREO* took a number of 3D images from the early phase of the mission when the two spacecraft were at optimal distance from each other. Unfortunately, the prime time of truly stereoscopic *STEREO* observations took place during the particularly quiet solar minimum after cycle 23.

The active, or indeed violent, processes in the solar corona are essential elements of space storms. Of particular importance to space storms are the *solar flares* and *coronal mass ejections (CMEs)*, which will be discussed in detail in Chap. 12.

In the past the corona was possible to observe during solar eclipses only. The early observations indicated two distinct components in the white-light corona: the *K corona* and the *F corona*. *K* comes from the German word *Kontinuum* and *F* from the dark *Fraunhofer lines*. The spectra of both components resemble the photospheric spectrum but in the *K* corona the Fraunhofer lines are absent. The *K* component is also strongly polarized, which indicates that it arises from *Thomson scattering* on free electrons (Chap. 9). Actually, there are weak dips corresponding to the strongest Fraunhofer lines (*H* and *K*) also in the *K* corona. The explanation for the filling of the lines is Doppler broadening due to the high temperature of the scattering electrons. This was an early hint that the corona might be hot, as first suggested by Grotrian as early as in 1931. Note that the white-light observations of coronal mass ejections extending far beyond 3 $R_\odot$ are also based on Thomson scattering on electrons in the dense plasma cloud.

The *F* corona shows the photospheric continuum with the Fraunhofer lines. The light is unpolarized and it is explained as photospheric light scattered on dust particles. The *K* corona decays faster than the *F* corona and the latter dominates beyond 2–3 $R_\odot$. The *K* corona is, in fact, the same phenomenon as the *zodiacal light* observed deep in interplanetary space.

The coronal structure is closely linked to the solar magnetism and illustrates the large-scale structure of the magnetic field. At the solar minimum the poloidal component dominates the large-scale structure of the magnetic field. Within the polar regions *polar plumes* emerge from large *coronal holes* and represent the plasma flowing out with the solar wind. At the solar maximum the polar coronal holes are not as easy to recognize because the actual magnetic field is dominated by the irregular contributions from the toroidal component. There can be several coronal holes and the magnetically closed regions often resemble Prussian helmets, and are called illustratively *helmet streamers*. Note that the word streamer refers to the visible closed structures. It is not directly associated with the stream of escaping plasma, the *solar wind*, which originates mostly, if not completely, from the coronal holes.

The high temperature of the corona was not known at the time of first spectroscopic observations, and the observed spectrum caused quite a lot of confusion. Recalling that helium was once found for the first time in the Sun, a new element, *coronium*, was suggested to explain some of the abundant but thus far unknown spectral features. In the years 1939–1941 Grotrian and Edlén, however, correctly identified several of the coronal lines to be those of highly ionized atoms. Three of the most conspicuous visible lines represent strong transitions of Fe XIV (530.3 nm), Ca XV (569.5 nm), and Fe X (637.5 nm). Of these Fe X is formed at $10^6$ K and Fe XIV at $2 \times 10^6$ K. Thus it is evident that a cool star of a temperature of about 6000 K can support a hot corona of millions of degrees.

The coronal spectrum is very rich in UV and X-ray lines. While the white-light observations require *coronagraphs*, i.e, devices where an occulting disk creates an artificial eclipse, many of the short wavelength emissions can also be observed against the solar surface as they emerge from the much hotter coronal gas. For example, the X-ray detector onboard the Japanese *Yohkoh* satellite was first to observe ionized iron up to Fe XXVI during solar flares. The emission is the Lyman $\alpha$ line of an iron ion with only one electron. Its wavelength is 178 pm and the required temperature is about $2 \times 10^7$ K. Such, and even higher, temperatures are not uncommon in solar flares. The tenuous corona is not in local thermodynamic equilibrium and particle populations of very different temperatures are produced by the rapidly varying magnetic field configurations.

The X-ray images of the Sun have revealed the very active behavior of the corona. The coronal holes are clearly seen as dark regions whereas the hot plasma radiating the X-rays is confined in the magnetic bottles of the closed field lines. In addition there are numerous small X-ray bright points arising from the bremsstrahlung of electrons being decelerated by the surrounding plasma (Chap. 9). The coronal holes remain colder because they are on open field lines, from which the plasma escapes as the solar wind before it is heated to the same temperatures as plasmas in the closed field line regions.

Also radio waves reveal important information on the magnetically active corona. They are emitted by electrons gyrating in the strong magnetic field (Chap. 9), and are particularly important in studies of radio flares associated with solar activity (Chap. 12).

That there is *some* temperature increase in the chromosphere is not so difficult to understand. The rarefied gas starts to deviate from local thermodynamic equilibrium and it does not need to find equilibrium with the lower atmospheric levels if some processes keep on heating it. There are two rich energy sources for the heating: the acoustic fluctuations and the magnetic network. The energy flux density of the sound waves in the chromosphere has been estimated to about $10\,\mathrm{W\,m^{-2}}$. This would be sufficient to heat the chromosphere up to 10 000 K, but this is not nearly enough for the coronal temperatures, which also require practically continuous heating. If the heating were turned off, the chromosphere would cool down in about 20 minutes.

The high temperature of the corona was once a great surprise and its heating is still among the toughest problems in solar physics. The acoustic fluctuations do not reach the coronal altitudes, but in principle there is no lack of energy. The energy flux needed to power the magnetically active regions is of the order of $10^4\,\mathrm{W\,m^{-2}}$, which on the average is only a fraction of $10^{-4}$ of the power in electromagnetic radiation. But the corona is optically very thin and there is no known mechanism to absorb the electromagnetic radiation. Thus the heating must be related to the magnetic field. Fortunately, there is enough energy also in the solar magnetic field. The problem is how to convert it into heat, in particular in the narrow transition region but also higher up where the mean temperature still increases from $10^6$ K to $2 \times 10^6$ K.

We can think of several mechanisms to dissipate the magnetic energy as heat, e.g., waves, instabilities, current sheet dissipation, and reconnection, which will all be discussed in the later chapters. The energy balance in MHD (Chap. 6) can be expressed writing the *Poynting theorem* in the form

$$-\oint_{\partial\mathscr{V}} \mathbf{E}\times\mathbf{H}\cdot d\mathbf{a} = \frac{\partial}{\partial t}\int_{\mathscr{V}} \frac{B^2}{2\mu_0}\,d\mathscr{V} + \int_{\mathscr{V}} \frac{J^2}{\sigma}\,d\mathscr{V} + \int_{\mathscr{V}} \mathbf{V}\cdot\mathbf{J}\times\mathbf{B}\,d\mathscr{V}\,. \tag{1.12}$$

The LHS describes the magnetic energy entering as Poynting flux through the surface $\partial\mathscr{V}$ of the volume $\mathscr{V}$ where the energy may show up as increasing magnetic energy (first term on the RHS) and be dissipated through ohmic heating (second term on the RHS) and mechanical work (acceleration, third term on the RHS) by the magnetic force ($\mathbf{J} \times \mathbf{B}$). Note that the ohmic, or more accurately resistive, term does not need to be determined by classical collisional resistivity but may rise from turbulence and/or wave–particle interactions.

MHD waves (known also as *Alfvén waves*, see Chap. 6) are excited by the motion of magnetic and acoustic disturbances in or near the photosphere. Spectral features in the transition region are wider than could be expected for the hot gas. The excess Doppler widening has been estimated to correspond to the velocity $10^4\,\mathrm{m\,s^{-1}}$, which may be a signature of upgoing Alfvén waves. When these waves propagate outward they are damped and part of their energy is transformed to heat. The linear damping of the Alfvén waves is, however, a very slow process. Nevertheless, within the diverging coronal holes the wave heating may be the only alternative, because there are no unstable flux tubes nor current sheets. One proposal how the heating could take place has been *phase-mixing* of waves of different wavelengths and speeds propagating in the same spatial volume. This can lead to large spatial gradients where the effective resistivity increases and shows as ohmic dissipation of the wave energy in the Poynting theorem. Phase-mixing is an example of *turbulent phenomena* in space plasmas.

Another proposed explanation for damping of Alfvén waves is that the waves have high enough frequencies to be damped by the cyclotron resonance with the plasma ions. Alfvén waves become electromagnetic ion cyclotron waves at frequencies close to the local ion cyclotron frequency (Chap. 4), and these waves are very efficiently damped by resonant interaction with ions (Chap. 5). As the magnetic field and, therefore, the cyclotron frequencies decrease with increasing radial distance, waves created at or near the solar surface by micro-flaring and/or turbulent motions can propagate without damping until they reach the distance at which the cyclotron frequency becomes comparable to the wave frequency.

Observations of ion temperatures in coronal holes indicate that minor ion populations (e.g., oxygen) can be very hot (up to $10^8$ K) and that their temperatures are anisotropic, being larger in the perpendicular direction relative to the magnetic field. This is a signature of cyclotron heating because the ions with the lowest cyclotron frequencies should be heated most efficiently and because the heating is due to wave electric fields directed perpendicularly to the ambient magnetic field. However, some theoretical calculations predict even too efficient wave damping by the heavy ions with lower cyclotron frequencies than the proton gyro frequency, leaving almost no wave energy to heat the major species. As a summary, cyclotron heating in the solar corona is not yet completely understood.

Even if the waves generated near the solar surface have small frequencies, the phenomenon called *turbulent cascading* may allow short wavelength fluctuations to be generated from the long wavelegth ones. The large wavenumber fluctuations may again be efficiently damped at scales close to the ion gyro radii. This turbulent heating mechanism in a way combines the ideas of cyclotron heating and phase mixing.

We know from observations that flux tubes in different scales, such as coronal loops, are continuously created and disrupted through various instabilities. The disrupting flux tubes convert magnetic energy into heat and acceleration whenever the disruptions take place, but the disruptions may be too sparse and localized to explain the heating of the whole corona. These processes may be important during strong solar activity, but the corona is hot also during quiet periods.

The *Skylab* mission revealed in 1973 that there are *X-ray bright points* everywhere on the Sun. Later it was demonstrated that their distribution is uniform over the whole Sun and that they exist also during quiet phases of the solar activity. They resemble small flares and the underlying particle acceleration is most likely due to continuous reconnection processes of the ever-changing magnetic field structures in the low corona.

While large flares can release some $10^{25}$ J of energy in the time scale of 10 minutes, they are too infrequent and can account for at most 1% of the heat to sustain the $10^6$ K temperature of the corona. Thus if small flares should explain the heating, they would need to be very abundant, indeed. A direct scaling down from the large flares may not be straightforward and the small flares may be relatively more dissipative.

The EUV observations at temperatures of $10^5$ K (i.e., in the transition region) have shown that there are localized hot spots that explode and shoot material upward at the speeds of hundreds of km $s^{-1}$. These hot upward plasma jets occur above the lanes of the magnetic network. It has been claimed that the jets would carry enough energy to heat the corona but the observations are inconclusive.

The UV and EUV observations of the *SOHO* and *TRACE* satellites have finally shown that there are even larger numbers of (relatively) small explosive events than was previously thought all over the Sun, perhaps some 20 000 events per minute. The inner solar atmosphere is very active also during the quiet phases of the solar cycle. The small activations have been dubbed *microflares* or *nanoflares*. Although this terminology is a bit inexact, "micro" can be associated with events of the order of $10^{19}$ J, which you need about one million to correspond to a flare, and "nano" with events of $10^{16}$ J, which you need one billion to one flare.

The brightest micro/nanoflares lie above regions of enhanced magnetic fields of the magnetic network and the stronger events correspond to greater fluctuations. This suggests that the lower corona is not only heated but continuously replenished by chromospheric material that has been heated to coronal temperatures (see Fig. 1.3). Thus a substantial part of the energy may come with the heated plasma from below. One scenario is that the new magnetic field emerges from the Sun in the centers of supergranular cells and is carried to their edges by the convective motion and finally reconnected with the magnetic field from the neighboring cells. In this scenario the energy released by the reconnection powers the microflares observed in the overlying low corona.

There has been some discussion whether the small-scale flares are abundant enough, or not, to account for the coronal heating. Some observations support this interpretation, others do not. However, observations have conclusively shown that there is a correlation between the solar magnetic field and coronal heating. The variability of the small-scale magnetic elements observed in the photosphere (so-called magnetic carpet) has been found to correlate with temperature fluctuations in the corona. Furthermore, observations of the temperature distribution of forming polar plumes within the coronal holes seem to correlate with photospheric fine-structure associated also with the supergranular structure and magnetic network.

## 1.2 The Carrier to the Earth: the Solar Wind

Toward the end of the 19th century it had become evident that there must be a connection between the solar activity and magnetic disturbances on the Earth, which is not mediated by electromagnetic radiation. There were still some very prominent sceptics, in particular Lord Kelvin, because it was very difficult to explain how such a connection could be established.

Lindemann [1919] seems to have been the first to suggest that quasi-neutral charged particle ejections related to solar activity were responsible for non-recurrent magnetic storms at the Earth. In 1929 Chapman proposed that the solar flares would emit plasma clouds and if such a cloud were to hit the Earth's magnetic field, it would cause magnetic disturbances. But how could these clouds escape from the strong gravitational field of the Sun? After all, the escape velocity on the solar surface is 618 km s$^{-1}$. The kinetic energy of a proton at that speed is 2 keV, corresponding to a temperature of $2 \times 10^7$ K, which was too much to be believed in those days. Today we know that such temperatures really do occur in coronal loops and flares, and the escape is no longer such a big mystery, although we do not yet know the details of how the plasma is heated and accelerated.

During the 1950s Biermann [1951, 1957] demonstrated that the structures of cometary tails were consistent with a *continuous* corpuscular outflow from the Sun, unrelated to large flares. Later Alfvén pointed out that the flow must be magnetized plasma. The first direct in situ observations of the solar wind came from the Russian *Lunik* III and *Venus* I spacecraft in 1959, and the definitive proof of its continuous nature was provided by the U.S. *Mariner* II in 1962–1967.

Today we know that there are two main types of solar wind, a fast (about 750 km s$^{-1}$), tenuous, and a denser but slower (about 350 km s$^{-1}$) wind. The details of the source regions and mechanisms are still under investigation, but the general view is that the fast wind originates from large coronal holes at high solar latitudes whereas the slow wind emerges from smaller and less permanent structures at lower latitudes. In addition to these, the CME-related outflow can be considered as a third independent solar wind type. Solar wind has never disappeared during the more than three decades it has been monitored. On May 11, 1999, the slow (300 km s$^{-1}$) wind had for a short while an extremely low density of 0.2 cm$^{-3}$ near the Earth.

### 1.2.1 Elements of solar wind expansion

Before direct spacecraft observations Chapman [1957] presented a static model to describe the existence of the continuous solar wind. He considered a sphere around the Sun and assumed that the thermal flux through the surface was constant. Assuming that $T \to 0$ when $r \to \infty$ he found the solution

$$T = T_0 (R_\odot / r)^{2/7} . \tag{1.13}$$

For a coronal temperature of $T_0 = 10^6$ K this predicts a temperature of $10^5$ K at 1 *AU*, which is quite good, although it was not known in 1957. An evident drawback of the model was that far from the Sun the pressure approaches a constant that is much larger than the pressure of the interstellar gas. As the temperature decreases toward zero with increasing distance, the density would have to increase without bound, which of course is unphysical.

One year later Parker [1958] presented another solution to the problem. While this solution was also based on strong simplifications, its basic idea is important. Parker noted that the corona cannot be in static equilibrium; it must either expand or collapse. Guided by this insight, he succeeded in predicting a supersonic solar wind just before the first satellite observations showed that he was essentially right. Parker's argumentation was the following.

Assume time-independent spherically symmetric outward-directed flow. Neglect the magnetic effects and write the continuity equation, momentum equation and equation of state as

$$4\pi r^2 nV = \text{const} \tag{1.14}$$

$$nmV\frac{dV}{dr} = -\frac{dP}{dr} - \frac{Gm_\odot mn}{r^2} \tag{1.15}$$

$$P = nk_BT\,. \tag{1.16}$$

Let the expansion be isothermal. This is clearly not true, but it is interesting to see where it leads. The solutions are of the form

$$\left(V - \frac{v_c^2}{V}\right)\frac{dV}{dr} = \frac{2v_c^2}{r} - \frac{Gm_\odot}{r^2}\,, \tag{1.17}$$

where $v_c = \sqrt{k_BT/m}$ is the isothermal sound speed, i.e., the polytropic index is set $\gamma = 1$. This equation has a *critical point*: $V = v_c$ , $r = r_c = Gm_\odot/(2v_c^2)$. Integration gives a family of curves

$$\left(\frac{V}{v_c}\right)^2 - \ln\left(\frac{V}{v_c}\right)^2 = 4\ln\frac{r}{r_c} + \frac{2Gm_\odot}{rv_c^2} + C\,. \tag{1.18}$$

Figure 1.8 illustrates these solutions. Solutions in regions I and II are unphysical and those in III have too high (supersonic) a velocity at the source. The solution IV crossing the critical point is Parker's solution for the supersonic solar wind. The critical point fixes the constant of integration to $C = -3$. Also V is a physically valid solution, called *stellar breeze*. There are stars that produce subsonic stellar breezes.

**Train your brain** by calculating the details of Parker's solution.

While elegant, Parker's solution is too simple for the real solar wind. In fact, the isothermal polytropic index $\gamma = 1$ leads to a diverging enthalpy (see Eq. (1.25) below), whereas for $\gamma = 5/3$ there is no critical point and thus the supersonic flow is not described correctly. The wind cools, as it expands, and thus thermal conduction must be taken into account. Because the solar wind plasma is effectively collisionless, ions and electrons cool with

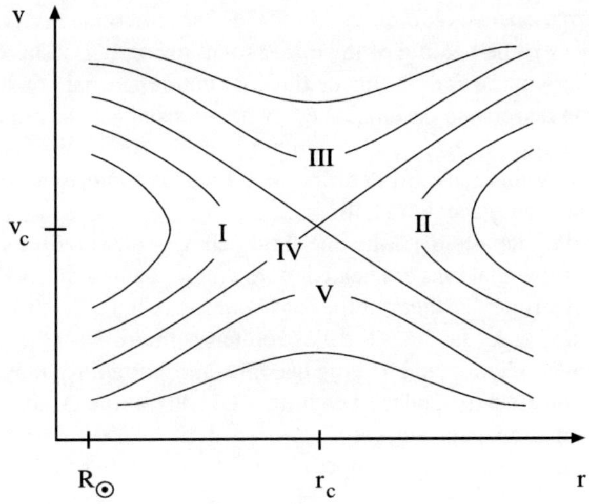


**Fig. 1.8** Solutions of (1.18).

different cooling rates and the interaction of plasma with magnetic fluctuations plays a different role in electron and ion expansion. Also the details of the fast and slow solar wind are different because the physical processes in their source regions are different. The observed solar wind properties at 1 *AU* are summarized in Table 1.1.

**Table 1.1** Typical solar wind parameters at 1 *AU*. $v_A = B/\sqrt{\mu_0 \rho_m}$ is the Alfvén velocity.

| | slow wind | fast wind |
|---|---|---|
| $V$ (km s$^{-1}$) | 350 | 750 |
| $n_e$ (m$^{-3}$) | $1\times10^7$ | $3\times10^6$ |
| $T_e$ (K) | $1.3\times10^3$ | $1\times10^5$ |
| $T_p$(K) | $3\times10^4$ | $2\times10^5$ |
| $B$ (nT) | 3 | 6 |
| $v_A$ (km s$^{-1}$) | 20 | 70 |

The solar wind transfers energy from the Sun. In the corona we must consider kinetic energy, internal energy, gravitational energy, thermal conduction, radiation, and heating. Most of these must also be taken into account in the description of the solar wind acceleration beyond the sound and Alfvén velocities.

In a steady state the divergence of the total energy flux must be zero

$$\nabla \cdot \left[ \mathbf{V} \left( \frac{1}{2}\rho V^2 + H - \frac{G m_\odot \rho}{r} \right) - \kappa \nabla T + \mathbf{F}_R + \mathbf{F}_H \right] = 0 \tag{1.19}$$

Here $H$ is the internal energy (*enthalpy*) and $\kappa$ the thermal conductivity. $\nabla \cdot \mathbf{F}_R$ describes the radiation and $\nabla \cdot \mathbf{F}_H$ the heating of the upper solar atmosphere. There is a temperature maximum somewhere in the corona. Inside this maximum thermal conduction is inwards, toward the transition region and chromosphere, where it balances the radiative loss through the strong Lyman $\alpha$ line. Outside the maximum thermal conduction is outwards. Chapman's model had only this outward contribution, whereas the original isothermal Parker solution did not take it into account at all.

The real solar wind departs from the one-fluid behavior already in the corona. Modern model calculations show that ions are heated more efficiently and reach a higher maximum temperature than electrons. Further out the ions cool faster than electrons, and at 1 *AU* the ion temperature is no longer far from the electron temperature.

In the outer corona radiation and heating become unimportant for plasma dynamics, but the internal energy of the plasma deserves further consideration. Assume that the coronal gas consists of protons and electrons only, let $n = n(r)$, $T = T(r)$, $n_e \approx n_i \approx n$, and neglect, for simplicity, the differences in the temperatures. Then the pressure is

$$P = n_e k_B T + n_i k_B T = 2nk_B T \tag{1.20}$$

and the thermal energy of the gas in a volume $\mathcal{V}$ is

$$U = \frac{3}{2}(n_e + n_i)k_B T \mathcal{V} = 3nk_B T \mathcal{V} \,. \tag{1.21}$$

The gravitational potential is given by

$$\Phi = -\frac{G m_\odot m n \mathcal{V}}{r} \,. \tag{1.22}$$

The thermal energy lifts the gas up when the volume $\mathcal{V}$ expands. At the same time the internal pressure pushes new gas into this volume performing the work $P\mathcal{V}$. The free energy is the enthalpy

$$H = U + P\mathcal{V} = 5nk_B T \mathcal{V} \,. \tag{1.23}$$

Assuming a temperature of $T = 2 \times 10^6$ K we find

$$\frac{H}{|\Phi|} \approx 0.5 \,. \tag{1.24}$$

This means that the heating of the corona to this "classical" temperature does not provide enough free energy to exceed the gravitational potential and the corona should collapse, not expand. Thus there must be some mechanism(s) doing extra work $Q$ on the gas. Using the actual solar wind observations the required energy can be estimated to be about $H + Q = 1.25|\Phi|$. There is no generally accepted theory yet to explain what powers the escape. Most likely it is of magnetic origin and associated to the heating of the ions in the corona. Once the ions escape, the more mobile electrons follow.

Assuming that there is enough energy available for the solar wind expansion and neglecting details of $\nabla \cdot (\mathbf{F}_R + \mathbf{F}_H)$, the energy transport equation can be written as

$$nmVr^2\left(\frac{1}{2}V^2+\frac{\gamma}{\gamma-1}\frac{P}{nm}-\frac{Gm_\odot}{r}\right)=r^2\kappa\frac{dT}{dr}+F_\infty\,. \tag{1.25}$$

Here $\kappa = \kappa_0 T^{5/2}$ ($\kappa_0 \approx 10^{-11}\,\mathrm{W\,m^{-1}\,K^{-1}}$) and $F_\infty$ is the energy flux far from the Sun. This equation also takes the cooling of the expanding (single fluid) wind into account. The internal energy is written in the form showing the polytropic index $\gamma$ and we immediately recognize the enthalpy problem with Parker's isothermal ($\gamma = 1$) solution in the second term of the LHS of (1.25).

There are three basically different classes of solutions, depending on the asymptotic behavior of the temperature:

(1) $T \sim r^{-2/7}$ heat conduction dominates in the far region
(2) $T \sim r^{-2/5}$ kinetic flux dominates in the far region
(3) $T \sim r^{-2/3}$ adiabatic expansion

In the fluid picture stellar winds belong either to class 1 (cold, tenuous winds) or class 3 (hot, dense winds), whereas class 2 is a limiting case between these two. However, the different particle species may fall into different categories. According to observations the proton temperature at 1 $AU$ behaves roughly as $T_p \sim r^{-2/3}$ being in the adiabatic class, whereas $T_e \sim r^{-1/3}$, which is closer to thermal conduction.

There are several reasons for the different cooling rates. The electrons are bound more tightly to the magnetic field of the solar wind and electrons and ions react in different ways to turbulence and plasma waves. Note that while any of these effects may be slow, the spatial and temporal scales are vast compared, e.g., to gyro radii or gyro periods.

### 1.2.2 The interplanetary magnetic field

A critical element to carry the effects of solar activity to the heliosphere is the magnetic field of the solar wind, the *interplanetary magnetic field* (IMF). In addition to its effects on the local properties of the solar wind, the IMF also breaks the solar rotation and it is critical to the dynamics of plasma environments of solar system bodies.

The observed structure of the IMF varies considerably from the ecliptic to the poles. To begin, let us consider a cylindrically symmetric case in the equatorial plane. Assume that the flow is radial and let $\Omega$ be the angular speed of the solar rotation. Let the angle between the radial direction and the magnetic field be $\psi$ and assume that the IMF is frozen into the expanding solar wind. (The frozen-in concept will be introduced in Chap. 6.) Close to the Sun the plasma rotates with the body but with the solar wind expansion the field is wound to a spiral. Let $\mathbf{V}$ be the flow velocity assumed to be radial, for simplicity. Its component perpendicular to the IMF is $V_\perp = V\sin\psi$. This can be imagined as the speed of the field line in this direction. The high conductivity ties the field line to the surface of the Sun, actually to the so-called *source surface*, where the magnetic field is, in the first approximation, radial. Thus the speed of the field line perpendicular to the radial direction is $\Omega(r - R_\odot)$, and

$$V\sin\psi = \Omega(r-R_\odot)\cos\psi \tag{1.26}$$

⇒

$$\tan\psi = \frac{\Omega(r - R_\odot)}{V} . \tag{1.27}$$

When $r$ increases, this approaches the Archimedean spiral. In this context it is known as the *Parker spiral.*

**Feed your brain**

With the help of literature discuss the description of the magnetic field in the solar atmosphere in terms of spherical harmonics. What is the role of the source surface in this description? What makes the magnetic field radial at the source surface?

We can calculate the magnetic field behavior as a function of distance from the Sun in a simple way. Let **B** be radial and constant on the surface of the Sun and write **B** and **V** in spherical coordinates with the origin in the center of the Sun $(r, \theta, \phi)$, where $\theta$ is the polar angle and $\phi$ the azimuthal angle,

$$\mathbf{B} = (B_r, 0, B_\phi) , \quad \mathbf{V} = (V_r, 0, V_\phi) . \tag{1.28}$$

Note that the components of the vectors are functions of $r$. From $\nabla \cdot \mathbf{B} = 0$ we get

$$B_r = B_0 (R_\odot / r)^2 . \tag{1.29}$$

Thus the radial component of the field decreases as $r^{-2}$. To find the azimuthal behavior we can write the steady state azimuthal force balance (see Chap. 2, Eq. (2.145)) as

$$\rho(\mathbf{V} \cdot \nabla \mathbf{V})_\phi = (\mathbf{J} \times \mathbf{B})_\phi , \tag{1.30}$$

where the plasma pressure is assumed to be azimuthally symmetric $(\nabla P)_\phi = 0$. Using Ampère's law and multiplying by $r^3$ we obtain

$$r^2 \rho V_r \frac{d}{dr}(r V_\phi) = \frac{1}{\mu_0} r^2 B_r \frac{d}{dr}(r B_\phi) . \tag{1.31}$$

The mass flux $r^2 \rho V_r$ and the magnetic flux $r^2 B_r$ are constants and we can integrate this equation to get

$$L = r V_\phi - \frac{r B_r B_\phi}{\mu_0 \rho V_r} . \tag{1.32}$$

In the constant of integration $L$ the first term is the *angular momentum per unit mass* and the second term describes the integral of the torque corresponding to the change in the angular momentum, known as *magnetic braking.*

To express $B_\phi$ in terms of $B_r$ we consider the frame that rotates with the angular speed $\Omega$. In this frame the velocity vector is $(V_r, 0, V_\phi - r\Omega)$. This vector is parallel to **B** and thus

$$B_\phi = \frac{V_\phi - r\Omega}{V_r} B_r . \tag{1.33}$$

At large distances $B_\phi \propto r^{-1}$, i.e., it decreases more slowly than the radial component, which explains the spiral formation.

Define now the radial *Alfvén Mach number* $M_A$

$$M_A = \frac{V_r}{v_A} = \frac{V_r\sqrt{\mu_0\rho}}{B_r} \tag{1.34}$$

$\Rightarrow$

$$V_\phi = \Omega r \frac{M_A^2\left(\frac{L}{r^2\Omega}\right) - 1}{M_A^2 - 1} \tag{1.35}$$

According to observations $M_A$ increases from ~0.1 in the corona to ~10 at $1\,AU$. Thus there is a critical point in the expression of $V_\phi$ where $M_A = 1$ at a certain distance $r = r_A$, which is called the *Alfvén radius*. At this distance (about $12R_\odot$) the flow becomes superalfvénic. As the azimuthal speed cannot be singular at that point we find the angular momentum per unit mass

$$L = \Omega r_A^2 \,. \tag{1.36}$$

This is equal to the angular momentum for a solid body with the radius $r_A$.

We can now write the azimuthal velocity as

$$V_\phi = \frac{V_r/v_A - 1}{(V_r r^2)/(v_A r_A^2) - 1}\Omega r \,. \tag{1.37}$$

Close to the Sun this reduces to

$$V_\phi \simeq r\Omega \tag{1.38}$$

corresponding to rigid rotation with the Sun. On the other hand at large distances

$$V_\phi \simeq r_A^2\Omega/r \,, \tag{1.39}$$

which expresses the conservation of angular momentum from the Alfvén radius outward. Thus $r_A$ can be interpreted as a lever arm, with which the solar wind brakes the solar rotation.

Out of the equatorial plane ($\theta \neq \pi/2$) the calculation is more complicated. The azimuthal component of the field turns out to be

$$B_\phi \approx -\frac{B_0 R_\odot^2 \Omega \sin\theta}{rV_r} \,. \tag{1.40}$$

Thus, far from the Sun the total magnetic field behaves as

- $B \to r^{-1}$ in the equatorial plane (the spiral becomes tightly wound)
- $B \to r^{-2}$ in the direction of the poles

Between the equatorial plane and the polar direction the field has a helical structure. At $1\,AU$ the equatorial spiral angle is typically about 44°.

**Train your brain**

1. Derive (1.40).

2. Show that the mass and angular momentum losses are related by

$$\frac{dJ}{dt} = \frac{2}{3}\Omega r_A^2 \frac{dm}{dt} \tag{1.41}$$

Calculate this for the present Sun and estimate how much time the present magnetic braking would need to stop the rotation. Compare the efficiency of the magnetic braking for $r_A = 12\,R_\odot$ to $r_A = R_\odot$.

### 1.2.3 The observed structure of the solar wind

The real solar wind is much more structured in space and time than the simple model calculations in the previous section suggest. The escaping flow originates from the coronal holes whose shapes and locations change all the time. Space-borne observations looking through the holes to the photosphere show further that the escape is highly structured within individual holes. In addition to this variability the solar eruptions eject faster or slower plasma and magnetic clouds to the background solar wind flow. These structures can drive various shock phenomena in the wind to large distances beyond the Earth's orbit.

When the solar activity is at its minimum, the solar magnetic field is as poloidal as it ever gets and the coronal structure is dominated by two large polar holes with opposite magnetic polarities. The almost radial solar wind flow escapes mostly from these holes and drags the frozen-in magnetic field in such a way that a *heliospheric current sheet* forms near the equatorial plane. However, as the holes have asymmetric shapes, the current sheet is asymmetric as well (Fig. 1.9). When the Sun rotates, the current sheet moves up and down, which led Alfvén to call this structure a ballerina's skirt. The Earth is either above or below the skirt. Depending on whether the field is pointing mostly toward or away from the Sun, the Earth is said to be either in the *toward sector* or the *away sector*. Superposed to this large-scale structure there are large variations in all components of the IMF.

Around the time of solar maximum the solar magnetic field structure is much less regular and the polar coronal holes are reduced in size. On the other hand, there are more smaller-scale opening and closing structures at lower latitudes. This also makes the solar wind structure more variable, which in turn drives magnetic activity in the terrestrial environment.

While the structure and magnetic field behavior in the polar directions can be inferred theoretically and even seen in pictures of polar plumes and in coronagraph images, it was not until the 1990s when first direct observations of the off-ecliptic solar wind behavior became available through the joint ESA and NASA spacecraft *Ulysses*, which was the first spacecraft on a high-inclination orbit around the Sun. Jupiter's gravitational field was used to insert the spacecraft into a trajectory with the aphelion at 5.3 $AU$, the perihelion at 1.3 $AU$, and the highest heliographic latitude 80°. *Ulysses* reached this point for the

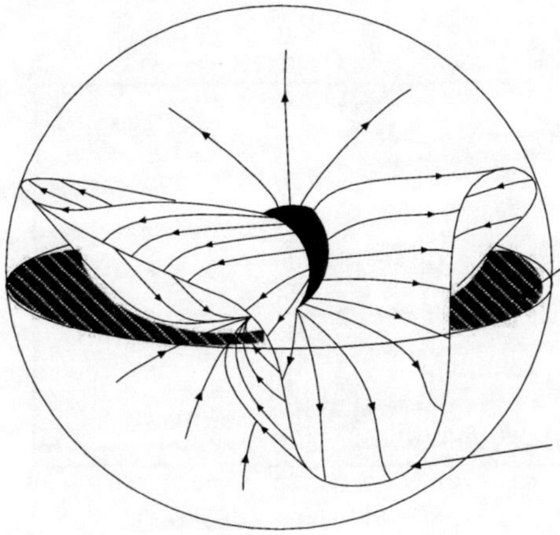


**Fig. 1.9** The ballerina skirt formation of the solar wind follows the shape and location of the dominating polar coronal holes.

first time above the southern solar hemisphere in September 1994 and above the northern hemisphere in March 1995. The next polar passages took place in 2000 and 2001. Early in 2004 *Ulysses* was again at its aphelion and began its third and last orbit until its radioisotope generators had decayed so much that the spacecraft practically froze to death in the summer of 2009.

Note that the large variability of the solar wind speed at 1 *AU* is partially due to the variable vertical distance from the heliospheric current sheet. The slowest speeds of the solar wind arise near the edges of the polar coronal holes and from intermittent coronal holes at lower latitudes. This is nicely illustrated in the observations by the *Ulysses* spacecraft during its first passage from high southern heliographic latitudes through the ecliptic plane to high northern latitudes (Fig. 1.10). When the spacecraft was within $\pm 20^\circ$ of the ecliptic, it observed both slow and fast solar wind, but at higher latitudes it encountered only fast, tenuous solar wind from the polar coronal holes.

### 1.2.4 Perturbed solar wind

While the steady fast solar wind with a sufficiently strong southward IMF component can drive significant activity in the magnetosphere, strongly perturbed solar wind is of particular interest to space storms. We will later discuss shocks (Chap. 11) and CMEs (Chap. 12) in the solar wind in greater detail, but for the completeness of the discussion on the solar wind a few words should be said here.

There are several types of shocks in the solar wind. Once a CME has left the vicinity of the Sun, it is customary to rename it an *interplanetary CME* (ICME). High-speed ICMEs drive shocks, the interaction regions between sectors of fast and slow solar wind evolve to shock structures, planets are obstacles to the solar wind flow causing shocks, and fi-

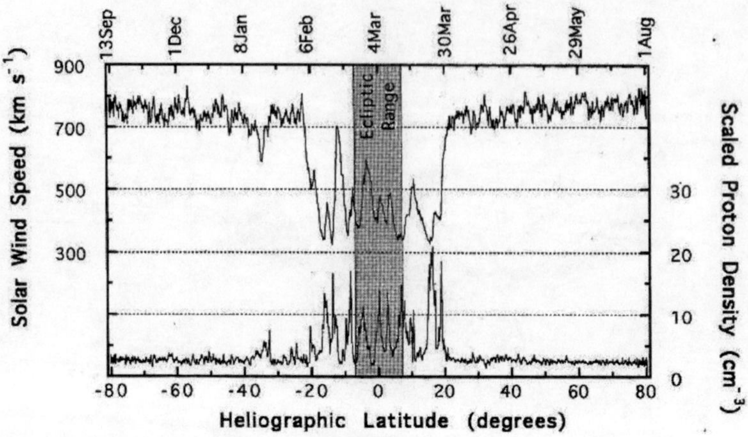


**Fig. 1.10** *Ulysses* observations of solar wind speed (upper curve) and proton density (lower curve) as a function of the heliographic latitude (From Phillips et al [1995].)

nally when the solar wind meets the interstellar plasma, it again becomes subsonic and a *termination shock* structure is formed somewhere inward from the heliopause.

For space storms the most important class of solar wind shocks are those driven by the fast ICMEs (Fig. 1.11). The ICMEs originate with different speeds, ranging from a few tens of $\mathrm{km\,s^{-1}}$ up to about $2000\,\mathrm{km\,s^{-1}}$. The slowest ICMEs are soon accelerated close to the speed of the ambient solar wind flow, whereas the fast ICMEs are decelerated. In order to drive a shock ahead of it the ICME must have a supersonic, or actually super-Alfvénic, velocity relative to the ambient plasma flow. Thus a slow ICME does not drive a shock, except perhaps close to the Sun, whereas a fast ICME does, as is clearly the case with a large number of ICMEs observed at $1\,AU$.

Close to the Sun the CME-related shocks are important in the acceleration of solar energetic particles. When an ICME and the shock ahead of it hit the magnetosphere of the Earth, they shake the system and, depending on the magnetic structure of the ICME–shock system, they drive the most severe magnetic storms in the terrestrial environment.

Another important class of solar wind shocks developing during the outflow are the *corotating interaction regions* (CIR). Figure 1.12 illustrates their formation. Consider a given direction in the non-rotating frame. Assume that at first slow wind is blowing in this direction. As the Sun rotates, a source region of fast wind turns into the same direction and the faster and more tenuous flow catches the slower and denser flow. Because both flows consist of ideal MHD plasma, they do not easily mix. As discussed in Chap. 11, a steepening boundary structure begins to form. Close to the Sun the field lines are still nearly radial and the boundary is more like a tangential discontinuity. Further out the Parker spiral becomes wound more tightly in the slower flow ahead the structure than in the fast flow behind. A fully developed CIR shock exhibits a *forward shock* ahead the structure and a *reverse shock* behind it. Note that one must be careful with the frame of reference: In the frame of the fast flow the reverse shock propagates backward, but in the frame of, e.g., the Earth or a spacecraft making observations in the solar wind, it

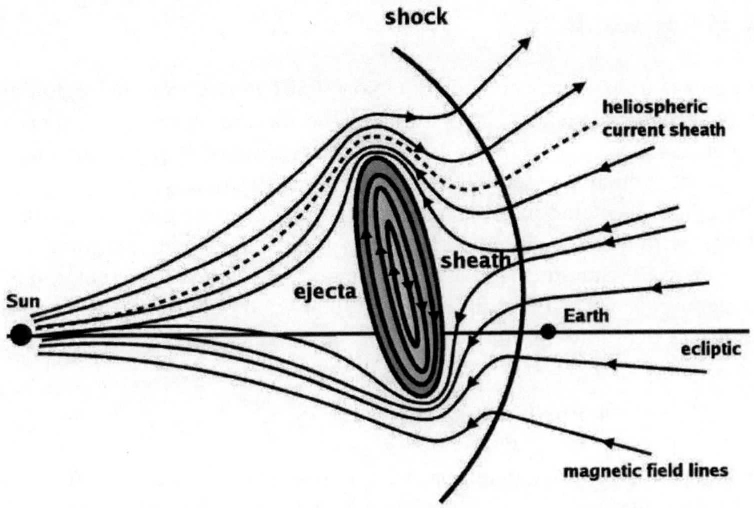


**Fig. 1.11** A sketch of a shock driven by an ICME. Note that the magnetic field of the ICME can rotate in different directions about the core, there can be a strong core field, and the whole structure can be strongly tilted. (Adaptation from Gosling and McComas [1987] by E. Kilpua.)

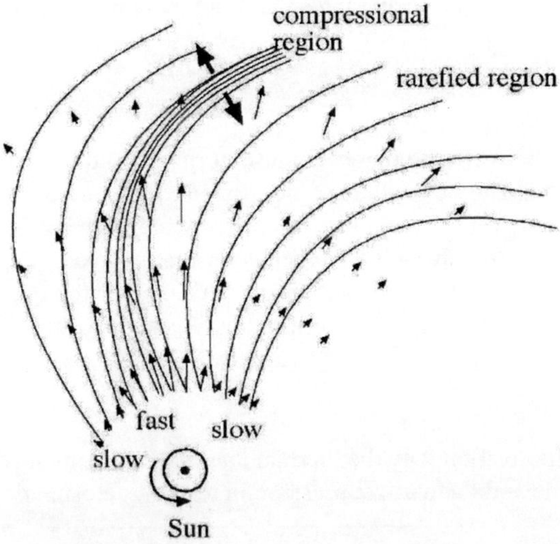


**Fig. 1.12** Formation of a corotating interaction region. The fast solar wind pushes toward slower wind and compresses the flow. The compression is observable at 1 $AU$, but does not usually form a shock structure until the compression has propagated beyond 2 $AU$.

propagates outward. The CIR-related shock formation usually takes place only beyond the Earth's orbit, whereas a CIR impinging upon the Earth's magnetosphere is a smoother structure of compressed plasma across which the speed changes from slow to fast.

## 1.3 The Magnetosphere

The term *magnetosphere* was coined by Gold [1959] to describe the region around the Earth where the geomagnetic field determines the motion of the charged particles. All magnetic planets (Mercury, the Earth, Jupiter, Saturn, Uranus, Neptune) are known to have a magnetosphere, which is essentially a magnetic cavity in the solar wind. The magnetic force deflects the solar wind particles around this cavity before they hit the surface of the planet. Planets with a dense enough atmosphere (Venus and Mars) and comets, when they are active close to the Sun, form structures that are called *induced magnetospheres*. In that case the deflection is due to the inability of the solar wind plasma to penetrate through the ionized atmosphere or ionized cometary gas.

### 1.3.1 Formation of the Earth's magnetosphere

When the supersonic solar wind approaches the magnetic field of the Earth, it pushes the magnetic field on the dayside and stretches it to a long tail on the nightside. In the first approximation the ideal solar wind and magnetospheric MHD plasmas cannot mix and a well-defined *magnetopause* forms. The distance from the center of the Earth to the dayside magnetopause can be estimated calculating the pressure balance between the magnetic pressure inside the magnetopause and the solar wind dynamic pressure

$$K\rho_{mSW}V_{SW}^2\cos^2\psi = \frac{B_{MS}^2}{2\mu_0}\,, \tag{1.42}$$

where $\psi$ is the angle between the magnetopause normal and the solar wind direction, $SW$ refers to the solar wind, and $MS$ to the magnetosphere. $K$ is a constant that would be 2 for an elastic collision (pure reflection) and 1 for a purely inelastic collision (absorption). For a fluid deflected around the obstacle $K$ depends on the upstream Mach number and is in the case of the Earth about 0.9. Typical subsolar distance of the magnetopause from the center of the Earth is about $10\,R_E$ (the Earth radius, $R_E \approx 6370\,\text{km}$).

**Train your brain**

Present a physical motivation why the thermal and magnetic pressures can be neglected in the solar wind side and the particle pressure in the magnetospheric side of (1.42).

As the solar wind flow in the frame of reference of the Earth is supersonic and super-Alfvénic, actually supermagnetosonic, a collisionless shock front called the *bow shock* is formed upstream of the magnetosphere (Fig. 1.13). For typical solar wind parameters the nose of the shock in the solar direction is about $3\,R_E$ upstream from the nose of the magnetopause. The shock is a fast MHD shock (Chap. 11) and it converts a considerable amount of solar wind kinetic energy to heat and electromagnetic energy. The region between the bow shock and the magnetopause is called the *magnetosheath*.

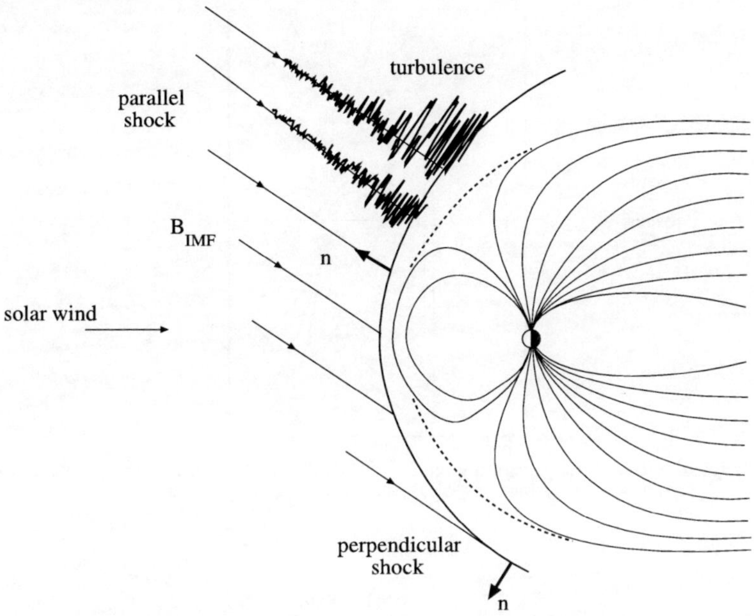


**Fig. 1.13** A magnetosphere and its bow shock. Concepts of parallel and perpendicular shocks refer to the angle between the shock normal and the direction of the upstream magnetic field (IMF). They will be discussed in Chap. 11.

As the magnetopause shields the planetary magnetic field from the solar wind, the magnetopause is a current layer where the current is determined by Ampère's law. Similarly the stretching of the long magnetotail requires a strong current inside the magnetosphere. Thus the solar wind–magnetosphere interaction must drive currents in the system. These current systems and their stability belong to the key issues in magnetospheric physics.

The first description of the magnetic cavity shielded from the solar wind by a current sheet was given by Chapman and Ferraro [1931] in their attempt to explain how magnetic storms would be driven by corpuscular radiation from the Sun. They essentially solved an image dipole problem of magnetostatics where the real dipole is inside the magnetosphere and the image dipole is placed in the infinitely conductive medium (Fig. 1.14).

Using modern terminology a diamagnetic current (see Eq. 6.48))

$$\mathbf{J}_{CF} = \frac{\mathbf{B}_{MS}}{B_{MS}^2} \times \nabla P_{SW} \tag{1.43}$$

separates the vacuum dipole from the conductive medium. This current is known as the *Chapman–Ferraro current* (Fig. 1.15). Because the IMF at 1 *AU* is only a few nanoteslas, the magnetopause current must shield the magnetospheric field to almost zero just outside the current sheet. Consequently, the field immediately inside the magnetopause increases so that about one half of it comes from the Earth's dipole and the other half from the current sheet, as illustrated in Fig. 1.16.

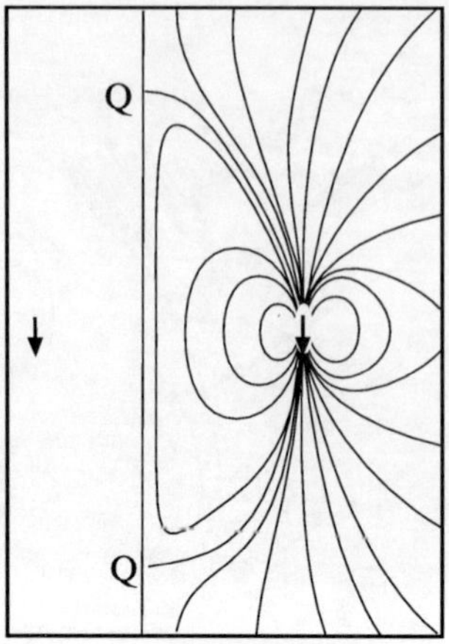


**Fig. 1.14** Image dipole solution illustrating the formation of two magnetic neutral points, cusps (Q), discussed in the next section.

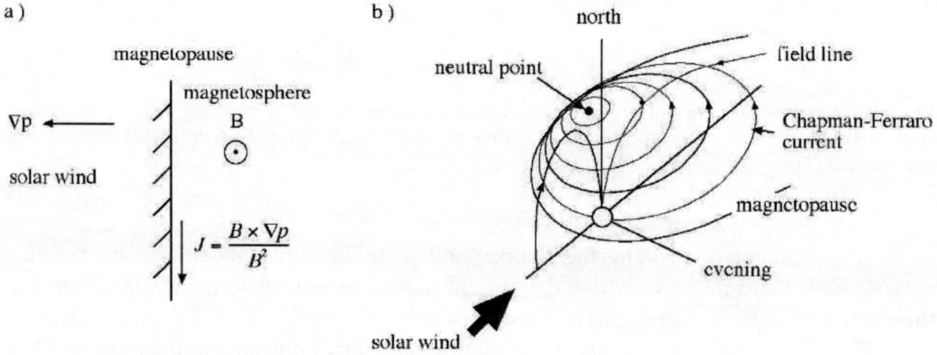


**Fig. 1.15** a) Principle of the Chapman–Ferraro current formation in two dimensions. b) Three-dimensional closure of the Chapman–Ferraro current.

The Chapman–Ferraro model describes a teardrop-shaped closed magnetosphere compressed on the dayside and stretched on the nightside, but actually not very far. However, in the 1960s spacecraft observations soon showed that the nightside magnetosphere, the *magnetotail* is very long. This requires a mechanism to transfer energy from the solar wind into the magnetosphere to keep up a current system that sustain the tail-like configuration. The magnetospheric energy budget will be discussed in Sect. 13.6.

### 1.3.2 The outer magnetosphere

Figure 1.17 is a sketch of the magnetosphere and some of the large-scale magnetospheric current systems. The overwhelming fraction of the volume consists of magnetic flux tubes connected to the polar region ionospheres. We call these regions *tail lobes*. In the northern

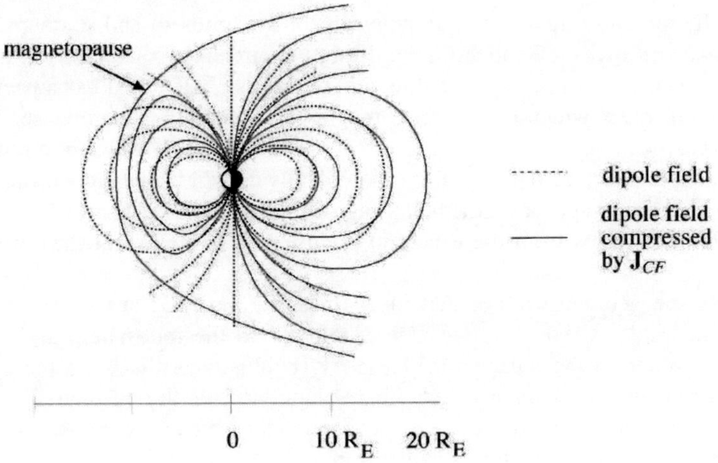


**Fig. 1.16** Sketch of the dipole field modified by $\mathbf{J}_{CF}$.

lobe the magnetic field points toward the Earth, in the southern away from the Earth. Consequently, there must be a current sheet between the lobes, where the *cross-tail current* points from the dawn to the dusk. The current is embedded within the *plasma sheet*. The current sheet can, in the first approximation, be described as the Harris sheet introduced in Chap. 3. The cross-tail current closes around the tail lobes forming the nightside part of the the *magnetopause current*. On the dayside magnetopause the magnetopause current is the same as the Chapman–Ferraro current and the two current systems join each other smoothly.

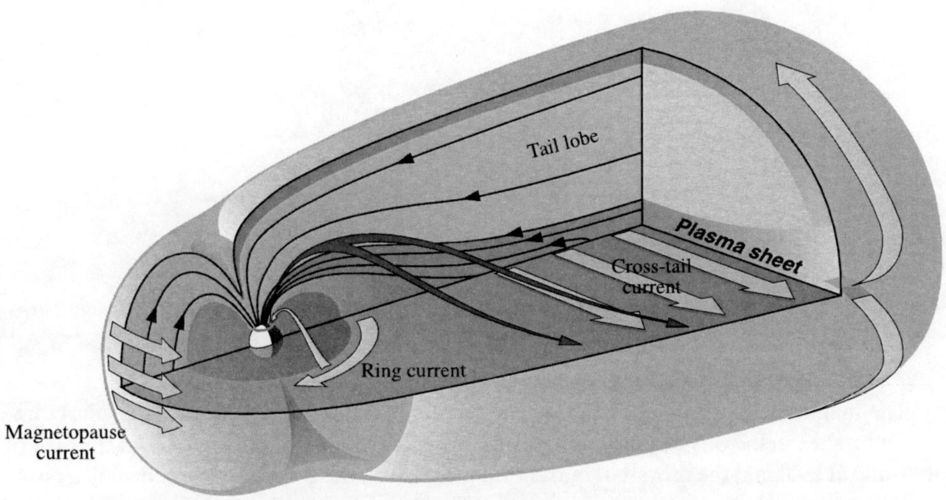


**Fig. 1.17** The magnetosphere and the large scale magnetospheric current systems. (Figure by courtesy of T. Mäkinen.)

Practically the entire magnetic flux poleward of the northern and southern auroral regions, the *auroral ovals*, extends to the tail lobes encircled by the cross-tail and magnetopause currents. At noon, i.e., in the direction of the Sun, each oval has a peculiar point, called the *polar cusp*, which is connected magnetically to the magnetopause. The formation of the cusp can be understood in terms of the image dipole description of Chapman and Ferraro. In this idealized picture it is topologically unavoidable that two singular points of zero magnetic field appear on the bounding surface in Figs. 1.14 and 1.15. If we follow the magnetic field lines from these neutral points, they indeed meet the auroral oval at noon.

In reality the geometry is not that ideal. Instead, the polar cusps are finite regions through which solar wind plasma can flow directly to the ionosphere and ionospheric plasma can escape to the solar wind. Figure 1.18 illustrates that both the cusp region and the magnetospheric boundary layers immediately inside the magnetopause are filled mostly by solar wind plasma that has entered through the cusps or across the magnetopause as a consequence of reconnection and diffusion.

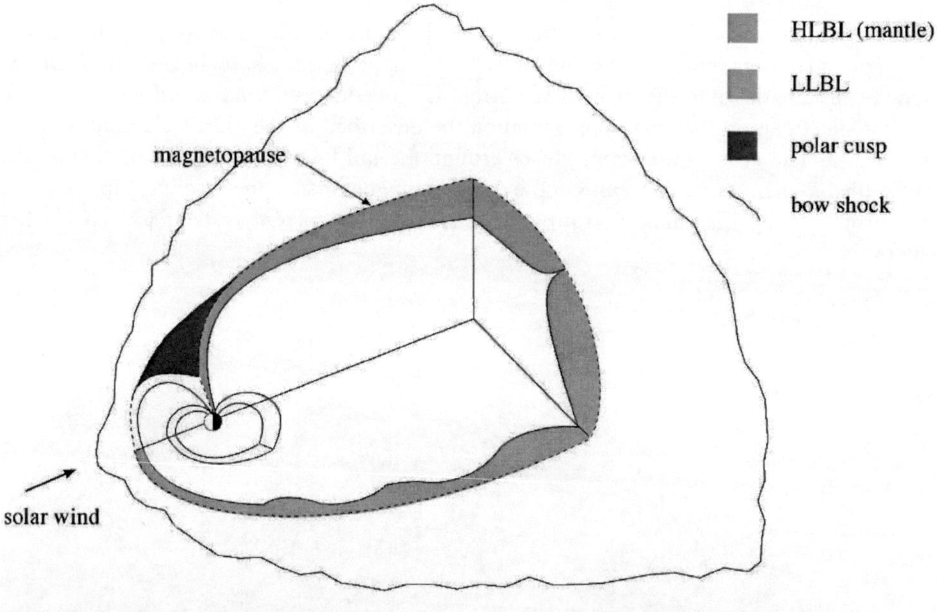


**Fig. 1.18** Sketch of magnetospheric boundary layers. HLBL stands for the *high-latitude boundary layer* and LLBL the *low-latitude boundary layer*.

The magnetotail, its stability and its connection to the auroral oval are particularly important issues in the physics of space storms. We can make a simple analysis of the size of the auroral oval and the cross-tail current intensity. Assume that the auroral oval is a circle around a region that we call the *polar cap* (PC). The magnetic flux in the polar cap is

$$\Phi_{PC} = \pi (R_E \sin \theta_{PC})^2 B_{PC} \,, \tag{1.44}$$

where $\theta_{PC}$ is the co-latitude of the boundary, say, 15°. The ionospheric magnetic field in the polar region is about 60 µT. Thus $\Phi_{PC} \approx 5 \times 10^8$ Wb. This must be the same as the magnetic flux in the tail lobe

$$\Phi_T = \frac{1}{2}\pi R_T^2 B_T \,, \tag{1.45}$$

where the tail lobe has been assumed to be a semi-circle with the radius $R_T$, and the average field in the lobe $B_T$. Equating the fluxes we get

$$\frac{R_T}{R_E} = \left(\frac{2B_{PC}}{B_T}\right)^{1/2} \sin\theta_{PC} \,. \tag{1.46}$$

In the central tail the magnetic field is about 20 nT, yielding tail radius of 20 $R_E$. Far in the tail the field is only 10 nT, giving a radius of 28 $R_E$. If the tail lobe magnetic flux increases through energy transfer from the solar wind into the magnetosphere, the oval must expand, because close to the surface of the Earth the magnetic flux density is determined by the, in this time scale, constant geomagnetic field. Consequently, the changes in the polar cap size are indicators of magnetospheric dynamics.

As the current sheet is embedded within a plasma sheet that is much more dense than the tail lobes, we can go further and estimate the cross-tail current applying the simple one-dimensional Harris model for the current sheet (Chap. 3). A rough balance between the lobe magnetic pressure and plasma pressure in the central current sheet, where the magnetic field changes sign, is

$$\frac{B_T^2}{2\mu_0} = nk_B(T_e + T_i) \,. \tag{1.47}$$

Now a 20-nT lobe field corresponds to a pressure of 0.16 nPa, which is consistent with typical observations in the tail ($n = 0.1 - 0.3\,\mathrm{cm}^{-3}$, proton temperature about 5 keV and electron temperature about 1 keV). Note that the plasma sheet is not homogeneous and these are order of magnitude estimates only.

Ampère's law across the current sheet is $2B_T = \mu_0 I$, where $I$ is the total current per unit length (units $\mathrm{A\,m^{-1}}$). Thus turning a 20-nT field to the opposite direction requires a current of 30 $\mathrm{mA\,m^{-1}}$ (i.e., 30 $\mathrm{A\,km^{-1}}$ or $2 \times 10^5\,\mathrm{A}\,R_E^{-1}$). Consequently, a piece of tail with a length of 5 $R_E$ carries a total current of 1 MA across the tail. At the magnetospheric boundaries this current splits to two parts encircling the lobe. Because the tail is very long, the total tail current is larger than 10 MA.

The plasma parameters in the tail vary with distance and magnetospheric activity. At mid-tail (30 – 40 $R_E$) typical numbers are given in Table 1.2.

### 1.3.3 The inner magnetosphere

Figure 1.19 is one more sketch of different plasma domains in the magnetosphere. The acronym PSBL stands for *plasma sheet boundary layer*. It is a transition layer between the almost empty tail lobe and the dense plasma sheet. Mapped along the magnetic field to the ionosphere, the PSBL forms a very thin strip at the poleward edge of the auroral oval,

**Table 1.2** Typical values of plasma parameters in the mid-tail. Plasma beta ($\beta$) is the ratio between magnetic and kinetic pressures.

| | magneto-sheath | tail lobe | plasma sheet boundary | central plasma sheet |
|---|---|---|---|---|
| $n\,(\mathrm{cm}^{-3})$ | 8 | 0.01 | 0.1 | 0.3 |
| $T_i$ (eV) | 150 | 300 | 1000 | 4200 |
| $T_e$ (eV) | 25 | 50 | 150 | 600 |
| $B$ (nT) | 15 | 20 | 20 | 10 |
| $\beta$ | 2.5 | $3 \cdot 10^{-3}$ | 0.1 | 6 |

whereas the rest of the oval maps to the plasma sheet, except at noon where the field lines lead to the cusp.

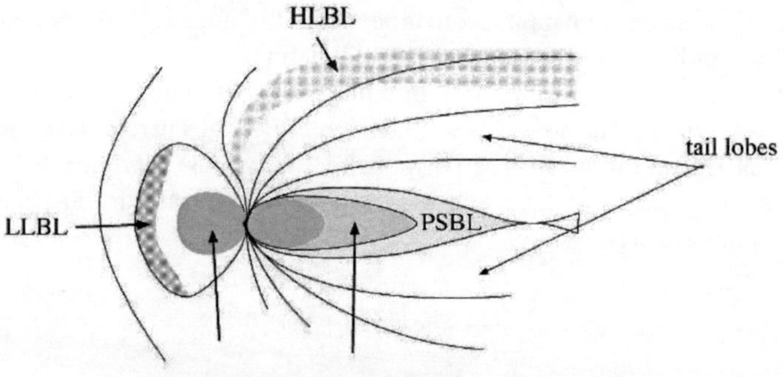


**Fig. 1.19** Sketch of magnetospheric plasma regions.

The inner magnetosphere is characterized by the corotation of the cool and dense upper ionized atmosphere with the Earth and energetic particles trapped within the magnetic bottle of the nearly dipolar magnetic field configuration (Chap. 3). The former is called the *plasmasphere* (Fig. 1.19), whereas the energetic particles form the *ring current* (RC) and the *radiation belts* (RB). The plasmasphere, ring current and radiation belts *are not* spatially distinct regions. They partially overlap and their mutual interactions are critical to the storm dynamics in the inner magnetosphere as discussed in detail in Chap. 14.

Approaching the inner magnetosphere from the tail the plasma sheet magnetic field changes from the stretched Harris-type configuration to a more dipolar form. This takes place somewhere near the *geostationary distance* ($6.6\,R_E$), but the transition is strongly dependent on the magnetospheric activity. During intense activity the cross-tail current sheet can be strongly intensified and the stretched plasmasheet can intrude deeply inside the geostationary distance. In the region of more dipolar configuration the tail current joins the ring current encircling the Earth in the westward direction.

The westward current is due to the westward drift of positively and the eastward drift of negatively charged energetic particles in the quasi-dipolar magnetic field (Chap. 3). As the drift currents are proportional to the energy density of the particles, the main ring current carriers are ions in the energy range 10–200 keV. Note that at the earthward edge of the ring current the negative pressure gradient may introduce a local eastward diamagnetic current contribution, but the net ring current is westward. During magnetospheric activity the ionosphere acts as a plasma source increasing the relative abundance of oxygen in the magnetosphere and during large storms a significant fraction of the ring current can be carried by oxygen ions.

Enhancement and decay of the ring current are the most characteristic elements in magnetospheric storm activity. The enhancement of the current requires efficient acceleration and transport of ions into the right location through radial diffusion. After the activity the current carriers slowly disappear through *charge exchange* with the low-energy neutral atoms of the Earth's *exosphere*, wave–particle interactions, and Coulomb collisions. These issues will be discussed in detail in Chap. 14.

The radiation belts are partly co-located with the ring current. While not a complete surprise, the detection of the radiation belts in 1958 was the first major discovery of the satellite era. *Explorer* I carried a simple Geiger counter of James Van Allen. The instrument was saturated when the satellite crossed the radiation belt. The observation was interpreted to be due to high-energy particles trapped in the magnetic bottle formed by the geomagnetic field (Chap. 3). To honor this observation the radiation belts are also known as *Van Allen belts*. In the process of analyzing the data Carl McIlwain introduced the $L$ parameter to label the field lines crossing the equator at a given distance in the units of $R_E$ (Chap. 3). In the *inner belt* ($L \approx 1.5 - 3$) the energetic population is dominated by protons in the energy range 0.1 MeV – 40 MeV with a substantial contribution of energetic electrons, whereas in the *outer belt* ($L > 4$) the energetic component is mostly electrons in the keV to MeV range. Note that while the energies of radiation belt ions are much higher than those of the ring current ions, their density is much smaller and thus the radiation belts do not contribute much to the total current around the Earth.

The electron belts, and also the *slot region* between them (Fig. 1.20) are highly variable. The storms can both increase and decrease the electron fluxes in the outer radiation belts in complicated ways that are not yet fully understood. The strongest storms may also inject large particle fluxes into the slot region. Because the dipole field at these distances (2–4 $R_E$) is a very stable magnetic magnetic bottle, it is very difficult to get particles there, but once the slot is filled, the loss of these electrons takes a long time. We will return to this important aspect of space storms in Chap. 14.

The plasmasphere is the innermost part of the magnetosphere. It consist of cold (~1 eV), dense (~ $10^3\,\mathrm{cm}^{-3}$) plasma of ionospheric origin. The existence of the plasmasphere was already known before the spaceflight era through the propagation studies of whistler mode waves (Chap. 4). The domain has a very steep outer edge, the *plasmapause* somewhere inside the geostationary distance. The location and fine structure of the plasmapause vary considerably as a function of magnetic activity. Figure 1.21 illustrates that during magnetospheric quiescence the density decreases rather smoothly at distances from 4–6 $R_E$, whereas during strong activity the plasmapause is steep and much closer to the Earth. This density gradient plays an important role in the generation and guidance of plasma waves that, in turn, interact with the energetic particles in the ring current and

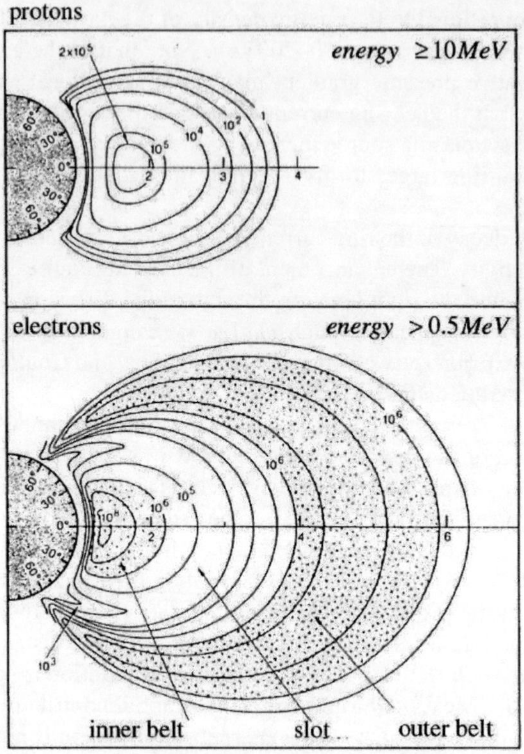


**Fig. 1.20** Fluxes of energetic protons and relativistic electrons in the radiation belts. The contours are given in units of particles per square centimeter and second ($cm^{-2}\,s^{-1}$). (Adapted from the textbook of Kivelson and Russell [1995].)

radiation belts. Thus the coldest and hottest components of the inner magnetosphere are intimately coupled to each other during the evolution of magnetospheric storms.

### 1.3.4 Magnetospheric convection

Magnetospheric plasma is in a continuous large-scale motion that is called *convection*.[2] The convection is driven by solar wind energy input into the magnetosphere. The convective motion is most directly observable in the polar ionosphere using scatter radar observations (Chap. 9), or by electric field measurements onboard polar orbiting satellites utilizing the fact that the motion-induced electric field **E** is related to the plasma flow velocity **V** by the simple equation

$$\mathbf{E} = -\mathbf{V} \times \mathbf{B}\,. \tag{1.48}$$

[2] Actually, advection would be a better description, as the motion is not driven by a thermal force. Sometimes it is wiser to conform with widely used inaccurate terminology than to try to change it.

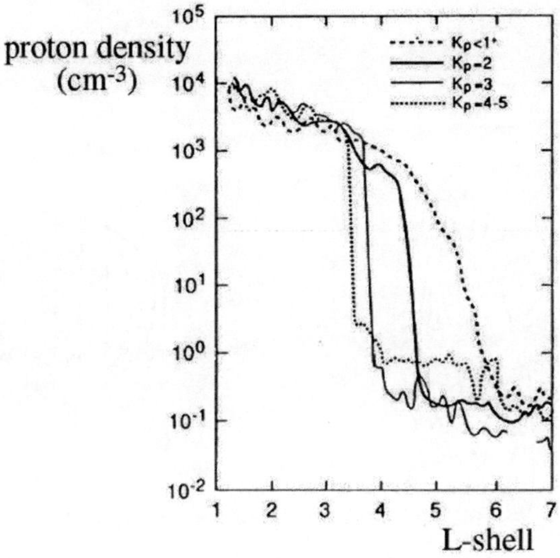


**Fig. 1.21** Plasma density in the night sector organized by the activity index $Kp$. $Kp < 1+$ corresponds to a very quiet magnetosphere, whereas $Kp = 4 - 5$ indicates a significant activity level, although not yet a large magnetic storm. (Adapted from Chappell [1972].)

Plasma flows from the dayside to the nightside across the polar cap, where it is returned back to the dayside through the morning and evening sectors. Convection is going on all the time. It weakens when the IMF points toward the north and is enhanced during southward-pointing IMF. Because ideal MHD (Chap. 6) is a very good description of the large-scale plasma motion above the resistive ionosphere, the magnetic field lines are electric equipotentials. Thus the convective motion, or alternatively the electric potential, in the ionosphere can be mapped to the tail lobes and the plasma sheet along the magnetic field lines.

If the magnetopause were fully closed, convection would circulate inside the magnetosphere so that the magnetic flux tubes crossing the polar cap from dayside to nightside would at some moment be reaching to the magnetospheric outer boundary where some kind of *viscous interaction* with the solar wind flow would sustain the circulation. This is actually the picture proposed by Axford and Hines [1961] to explain the convection illustrated in Fig. 1.22. The classical (collisional) viscosity on the magnetopause is extremely weak, but finite gyro radius effects and wave–particle interactions give rise to some level of "anomalous" viscosity. It is estimated to provide about 10% of the momentum transfer from the solar wind to the magnetosphere.

The magnetosphere is, however, not fully closed. In the same year when Axford and Hines presented with the viscous interaction model, Dungey [1961] explained the convection in terms of reconnection (Chap. 8). His idea is illustrated in Fig. 1.23.

In this picture a magnetic field line in the solar wind is cut and reconnected with a terrestrial field line on the dayside magnetopause. The solar wind flow drags the newly-connected field line to the nightside and the part of the field line that is inside the magneto-

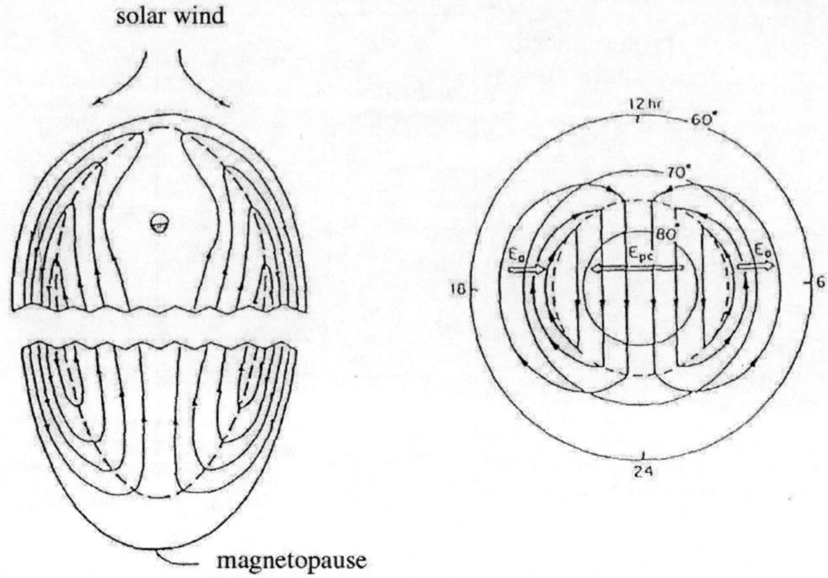


**Fig. 1.22** Convection in the equatorial plane of a closed magnetosphere. On the left the so-called Axford–Hines model. On the right the mapping of the motion to the polar cap with open arrows indicating the polar cap electric field $\mathbf{E} = -\mathbf{V} \times \mathbf{B}$.

sphere becomes a field line in the tail lobe. Consequently, more and more magnetic flux is piling up in the lobe and pushing the flow toward the cross-tail current layer. Somewhere 100–200 $R_E$ down the tail the field lines piling up in the northern and southern lobes reconnect again across the tail current layer. At this point the ionospheric end of the field line has reached the nightside oval near midnight. Now the earthward outflow from the reconnection site in the tail drags the newly-closed field line toward the Earth. The return flow cannot penetrate to the corotating plasmasphere and must go around the Earth to the dayside. The ionospheric end of the field line returns toward the dayside along either the dawnside or the duskside auroral oval. Once approaching the dayside magnetopause the magnetospheric plasma provides the inflow to the dayside reconnection from the inside.

If the dayside and nightside reconnection rates balance each other, a steady-state convection may arise (Sect. 13.3.1). More typically the changes in the driver (solar wind) and in the magnetospheric response are faster than the circulation time scale of a few hours. Thus reconnection may cause significant erosion of the dayside magnetospheric magnetic field pushing the magnetopause closer to the Earth than a simple pressure balance calculation would indicate. Furthermore, the changing magnetic flux in each tail lobe causes expansion and contraction of the polar caps.

The reconnection is most efficient when the solar wind magnetic field (IMF) has a due southward-pointing orientation. The increase in the tail lobe magnetic field and strengthening of plasma convection inside the magnetosphere during southward IMF has a strong observational basis. If we calculate the (rectified) east-west component of the solar wind

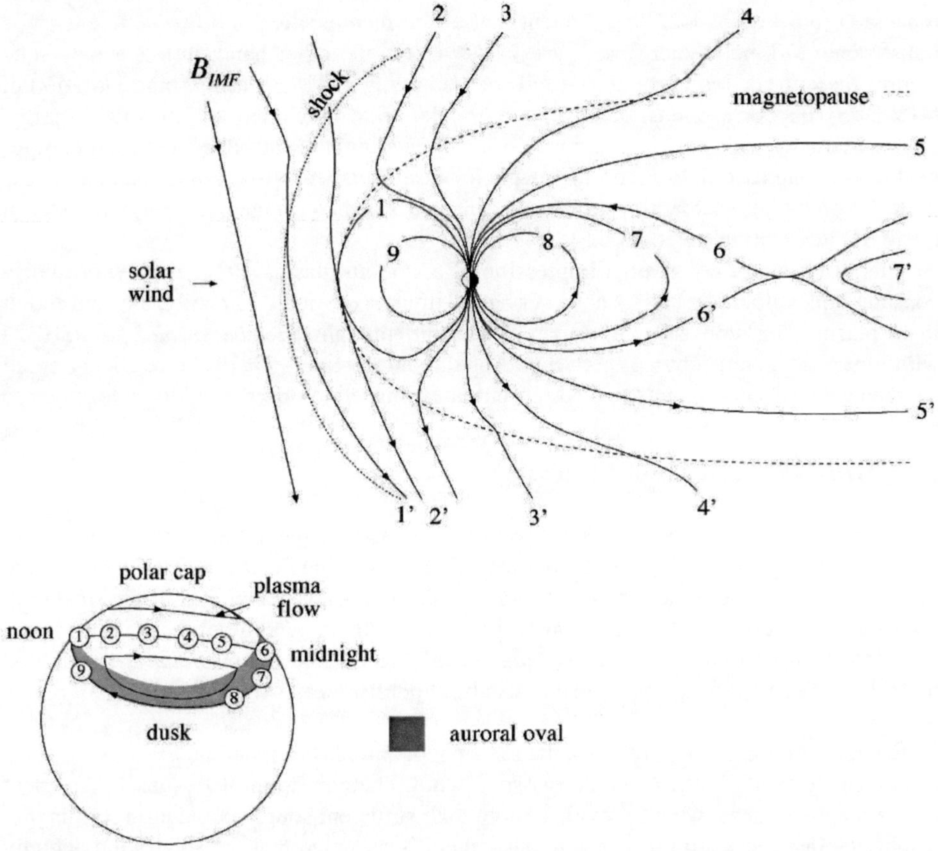


**Fig. 1.23** Convection in the magnetosphere opened by reconnection. The lower picture illustrates the motion of the ionospheric end of the magnetic field line assuming that plasma and magnetic field are frozen-in to each other. Note that the tail in this picture is strongly compressed in the horizontal direction. In reality the far-tail neutral line is located somewhere at $100\,R_E$ or even further. (Adapted from the textbook of Kivelson and Russell [1995].)

electric field ($E = VB_{south}$) incident on the magnetopause and the corresponding potential drop over the magnetosphere, we find that some 10% of the electric field "penetrates" into the magnetosphere corresponding to the convection electric field. Note, however, that in the relationship $\mathbf{E} = -\mathbf{V} \times \mathbf{B}$ there is no causal information, whether it is the electric field that drives the magnetospheric convection, or convection that gives rise to the motion-induced electric field. Of course, the ultimate driver is the solar wind flow against the magnetosphere.

The increase of the magnetic flux in the lobes is a bit more sophisticated than the frequently used sloppy phrase that reconnection transports solar wind magnetic flux to the lobe. It is more appropriate to describe the process as energy transfer where solar wind kinetic energy is converted to the magnetospheric magnetic field energy on the magnetopause. From Figure 1.23 it is evident that at the reconnection site magnetic energy is transformed to kinetic energy as there $\mathbf{J} \cdot \mathbf{E} > 0$. On the other hand, in this steady state picture the current loop around the tail lobes implies $\mathbf{J} \cdot \mathbf{E} < 0$ at the high-latitude tail boundary. This corresponds to conversion of solar wind flow energy to magnetic energy, i.e, a dynamo (Chap. 8). The main role of reconnection is to introduce a normal component of the magnetic field $B_n$ on the magnetopause. This leads to a finite magnetic stress ($\propto B_n B_t$) on the magnetopause surface and this stress is the key agent of energy transfer in the MHD description (Sect. 13.6.5).

This discussion may give an impression of a smooth plasma circulation with a fairly constant bulk velocity, which is not a very good impression at all. In reality, the convection in the plasma sheet consists in large part of intermittent high-speed *bursty bulk flows* (BBF) with almost stagnant flows in between [Angelopoulos et al, 1992]. The relatively small average velocity corresponds to the high-latitude convection observed in the ionosphere.

### 1.3.5 Origins of magnetospheric plasma

Thus far we have discussed the magnetosphere from the magnetic field viewpoint without addressing the question of the origin of the plasma convecting in concert with the magnetic field. The origins and losses of magnetospheric plasma is a vast complex of physical phenomena. A comprehensive discussion of the status of understanding of these issues in the late 1990s can be found in the book *Magnetospheric Plasma Sources and Losses* edited by Hultqvist et al [1999]. The book has also been published as vol. 88 (Nos. 1–2) of *Space Science Reviews*, 1999.

Except for its innermost regions, the ionosphere and plasmasphere, the magnetosphere is a magnetic cavity in the much denser solar wind. There are some $10^{29}$ ions s$^{-1}$ incident on the magnetopause, which provide a more than sufficient source population for magnetospheric plasmas. Until the 1980s it was generally assumed that the solar wind actually was the main source. A good reason to believe so was the fact that the solar wind ion energies are in the keV range, which is not too far from the plasma sheet temperature, whereas in the near-Earth plasma reservoir, the ionosphere, the ion energies range from below 1 eV to a few tens of eV only.

The first indications that ionospheric plasma might escape in large amounts to the magnetosphere came with observations of heavy ($m/q = 16$) energetic (up to 17 keV) ions by the polar orbiting satellite 1971-089A in the 1970s [Shelley et al, 1972]. These were presumed to be $O^+$ ions, which could only come from the ionosphere, as the oxygen ions in the solar wind have much higher charge states, typically $O^{6+}$, as a consequence of their origin in the hot solar corona. The first observations were made during magnetospheric storms, but the subsequent satellite observations confirmed the existence of ionospheric plasma in the magnetosphere also during magnetically quiet times. Chappell et al [1987] finally suggested that the ionosphere is capable of supplying all magnetospheric plasma under any magnetic conditions.

As so often, the truth lies somewhere between these two extremes. Both the solar wind and ionospheric sources are highly variable. There is always some diffusion through the magnetopause, but the rate at which solar wind plasma penetrates to the magnetosphere depends on how efficiently reconnection opens the magnetopause and, consequently, on the direction of the IMF. The estimates of the dayside magnetopause source are in the range $10^{26}$–$10^{27}$ ions s$^{-1}$, perhaps reaching $10^{28}$ ions s$^{-1}$ during strong solar wind driving. The ionospheric supply is somewhat smaller, peaking during strong geomagnetic activity, when roughly the same amount of $O^+$ ions and protons, $10^{26}$ each, escape per second. Most of the ion upflow takes place from the auroral region including the polar cusp.

Note that the strong solar wind inflow when the magnetopause is most open does not necessarily imply the most efficient net gain of plasma because the open magnetosphere is at the same time most leaky. As discussed in Sect. 13.5.2 there are strong indications that more plasma can accumulate in the plasma sheet during periods of northward than southward IMF. The estimates of plasma outflow in the far tail indicate that some $10^{28}$ ions s$^{-1}$ escape downstream. As neither the dayside magnetopause nor the ionosphere seem to be able to provide that much plasma, most of the total solar wind plasma entrance likely takes place along the tail magnetopause. Part of this plasma flows directly downwind but some fraction of it is first convected to the plasma sheet earthward of the distant neutral line and thereafter circulated toward the Earth.

### 1.3.6 Convection and electric fields

In ideal MHD the macroscopic plasma motion $\mathbf{V}$ and the electric field are coupled to each other through $\mathbf{E} = -\mathbf{V} \times \mathbf{B}$. This electric field is always perpendicular to the magnetic field. If the magnetic field is time-independent, the electric field is also curl-free and can be expressed as the gradient of scalar potential

$$\mathbf{E} = -\nabla\varphi \,. \tag{1.49}$$

Of course, these assumptions are not always fulfilled in the magnetosphere and the inductive fields given by Faraday's law

$$\frac{\partial \mathbf{B}}{\partial t} = -\nabla \times \mathbf{E} \tag{1.50}$$

must be taken into account during rapid changes of the magnetic field, which often occur during space storms. Let us leave such intricacies, as well as the disturbing properties of BBFs, aside for the time being and consider convection of plasma consisting of low-energy particles in the equatorial plane within the plasma sheet and plasmasphere.

For simplicity we assume that the magnetospheric magnetic field points upward (north) everywhere in the equatorial plane. Thus the return convection in the plasmasheet is equivalent to a dawn-to-dusk directed electric field $E_0\mathbf{e}_y$ that we assume to be constant (we select the coordinates such that the $x$-axis is toward the Sun, the $y$-axis toward the dusk, and thus the magnetic field is in the direction of the $z$-axis). Let $r$ be the distance to the center of the Earth and $\phi$ the angle from the direction of the Sun. Then the electric field is given by

$$\mathbf{E}_{conv} = -\nabla(-E_0 r \sin\phi) \tag{1.51}$$

and its potential is

$$\varphi_{conv} = -E_0 r \sin\phi \,. \tag{1.52}$$

The Earth with its atmosphere rotates in this frame of reference. The corotation extends in the equatorial plane roughly up to the plasmapause. The angular velocity to the east is evidently $\Omega_E = 2\pi/24\,\text{h}$. In the fixed frame plasma thus moves with the velocity

$$\mathbf{V}_{rot} = \Omega_E r \mathbf{e}_\phi \,, \tag{1.53}$$

where $\mathbf{e}_\phi$ is the unit vector pointing toward the east. $\mathbf{V}_{rot} = \mathbf{E}_{rot} \times \mathbf{B}/B^2$ is the convection velocity associated with the corotation electric field $\mathbf{E}_{rot}$, the potential of which is

$$\varphi_{rot} = \frac{-\Omega_E k_0}{r} = \frac{-\Omega_E B_0 R_E^3}{r} \,. \tag{1.54}$$

Here $k_0 = 8 \times 10^{15}\,\text{Tm}^3$ is the Earth's dipole moment and $B_0$ the dipole field on the surface of the Earth at the equatorial plane $\approx 30\,\mu\text{T}$. The convection and corotation electric fields are illustrated in Fig. 1.24. The equipotential curves of the always earthward pointing corotation field are circles.

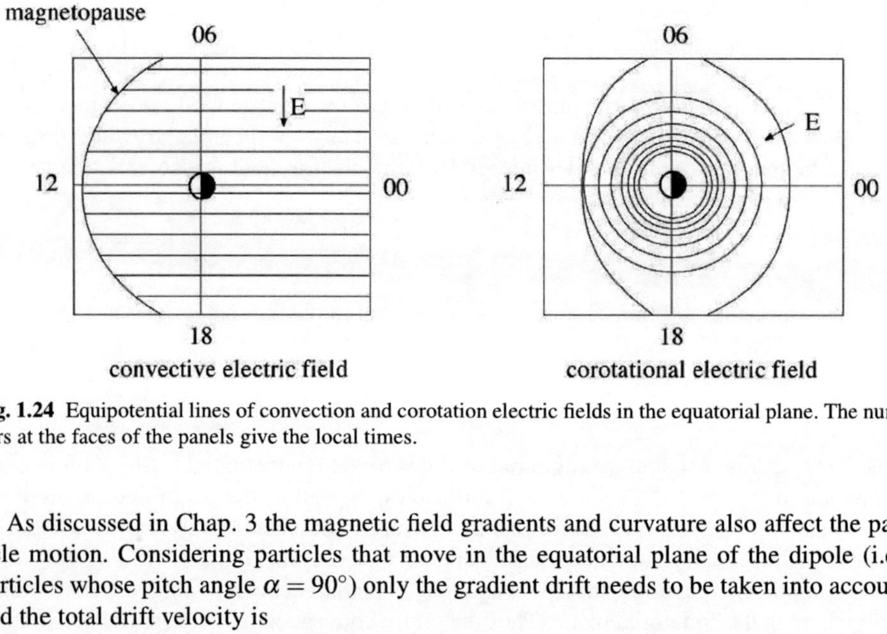


**Fig. 1.24** Equipotential lines of convection and corotation electric fields in the equatorial plane. The numbers at the faces of the panels give the local times.

As discussed in Chap. 3 the magnetic field gradients and curvature also affect the particle motion. Considering particles that move in the equatorial plane of the dipole (i.e., particles whose pitch angle $\alpha = 90°$) only the gradient drift needs to be taken into account and the total drift velocity is

$$\mathbf{v}_D = \frac{1}{B^2}\left[\mathbf{E}_{conv} + \mathbf{E}_{rot} - \nabla\left(\frac{\mu B}{q}\right)\right] \times \mathbf{B} = \frac{1}{B^2}\mathbf{B} \times \nabla\varphi_{eff} \,, \tag{1.55}$$

where $\mu$ is the magnetic moment of the particles and the effective potential is

$$\varphi_{eff} = -E_0 r \sin\phi - \frac{\Omega_E B_0 R_E^3}{r} + \frac{\mu B_0 R_E^3}{q r^3} \,. \tag{1.56}$$

The particles move along streamlines $\varphi_{eff}$ = constant. These streamlines depend on both the charge and energy of the particles through their magnetic moments. For cold particles ($\mu = 0$) the streamlines are equipotential lines of the combined convection and corotation fields (Fig. 1.25). In this case the motion is a pure E×B-drift and all particles move with the same velocity.

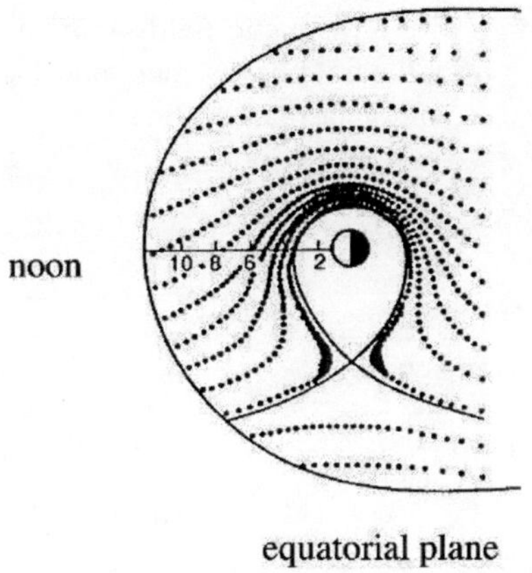


**Fig. 1.25** Orbits of low-energy particles (i.e, magnetic moment $\mu \approx 0$) in the equatorial plane assuming $E_0 = 0.3\,\mathrm{mV\,m^{-1}}$. The distance between consecutive points is 10 min. (Adapted from Kavanagh et al [1968].)

Figure 1.25 illustrates the formation of a *separatrix* that separates the cold corotating plasmaspheric plasma from the cold plasma outside. In this model the separatrix is thus the plasmapause. The separatrix has an electric neutral point at the distance

$$r = \sqrt{\frac{\Omega_E B_0 R_E^3}{E_0}} \tag{1.57}$$

in the direction of 18 h local time. The plasmasphere thus has a bulge in the evening sector. There is a corresponding bulge in the real plasmapause, but its orientation and size depend on the strength of the convection electric field. We will revisit the bulge when discussing Fig. 14.4.

While this model for the plasmasphere is a strong simplification, it nevertheless explains why the plasmasphere is compressed during strong magnetic activity: the enhanced energy input enhances the convection velocity and thus the dawn-to-dusk electric field. The rotation of the Earth is constant and the corotation electric field is always the same. Consequently the separatrix is pushed toward the Earth when the convection enhances. Note that the real plasmapause reacts to the changing electric field with some delay, which leads to observations of detached clouds of plasmaspheric plasma outside the plasmapause.

When the magnetic moments of the particles are increased, the magnetic gradients start to separate the motion of positive and negative charges. To illustrate this effect consider particles whose magnetic moment is so strong that it supersedes corotation. Now the effective potential (still in the equatorial plane only) is

$$\varphi_{eff} = -E_0 r \sin\phi + \frac{\mu B_0 R_E^3}{q r^3} \,. \tag{1.58}$$

This implies that far from the Earth the particles follow the convection electric field, but closer in they drift according to the magnetic field gradient. This way the dipole field shields the near-Earth space from the hot plasma sheet plasma and the cold plasmasphere and hot plasmasheet are two separate plasma domains.

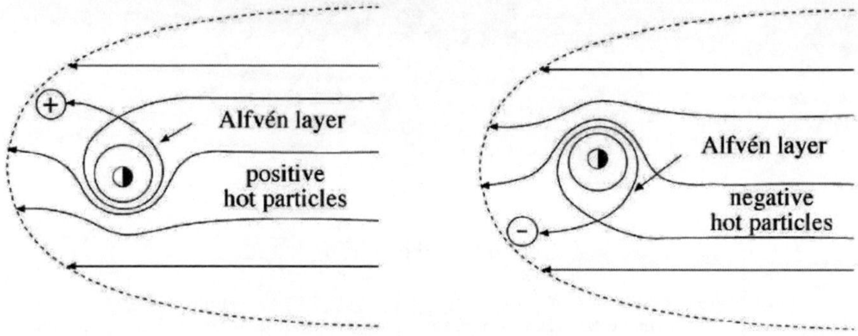


**Fig. 1.26** The formation of Alfvén layers.

The positive and negative charges drift according to Fig. 1.26 and their separatrices, called the *Alfvén layers*, are different. Because the plasma sheet is a finite particle source, a larger fraction of positive charges pass the Earth in the evening sector and a larger fraction of negative charges in the morning sector. This leads to piling of positive space charge in the evening sector and negative charge in the morning sector. These charge accumulations are discharged by magnetic field-aligned currents to the ionosphere from the evening sector and from the ionosphere to the magnetosphere in the morning sector.

This picture also gives a qualitative explanation how very high-energy particles can get access and become trapped in the magnetic bottle of the Earth's dipole field during strong magnetic activity. As the convective electric field grows rapidly, the particles E×B-drift deeper into the ring current and radiation belts than in quiet times. Once the activity ceases, the trapping boundary (i.e. the Alfvén layer) moves outward and thus particles that were originally on open drift paths past the Earth find themselves trapped into the expanding magnetic bottle.

## 1.4 The Upper Atmosphere and the Ionosphere

An ionosphere is formed around all planets having a neutral atmosphere. It is mainly produced by photoionization due to solar EUV radiation. Additional collisional ionization is provided by particle precipitation from the magnetosphere. At high enough energies electrons produce X-rays through *bremsstrahlung*, when they are stopped in the atmosphere. This leads to weak ionization at also lower altitudes. Another observable but dynamically rather unimportant source of ionization associated with space storms is the so-called *solar flare effect* caused by X-ray and EUV radiation from a large enough flare. It has a crochet-like appearance in the ground-based magnetograms, its shape following the flare evolution. The phenomenon has historical interest because it can be seen as a 110-nT perturbation in the Greenwich magnetogram during the Carrington flare in 1859 (e.g., Cliver and Svalgaard [2004]).

Due to its origin in solar EUV radiation and magnetospheric particle precipitation the ionospheric ion density depends strongly on the time of the day, the season, and solar and magnetospheric activity. Although some low-latitude processes, e.g., the *equatorial spread-F* caused by the Rayleigh–Taylor instability (Chap. 7), have some correlation with space storms, for our theme the high-latitude ionosphere and its coupling to the magnetosphere are of the greatest interest. Therefore we limit the discussion here to some of the key elements in high-latitude ionospheric electrodynamics. A thorough treatise on ionospheric physics is the textbook by Kelley [1989]. The textbook by Kivelson and Russell [1995] provides a reader-friendly introduction to the formation of the ionosphere.

### 1.4.1 The thermosphere and the exosphere

The Earth's atmosphere behaves as a collision-dominated gas up to altitudes of about 400–500 km. The ionosphere is formed in the *thermosphere* at altitudes above 80–85 km, where the neutral gas is in *hydrostatic equilibrium*

$$n_n m_n g = -\frac{d}{dh}(n_n k_B T_n) \,. \tag{1.59}$$

Here $m_n$ is the mass of the neutral gas molecules or atoms and $h$ is the altitude. If the temperature $T_n$ of the gas is assumed to be altitude-independent, the density profile of the atmosphere is

$$n_n = n_0 \exp\left(\frac{-(h-h_0)}{H_n}\right) , \tag{1.60}$$

where

$$H_n = \frac{k_B T_n}{m_n g} \tag{1.61}$$

is the *density scale height*. The scale height is different for different molecules and atoms, which thus have different density profiles. While the collisions bring all constituents into the same temperature, they do not homogenize the composition of the gas. In reality also the temperature is altitude-dependent and thus our simple discussion is not fully accurate. Furthermore, strong solar and magnetospheric activity lead to heating of the thermosphere and thus to enhanced scale height.

**Train your brain**

Instead of density scale height, *pressure scale height* is often used. Find an expression for it in terms of the *gas constant* $\mathscr{R} = P\mathscr{V}/nT$.

The nearly collisionless gas above the thermosphere is called the *exosphere*. The bottom of the exosphere, the *exobase*, can be defined either as the altitude where the collisions become negligible, or above which the constituents of the gas are on purely ballistic trajectories. At the exobase the particle mean free path and the pressure scale height are equal.

The exosphere has a particular role in the physics of space storms. It extends as the *geocorona* far into the near-Earth space and is a key element in the loss of ring current carriers (Chap. 14).

### 1.4.2 Structure of the ionosphere

The existence of the ionosphere was revealed early in the 20th century by the first long-distance radio communication experiments, including Marconi's famous transmission of electromagnetic signals across the Atlantic. During the years 1924–1926 Appleton and Barnett and, independently, Breit and Tuve demonstrated mathematically and experimentally that there is an ionized electrically conductive layer in the upper atmosphere from which radio waves are reflected. This layer became to be called the *E layer* (or *E region*; E for electric). Today we know that the E layer is within the altitude range 90–120 km and it is ionized mostly by precipitating electrons. The global ionization maximum due to the solar EUV radiation is higher up at about 250 km. The altitude range above the E layer, reaching to about 800 km is called the *F layer* (or sometimes the Appleton layer). Later also an ion density enhancement below the E layer was identified and became to be called (logically?) the *D layer*. Figure 1.27 illustrates the altitude profiles of electron and major ion and neutral atom densities in the ionosphere. Note that the ionization degree of the ionosphere is very low at low altitudes, but nevertheless the gas behaves like a collisional plasma, where the dominant collisions are with the thermospheric neutral atoms and molecules.

As already discussed, the ionosphere is a significant source of magnetospheric plasma. From the polar cap enclosed by the auroral oval a tenuous *polar wind* flows continuously upward. Its escape resembles the outflow of solar wind from the Sun, as the outflow starts as subsonic and is transformed to supersonic at higher altitudes. The strongest outflow, however, takes place on magnetic field lines attached to the auroral oval. Plasma processes associated with the electrodynamic coupling between the ionosphere and magnetosphere heat ionospheric plasma, which starts to lift up, partly due to thermal energy, partly due to the mirror force (Chap. 3). The acceleration and heating from the cold ionosphere up to keV energies in the magnetosphere most likely takes place in several steps and involves both quasi-static acceleration and wave–particle interaction mechanisms. When leaving the ionosphere the typical ion energies are of the order of 1 eV, but already above 10 000 km they may exceed 10 keV. Although there is also downward plasma motion, the ionosphere and thus the atmosphere experience a net loss of matter. The estimates are uncertain, but

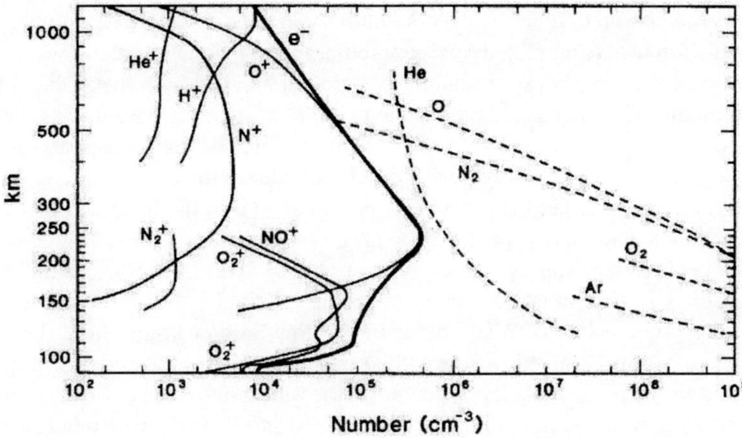


**Fig. 1.27** Dayside ionospheric electron (thick line), ion (thin lines) and neutral atom (dashed lines) density profiles according to the definition of "International quiet solar year". (From Johnson [1969].)

total upward flux is of the order of $2\,\mathrm{kg\,s^{-1}}$ or more (see, e.g., Chap. 2 of Hultqvist et al [1999]).

### 1.4.3 Electric currents in the polar ionosphere

For the physics of space storms the most important property of the ionosphere is its finite electric conductivity. Due to the anisotropy caused by the strong background magnetic field and the vastly different relative gyro and collision frequencies between the ions and electrons the conductivity is a tensor. The ionospheric Ohm's law can be written in the form

$$\mathbf{J} = \begin{pmatrix} \sigma_P & \sigma_H & 0 \\ -\sigma_H & \sigma_P & 0 \\ 0 & 0 & \sigma_\parallel \end{pmatrix} \cdot \mathbf{E}\,. \tag{1.62}$$

Here the elements of the conductivity tensor, assuming for simplicity only one ion population, are given by

$$\begin{aligned} \sigma_P &= \left[ \frac{1}{m_e \nu_{en}} \left( \frac{\nu_{en}^2}{\nu_{en}^2 + \omega_{ce}^2} \right) + \frac{1}{m_i \nu_{in}} \left( \frac{\nu_{in}^2}{\nu_{in}^2 + \omega_{ci}^2} \right) \right] n_e e^2 \\ \sigma_H &= \left[ \frac{1}{m_e \nu_{en}} \left( \frac{\omega_{ce} \nu_{en}}{\nu_{en}^2 + \omega_{ce}^2} \right) - \frac{1}{m_i \nu_{in}} \left( \frac{\omega_{ci} \nu_{in}}{\nu_{in}^2 + \omega_{ci}^2} \right) \right] n_e e^2 \\ \sigma_\parallel &= \left[ \frac{1}{m_e \nu_{en}} + \frac{1}{m_i \nu_{in}} \right] n_e e^2\,. \end{aligned} \tag{1.63}$$

$\nu_{\alpha n}$ are the electron and ion collision frequencies with neutrals and $\omega_{c\alpha}$ are the angular frequencies of electron and ion gyro motions.

The *Pedersen conductivity* $\sigma_P$ is the conductivity in the direction of the ambient electric field $\mathbf{E}_\perp$, which in turn is practically perpendicular to the magnetic field in the ionosphere. The *Hall conductivity* $\sigma_H$ is the conductivity perpendicular to both the ambient magnetic and electric fields. The magnetic field-aligned conductivity $\sigma_\parallel$ is the same as the classical collisional conductivity in the absence of magnetic field. In the ionosphere it is several orders of magnitude larger than the perpendicular conductivities. Consequently, the quasi-static ionospheric electric field is practically perpendicular to the magnetic field.

The Pedersen conductivity peaks in a narrow layer above the altitude of 150 km, whereas the peak of the Hall conductivity is at about 120 km. Due to diurnally and seasonally variable solar illumination conditions at these altitudes the peak conductivities can vary more than two orders of magnitude. In the low-latitude ionosphere the day–night asymmetry is most pronounced, whereas the polar ionospheres have very strong seasonal variability. From the magnetospheric viewpoint the ionospheric current layers are thin and often treated as two-dimensional current layers in studies of magnetosphere–ionosphere coupling. From the viewpoint of ionospheric processes the structure is, however, three-dimensional.

**Feed your brain**

With the help of literature find out how the elements of the conductivity tensor (1.63) are derived.

We have already encountered the plasma convection across the polar cap. Above the dense ionosphere plasma is collision-free and both positive and negative charges E×B-drift with the same velocity causing no net electric current perpendicular to the magnetic field. In the E layer the ions are so strongly collisional that they cannot make full gyro orbits between collisions with neutrals. Thus they drift predominantly in the direction of the electric field and carry most of the *Pedersen current* ($\sigma_P\mathbf{E}$). Electrons, on the other hand, are still strongly magnetized and follow the E×B-drift, i.e., the convection, and carry most of the *Hall current* ($\sigma_H\mathbf{E}$), which thus is directed opposite to electron drift motion. In the polar cap the current is distributed over a wide area, but in the evening and morning sectors the current is squeezed into narrow channels, in which the current density is much higher. These currents are called *electrojets*. In the evening sector the electrojet current is eastward and in the morning sector westward. The currents can be monitored with ground-based magnetometers. The eastward current gives a positive contribution to the northward component of the magnetic field measured below the electrojet and the westward current a negative contribution.

The high parallel conductivity allows for large electric current along the magnetic field even for a very small parallel electric field. In fact, the ionospheric electrodynamics is intimately coupled to the magnetospheric current systems through magnetic *field-aligned currents* (FAC). We postpone the details to Sect. 6.5, as this discussion relies heavily on concepts to be covered later in this book. Here we just illustrate the current systems with the aid of two figures.

Figure 1.28 is a classic statistical presentation of upward and downward flowing FACs during weak auroral activity. If we follow the magnetic field from the equatorward slices

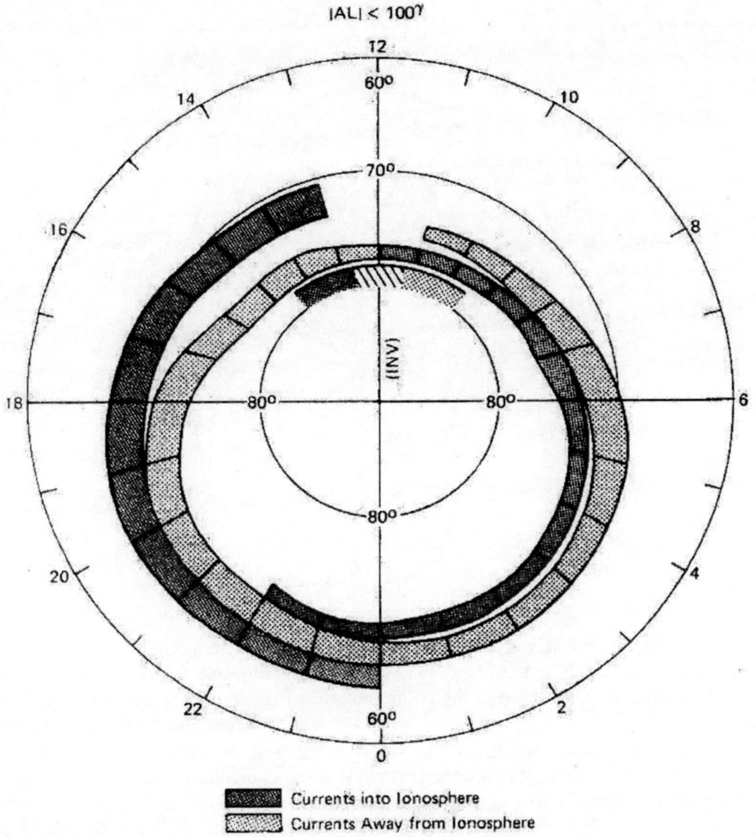


**Fig. 1.28** Statistical pattern of FACs above the northern auroral oval during weak auroral activity originally presented by Iijima and Potemra [1976]. The grey domains illustrate the current away from the ionosphere and the black areas the downward return current back to the ionosphere.

of the current pattern we end up at the ring current region in the magnetosphere. In the evening sector this FAC flows from the magnetosphere to the ionosphere and in the morning sector from the ionosphere to the magnetosphere. This is the same sense of the currents that is needed to discharge the excessive space charge of the Alfvén layers (Fig. 1.26). This FAC system is called the *Region 2* current system. *Region 1* currents flow in turn in the poleward slices in Fig. 1.28. They are directed opposite to Region 2 currents. The current thus comes into the ionosphere in the morning sector and leaves it in the evening sector. Region 1 is located close to the boundary between open and closed field lines. Consequently a magnetic field-aligned mapping from this strip in the ionosphere leads to the magnetospheric boundary layers. As the circuit closes through the resistive ionosphere, the maintenance of the current requires an existence of a dynamo somewhere in the magnetospheric boundary. Figure 1.29 is a summary of the most important field-aligned currents and their closures in the ionosphere and magnetosphere.

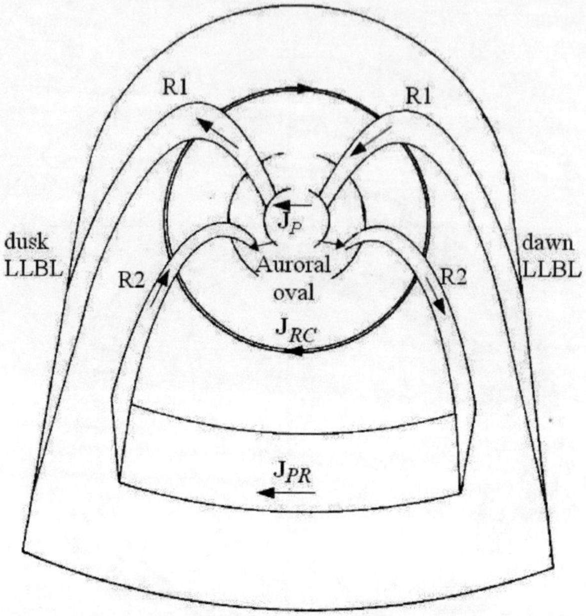


**Fig. 1.29** Sketch of the major FACs. The morning sector Region 1 current originates near the outer boundary of the magnetosphere, likely in the LLBL, and flows to the poleward side of the auroral oval. There it closes as Pedersen current partly through the polar cap and partly across the auroral oval. Current crossing the polar cap rises in the evening sector again as Region 1 current and flows to the magnetospheric boundary. Current crossing the morning sector oval as a Pedersen current continues as Region 2 current to the inner magnetosphere, where it connects again to the perpendicular current flowing westward to the evening sector and joins there to the evening sector Region 2 current. This current loop in the tail is called *partial ring current* ($J_{PR}$). The evening sector Region 2 FAC reaches the equatorward side of the oval and closes through the oval to the evening side Region 1 as a Pedersen current. As discussed in Sect. 6.5, if the height-integrated conductivities were constant, all FAC would connect to Pedersen currents. In reality the Hall conductivity has gradients implying that also a part of the ionospheric Hall currents are involved in the closure of FACs.

The overlapping current system in the time sector 22–24 of Fig. 1.28 is one of the key regions in magnetosphere–ionosphere coupling. It is known as the *Harang discontinuity* and it plays a special role in the dynamics of magnetospheric substorms to which we will return in Chap. 13. Also in the noon sector there is a special current system poleward of Region 1. This current system is observed during northward IMF and thought to be connected to the high-latitude magnetopause tailward of the cusp region.

## 1.5 Space Storms Seen from the Ground

Effects of space storms reach all the way down to the surface of the Earth where the storm development can be followed using various ground-based instruments, in particular, magnetometers. The Earth is not only an auditorium where we can watch the storms, but the conductive ground is a part of the global space storm system.

### 1.5.1 Measuring the strength of space storms

Scientists use several different methods to characterize the strength of space storms. It is understandable that those who are mostly interested in the storms on the Sun use different metrics than those studying magnetic storms in the near-Earth space.

The most traditional method of describing solar activity is the sunspot number (e.g., Fig. 1.6). It does not actually tell us anything of individual storms, but it describes very well the progress of the most important space climate cycle. Another widely used indicator of solar activity is the *10.7-cm radio flux* (F10.7). Radio emissions are created high in the solar atmosphere by electrons gyrating in the solar magnetic field (Chap. 12) and are strongly enhanced during strong magnetic activity. F10.7 follows the sunspot activity and it has also been found to be a very good proxy for energy input to the upper atmosphere of the Earth although the variations in the energy input itself mostly depends on the variations in the EUV irradiance.

The solar flares to be discussed in Chap. 12 give rise to strongly enhanced X-ray emissions. The emissions are today monitored regularly by geostationary satellites and their intensity is readily available through the internet. The intensity is indexed into different classes (A,B,C,M,X) according to the X-ray flux as given in Table 1.3. Within each class the intensity is given in decimals, e.g., M7.5 indicates the flux $7.5 \times 10^{-5}\,\mathrm{W\,m^{-2}}$. The largest measured X-ray flare (till the end of cycle 23) took place on November 4, 2003. It was classified as X27. This was close to the upper sensitivity limit of the *GOES* satellites measuring the intensity and thus not necessarily fully accurate.

**Table 1.3** Solar X-ray emission classes.

| | |
|---|---|
| A | $10^{-8} - 10^{-7}\,\mathrm{W\,m^{-2}}$ |
| B | $10^{-7} - 10^{-6}\,\mathrm{W\,m^{-2}}$ |
| C | $10^{-6} - 10^{-5}\,\mathrm{W\,m^{-2}}$ |
| M | $10^{-5} - 10^{-4}\,\mathrm{W\,m^{-2}}$ |
| X | $\geq 10^{-4}\,\mathrm{W\,m^{-2}}$ |

In order to characterize the storms in the magnetosphere several activity indices have been developed to measure the strength of the magnetic perturbations [Mayaud, 1980]. The large number of useful indices illustrates the large variety of storm features; sometimes the effects are stronger at high latitudes, sometimes at low; sometimes the background current systems are already strong before the main perturbation; different current systems may decay at different rates, etc. Furthermore, different time scales from minutes to annual activity levels require different indexing methods. Instead of penetrating to the details of the great variety of indices, we discuss briefly the most widely used indices for global storm levels, *Dst* and *Kp*, and for the activity at auroral latitudes, *AE*, which we will be using in later chapters.

The *Dst* index is a weighted average of the deviation from the quiet level of the horizontal ($H$) magnetic field component measured at four low-latitude stations around the globe. The westward ring current flowing around the Earth at the distance of about 3–4 $R_E$ is the main source of the *Dst* index. During a magnetospheric storm the ring current is enhanced, which causes a negative deviation in $H$. Consequently, the more negative the peak *Dst* index is, the stronger the storm is said to be. The threshold between weak and moderate storms is typically set to −40 or −50 nT, moderate storms range from −50 to −100 nT. Storms stronger than −100 nT can be called intense and those stronger than −200 nT big. The *Dst* index is calculated once an hour. A similar 1-minute index derived from a partly different set of six low-latitude stations (*SYM–H*) is also in use.

A magnetometer reacts to all current systems, including the magnetopause current, cross-tail current and induced currents in the ground due to rapid changes in the ionospheric currents. Furthermore, high solar wind pressure pushes the magnetopause closer to the Earth forcing the magnetopause current to increase because it must shield more of the geomagnetic field from the solar wind. The effect is strongest on the dayside where the geomagnetic field just inside the magnetopause is strongest. Here the magnetopause current flows in the direction opposite to the ring current. Thus a pressure pulse causes a positive deviation in the $H$-component measured on ground. In fact, this is an excellent signature of an interplanetary shock hitting the magnetopause. If the solar wind parameters are known, the pressure effect can be cleaned away from the *Dst* index. The so-called *pressure-corrected Dst* index can be defined as

$$Dst^* = Dst - b\sqrt{P_{dyn}} + c \,, \tag{1.64}$$

where $P_{dyn}$ is the solar wind dynamic pressure and $b$ and $c$ are empirical parameters. Owing to different statistical methods and different data sources, somewhat different values of these parameters have been determined. For example, O'Brien and McPherron [2000] obtained $b = 7.26\,\text{nT}\,\text{nPa}^{-1/2}$ and $c = 11\,\text{nT}$.

The contribution from the dawn-to-dusk directed tail current is more difficult to compensate. During strong activity this current intensifies and moves closer to the Earth, enhancing thus the *Dst* index. How to handle this effect is still a controversial issue. The estimates of the effect vary in the range 25–50% [Turner et al, 2000; Alexeev et al, 1996]. During the fastest evolution of the storm main phase (Chap. 13) the temporal changes in the ionospheric currents lead to induction currents in the ground, which may contribute up to 25% to the *Dst* index [Langel and Estes, 1985; Häkkinen et al, 2002].

Another widely used index is the planetary $K$ index, $Kp$. Each magnetic observatory has its own $K$ index and $Kp$ is an average of $K$ indices from 13 mid-latitude stations. It is a quasi-logarithmic range index expressed in a scale of one-thirds: 0, 0+, 1−, 1, ..., 8+, 9−, 9. As $Kp$ is based on mid-latitude observations, it is more sensitive to high-latitude auroral current systems and substorm activity than the *Dst* index. As $Kp$ is a 3-hour index, it does not reflect short-term changes in auroral activity.

The fastest variations in the current systems take place at auroral latitudes. To investigate the strength of the auroral currents the use of *auroral electrojet indices* (*AE*) is a common method. The standard *AE* index is calculated from 11 or 12 magnetometer stations located under the average auroral oval on the northern hemisphere. It is derived from the magnetic north component of each station by taking the envelope of the largest negative deviation from the quiet time background, called the *AL* index, and the largest positive deviation, called the *AU* index. The *AE* index itself is calculated as $AE = AU - AL$ (all in nT). Thus *AL* is the measure of the strongest westward current in the auroral oval, *AU* is the measure of the strongest eastward current, and *AE* characterizes the total electrojet activity. These indices are typically given with 1-minute time resolution, but for long-term statistical studies longer cadences are also used.

As the auroral electrojets flow at much much lower altitudes than other magnetospheric currents, the magnetic deviations due to auroral currents are larger than those used to calculate *Dst*. For example, during typical substorm activations *AE* is about 200–400 nT and during strong storms the deviations can exceed 2000 nT, whereas the equatorial perturbations exceed −200 nT only during the largest storms.

There are some issues with the *AE* indices that their user must be aware of. During the strongest activity the peak ionospheric currents move well equatorward of the *AE* stations and thus the indices do not capture the real strength of the auroral currents. The same applies to times of quiescence during prolonged northward IMF, when the oval contracts to very high latitudes. Another problem is sparse, and during some periods of time, lacking coverage in the Siberian sector. For case studies magnetometer data can be collected from some 70 or 80 high- and mid-latitude stations giving a much better coverage during strong magnetospheric activity. Furthermore, if the study is limited to a given local time sector, long meridional magnetometer chains can be utilized, e.g., the IMAGE magnetometers of the MIRACLE network in Fennoscandia (cf., Kauristie et al [1996]).

We will discuss the ground-based observations of magnetospheric storms, including the auroras, and the evolution of different current systems in much greater detail in Chap. 13.

### 1.5.2 Geomagnetically induced currents

The rapidly varying ionospheric currents cause rapid time variations in the magnetic field on the surface of the Earth. These give rise to the induced *geoelectric field* according to Faraday's law $\partial \mathbf{B}/\partial t = -\nabla \times \mathbf{E}$. This electric field drives electric currents in any conductive system upon which it is applied. When these currents flow in man-made conductive networks they are called *geomagnetically induced currents* (GIC). As the electric field penetrates into the soil and water, the induction effects are also felt in gas pipelines buried under ground and in undersea telecommunication cables.

While the basic idea of current induction is elementary, its consequences are complicated. The actual induced current depends on the conductivity structure of the medium where the induced electric field is driving the current. Tanskanen et al [2001] concluded from a study of 77 substorms that at the time of the substorm onset, i.e., when the magnetic field variation is most rapid, about 40% of the *AL* index comes from the currents underneath and thus the index does not describe the real ionospheric currents correctly at substorm onset. Furthermore, at the stations surrounded by the Arctic Ocean the deviations are 10–20% larger than deviations at inland stations. In order to remove the induction effect from the *AL* index at other times than the onset the average correction needed for the inland stations is 15–20% and for the stations close to the ocean 25–30%.

There is considerable practical interest in the GIC effects, as the currents sometimes reach harmful levels. In fact, the first reported space-weather-related problems on technological systems are from events around the year 1850 when electric telegraph communications were disturbed and in some cases completely stopped during strong auroral activations. As expected, the great magnetic storm following Carrington's flare in 1859 also caused troubles to telegraph connections [Prescott, 1860]. For a long time telegraph and, later, telephone communication lines were the most space-weather-sensitive systems. The first reported effect on a power transmission network took place on March 24, 1940, when a great geomagnetic storm caused voltage dips, large swings in reactive power, and tripping of transformer banks in the United States and Canada [Davidson, 1940]. The effects of the storm were also felt on telephone lines. For example, 80% of long-distance telephone connections from Minneapolis, Minnesota, were out of operation. As our focus is in the physics of space storms themselves and not in their effects, we refer the interested reader to the more thorough discussion of space weather effects in Bothmer and Daglis [2007]. However, we will discuss some of the physics aspects of geomagnetic current induction in Sect. 15.2.

# 2. Physical Foundations

Physics of space storms is founded on physics of hot tenuous space plasmas. While the reader is assumed to be familiar with the basic concepts of plasma physics and master the classical electrodynamics, the motivation for this chapter is to review some of the main concepts, to introduce definitions and the notation to be used elsewhere in the book, and to highlight some aspects that are specific to space plasma physics.

## 2.1 What is Plasma?

There is no rigorous way to define the plasma state. A good practical description for our purposes is:

> Plasma is *quasi-neutral* gas with so many *free charges* that *collective electromagnetic phenomena* are important to its physical behavior.

In this treatise we discuss quasi-neutral plasmas only. This means that in a given plasma element there is an equal amount of positive and negative charges. There is no clear threshold for the required degree of ionization. Roughly 0.1% ionization already makes the gas look like plasma, and 1% is sufficient for almost perfect conductivity.

Plasma is sometimes called the fourth state of matter because it arises as the next natural step in the sequence from solid to liquid to gas, when the temperature is increased. There are two natural ways to produce plasma in space. The most common is to heat the gas to a high enough temperature. Usually $10^5$–$10^6$ K (10–100 eV) is sufficient (1 eV $\leftrightarrow$ 11 600 K). Also ionizing radiation is important because it creates and sustains the photospheric and ionospheric plasmas at lower temperatures where the electrons and ions recombine if the radiation stops. The transition from gas to plasma is gradual and thus different from, e.g., the phase transition from liquid to gas. The collective electromagnetic behavior gives plasma liquid-like properties. We speak of *fluid* description of plasmas when dealing with macroscopic plasma properties.

H.E.J. Koskinen, *Physics of Space Storms: From the Solar Surface to the Earth*,
Springer Praxis Books, DOI 10.1007/978-3-642-00319-6_2,

Three key concepts *Debye shielding*, *plasma oscillations*, and *gyro motion* of charged particles in the magnetic field, lie at the heart of plasma physics. Let us review them briefly.

### 2.1.1 Debye shielding

The electrostatic Coulomb potential of charge $q$ is $\varphi = q/(4\pi\varepsilon_0 r)$. In a fully ionized plasma individual particles either attract or repel each other by the force due to the gradient of this potential.

Quasi-neutrality implies that in equilibrium there is no net charge in a "large enough" volume. If we introduce an extra test charge $q_T$ into the equilibrium plasma, the charges must be redistributed to maintain the quasi-neutrality within certain volume around $q_T$. Let us denote the different plasma populations (e.g., ions and electrons) by $\alpha$ and assume that each population is in a *Boltzmann equilibrium*

$$n_\alpha(\mathbf{r}) = n_{0\alpha} \exp\left(-\frac{q_\alpha \varphi}{k_B T_\alpha}\right), \tag{2.1}$$

where $k_B$ is the *Boltzmann constant* ($k_B = 1.38 \times 10^{-23}\,\mathrm{J K^{-1}}$) and $T_\alpha$ is the temperature of population $\alpha$. The potential of $q_T$ becomes the *shielded potential*

$$\varphi = \frac{q_T}{4\pi\varepsilon_0 r} \exp\left(-\frac{r}{\lambda_D}\right), \tag{2.2}$$

where

$$\lambda_D^{-2} = \frac{1}{\varepsilon_0} \sum_\alpha \frac{n_{0\alpha} q_\alpha^2}{k_B T_\alpha} \tag{2.3}$$

defines the *Debye length* $\lambda_D$. The rearrangement of the charges is called *Debye shielding* and it is the most fundamental manifestation of the collective behavior of the plasma. Intuitively $\lambda_D$ is the limit beyond which the thermal speed of the plasma particles is high enough to escape from the Coulomb potential of $q_T$. Often the electron and ion Debye lengths are given separately. Numerically the electron Debye length is

$$\lambda_D(\mathrm{m}) \approx 7.4 \sqrt{\frac{T(\mathrm{eV})}{n(\mathrm{cm}^{-3})}}. \tag{2.4}$$

Using the Debye length we can redefine the plasma state in a slightly more quantitative way. That the collective properties really dominate the plasma behavior there must be a large number of particles in the *Debye sphere* of radius $\lambda_D$, i.e., $(4\pi/3)n_0\lambda_D^3 \gg 1$. The factor $4\pi/3$ is often neglected and we call $\Lambda = n_0\lambda_D^3$ the *plasma parameter*. Because plasma must also be quasi-neutral, its size $L = V^{1/3}$ must be larger than $\lambda_D$. Thus for a plasma

$$\frac{1}{\sqrt[3]{n_0}} \ll \lambda_D \ll L. \tag{2.5}$$

Note that many sources [e.g., Boyd and Sanderson, 2003] call $g = 1/n_0\lambda_D^3$ plasma parameter.

**Train your brain**

Derive (2.2) for the shielded potential of a test charge $q_T$ in a plasma with Boltzmann's density distribution.

Hints:
(i) Use $e^{-x} \simeq 1 - x$ when substituting the densities into Coloumb's law and make use of quasi-neutrality.
(ii) Make also use of spherical symmetry to write

$$\nabla^2 \varphi = \frac{1}{r^2}\frac{d}{dr}\left(r^2\frac{d\varphi}{dr}\right) .$$

(iii) After solving the differential equation require that the solution approaches the Coulomb potential of $q_T$ when $r \to 0$ and remains finite at all distances.

### 2.1.2 Plasma oscillations

If plasma equilibrium is disturbed by a small perturbation, plasma starts to oscillate. Much of space plasma physics concerns the great variety of plasma responses to perturbations. The most fundamental example is the *plasma oscillation*.

Considering freely moving cold ($T_e = 0$) electrons and fixed background ions it is an easy exercise to show that a small perturbation in the electron density causes the plasma oscillation at the *plasma frequency*

$$\omega_{pe}^2 = \frac{n_0 e^2}{\varepsilon_0 m_e} . \tag{2.6}$$

Note that both the angular frequency $\omega_{pe}$ and the corresponding oscillation frequency $f_{pe} = \omega_{pe}/2\pi$ are usually called plasma frequency. So, be careful!

Plasma frequency is inversely proportional to the square root of the mass of the moving particles, here electrons. Thus the ion plasma frequency is a much smaller quantity than the electron plasma frequency. When we speak of plasma frequency, we usually mean the electron plasma frequency. A useful rule of thumb is

$$f_{pe}(\mathrm{Hz}) \approx 9.0\sqrt{n(\mathrm{m}^{-3})} .$$

The plasma oscillation determines a natural length scale in the plasma known as the *electron inertial length* $c/\omega_{pe}$, where $c$ is the speed of light. Physically it gives the attenuation length scale of an electromagnetic wave with the frequency $\omega_{pe}$ when it penetrates to plasma (wave propagation in plasmas will be discussed in detail in Chaps. 4 and 5). It is analogous to the *skin depth* in classical electromagnetism defined in (4.26) and is thus often called *electron skin depth*.

Similarly, the *ion plasma frequency* is defined by

$$\omega_{pi}^2 = \frac{n_0 e^2}{\varepsilon_0 m_i} \, . \tag{2.7}$$

The corresponding *ion inertial length* is $c/\omega_{pi}$. It is associated with damping of fluctuations near the ion plasma frequency.

### 2.1.3 Gyro motion

Space plasmas are practically always embedded in a magnetic field. The magnetic field may be due to external or internal current systems. The known magnetic flux densities in space vary by more than 20 orders of magnitude. The interstellar magnetic field is typically less than 1 nT, the magnetic field of the solar wind at the distance of the Earth (1 *AU*) is a few nT, the field on the surface of the Earth varies $3–6\times10^{-5}$ T (0.3–0.6 gauss) and in large fusion devices the fields are several teslas. The largest known fields, exceeding $10^8$ T, are found at the rapidly rotating neutron stars (pulsars). Observations of slowly decelerating pulsars emitting X- and soft gamma rays indicate even stronger magnetic fields, exceeding $10^{11}$ T.

A charged particle in a magnetic field performs a circular motion perpendicular to the field. The angular frequency of this gyro motion for particle species $\alpha$ is

$$\omega_{c\alpha} = \frac{|q_\alpha| B}{m_\alpha} \, . \tag{2.8}$$

This is called the *gyro frequency* (or cyclotron frequency, Larmor frequency). The corresponding oscillation frequencies $f_{c\alpha} = \omega_{c\alpha}/(2\pi)$ of electrons and protons are given by

$$\begin{aligned} f_{ce}(\mathrm{Hz}) &\approx 28\, B(\mathrm{nT}) \\ f_{cp}(\mathrm{Hz}) &\approx 1.5 \times 10^{-2}\, B(\mathrm{nT}) \, . \end{aligned}$$

Again the same term is used for both $\omega_c$ and $f_c$.

As discussed later in this chapter the gyro motion determines another important length scale, the *electron* or *ion gyro radius*, also known as cyclotron, or Larmor radius

$$r_{L\alpha} = \frac{v_{\perp\alpha}}{|q_\alpha|} \, , \tag{2.9}$$

where $v_{\perp\alpha}$ is the speed of the particle perpendicular to the magnetic field.

### 2.1.4 Collisions

Most of the volume where space storms take place is filled by fully ionized plasmas that behave in a "collisionless" manner. However, there are two important exceptions: in the solar photosphere and in the ionosphere collisions between charged particles and neutrals have a strong influence on the plasma properties, determining, e.g., the ionospheric Ohm's law. Furthermore, the charge exchange collisions between charged particles and the Earth's ring current are important to the dynamics of storms in the inner magnetosphere (Chap. 14).

For the collisionless behavior of fully ionized plasmas the *Coulomb interaction* (Coulomb collisions) between charged particles is essential. In a plasma the finite Debye length limits the Coulomb interaction within the Debye sphere, but yet each particle sees $\Lambda$ other charges. If we can calculate the collisional *cross-section* $\sigma$, we can determine the *mean free path*

$$l_{mfp} = 1/(n\sigma) \tag{2.10}$$

of the particles and their *collision frequency*

$$\nu_c = n\sigma\langle v\rangle \,, \tag{2.11}$$

where $\langle v\rangle$ is the average speed of the particles.

For Coulomb collisions it is sufficient to consider small-angle collisions, in which the particles are just slightly deflected. The reason for this is that each particle interacts with a large number of particles at long distance, whereas the probability for nearby collisions with large deflection angles is much smaller. The rigorous calculation of collisional cross-sections is rather challenging. For electron–ion collisions $\sigma \propto v_0^{-4}$ and

$$\nu_c = \nu_{ei} = \frac{2n_0(Ze^2)^2 \ln\Lambda}{\varepsilon_0^2 m_e^2 v_0^3} \,, \tag{2.12}$$

where $v_0$ is the particle speed far from the collision and $\ln\Lambda$ is called the *Coulomb logarithm*. Typical values of the Coulomb logarithm are in the range 10–20.

When the temperature of the plasma increases or the density decreases, $g = \Lambda^{-1}$ decreases. At the limit $g \rightarrow 0$ plasma becomes *collisionless*. Physically this means that the time between individual collisions, or the mean free path, becomes longer than the temporal or spatial scales of the problems under study. *This does not mean* that the electromagnetic interaction between plasma particles would become negligible. At the collisionless limit it is, however, sufficient to consider the effect of average electromagnetic fields on the particles instead of individual collisions.

**Train your brain**

Show that in a fully ionized plasma the frequency of small-angle Coulomb collisions is much larger than the frequency of large-angle collisions. To what plasma parameter the ratio of these frequencies is related?

**Feed your brain**

Derive Equation (2.12). The derivation can be found in many textbooks, including some listed in the References section of this book.

## 2.2 Basic Electrodynamics

In this section we review some of the basic concepts of classical electrodynamics that are most important in plasma physics.

### 2.2.1 Maxwell's equations

In plasma physics we usually write Maxwell's equations in the vacuum form

$$\nabla\cdot\mathbf{E} = \rho/\varepsilon_0 \tag{2.13}$$

$$\nabla\cdot\mathbf{B} = 0 \tag{2.14}$$

$$\nabla\times\mathbf{E} = -\frac{\partial\mathbf{B}}{\partial t} \tag{2.15}$$

$$\nabla\times\mathbf{B} = \mu_0\mathbf{J} + \frac{1}{c^2}\frac{\partial\mathbf{E}}{\partial t}\,, \tag{2.16}$$

where the source terms *charge density* $\rho$ and *current density* $\mathbf{J}$ are determined by the particle distribution functions (Sect. 2.3.3). We call $\mathbf{E}$ the *electric field* ([E] = $\mathrm{V\,m^{-1}}$) and $\mathbf{B}$ *magnetic field* ([B] = $\mathrm{V\,s\,m^{-2}} \equiv \mathrm{T}$). It would be more orthodox to call $\mathbf{B}$ magnetic induction, or more descriptively magnetic flux density, as the *magnetic flux* through a surface $S$ is

$$\Phi = \int_S \mathbf{B}\cdot d\mathbf{S}\,. \tag{2.17}$$

The SI units of the source terms in Maxwell's equations are $[\rho] = \mathrm{A\,s\,m^{-3}} = \mathrm{C\,m^{-3}}$ and $[J] = \mathrm{A\,m^{-2}}$. The natural constants in SI units are

$$\varepsilon_0 \approx 8.854\times10^{-12}\,\mathrm{A\,s\,V^{-1}m^{-1}}\,, \quad \text{vacuum permittivity}$$

$$\mu_0 = 4\pi\times10^{-7}\,\mathrm{V\,s\,A^{-1}m^{-1}}\,, \quad \text{vacuum permeability}$$

$$c = 1/\sqrt{\varepsilon_0\mu_0} = 299\,792\,458\ \mathrm{m\,s^{-1}} \quad \text{definition of the speed of light.}$$

In studies of electromagnetic media the *electric displacement* $\mathbf{D}$ and the *magnetic field intensity* $\mathbf{H}$ (the "magnetic field" of engineering physics) are useful and Maxwell's equations are written as

$$\nabla\cdot\mathbf{D} = \rho_f \tag{2.18}$$

$$\nabla \cdot \mathbf{B} = 0 \tag{2.19}$$

$$\nabla \times \mathbf{E} = -\frac{\partial \mathbf{B}}{\partial t} \tag{2.20}$$

$$\nabla \times \mathbf{H} = \mathbf{J}_f + \frac{\partial \mathbf{D}}{\partial t}, \tag{2.21}$$

where $\rho_f$ and $\mathbf{J}_f$ are the source terms due to "free" charges. If the properties of the medium can be described in terms of *electric polarization* **P** and *magnetization* **M**, fields **D** and **H** are given by the *constitutive equations*

$$\mathbf{D} = \varepsilon_0 \mathbf{E} + \mathbf{P} \tag{2.22}$$

$$\mathbf{H} = \mathbf{B}/\mu_0 - \mathbf{M}. \tag{2.23}$$

In plasma physics the use of **D** and **H** is sometimes convenient notation, but the constitutive relations may pose a problem. There is no unique way to define the polarization field in a medium of free charges, although sometimes a useful **P** can be introduced formally (e.g., Eq. 9.73). However, the change of polarization is a real plasma phenomenon and the corresponding *polarization current*

$$\mathbf{J}_P = \frac{\partial \mathbf{P}}{\partial t} \tag{2.24}$$

is well-defined (see., e.g., Sect. 3.5.1). Also the *magnetization current*

$$\mathbf{J}_M = \nabla \times \mathbf{M} \tag{2.25}$$

is a useful concept in plasma physics.

The Maxwell equations form a set of 8 partial differential equations. If we know the source terms, we have more than enough equations to calculate the six unknown field components. If we, however, want to treat all 10 variables (**E**, **B**, **J**, $\rho$) self-consistently, we need more equations. In a conductive medium it is customary to use *Ohm's law*

$$\mathbf{J} = \sigma \cdot \mathbf{E}, \tag{2.26}$$

where the *conductivity* $\sigma$ ($[\sigma] = \mathrm{A\,(V\,m)^{-1}} = (\Omega\,\mathrm{m})^{-1}$) is, in general, a tensor and may also depend on **E** and **B**.

Recall that Ohm's law is not a fundamental law in the same sense as Maxwell's equations but merely an empirical relationship to describe the conductivity of the medium similarly to the constitutive relations $\mathbf{D} = \varepsilon \cdot \mathbf{E}$ and $\mathbf{B} = \mu \cdot \mathbf{H}$ where $\varepsilon$ and $\mu$ are, in general, tensors. The medium is called *linear* if $\varepsilon$, $\mu$, and $\sigma$ are scalars and constant in space and time. Note that also in the linear media they usually are functions of the wave number and frequency of electromagnetic fields penetrating into the medium. Much of plasma physics deals with the properties of $\varepsilon(\omega, \mathbf{k})$.

### 2.2.2 Lorentz force

Experimental determination of **E** and **B** is based on the *Lorentz force*

$$\mathbf{F} = \frac{d\mathbf{p}}{dt} = q(\mathbf{E} + \mathbf{v} \times \mathbf{B}) \tag{2.27}$$

on a particle with charge $q$ and velocity **v**. Close to a body with strong gravity (e.g., the Sun) also the gravitational force ($m\mathbf{g}$) must be taken into account. In principle, a complete description of plasma would mean solving the equation of motion (with gravitation if needed) for all plasma particles. In practice, this is impossible.

Often it is useful, and in many problems sufficient, to trace the motion of individual charges in a given electromagnetic field. Examples of this are the motion of cosmic rays, or high-energy particles in the Earth's radiation belts. These problems are often relativistic

$$\mathbf{F} = \frac{d}{dt}(\gamma m \mathbf{v}) = q(\mathbf{E} + \mathbf{v} \times \mathbf{B}) , \tag{2.28}$$

where $\gamma = (1 - \beta^2)^{-1/2}$ is the *Lorentz factor* with $\beta = v/c$. The time component of the underlying four-force gives the power

$$\frac{dW}{dt} = \frac{d}{dt}(\gamma m c^2) = q\mathbf{E} \cdot \mathbf{v} . \tag{2.29}$$

Because the magnetic part of the Lorentz force is perpendicular to **v**, only the electric field performs work ($W$). Thus any "magnetic" acceleration of charged particles requires the change in the magnetic field, which induces an electric field in the frame of reference where the acceleration is observed.

### 2.2.3 Potentials

Equation $\nabla \cdot \mathbf{B} = 0$ implies that there is a *vector potential* **A**, for which $\mathbf{B} = \nabla \times \mathbf{A}$. Inserting **A** into Faraday's law we find

$$\nabla \times (\mathbf{E} + \partial \mathbf{A}/\partial t) = 0 \tag{2.30}$$

$\Rightarrow$

$$\mathbf{E} = -\partial \mathbf{A}/\partial t - \nabla \varphi , \tag{2.31}$$

where $\varphi$ is the *scalar potential.*

Thus we have expressed six variables (**E**, **B**) using four functions (**A**, $\varphi$). For this we needed four components of Maxwell's equations. The remaining four equations are now

$$\nabla^2 \varphi + \frac{\partial (\nabla \cdot \mathbf{A})}{\partial t} = -\rho/\varepsilon_0 \tag{2.32}$$

$$\nabla^2 \mathbf{A} - \frac{1}{c^2}\frac{\partial^2 \mathbf{A}}{\partial t^2} - \nabla(\nabla \cdot \mathbf{A} + \frac{1}{c^2}\frac{\partial \varphi}{\partial t}) = -\mu_0 \mathbf{J} . \tag{2.33}$$

At first these look more complicated than the original equations, but they are much easier to solve analytically. The point is that **E** and **B** are derivatives of the scalar and vector potentials and there is quite a lot of freedom to transform the potentials keeping their derivatives unchanged. Such transformations are called *gauge transformations*. There are several *gauge functions* $\Psi$ to define the transformations

$$\mathbf{A} \rightarrow \mathbf{A}' = \mathbf{A} + \nabla\Psi \tag{2.34}$$

$$\varphi \rightarrow \varphi' = \varphi - \partial\Psi/\partial t\,. \tag{2.35}$$

The *Lorenz*[1] *gauge* is defined by

$$\nabla\cdot\mathbf{A}' + \frac{1}{c^2}\frac{\partial\varphi'}{\partial t} = 0\,. \tag{2.36}$$

This gauge always exists but is not unique. It transforms the Maxwell equations to inhomogeneous wave equations

$$(\nabla^2 - \frac{1}{c^2}\frac{\partial^2}{\partial t^2})\varphi = -\rho/\varepsilon_0 \tag{2.37}$$

$$(\nabla^2 - \frac{1}{c^2}\frac{\partial^2}{\partial t^2})\mathbf{A} = -\mu_0\mathbf{J}\,. \tag{2.38}$$

The solutions of which are the *retarded potentials*

$$\varphi(\mathbf{r},t) = \frac{1}{4\pi\varepsilon_0}\int\frac{\rho(\mathbf{r}',t-R/c)}{R}\,d^3r' \tag{2.39}$$

$$\mathbf{A}(\mathbf{r},t) = \frac{\mu_0}{4\pi}\int\frac{\mathbf{J}(\mathbf{r}',t-R/c)}{R}\,d^3r'\,, \tag{2.40}$$

where $R = |\mathbf{r}-\mathbf{r}'|$ and integrations are over the volume where the source terms are not zero. Thus we have solved Maxwell's equations for given $\rho$ and **J**.

In terms of special relativity the wave equations are actually the time and space components of the wave equation for the four-vector $A^\alpha(\varphi/c,\mathbf{A})$

$$\partial^2 A^\alpha \equiv \left(\nabla^2 - \frac{1}{c^2}\frac{\partial^2}{\partial t^2}\right)A^\alpha = -\mu_0 j^\alpha\,, \tag{2.41}$$

where $j^\alpha = (c\rho,\,\mathbf{J})$ is the four-current.

**Feed your brain** by deriving the expressions for the retarded potentials

[1] This is not a spelling error. The first person to apply this method was *Ludvig V. Lorenz* (1829–1891) in 1867, not the much more famous *Hendrik A. Lorentz* (1853–1928).

***Example: The radiation terms of the electromagnetic fields***

Denote the retarded quantities by brackets $[f] = f(\mathbf{r}', t - R/c)$ and calculate the fields from the potentials. This results in

$$\mathbf{E} = \frac{1}{4\pi\varepsilon_0}\left\{\int \frac{[\rho]\mathbf{R}}{R^3}\, d^3r' + \frac{1}{c}\int\left(\frac{2[\mathbf{J}]\cdot\mathbf{RR}}{R^4} - \frac{[\mathbf{J}]}{R^2}\right) d^3r' \right.$$
$$\left. + \frac{1}{c^2}\int\left(\frac{([\dot{\mathbf{J}}]\times\mathbf{R})\times\mathbf{R}}{R^3}\right) d^3r'\right\} \tag{2.42}$$

$$\mathbf{B} = \frac{\mu_0}{4\pi}\left\{\int \frac{[\mathbf{J}]\times\mathbf{R}}{R^3}\, d^3r' + \frac{1}{c}\int \frac{[\dot{\mathbf{J}}]\times\mathbf{R}}{R^2}\, d^3r'\right\}, \tag{2.43}$$

where the dot above $\mathbf{J}$ denotes the time derivative. Far from the sources ($R \to \infty$) the *radiation terms* dominate

$$\mathbf{E}_{rad} = \frac{1}{4\pi\varepsilon_0 c^2}\int \frac{([\dot{\mathbf{J}}]\times\mathbf{R})\times\mathbf{R}}{R^3}\, d^3r' \tag{2.44}$$

$$\mathbf{B}_{rad} = \frac{1}{4\pi\varepsilon_0 c^3}\int \frac{[\dot{\mathbf{J}}]\times\mathbf{R}}{R^2}\, d^3r'\,. \tag{2.45}$$

$\mathbf{E}_{rad}$ and $\mathbf{B}_{rad}$ vanish as $1/R$. The fields due to static currents and charges vanish as $1/R^2$ or faster. Radiation requires temporal variation of $\mathbf{J}$ and a charge moving with a constant velocity does not radiate. We will discuss the electromagnetic radiation in more detail in Chap. 9.

---

Another important gauge is the *Coulomb gauge*

$$\nabla\cdot\mathbf{A}' = 0\,. \tag{2.46}$$

The vector potential is found by transformation

$$\nabla^2\Psi = -\nabla\cdot\mathbf{A}\,, \tag{2.47}$$

which defines $\Psi$ uniquely (to an additive constant) when $\mathbf{A}$ and $\varphi \to 0$ for $r \to \infty$.

Now the scalar potential

$$\varphi = \frac{1}{4\pi\varepsilon_0}\int \frac{\rho(\mathbf{r}', t)}{R}\, d^3r' \tag{2.48}$$

is *not retarded* but determined by the instantaneous value of $\rho$ everywhere. Thus the Coulomb gauge is not Lorentz[2] covariant and one must be careful when transforming between moving coordinate systems.

---

[2] Now the credit goes to the right Lorentz

The vector potential is obtained from the wave equation

$$\nabla^2 \mathbf{A} - \frac{1}{c^2}\frac{\partial^2 \mathbf{A}}{\partial t^2} = \frac{1}{c^2}\nabla\frac{\partial \varphi}{\partial t} - \mu_0 \mathbf{J} \,. \tag{2.49}$$

The first term on the RHS is curl-free. Applying the Helmholtz theorem of vector calculus we can divide the current to curl-free and source-free components

$$\mathbf{J} = \mathbf{J}_l + \mathbf{J}_t \; ; \;\; \nabla \times \mathbf{J}_l = 0 \; ; \;\; \nabla \cdot \mathbf{J}_t = 0 \,,$$

where $l$ stands for *longitudinal* (curl-free) and $t$ for *transversal* (source-free). The continuity equation $\partial \rho / \partial t + \nabla \cdot \mathbf{J} = 0$ reduces (2.49) to

$$\nabla^2 \mathbf{A} - \frac{1}{c^2}\frac{\partial^2 \mathbf{A}}{\partial t^2} = -\mu_0 \mathbf{J}_t \,. \tag{2.50}$$

Consequently, the Coulomb gauge is called *transversal gauge*. It is also called *radiation gauge* because the vector potential calculated from the transversal current

$$\mathbf{A}(\mathbf{r},t) = \frac{\mu_0}{4\pi}\int \frac{\mathbf{J}_t(\mathbf{r}',t-R/c)}{R}\, d^3r' \tag{2.51}$$

is sufficient for the calculation of the radiation fields. The Coulomb gauge separates the electric field to its static ($s$) and inductive ($i$) parts

$$\mathbf{E}_s = -\nabla\varphi \; ; \; \mathbf{E}_i = -\partial \mathbf{A}/\partial t \,, \tag{2.52}$$

but this separation is not Lorentz covariant.

The Coulomb gauge is technically easier to use than the Lorenz gauge. It is particularly useful when no sources are present. Then $\varphi = 0$ and

$$\mathbf{E} = -\partial \mathbf{A}/\partial t \; ; \; \mathbf{B} = \nabla \times \mathbf{A} \,. \tag{2.53}$$

This is sometimes called the *temporal gauge*. It is useful, e.g., in studies of Alfvén waves and wave–wave interactions.

For specific purposes there are several other useful potential presentations. Plasmas are often embedded in a background magnetic field created by external currents ($\nabla \times \mathbf{B} = 0$, e.g., the intrinsic magnetic field of a planet). Then the magnetic field can be expressed in terms of the magnetic scalar potential as

$$\mathbf{B} = -\nabla\psi \,. \tag{2.54}$$

Because $\nabla \cdot \mathbf{B} = 0$, $\psi$ can be solved from the Laplace equation

$$\nabla^2\psi = 0 \tag{2.55}$$

using familiar potential theory methods.

Another representation of the magnetic field is in terms of *Euler potentials* $(\alpha, \beta, \chi)$ as

$$\mathbf{A} = \alpha \nabla \beta + \nabla \chi \tag{2.56}$$

$\Rightarrow$

$$\mathbf{B} = \nabla \times \mathbf{A} = \nabla \times (\alpha \nabla \beta + \nabla \chi) = \nabla \alpha \times \nabla \beta \,. \tag{2.57}$$

Note that $\mathbf{B}$ is perpendicular to both $\nabla\alpha$ and $\nabla\beta$, and $\alpha$ and $\beta$ are constants along the magnetic field. Thus the magnetic field line can be visualized as the intersection line of $\alpha =$ const. and $\beta =$ const. This presentation is particularly useful in problems where tracing of magnetic field lines is required.

### 2.2.4 Energy conservation

The energy conservation of electromagnetic fields is expressed by the *Poynting theorem.* In a linear medium the energy densities of electric and magnetic fields are given by

$$w_E = \frac{1}{2}\mathbf{E}\cdot\mathbf{D} \tag{2.58}$$

$$w_M = \frac{1}{2}\,\mathbf{H}\cdot\mathbf{B} = \frac{1}{2}\,\mathbf{J}\cdot\mathbf{A}\,. \tag{2.59}$$

Define the *Poynting vector* as $\mathbf{S} = \mathbf{E} \times \mathbf{H}$. From Maxwell's equations we find

$$\nabla \cdot \mathbf{S} = -\mathbf{E}\cdot\mathbf{J} - \mathbf{E}\cdot\frac{\partial \mathbf{D}}{\partial t} - \mathbf{H}\cdot\frac{\partial \mathbf{B}}{\partial t}\,. \tag{2.60}$$

The Poynting theorem is the integral of this expression over volume $\mathscr{V}$

$$-\int_{\mathscr{V}} \mathbf{J}\cdot\mathbf{E}\,d^3r = \int_{\mathscr{V}} \nabla\cdot\mathbf{S}\,d^3r + \int_{\mathscr{V}} \frac{\partial}{\partial t}(w_E + w_M)\,d^3r\,. \tag{2.61}$$

The LHS is the work performed by the electromagnetic field per unit time (i.e., power) in volume $\mathscr{V}$. The first term on the RHS is $\oint_{\partial\mathscr{V}} \mathbf{S}\cdot d\mathbf{a}$, i.e., the energy flux per unit time through the surface $\partial\mathscr{V}$. Thus the Poynting vector gives the flux of electromagnetic energy density. The last term on the RHS expresses the rate of change of the electromagnetic energy in volume $\mathscr{V}$.

In the following we often assume that the fields have harmonic time or space dependence ($\propto \exp(-i\omega t)$, $\exp(i\mathbf{k}\cdot\mathbf{r})$), or both in the case of *plane waves*. For complex fields one must be careful with products. We interpret the real part of the complex vector as the physical field. For example, consider an electric field with harmonic time dependence

$$\mathbf{E}(\mathbf{r},t) = \mathrm{Re}\{\mathbf{E}(\mathbf{r})\exp(-i\omega t)\} = \frac{1}{2}\left[\mathbf{E}(\mathbf{r})\exp(-i\omega t) + \mathbf{E}^*(\mathbf{r})\exp(i\omega t)\right].$$

Denote the complex conjugate by cc. The product of $\mathbf{E}$ and $\mathbf{J}$ is

$$\mathbf{J}\cdot\mathbf{E} = \frac{1}{4}[\mathbf{J}(\mathbf{r})\exp(-i\omega t)+\text{cc}]\cdot[\mathbf{E}(\mathbf{r})\exp(-i\omega t)+\text{cc}]$$
$$= \frac{1}{2}\text{Re}\{\mathbf{J}^*(\mathbf{r})\cdot\mathbf{E}(\mathbf{r})+\mathbf{J}(\mathbf{r})\cdot\mathbf{E}^*(\mathbf{r})\exp(-2i\omega t)\}\,. \tag{2.62}$$

The time average of this is

$$\langle\mathbf{J}\cdot\mathbf{E}\rangle = \frac{1}{2}\text{Re}\{\mathbf{J}^*\cdot\mathbf{E}\}\,. \tag{2.63}$$

The Poynting theorem now reads as

$$\frac{1}{2}\int_{\mathscr{V}}\mathbf{J}^*\cdot\mathbf{E}\,d^3r+\oint_{\partial\mathscr{V}}\mathbf{S}\cdot d\mathbf{a}+2i\omega\int_{\mathscr{V}}(w_E+w_M)\,d^3r=0\,. \tag{2.64}$$

Note that $\mathbf{S}=\frac{1}{2}\mathbf{E}\times\mathbf{H}^*$ ; $w_E=\frac{1}{4}\mathbf{E}\cdot\mathbf{D}^*$ ; $w_M=\frac{1}{4}\mathbf{H}\cdot\mathbf{B}^*$ .

Using the Poynting vector we can express the momentum density of the electromagnetic field as

$$\hat{\mathbf{p}}=\mathbf{D}\times\mathbf{B}=\mu_0\varepsilon_0\mathbf{S} \tag{2.65}$$

when the momentum of the field is

$$\mathbf{p}_{field}=\int_{\mathscr{V}}\mathbf{D}\times\mathbf{B}\,d^3r\,. \tag{2.66}$$

The elements of the *Maxwell stress tensor* are

$$T_{ij}=E_iD_j+B_iH_j-\frac{1}{2}(\mathbf{E}\cdot\mathbf{D}+\mathbf{B}\cdot\mathbf{H})\delta_{ij}\,. \tag{2.67}$$

With this we can express the conservation of momentum as

$$\frac{d}{dt}(\mathbf{p}_{mech}+\mathbf{p}_{field})_i=\sum_j\int_{\mathscr{V}}\frac{\partial}{\partial x_j}T_{ij}\,d^3r=\oint_{\partial\mathscr{V}}\sum_j T_{ij}n_j\,da; \tag{2.68}$$

where the mechanical force is the Lorentz force

$$\frac{d\mathbf{p}_{mech}}{dt}=\int_{\mathscr{V}}(\rho\mathbf{E}+\mathbf{J}\times\mathbf{B})\,d^3r\,. \tag{2.69}$$

### 2.2.5 Charged particles in electromagnetic fields

In a homogeneous static magnetic field in absence of an electric field the *equation of motion* of a charged particle

$$m\frac{d\mathbf{v}}{dt}=q(\mathbf{v}\times\mathbf{B}) \tag{2.70}$$

has a solution with constant speed along the magnetic field and circular motion around the magnetic field line with the angular frequency

$$\omega_c=\frac{qB}{m}\,. \tag{2.71}$$

The radius of the circular motion (*Larmor radius*, cyclotron radius, gyro radius) is

$$r_L = \frac{v_\perp}{|\omega_c|} = \frac{mv_\perp}{|q|B} , \tag{2.72}$$

where $v_\perp = \sqrt{v_x^2 + v_y^2}$ is the velocity perpendicular to the magnetic field. The gyro period is

$$\tau_L = \frac{2\pi}{|\omega_c|} . \tag{2.73}$$

Looking along the magnetic field, the particle rotating clockwise has a negative charge. In plasma physics this is *the convention of right-handedness*.

This way we have decomposed the velocity to a constant speed $v_\parallel$ along the field and circular velocity $v_\perp$ perpendicular to the field. The sum of these components is a helical motion with the *pitch angle* $\alpha$ defined as

$$\tan\alpha = v_\perp / v_\parallel . \tag{2.74}$$

Hannes Alfvén realized that this decomposition is convenient even in temporally and spatially varying fields if the variations are slow compared to the gyro motion. The method is called *guiding center approximation*. The center of the gyro motion is the *guiding center* (GC) and the frame of reference where $v_\parallel = 0$ is the *guiding center system* (GCS).

In the GCS the charge gives rise to a current $I = q/\tau_L$ with the associated *magnetic moment*

$$\mu = I\pi r_L^2 = \frac{1}{2}\frac{q^2 r_L^2 B}{m} = \frac{1}{2}\frac{mv_\perp^2}{B} = \frac{W_\perp}{B} . \tag{2.75}$$

The magnetic moment is actually a vector

$$\boldsymbol{\mu} = \frac{1}{2} q \mathbf{r}_L \times \mathbf{v}_\perp , \tag{2.76}$$

which is always *opposite to the ambient magnetic field*. Charged particles tend to weaken the magnetic field and thus plasma can be considered as a *diamagnetic* medium.

If there is also a constant electric field, the GC drifts perpendicular to both the electric and magnetic fields with the velocity

$$\mathbf{v}_E = \frac{\mathbf{E} \times \mathbf{B}}{B^2} . \tag{2.77}$$

This is called *electric drift* or E×B drift. The drift velocity is independent of the charge and mass of the particle.

The E×B drift corresponds to the Lorentz transformation to the frame co-moving with the GC

$$\mathbf{E}' = \mathbf{E} + \mathbf{v} \times \mathbf{B} . \tag{2.78}$$

In this frame $\mathbf{E}' = 0 \Rightarrow \mathbf{E} = -\mathbf{v} \times \mathbf{B}$, from which we find the solution (2.77) for $\mathbf{v}$. This coordinate transformation is possible for all sufficiently weak forces $\mathbf{F}_\perp$ resulting in a general expression for the drift velocity

$$\mathbf{v}_D = \frac{\mathbf{F}_\perp \times \mathbf{B}}{qB^2} . \tag{2.79}$$

This requires $F/qB \ll c$. If $F \gtrsim qcB$, the GC approximation cannot be used.

From (2.79) we readily find the *gravitational drift* velocity

$$\mathbf{v}_g = \frac{m\mathbf{g} \times \mathbf{B}}{qB^2} \propto \frac{m}{q} . \tag{2.80}$$

Gravity separates particles according to their $m/q$, not in the direction of the gravitational force but perpendicular to it and to **B**.

The same formalism applies to a slowly time varying electric field if we assume the magnetic field to be constant. This results in the *polarization drift*

$$\mathbf{v}_P = \frac{1}{\omega_c B} \frac{d\mathbf{E}_\perp}{dt} . \tag{2.81}$$

We will discuss inhomogeneous magnetic fields and rapidly time varying electric fields in Chap. 3.

## 2.3 Tools of Statistical Physics

Plasma physics is sometimes considered as applied electrodynamics. Equally well it could be characterized as statistical physics of charged particles. The computation of the motion of all plasma particles from Maxwell's equations and the Lorentz force is an impossible task. Fortunately, we do not always need to know the details of individual particles, but we are interested in the macroscopic properties of the gas or fluid (density, flux, flow velocity, temperature, pressure, heat flux, etc.) and their evolution in space and time. To handle this we need tools of statistical physics.

### 2.3.1 Plasma in thermal equilibrium

There are different ways to find the fundamental plasma equations. Here we start from *equilibrium statistical mechanics*. Let there be $N$ particles in the plasma ($N/2$ electrons, $N/2$ singly-charged ions). Assume that the plasma is in *thermal equilibrium* at the temperature $T$. The probability of finding the particles in locations $(\mathbf{r}_1, ..., \mathbf{r}_N)$ is given by the *Gibbs distribution*

$$D(\mathbf{r}_1, ..., \mathbf{r}_N) = \frac{1}{Z} \exp\left( - \frac{\sum_k \sum_{i>k} W_{ik}}{k_B T} \right) , \tag{2.82}$$

where

$$W_{ik} = \frac{q_i q_k}{4\pi\varepsilon_0 |\mathbf{r}_i - \mathbf{r}_k|} + \varphi_{ext}$$

and

$$Z = \int \exp\left(-\frac{\sum_k \sum_{i>k} W_{ik}}{k_B T}\right) d^3r_1 ... d^3r_N \, .$$

$Z$ is the *partition function* and $\varphi_{ext}$ describes the potential energy of all external fields.

The probability of finding particle 1 at $\mathbf{r}_1$ is

$$F_1(\mathbf{r}_1) = \int D \, d^3r_2 ... d^3r_N \, . \tag{2.83}$$

If there are no external forces, $F_1 = 1/\mathcal{V}$ ($\mathcal{V}$ is the volume). Correspondingly, the probability of finding particle 1 at $\mathbf{r}_1$ and particle 2 at $\mathbf{r}_2$ is

$$F_2(\mathbf{r}_1, \mathbf{r}_2) = \int D \, d^3r_3 ... d^3r_N \tag{2.84}$$

and so on

$$F_s(\mathbf{r}_1, ..., \mathbf{r}_s) = \int D \, d^3r_{s+1} ... d^3r_N \, . \tag{2.85}$$

Functions $F_1, ..., F_s$ are called *reduced distributions*. At the limit of non-interacting particles ($W_{ik} \to 0$)

$$F_s \to F_1(\mathbf{r}_1) F_1(\mathbf{r}_2) \cdots F_1(\mathbf{r}_s) = 1/\mathcal{V}^s \, . \tag{2.86}$$

The reduced distributions can be written using the *Mayer cluster expansion* (we use the notation: $\mathbf{r}_1 \to 1$ when there is no risk of confusion):

$$\begin{aligned} F_2(1,2) &= [1 + P_{12}(1,2)] F_1(1) F_1(2) \\ F_3(1,2,3) &= [1 + P_{12}(1,2) + P_{12}(2,3) + P_{12}(1,3) + T_{123}(1,2,3)] \times \\ & \quad F_1(1) F_1(2) F_1(3) \end{aligned} \tag{2.87}$$

and so on. $P_{12}$ is the *two-particle* (or *pair*) *correlation function* and $T_{123}$ is the *three-particle correlation function*. At the plasma limit ($\Lambda \gg 1$) the Coulomb interaction is weak and $T_{123} \ll P_{12} \ll 1$. Thus it is usually sufficient to consider pair correlations only. Note that $P$ is symmetric: $P_{12}(1,2) = P_{12}(|\mathbf{r}_1 - \mathbf{r}_2|)$.

The complete Gibbs distribution depends also on velocity:

$$D^*(\mathbf{r}_1, ..., \mathbf{r}_N, \mathbf{v}_1, ..., \mathbf{v}_N) = \frac{1}{Z^*} \exp\left(-\frac{\sum_k \sum_{i>k} W_{ik}}{k_B T}\right) \exp\left(-\frac{\sum_i \frac{1}{2} m_i v_i^2}{k_B T}\right) . \tag{2.88}$$

In this book we will consider non-relativistic plasmas only and can neglect the velocity correlations. The relativistic particles encountered in radiation belts or in solar energetic particle events can be treated as test particles and are not assumed to have significant effects on the macroscopic quantities. Of course, there are relativistic plasmas in the universe. For example, in the magnetospheres of pulsars not only relativistic but also quantum effects become important. Quantum fluctuations produce electron–positron pairs, which annihilate and radiate 511-keV gamma rays.

Differentiating $F_s$, setting $s = 2$, and assuming $T_{123} \ll P_{12}$ we can derive the equation for $P_{12}$

$$\frac{\partial P_{12}}{\partial \mathbf{r}_1} + \frac{1}{4\pi\varepsilon_0 k_B T}\frac{\partial}{\partial \mathbf{r}_1}\left(\frac{q_1 q_2}{|\mathbf{r}_1 - \mathbf{r}_2|}\right) + \tag{2.89}$$
$$\frac{1}{4\pi\varepsilon_0 k_B T}\sum_\alpha \frac{N_\alpha}{V}\int [P_{12}(2,\alpha) + P_{12}(1,\alpha)]\frac{\partial}{\partial \mathbf{r}_1}\left(\frac{q_1 q_\alpha}{|\mathbf{r}_1 - \mathbf{r}_\alpha|}\right) d^3 r_\alpha = 0\,,$$

where $\alpha$ indexes the particle species. This equation can be solved by Fourier transformation. The result is

$$P_{12}(|\mathbf{r}_1 - \mathbf{r}_2|) = -\frac{q_1 q_2}{4\pi\varepsilon_0 k_B T}\frac{\exp(-|\mathbf{r}_1 - \mathbf{r}_2|/\lambda_D)}{|\mathbf{r}_1 - \mathbf{r}_2|}\,, \tag{2.90}$$

where we again encounter the Debye shielding. The assumption $P_{12} \ll 1$ is valid if $|\mathbf{r}_1 - \mathbf{r}_2| > \lambda_D$. The Mayer expansion is valid also inside the Debye sphere, where $P_{12} \propto 1/|\mathbf{r}_1 - \mathbf{r}_2|$ as long as the distance $|\mathbf{r}_1 - \mathbf{r}_2|$ remains larger than the average distance between particles in temperature $T$.

From this description it is possible to derive equilibrium thermodynamic properties of the plasma. For example, in the plasma approximation ($\Lambda \gg 1$) the equation of state is practically that of the *ideal gas*

$$P = nk_B T + O\left(\frac{1}{\Lambda}\right). \tag{2.91}$$

Unfortunately, due to the small collision rates space plasmas seldom are in thermal equilibrium and we must look for a more general approach.

### 2.3.2 Derivation of Vlasov and Boltzmann equations

There are two main roads to the Boltzmann equation for a plasma. Consider first the *Klimontovich approach.* It starts from the exact density of particles in the six-dimensional phase space $(\mathbf{r}, \mathbf{v})$. Consider a single particle whose orbit in this space is $(\mathbf{R}_1(t), \mathbf{V}_1(t))$. The "density" of this particle is

$$N(\mathbf{r}, \mathbf{v}, t) = \delta[\mathbf{r} - \mathbf{R}_1(t)]\delta[\mathbf{v} - \mathbf{V}_1(t)]\,, \tag{2.92}$$

where $\delta$ is Dirac's delta function.[3]

Summing over all particles of a given species $\alpha$ we get the density function $N_\alpha$ for the species. Writing the equation of motion under the Lorentz force for each particle and summing over particles of a given species leads to the *Klimontovich equation* for $N_\alpha$

$$\frac{\partial N_\alpha}{\partial t} + \mathbf{v}\cdot\frac{\partial N_\alpha}{\partial \mathbf{r}} + \frac{q_\alpha}{m_\alpha}(\mathbf{E} + \mathbf{v}\times\mathbf{B})\cdot\frac{\partial N_\alpha}{\partial \mathbf{v}} = 0\,. \tag{2.93}$$

[3] Dirac's delta is not really a function, being infinite at one point and zero elsewhere, but we prefer to use in this context the sloppy language of physicists.

This is still a very detailed equation containing exact information of the orbits of all particles. $N_\alpha$ is composed of sums of $\delta$-functions, which makes practical calculations cumbersome. Because we are not interested in the orbits of individual particles, we can take *ensemble averages* of $N_\alpha$ and of equation (2.93). Denoting the average of $N_\alpha(\mathbf{r},\mathbf{v},t)$ by $f_\alpha(\mathbf{r},\mathbf{v},t)$ and neglecting the particle collisions, the ensemble averaging of (2.93) leads to the *Vlasov equation* for $f_\alpha$

$$\frac{\partial f_\alpha}{\partial t} + \mathbf{v} \cdot \frac{\partial f_\alpha}{\partial \mathbf{r}} + \frac{q_\alpha}{m_\alpha}(\mathbf{E} + \mathbf{v} \times \mathbf{B}) \cdot \frac{\partial f_\alpha}{\partial \mathbf{v}} = 0. \tag{2.94}$$

Another route is the *Liouville approach*. It starts from distribution functions and avoids $\delta$-functions and ensemble averaging. Consider a general distribution of $N$ particles $F(\mathbf{r}_1,...,\mathbf{r}_N;\mathbf{v}_1,...,\mathbf{v}_N;t)$, which is normalized as $\int F\, d^3r_1 \cdots d^3r_N\, d^3v_1 \cdots d^3v_N = 1$. For a plasma of $N/2$ ions and $N/2$ electrons in thermodynamic equilibrium $F = D$, where $D$ is the Gibbs distribution of the previous section.

The penalty of avoiding $\delta$-functions is to deal with a $6N$-dimensional phase space. $F$ contains information of all particles and is again much too detailed for practical use. A set of *reduced distribution functions* can be defined as follows. The *one-particle distribution function* $f_\alpha^{(1)}$ for species $\alpha$ is

$$f_\alpha^{(1)}(\mathbf{r}_1,\mathbf{v}_1,t) = \mathcal{V} \int F\, d^3r_2 \cdots d^3r_N d^3v_2 \cdots d^3v_N\,. \tag{2.95}$$

$\mathcal{V}$ is the finite spatial volume where $F$ is nonzero for all $\mathbf{r}_1,\mathbf{r}_2,...,\mathbf{r}_N$. The *two-particle distribution function* is

$$f_{\alpha\beta}^{(2)}(\mathbf{r}_1,\mathbf{r}_2,\mathbf{v}_1,\mathbf{v}_2,t) = \mathcal{V}^2 \int F\, d^3r_3 \cdots d^3r_N d^3v_3 \cdots d^3v_N \tag{2.96}$$

and so on. Statistical physics tells us that $F$ fulfills the *Liouville equation*

$$\frac{\partial F}{\partial t} + \sum_{i=1}^{N} \left( \frac{\partial F}{\partial \mathbf{r}_i} \cdot \mathbf{v}_i + \frac{\partial F}{\partial \mathbf{v}_i} \cdot \mathbf{a}_i^T \right) = 0\,, \tag{2.97}$$

where $\mathbf{a}_i^T$ is the acceleration by all interactions, including collisions.

The equation of motion for $f_\alpha^{(1)}$ is found by integrating (2.97) over all coordinates except $(\mathbf{r}_1,\mathbf{v}_1)$

$$\frac{\partial f_\alpha^{(1)}}{\partial t} + \mathbf{v}_1 \cdot \frac{\partial f_\alpha^{(1)}}{\partial \mathbf{r}_1} + \mathcal{V} \int \mathbf{a}_i^T \cdot \frac{\partial F}{\partial \mathbf{v}_1} d^3r_2 \cdots d^3r_N d^3v_2 \cdots d^3v_N = 0\,. \tag{2.98}$$

Here the total number of particles was assumed to be conserved.

If there are external forces $(\mathbf{a}_1^E)$ only, we again get the Liouville equation

$$\frac{\partial f_\alpha^{(1)}}{\partial t} + \mathbf{v}_1 \cdot \frac{\partial f_\alpha^{(1)}}{\partial \mathbf{r}_1} + \mathbf{a}_1^E \cdot \frac{\partial f_\alpha^{(1)}}{\partial \mathbf{v}_1} = 0\,. \tag{2.99}$$

Denote the interactions between particles by $\mathbf{a}_{ij}$. Now the third term of (2.98) reduces to

$$\mathbf{a}_1^E \cdot \frac{\partial f_\alpha^{(1)}}{\partial \mathbf{v}_1} + \sum_\beta \int \mathbf{a}_{1\beta} \cdot \frac{\partial}{\partial \mathbf{v}_1} f_{\alpha\beta}^{(2)}(\mathbf{r}_1, \mathbf{r}_\beta, \mathbf{v}_1, \mathbf{v}_\beta, t) d^3 r_\beta d^3 v_\beta \, .$$

Note that (2.98) is not a closed equation for $f^{(1)}$, as it depends on $f^{(2)}$. We could write a similar equation for $f^{(2)}$, which then depends on $f^{(3)}$, and so on. This is called the *BBGKY hierarchy* (after Bogoliubov, Born, Green, Kirkwood, and Yvon). In higher orders this hierarchy becomes intractable and the series must be truncated with physical arguments. We do it by approximating $f^{(2)}$.

If the interactions between particles were *strong* and of *short-range* (as in ordinary gases) we would end up with the *Boltzmann equation*

$$\frac{df_\alpha^{(1)}}{dt} \equiv \frac{\partial f_\alpha^{(1)}}{\partial t} + \mathbf{v}_1 \cdot \frac{\partial f_\alpha^{(1)}}{\partial \mathbf{r}_1} + \mathbf{a}_1^E \cdot \frac{\partial f_\alpha^{(1)}}{\partial \mathbf{v}_1} = \left( \frac{\partial f_\alpha^{(1)}}{\partial t} \right)_c \, . \tag{2.100}$$

However, in a plasma the dominating interaction is the *long-range* Coulomb force, which is, in this context, *weak*. Fortunately, in a plasma the combined effect of remote charges is, on the average, stronger than the acceleration due to the nearest neighbor. The average acceleration $\langle \mathbf{a}^{int} \rangle$ is from the viewpoint of a single particle the same as the acceleration by the external Coulomb force $\mathbf{a}^E$. Thus we can replace $\mathbf{a}_1 = \mathbf{a}_1^E + \langle \mathbf{a}^{int} \rangle$. The effect of binary collisions is

$$\left( \frac{\partial f_\alpha^{(1)}}{\partial t} \right)_c = -\sum_\beta \int \left( \mathbf{a}_{1\beta} - \langle \mathbf{a}_{1\beta}^{int} \rangle \right) \cdot \frac{\partial}{\partial \mathbf{v}_1} f_{\alpha\beta}^{(2)} d^3 r_\beta d^3 v_\beta \, . \tag{2.101}$$

Assuming that the only external force is the Lorentz force we have the Boltzmann equation for plasma

$$\frac{\partial f_\alpha^{(1)}}{\partial t} + \mathbf{v}_1 \cdot \frac{\partial f_\alpha^{(1)}}{\partial \mathbf{r}_1} + \frac{q_\alpha}{m_\alpha} \langle \mathbf{E} + \mathbf{v}_1 \times \mathbf{B} \rangle \cdot \frac{\partial f_\alpha^{(1)}}{\partial \mathbf{v}_1} = \left( \frac{\partial f_\alpha^{(1)}}{\partial t} \right)_c \, , \tag{2.102}$$

where the average fields $\langle \mathbf{E} \rangle$ and $\langle \mathbf{B} \rangle$ fulfill the average Maxwell equations

$$\nabla \cdot \langle \mathbf{E} \rangle = \frac{\rho}{\varepsilon_0} \; ; \; \nabla \times \langle \mathbf{B} \rangle = \mu_0 \mathbf{J} + \frac{1}{c^2} \frac{\partial \langle \mathbf{E} \rangle}{\partial t} \, . \tag{2.103}$$

Note that the normalization of $f_\alpha^{(1)}$ is different from the normalization of the distribution function $f_\alpha$ in the Vlasov equation (2.94). We retain the same plasma kinetic equation by substitution $f_\alpha = (N_\alpha / \mathscr{V}) f_\alpha^{(1)}$.

A thorough treatment of the collision term is a substantial task. The interested reader is encouraged to consult advanced text-books on Balescu–Lenard and Fokker–Planck equations. We will discuss some elements of the Fokker–Planck theory in Chap. 10. Note that the interparticle collisions may be of very variable nature. They may be elastic, but the kinetic energy of a colliding plasma particle may also be transferred to internal energy of neutral particles or molecular ions of the plasma. Furthermore, there are collisions leading to recombination, ionization, and charge exchange, which are important processes associated with space storms.

A simple and often sufficient first approximation for the collision term is the *relaxation time approximation*, also called the *Krook model* where the average collision frequency is approximated by a constant $\nu_c$ and

$$\left(\frac{\partial f_\alpha}{\partial t}\right)_c = -\nu_c(f - f_0) . \tag{2.104}$$

where $f_0$ is the equilibrium distribution and $|f - f_0| \ll f_0$. Note that the equilibrium here is a wider concept than a Maxwellian distribution. It is enough that $f_0$ is a stable solution of the Vlasov equation.

### 2.3.3 Macroscopic variables

The Vlasov and Boltzmann equations are equations of motion for the *single particle distribution function* $f(\mathbf{r},\mathbf{v},t)$. The function expresses the number density of particles in a volume element $dx\,dy\,dz\,dv_x\,dv_y\,dv_z$ of a six-dimensional phase space $(\mathbf{r},\mathbf{v})$ at the time $t$ (thus the SI units of $f$ are $\mathrm{m}^{-6}\,\mathrm{s}^3$). In the following we use the normalization

$$\int_{\mathscr{V}} \int_{\mathbf{V}} f(\mathbf{r},\mathbf{v},t) d^3r d^3v = N , \tag{2.105}$$

where $N$ is the number of all particles in the phase space volume considered.

The average density in volume $\mathscr{V}$ is $\langle n \rangle = N/\mathscr{V}$. However, the *particle density* is usually a function of space and time. It is defined as the zero order *velocity moment* of the distribution function

$$n(\mathbf{r},t) = \int f(\mathbf{r},\mathbf{v},t)\, d^3v . \tag{2.106}$$

We define the *macroscopic* quantities as velocity moments of the distribution function

$$\int f\, d^3v \;\; ; \;\; \int \mathbf{v} f\, d^3v \;\; ; \;\; \int \mathbf{v}\mathbf{v} f\, d^3v .$$

In a plasma different particle populations (labeled by $\alpha$) may have different distributions and thus have different velocity moments ($n_\alpha(\mathbf{r},t)$, etc.). If the particles of a species are charged with charge $q_\alpha$, the *charge density* of the species

$$\rho_\alpha = q_\alpha n_\alpha . \tag{2.107}$$

The first-order moment yields the *particle flux*

$$\Gamma_\alpha(\mathbf{r},t) = \int \mathbf{v} f_\alpha(\mathbf{r},\mathbf{v},t)\, d^3v . \tag{2.108}$$

Dividing this by particle density we get the *average velocity*

$$\mathbf{V}_\alpha(\mathbf{r},t) = \frac{\int \mathbf{v} f_\alpha(\mathbf{r},\mathbf{v},t)\, d^3v}{\int f_\alpha(\mathbf{r},\mathbf{v},t)\, d^3v}\,, \tag{2.109}$$

from which we can further determine the *current density*

$$\mathbf{J}_\alpha(\mathbf{r},t) = q_\alpha \Gamma_\alpha = q_\alpha n_\alpha \mathbf{V}_\alpha\,. \tag{2.110}$$

In the second order we find the *pressure tensor*

$$\mathcal{P}_\alpha(\mathbf{r},t) = m_\alpha \int (\mathbf{v}-\mathbf{V}_\alpha)(\mathbf{v}-\mathbf{V}_\alpha) f_\alpha(\mathbf{r},\mathbf{v},t)\, d^3v\,, \tag{2.111}$$

which in a spherically symmetric case reduces to the *scalar pressure*

$$P_\alpha = \frac{m_\alpha}{3}\int (\mathbf{v}-\mathbf{V}_\alpha)^2 f_\alpha(\mathbf{r},\mathbf{v},t)\, d^3v = n_\alpha k_B T_\alpha\,. \tag{2.112}$$

Here we introduce the concept of *temperature* $T_\alpha$. In the frame moving with the velocity $\mathbf{V}$ the temperature is given by

$$\frac{3}{2}k_B T_\alpha(\mathbf{r},t) = \frac{m_\alpha}{2}\frac{\int v^2 f_\alpha(\mathbf{r},\mathbf{v},t)\, d^3v}{\int f_\alpha(\mathbf{r},\mathbf{v},t)\, d^3v}\,, \tag{2.113}$$

which for a *Maxwellian distribution* is the temperature of classical thermodynamics. In collisionless plasmas equilibrium distributions may be far from Maxwellian. Thus temperature is a non-trivial concept in plasma physics.

**Train your brain**

Show that a spherically symmetric (in the velocity space) distribution function $f_\alpha(\mathbf{r},v,t)$ yields an isotropic pressure $P_{\alpha ij} = p_\alpha \delta_{ij}$. What kind of distribution function yields the diagonal gyrotropic form

$$P_{\alpha ij} = p_\perp \delta_{ij} + (p_\parallel - p_\perp)\delta_{3i}\delta_{3j}\ ?$$

What is the value of scalar pressure $p$ in this case? Here the "parallel" direction (e.g., the direction of background magnetic field) is assumed to be in the direction of the axis number 3.

The relation between the particle pressure and *magnetic pressure* (*magnetic energy density*) is the *plasma beta*

$$\beta = \frac{2\mu_0 \sum_\alpha n_\alpha k_B T_\alpha}{B^2}\,. \tag{2.114}$$

If $\beta > 1$, plasma governs the evolution of the magnetic field. If $\beta \ll 1$, the magnetic field determines the plasma dynamics. Values of beta are very different and highly variable in various landscapes of space storms. In the solar photosphere beta varies from 1 to 100. In the lower corona it is of the order of $10^{-4}$–$10^{-2}$ and higher up it starts rising again to be around 1 in the solar wind, but also there with large variations. In the Earth's magnetosphere the lowest beta values ($\beta \sim 10^{-6}$) are found in the auroral region magnetic field lines at altitudes of a few Earth radii. In the tail plasma sheet $\beta \sim 1$, but in the tail lobes it is some 4 orders of magnitude smaller.

The chain of moments continues to higher orders. The third order introduces the *heat flux* , i.e., temperature multiplied by velocity. It can usually be neglected in the magnetosphere but is very important at the solar end of space storms.

### 2.3.4 Derivation of macroscopic equations

Next we derive macroscopic equations by taking velocity moments of the Boltzmann equation. For the needs of many space applications we could start from the Vlasov equation, but retaining the collision term gives us a more complete macroscopic theory. When not needed, the collision effects can be dropped at the macroscopic level.

We start from the Boltzmann equation for species $\alpha$

$$\frac{\partial f_\alpha}{\partial t} + \mathbf{v} \cdot \frac{\partial f_\alpha}{\partial \mathbf{r}} + \frac{q_\alpha}{m_\alpha}(\mathbf{E} + \mathbf{v} \times \mathbf{B}) \cdot \frac{\partial f_\alpha}{\partial \mathbf{v}} = \left(\frac{\partial f_\alpha}{\partial t}\right)_c . \tag{2.115}$$

#### *Zeroth moment*

We first integrate (2.115) over the velocity space. For physical distributions $f_\alpha \to 0$, when $|v| \to \infty$, and the force term vanishes in the integration. If there are no ionizing nor recombining collisions, or charge-exchange collisions between ions and neutrals, the zero-order moment of the collision term is also zero. The integral of the first term of (2.115) yields the time derivative of density. The second term is of the first order in velocity

$$\int \mathbf{v} \cdot \frac{\partial f_\alpha}{\partial \mathbf{r}} d^3v = \nabla \cdot \int \mathbf{v} f_\alpha d^3v = \nabla \cdot (n_\alpha \mathbf{V}_\alpha) \tag{2.116}$$

and we have found the *equation of continuity*

$$\frac{\partial n_\alpha}{\partial t} + \nabla \cdot (n_\alpha \mathbf{V}_\alpha) = 0 . \tag{2.117}$$

Continuity equations for charge or mass densities are obtained by multiplying (2.117) by $q_\alpha$ or $m_\alpha$, respectively. The equation of continuity is an example of the general form of a *conservation law*

$$\frac{\partial F}{\partial t} + \nabla \cdot \mathbf{G} = 0 , \tag{2.118}$$

where $F$ is the density of a physical quantity and $\mathbf{G}$ the associated flux.

#### *First moment*

Multiply (2.115) by $m_\alpha \mathbf{v}$ and integrate over $\mathbf{v}$. This yields the *momentum transport equation,* which actually is the macroscopic *equation of motion*

$$n_\alpha m_\alpha \frac{\partial \mathbf{V}_\alpha}{\partial t} + n_\alpha m_\alpha \mathbf{V}_\alpha \cdot \nabla \mathbf{V}_\alpha - n_\alpha q_\alpha \langle \mathbf{E} + \mathbf{V}_\alpha \times \mathbf{B} \rangle + \nabla \cdot \mathscr{P}_\alpha$$
$$= m_\alpha \int \mathbf{v} \left( \frac{\partial f_\alpha}{\partial t} \right)_c d^3 v \,. \tag{2.119}$$

**Train your brain**

Make a careful derivation of Eq. (2.119). You need to apply the continuity equation.

The average electric and magnetic fields in (2.119) are determined by both internal and external sources and fulfill the average Maxwell equations

$$\nabla \cdot \langle \mathbf{E} \rangle = \sum_\alpha \frac{n_\alpha q_\alpha}{\varepsilon_0} + \rho_{ext}/\varepsilon_0 \tag{2.120}$$

$$\nabla \times \langle \mathbf{B} \rangle = \frac{1}{c^2} \frac{\partial \langle \mathbf{E} \rangle}{\partial t} + \mu_0 \sum_\alpha n_\alpha q_\alpha \mathbf{V}_\alpha + \mu_0 \mathbf{J}_{ext} \,. \tag{2.121}$$

Because collisions transport momentum between different plasma populations, the collision integral does not vanish, except for collisions between the same type of particles. The collision term is a complicated function of velocity. A useful approximation related to the Krook model (2.104) is

$$m_\alpha \int \mathbf{v} \left( \frac{\partial f_\alpha}{\partial t} \right)_c d^3 v = -\sum_\beta m_\alpha n_\alpha (\mathbf{V}_\alpha - \mathbf{V}_\beta) \langle \nu_{\alpha\beta} \rangle \,, \tag{2.122}$$

where $\langle \nu_{\alpha\beta} \rangle$ is the average collision between particles of type $\alpha$ and $\beta$.

The second-order contributions $\mathbf{V}_\alpha \cdot \nabla \mathbf{V}_\alpha$ and $\mathscr{P}_\alpha$ arise from terms containing products $\mathbf{vv}$ or $\mathbf{v} \cdot \mathbf{v}$. The divergence of $\mathscr{P}_\alpha$ contains information of inhomogeneity and viscosity of the plasma. Note that $\mathscr{P}_\alpha$ is not independent of the collisions. For example, if the collisions are frequent enough, the pressure tensor becomes diagonal, or even isotropic in which case $\nabla \cdot \mathscr{P} \to \nabla P$.

### *Second moment*

The second velocity moment yields the *energy* or *heat transport equation* (conservation law of energy). We can write the equation in the form

$$\frac{3}{2} n_\alpha k_B \left( \frac{\partial T_\alpha}{\partial t} + \mathbf{V}_\alpha \cdot \nabla T_\alpha \right) + P_\alpha \nabla \cdot \mathbf{V}_\alpha =$$
$$-\nabla \cdot \mathbf{H}_\alpha - (\mathscr{P}'_\alpha \cdot \nabla) \cdot \mathbf{V} + \frac{\partial}{\partial t} \left( \frac{n_\alpha m_\alpha V_\alpha^2}{2} \right)_c \,, \tag{2.123}$$

where the isotropic part of the pressure $P_\alpha$ is written on the LHS and the non-isotropic part $\mathscr{P}'_\alpha$ on the RHS. The relation between the scalar pressure $P_\alpha$ and temperature $T_\alpha$ is assumed to be that of an ideal gas $P_\alpha = n_\alpha k_B T_\alpha$ .

The third-order term $\mathbf{H}_\alpha$ describes the *heat flux*. An equation for it is found by taking the third moment. This contains fourth-order contributions, and so on. The chain of equations must again be truncated at some point, just as was done in the case of kinetic equations. In many practical problems this is made in the second order, either by neglecting the heat flux, or by substituting the energy equation by an equation of state. Here physical insight is essential. Krall and Trivelpiece [1973] state this: "The fluid theory, though of great practical use, relies heavily on the cunning of its user". In collisional and Maxwellian plasmas the truncation may be easy to motivate, but in collisionless space plasmas it is a more subtle issue.

### 2.3.5 Equations of magnetohydrodynamics

Now we have macroscopic equations for each plasma species. In a real plasma several species co-exist; in addition to electrons and protons, there may be a variety of heavier ions, as well as neutral particles, which may contribute to plasma dynamics through collisions, including charge-exchange processes (e.g., Sect. 14.1.4). Sometimes it is also necessary to consider different species of the same type of particles; e.g., in the same spatial volume there may be two electron populations of widely different temperatures or average velocities. Such situations often give rise to plasma instabilities to be discussed in Chap. 7.

As the first step toward a single-fluid theory it is useful to consider all electrons as one fluid and all ions as another. This is called a *two-fluid model*. The separate fluid components interact through collisions and electromagnetic interaction. In the following derivation of the single-fluid theory, it may be practical to think only two components although we have written the expressions for an arbitrary number of species.

*Magnetohydrodynamics* (MHD) is probably the most widely known plasma theory. In MHD the plasma is considered as a single fluid in the center-of-mass (CM) frame. This is a well-motivated approach in collision-dominated plasmas, where the collisions constrain the plasma particles to follow each other closely and thermalize the distribution toward a Maxwellian, which makes the interpretation of velocity moments straightforward. MHD works also remarkably well in collisionless tenuous space plasmas. However, great care should be exercised both with interpretation and approximations.

The *single-fluid* variables are defined as:

mass density

$$\rho_m(\mathbf{r},t) = \sum_\alpha n_\alpha m_\alpha \,, \tag{2.124}$$

charge density

$$\rho_q(\mathbf{r},t) = \sum_\alpha n_\alpha q_\alpha \tag{2.125}$$

$(= e(n_i - n_e)$ for singly charged ions and electrons$)$ , macroscopic velocity

$$\mathbf{V}(\mathbf{r},t) = \frac{\sum_\alpha n_\alpha m_\alpha \mathbf{V}_\alpha}{\sum_\alpha n_\alpha m_\alpha}, \tag{2.126}$$

current density

$$\mathbf{J}(\mathbf{r},t) = \sum_\alpha n_\alpha q_\alpha \mathbf{V}_\alpha, \tag{2.127}$$

and pressure tensor in the CM frame

$$\mathscr{P}_\alpha^{CM}(\mathbf{r},t) = m_\alpha \int (\mathbf{v}-\mathbf{V})(\mathbf{v}-\mathbf{V}) f_\alpha \, d^3v, \tag{2.128}$$

from which we get the total pressure

$$\mathscr{P}(\mathbf{r},t) = \sum_\alpha \mathscr{P}_\alpha^{CM}(\mathbf{r},t). \tag{2.129}$$

Summing the individual continuity and momentum transport equations over particle species yields the continuity equations

$$\frac{\partial \rho_m}{\partial t} + \nabla \cdot (\rho_m \mathbf{V}) = 0 \tag{2.130}$$

$$\frac{\partial \rho_q}{\partial t} + \nabla \cdot \mathbf{J} = 0 \tag{2.131}$$

and the momentum transport equation

$$\rho_m \left( \frac{\partial \mathbf{V}}{\partial t} + \mathbf{V} \cdot \nabla \mathbf{V} \right) = \rho_q \mathbf{E} + \mathbf{J} \times \mathbf{B} - \nabla \cdot \mathscr{P}. \tag{2.132}$$

The momentum equation corresponds to the Navier–Stokes equation of hydrodynamics (6.2) where the viscosity terms are written explicitly (here they are hidden in $\nabla \cdot \mathscr{P}$). At macroscopic level the deviations from charge neutrality are small and $\rho_q \mathbf{E}$ is usually negligible. The magnetic part of the Lorentz force $\mathbf{J} \times \mathbf{B}$ (sometimes called Ampère's force) is, however, essential in the theory of magnetic fluids.

Ohm's law in fluid description is a more complicated issue. In the particle picture the plasma current is the sum of all charged particle motions. In a single-fluid theory the current transport equation is derived by multiplying the momentum transport equations of each particle population by $q_\alpha / m_\alpha$ and summing over all populations. In the two-fluid case $(e,i)$ we get

$$\begin{aligned} \frac{\partial \mathbf{J}}{\partial t} + \nabla \cdot (\mathbf{VJ} + \mathbf{JV} - \mathbf{VV}\rho_q) &= \sum_\alpha \frac{n_\alpha q_\alpha^2}{m_\alpha} \mathbf{E} \\ + \left( \frac{e^2}{m_e} + \frac{e^2}{m_i} \right) \frac{\rho_m \mathbf{V} \times \mathbf{B}}{m_e + m_i} &- \left( \frac{e m_i}{m_e} - \frac{e m_e}{m_i} \right) \frac{\mathbf{J} \times \mathbf{B}}{m_e + m_i} \\ - \frac{e}{m_e} \nabla \cdot \left( \mathscr{P}_i^{CM} \frac{m_e}{m_i} - \mathscr{P}_e^{CM} \right) &+ \sum_\alpha \int q_\alpha \mathbf{v} \left( \frac{\partial f_\alpha}{\partial t} \right)_c d^3v, \end{aligned} \tag{2.133}$$

where the products **VJ**, etc., are cartesian tensors (dyads) with elements $V_iJ_k$, and the divergence of a dyad is a vector, e.g., with components $\sum_i \partial_i V_i J_k$. This equation expresses the relationship between the electric current and the electric field. Thus it can be called *generalized Ohm's law*.

The first step to simplify (2.133) is to approximate the collision integral introducing a constant collision frequency $\nu$

$$\sum_\alpha \int q_\alpha \mathbf{v} \left( \frac{\partial f_\alpha}{\partial t} \right)_c d^3v = -\nu \mathbf{J} . \tag{2.134}$$

Defining the *conductivity* by $\sigma = ne^2/\nu m_e$ and neglecting all derivatives and the magnetic field in (2.133) we get the familiar form of Ohm's law $\mathbf{J} = \sigma \mathbf{E}$.

Not all terms in the generalized Ohm's law are equally important. There are some that clearly are smaller than the others (e.g. $\propto m_e/m_i$). Furthermore, the derivatives of the second-order terms **VJ**, **JV** and **VV** can usually be neglected. At this level we have the generalized Ohm's law in the form that contains the most important terms for space plasmas:

$$\mathbf{E} + \mathbf{V} \times \mathbf{B} = \frac{\mathbf{J}}{\sigma} + \frac{1}{ne} \mathbf{J} \times \mathbf{B} - \frac{1}{ne} \nabla \cdot \mathcal{P}_e + \frac{m_e}{ne^2} \frac{\partial \mathbf{J}}{\partial t} . \tag{2.135}$$

Assume further so *slow temporal changes and large spatial gradient scales* that $|\mathbf{J} \times \mathbf{B}|$, $|\partial \mathbf{J}/\partial t|$, and $|\nabla \cdot \mathcal{P}|$ are all smaller than $|\mathbf{V} \times \mathbf{B}|$. This leaves us with the standard form of Ohm's law in MHD

$$\mathbf{J} = \sigma(\mathbf{E} + \mathbf{V} \times \mathbf{B}) , \tag{2.136}$$

which already familiar from elementary electrodynamics in cases when moving frames are taken into account. Here the moving frame is attached to the fluid flow with the velocity **V**. If the conductivity is very large, we find Ohm's law of the *ideal MHD*

$$\mathbf{E} + \mathbf{V} \times \mathbf{B} = 0 . \tag{2.137}$$

The road from the Liouville or Klimontovich equations to this simple equation is long and there are several potholes on the road. For example, while the ideal MHD is a reasonable starting point, it is not at all clear that the next term to take into account should be $\mathbf{J}/\sigma$. In many space applications the *Hall term* $\mathbf{J} \times \mathbf{B}/ne$ and the pressure term $\nabla \cdot \mathcal{P}/ne$ are more important.

There are effects that originate at the microscopic level, which are not due to actual interparticle collisions, but which may lead to "effective" resistivity or viscosity at the macroscopic level. Various wave–particle interactions and microscopic instabilities tend to inhibit the current flow. Often the macroscopic effect of these processes looks analogous to finite $\nu$ and is called *anomalous resistivity*.[4]

Another issue is that plasma does not need to exhibit a local Ohm's law at all. In tenuous space plasmas it may happen that there are not enough current carriers to satisfy $\nabla \cdot \mathbf{J} = 0$ without extra acceleration of the charges. An example is the magnetic field-aligned po-

[4] This is one more example of unfortunate terminology. There is nothing anomalous in the physics behind the non-collisional resistivity.

tential drop above the discrete auroras. The coupling between the ionosphere and magnetosphere requires more upward field-aligned current to be drawn through this region than there are electrons readily available from the magnetosphere. The *global plasma system* reacts to this by setting up an upward-directed electric field to accelerate electrons to so high velocities that the current continuity is maintained. This results in a *global current–voltage relationship*, which Knight [1973] derived into the form

$$J_{\|} = -en\sqrt{\frac{k_B T_e}{2\pi m_e}\frac{B_I}{B_E}}\left[1-\left(1-\frac{B_E}{B_I}\right)\exp\left(-\frac{e\triangle\varphi}{k_B T_e(B_I/B_E-1)}\right)\right] . \tag{2.138}$$

Here $B_I$ is the magnetic field in the ionosphere, $B_E$ in the equatorial plane in the magnetosphere and $\triangle\varphi$ the potential difference between them. At the limit $e\triangle\varphi/k_B T \ll (B_I/B_E-1)$ this reduces to

$$J_{\|} = K\left(\triangle\varphi + \frac{k_B T_e}{e}\right) , \tag{2.139}$$

which is often approximated as the direct linear relationship between the current and voltage of the form

$$J_{\|} = K\triangle\varphi . \tag{2.140}$$

This last form is known as the *Knight relation*. The coefficient $K$ is a function of plasma parameters and thus not a universal constant.

**Feed your brain**

The current–voltage relationship is actually not quite as simple as given above. Read carefully the paper by Janhunen and Olsson [1998] and fill in the gaps in their derivations.

The next equation in the velocity moment chain is the *energy transport equation*. After some tedious but straightforward calculation the energy equation can be written in the conservation form

$$\frac{\partial}{\partial t}\left[\rho_m\left(\frac{V^2}{2}+w\right)+\frac{B^2}{2\mu_0}\right] = -\nabla\cdot\mathbf{H} . \tag{2.141}$$

Here $w$ is the *enthalpy* that is related to the the internal free energy (per unit mass) of the plasma $u$ by $w = u + P/\rho_m$. The RHS is the divergence of the heat flux vector $\mathbf{H}$, which is a third-order moment. After some reasonable approximations it can be written as

$$\mathbf{H} = \left(\frac{V^2}{2}+u+\frac{P+B^2/\mu_0}{\rho_m}\right)\rho_m\mathbf{V} - \frac{\mathbf{B}}{\mu_0}\left(\mathbf{V}+\frac{\mathbf{J}}{ne}\right)\cdot\mathbf{B}$$
$$-\frac{\mathbf{J}\times\mathbf{B}}{\sigma\mu_0} + \frac{\mathbf{J}B^2}{\mu_0 ne} + \frac{m_e\mathbf{B}}{\mu_0 ne^2}\times\frac{\partial\mathbf{J}}{\partial t} . \tag{2.142}$$

When integrated over a finite volume $\mathcal{V}$ the LHS of (2.141) describes the temporal change of the energy of the MHD plasma in that volume and the RHS the the energy flux through the boundary $\partial\mathcal{V}$ and energy losses due to resistivity. Thus we have found the MHD equivalent of Poynting's theorem of elementary electrodynamics.

Because the energy equation depends on third-order terms, we do not get a closed set of MHD equations without some further approximations. Often the chain is cut by *selecting* an equation of state. After this the energy equation can be written in a simpler form. Another frequently adopted approach is to assume an isotropic pressure. We can start from the ideal gas law $P = nk_BT$ and use some of the following equations of state depending on what kind of processes we are considering:

- *adiabatic process*

$$T = T_0\left(\frac{n}{n_0}\right)^{\gamma-1} \;;\; P = P_0\left(\frac{n}{n_0}\right)^{\gamma}, \tag{2.143}$$

  where the *polytropic index* $\gamma = c_p/c_v$ is $5/3$ in a three-dimensional plasma and $c_p$ and $c_v$ are the specific heat constants for constant pressure and constant volume, respectively.

- *isothermal process*
  the above with $\gamma = 1 \Rightarrow P = nk_BT_0$

- *isobaric process*
  the above with $\gamma = 0$, i.e., constant pressure

- *isometric process*
  the above with $\gamma = \infty$, i.e., $P \approx 0$, e.g. the case of $\beta \ll 1$ .

Using the equation of state we can write the equations of MHD in the form

$$\frac{\partial\rho_m}{\partial t} + \nabla\cdot(\rho_m\mathbf{V}) = 0 \tag{2.144}$$

$$\rho_m\left(\frac{\partial}{\partial t} + \mathbf{V}\cdot\nabla\right)\mathbf{V} + \nabla P - \mathbf{J}\times\mathbf{B} = 0 \tag{2.145}$$

$$\mathbf{E} + \mathbf{V}\times\mathbf{B} = \mathbf{J}/\sigma \tag{2.146}$$

$$P = P_0\left(\frac{n}{n_0}\right)^{\gamma} \tag{2.147}$$

$$\frac{\partial\mathbf{B}}{\partial t} = -\nabla\times\mathbf{E} \tag{2.148}$$

$$\nabla\times\mathbf{B} = \mu_0\mathbf{J}\,. \tag{2.149}$$

### 2.3.6 Double adiabatic theory

Due to the presence of the magnetic field the particle distributions in space plasmas are not always isotropic and the pressure tensor does not even need to be diagonal. To fully appreciate the anisotropic effects we need to refer to some concepts to be investigated in Chap. 3, but their macroscopic consequences are useful to introduce here for completeness of the present discussion.

Consider the ideal MHD equations

$$\frac{\partial \rho_m}{\partial t} + \nabla \cdot (\rho_m \mathbf{V}) = 0 \tag{2.150}$$

$$\rho_m \left( \frac{\partial}{\partial t} + \mathbf{V} \cdot \nabla \right) \mathbf{V} + \nabla \cdot \mathscr{P} - \mathbf{J} \times \mathbf{B} = 0 \tag{2.151}$$

$$\mathbf{E} + \mathbf{V} \times \mathbf{B} = 0 \tag{2.152}$$

and assume that the pressure tensor is diagonal and *gyrotropic*

$$\mathscr{P} = \begin{pmatrix} P_\perp & 0 & \\ 0 & P_\perp & 0 \\ 0 & 0 & P_\parallel \end{pmatrix} . \tag{2.153}$$

Assume further that both the parallel and perpendicular pressures behave adiabatically and fulfill the ideal gas equation of state

$$P_\parallel = n k_B T_\parallel \tag{2.154}$$

$$P_\perp = n k_B T_\perp \, . \tag{2.155}$$

There are one parallel and two perpendicular dimensions. From thermodynamics we know that the polytropic index depends on the number of dimensions $d$ as $\gamma = (d+2)/d$. Setting $\gamma_\perp = 2$ and $\gamma_\parallel = 3$ is, however, *wrong* because the magnetic field not only breaks the symmetry of the pressure tensor but also couples the perpendicular motion to the parallel motion in inhomogeneous plasma (e.g, the mirror force, see Chap. 3).

Assume that the motion of the individual particles is adiabatic, which means that the magnetic moment $\mu = W_\perp / B$ is constant. Then the average magnetic moment $\langle \mu \rangle = k_B T_\perp / B = P_\perp / nB$ is also constant. This yields the perpendicular equation of state

$$\frac{d}{dt} \left( \frac{P_\perp}{\rho_m B} \right) = 0 \, . \tag{2.156}$$

The parallel direction is more difficult. Chew, Goldberger, and Low developed a theory [Chew et al, 1956] *assuming* that the heat flux parallel to the magnetic field is negligible. This leads to the equation of state

$$\frac{d}{dt} \left( \frac{P_\perp^2 P_\parallel}{\rho_m^5} \right) = \frac{d}{dt} \left( \frac{P_\parallel B^2}{\rho_m^3} \right) = 0 \, . \tag{2.157}$$

This anisotropic version of MHD is called *double adiabatic theory* or *CGL theory*. Now the pressure tensor is of the form $\mathscr{P} = P_\perp \mathscr{I} + (P_\parallel - P_\perp)\mathbf{bb}$, where $\mathbf{b} = \mathbf{B}/B$ and $\mathscr{I}$ is the unit tensor. The momentum equation separates into two equations

$$\rho_m \left( \frac{d\mathbf{V}}{dt} \right)_\perp + \nabla_\perp \left( P_\perp + \frac{B^2}{2\mu_0} \right) - \frac{(\mathbf{B} \cdot \nabla)\mathbf{B}}{\mu_0} \left( \frac{P_\perp - P_\parallel}{B^2/\mu_0} + 1 \right) = 0 \tag{2.158}$$

$$\rho_m \left(\frac{d\mathbf{V}}{dt}\right)_{\parallel} + \nabla_{\parallel} P_{\parallel} + (P_{\perp} - P_{\parallel}) \left(\frac{\nabla B}{B}\right)_{\parallel} = 0\,. \qquad (2.159)$$

In the CGL theory the parallel and perpendicular polytropic indices are not constant numbers. Assuming that $p_{\parallel} \propto n^{\gamma_{\parallel}}$ and $p_{\perp} \propto n^{\gamma_{\perp}}$ the following relations are found

$$\gamma_{\perp} = 1 + \frac{\ln(B/B_0)}{\ln(n/n_0)} \qquad (2.160)$$

$$\gamma_{\parallel} = 3 - 2\frac{\ln(B/B_0)}{\ln(n/n_0)}\,, \qquad (2.161)$$

from which

$$\gamma_{\parallel} + 2\gamma_{\perp} = 5\,. \qquad (2.162)$$

While being related to each other, $\gamma_{\perp}$ and $\gamma_{\parallel}$ are spatially varying functions in an inhomogeneous plasma.

In space physics the CGL equations (2.158, 2.159) are sometimes useful, e.g., in the studies of firehose and mirror instabilities (Chap. 7) related to shock waves. However, one has to be careful with the validity of the approach. For example, the CGL theory predicts that the temperature depends on the magnetic field as

$$T_{\perp} \propto B \;\; ; \;\; T_{\parallel} \propto (n/B)^2\,. \qquad (2.163)$$

For example, direct observations in the magnetic dipole field geometry above the auroral ionosphere show that the perpendicular temperature does not scale as $T_{\perp} \propto B$. Here, and in many other practical examples, the CGL heat flux argument is not valid. In the auroral case the particles precipitate to the upper atmosphere carrying energy (heat) with them. This is actually one of the major sinks of energy associated with space storms in the magnetosphere, as will be discussed in Chap. 13.

# 3. Single Particle Motion

In Chap. 2 we discussed the idea of the guiding center (GC) approximation and the solutions of

$$\frac{d\mathbf{p}}{dt} = q(\mathbf{E} + \mathbf{v} \times \mathbf{B}) + \mathbf{F}_{non-EM} \tag{3.1}$$

for homogeneous fields. Here we consider the motion in inhomogeneous fields, starting for simplicity, at the non-relativistic limit ($\gamma = 1,\ \mathbf{p} = m\mathbf{v}$).

## 3.1 Magnetic Drifts

If the inhomogeneities of the magnetic field ($\partial_t, \nabla$) are small as compared to the Larmor motion

$$|\partial\mathbf{B}/\partial t| \ll \omega_c B \ ; \ |\nabla B|_\perp \ll B/r_L \ ; \ |\nabla B|_\parallel \ll (\omega_c/v_\parallel)B \,,$$

we can use perturbation theory to solve the equation of motion [Northrop, 1963]. Note that, in addition to field geometry, the validity of these conditions depends on the energy and mass of the particles.

For weak inhomogeneities we can make a Taylor expansion around the GC. Let $\mathbf{B}_0$ be the field at the GC and $\mathbf{r}$ the particle's distance from it. Then

$$\mathbf{B}(\mathbf{r}) = \mathbf{B}_0 + \mathbf{r} \cdot (\nabla\mathbf{B})_0 + \ldots \,. \tag{3.2}$$

In general $\nabla\mathbf{B}$ is a tensor whose components form the matrix $(\partial_i B_j)$. The tensor describes two effects: the gradient of the field strength and the curvature of the field lines. These are tied to each other because a gradient of the field implies curvature of the field lines somewhere in the global magnetic field configuration. Here we follow the standard textbook approach and treat the gradients and the curvature separately.

To study the gradient effects we move to the frame of reference where $v_\parallel = 0$, which often is *not* an inertial frame. The equation of motion is

$$\frac{d\mathbf{v}}{dt} = \frac{q}{m}(\mathbf{v} \times \mathbf{B}_0) + \frac{q}{m}(\mathbf{v} \times [\mathbf{r} \cdot (\nabla\mathbf{B})_0]) + \ldots \,. \tag{3.3}$$

H.E.J. Koskinen, *Physics of Space Storms: From the Solar Surface to the Earth*, Springer Praxis Books, DOI 10.1007/978-3-642-00319-6_3,

Let $\mathbf{v}_0$ be the solution of the "unperturbed" equation and write $\mathbf{v} = \mathbf{v}_0 + \mathbf{u}$, where $\mathbf{u}$ is a small correction. Now (3.3) contains the second order term $\mathbf{u} \times [\mathbf{r} \cdot (\nabla \mathbf{B})_0]$. Because $\mathbf{r} \approx \mathbf{r}_L$, the first-order equation is

$$\frac{d\mathbf{v}}{dt} = \frac{q}{m}(\mathbf{v} \times \mathbf{B}_0) + \frac{q}{m}(\mathbf{v}_0 \times [\mathbf{r}_L \cdot (\nabla \mathbf{B})_0]) \,. \tag{3.4}$$

This looks formally similar to the zero-order equation (2.79) with the external force $\mathbf{F} = q(\mathbf{v}_0 \times [\mathbf{r}_L \cdot (\nabla \mathbf{B})_0])$, but now $\mathbf{F}$ is a function of $\mathbf{B}$ through $\mathbf{r}_L$ and $\nabla \mathbf{B}$.

We are looking for the drift of the GC and thus we have to find the average effect over one Larmor rotation, denoting the average by $\langle \ \rangle$. We use cylindrical coordinates, where $\mathbf{e}_z \parallel \mathbf{B}_0$ , $\mathbf{e}_\phi \parallel \mathbf{v}_0$, $\mathbf{B} = B_r \mathbf{e}_r + B_\phi \mathbf{e}_\phi + B_z \mathbf{e}_z$. The unperturbed Larmor radius vector is given by

$$\mathbf{r}_L = -\frac{m}{qB_0^2}(\mathbf{v}_0 \times \mathbf{B}_0) \,. \tag{3.5}$$

A brief exercise yields

$$\mathbf{F} = \left\langle -q\mathbf{v}_0 \times r_L \left( \frac{\partial \mathbf{B}}{\partial r} \right)_0 \right\rangle \,. \tag{3.6}$$

Hereafter we leave out the subscript 0. $\mathbf{F}$ has both perpendicular and parallel components

$$\mathbf{F}_\parallel = \left\langle q\mathbf{v} \times \mathbf{r}_L \left( \frac{\partial B_r}{\partial r} \right) \right\rangle \tag{3.7}$$

$$\mathbf{F}_\perp = \left\langle -q\mathbf{v} \times r_L \left( \frac{\partial B_z}{\partial r} \right) \mathbf{e}_z \right\rangle \,. \tag{3.8}$$

Calculate first $\mathbf{F}_\parallel$. By definition $\mathbf{v} \times \mathbf{r}_L = (2\mu/q)\,\mathbf{e}_z$. Thus

$$\mathbf{F}_\parallel = 2\mu \left\langle \frac{\partial B_r}{\partial r} \right\rangle \mathbf{e}_z = -\mu \left( \frac{\partial B_z}{\partial z} \right) \mathbf{e}_z = -\mu \nabla_\parallel B \,. \tag{3.9}$$

To calculate $\mathbf{F}_\perp$ we select the *xy*-plane as the plane of the gyro motion, when

$$\mathbf{v} \times \mathbf{e}_z = -\frac{\mathbf{r}_L}{r_L} v$$

$$\frac{\partial}{\partial r} = \cos\phi \frac{\partial}{\partial x} + \sin\phi \frac{\partial}{\partial y}$$

$$\mathbf{r}_L = -r_L(\cos\phi\, \mathbf{e}_x + \sin\phi\, \mathbf{e}_y) \,.$$

Noting that $\langle \cos^2\phi \rangle = \langle \sin^2\phi \rangle = 1/2$ and $\langle \sin\phi \cos\phi \rangle = 0$ we get

$$\mathbf{F}_\perp = -\frac{qvr_L}{2} \left\langle \frac{\partial B_z}{\partial x} \mathbf{e}_x + \frac{\partial B_z}{\partial y} \mathbf{e}_y \right\rangle \,. \tag{3.10}$$

As $\nabla_\perp = \mathbf{e}_x \dfrac{\partial}{\partial x} + \mathbf{e}_y \dfrac{\partial}{\partial y}$,

$$\mathbf{F}_{\perp} = -\mu \nabla_{\perp} B \,. \tag{3.11}$$

Thus the total force is

$$\mathbf{F} = -\mu \nabla B \,. \tag{3.12}$$

**Train your brain**

Write down the intermediate steps in the derivation of (3.12)

The force causes acceleration along the magnetic field

$$\frac{d\mathbf{v}_{\parallel}}{dt} = -\frac{\mu}{m} \nabla_{\parallel} B \,. \tag{3.13}$$

In the perpendicular direction we find a drift across the magnetic field using the same reasoning as in the zero-order case, i.e., the drift velocity $\mathbf{v}_G$ must balance the force term

$$\mathbf{v}_G = \frac{\mathbf{F}_{\perp} \times \mathbf{B}}{qB^2} \tag{3.14}$$

$\Rightarrow$

$$\mathbf{v}_G = \frac{\mu}{qB^2} \mathbf{B} \times (\nabla B) = \frac{W_{\perp}}{qB^3} \mathbf{B} \times (\nabla B) \,. \tag{3.15}$$

This is called the *gradient drift*. It depends both on the perpendicular energy and on the charge of the particle. Thus the drift contributes to the net plasma current.

We assumed that $\mathbf{v}_{\parallel} = 0$ but found $d\mathbf{v}_{\parallel}/dt \neq 0$. Thus, depending on the force, the reference frame may be non-inertial. In a curved magnetic field also the GC motion is curved. Denote the GC velocity by $\mathbf{w}$ (note that generally $\mathbf{w}_{\parallel} \neq \mathbf{v}_{\parallel}$). We let $\mathbf{v}_{\parallel} \neq 0$ and transform to a frame co-moving with the GC. Let the orthogonal basis $\{\mathbf{e}_i\}$ define the coordinate axes and choose $\mathbf{e}_3 \parallel \mathbf{v}_{\parallel} \parallel \mathbf{B}$. Now $\mathbf{v} = \sum v_i \mathbf{e}_i$, and $\{\mathbf{e}_i\}$ rotates when its origin moves with the GC. The acceleration is

$$\frac{d\mathbf{v}}{dt} = \sum_i \left( \frac{dv_i}{dt} \mathbf{e}_i + v_i \frac{d\mathbf{e}_i}{dt} \right) = \sum_i \left( \frac{dv_i}{dt} \mathbf{e}_i + v_i \left( \mathbf{w}_{\parallel} \cdot \nabla \right) \mathbf{e}_i \right) . \tag{3.16}$$

The term $\sum v_i (\mathbf{w}_{\parallel} \cdot \nabla) \mathbf{e}_i$ is due to the curvature and causes a centrifugal effect. Consider again the averages over one Larmor rotation

$$\mathbf{F}_C = -\left\langle m \sum_i v_i (\mathbf{w}_{\parallel} \cdot \nabla) \mathbf{e}_i \right\rangle . \tag{3.17}$$

Due to the assumption of weak curvature $(\mathbf{w}_{\parallel} \cdot \nabla)\, \mathbf{e}_i$ can be approximated to be constant in every point during one rotation. Because $v_1$ and $v_2$ oscillate, $\langle v_1 \mathbf{e}_1 \rangle = \langle v_2 \mathbf{e}_2 \rangle = 0$. Furthermore, during one rotation $v_{\parallel} \approx w_{\parallel}$ and thus

$$\mathbf{F}_C = -m w_{\parallel}^2 (\mathbf{e}_3 \cdot \nabla)\, \mathbf{e}_3 \,. \tag{3.18}$$

A little exercise in differential geometry yields

$$(\mathbf{e}_3 \cdot \nabla)\,\mathbf{e}_3 = \mathbf{R}_C / R_C^2\,, \tag{3.19}$$

where $\mathbf{R}_C$ is the *radius of curvature* vector, pointing inward. Now

$$\mathbf{F}_C = -mw_\parallel^2 \frac{\mathbf{R}_C}{R_C^2}\,. \tag{3.20}$$

Because $\mathbf{B} = B\,\mathbf{e}_3$,

$$(\mathbf{e}_3 \cdot \nabla)\mathbf{e}_3 = (\mathbf{B} \cdot \nabla \mathbf{B})/B^2 \tag{3.21}$$

and we can write the *curvature drift* velocity as

$$\mathbf{v}_C = \frac{-mw_\parallel^2}{qB^2} \frac{\mathbf{R}_C \times \mathbf{B}}{R_C^2} = \frac{mw_\parallel^2}{qB^4} \mathbf{B} \times (\mathbf{B} \cdot \nabla)\mathbf{B}\,. \tag{3.22}$$

Now we can again approximate $v_\parallel \approx w_\parallel$ and express the curvature drift in terms of the parallel energy $W_\parallel \approx (1/2)mw_\parallel^2$.

| **Train your brain** |
| --- |
| Fill in all steps leading to the curvature drift velocity (3.22) |

If there are no local currents ($\nabla \times \mathbf{B} = 0$), the expression for the curvature drift velocity simplifies to

$$\mathbf{v}_C = \frac{2W_\parallel}{qB^3} \mathbf{B} \times \nabla B \tag{3.23}$$

and $\mathbf{v}_G$ and $\mathbf{v}_C$ can be combined to

$$\mathbf{v}_{GC} = \frac{W_\perp + 2W_\parallel}{qB^3} \mathbf{B} \times \nabla B = \frac{W}{qBR_C} (1 + \cos^2 \alpha)\, \mathbf{n} \times \mathbf{t}\,, \tag{3.24}$$

where $\mathbf{t} \parallel \mathbf{B}$ and $\mathbf{n} \parallel \mathbf{R}_C$ are unit vectors.

Drifting particles are often relativistic. The above drift velocities are easy to cast into the relativistic form substituting $m$ by $\gamma m$.

The perturbation theory can be continued to higher orders. The recipe is the same as above: First determine the force due to the higher-order perturbation and then calculate the drift velocity to balance this effect.

## 3.2 Adiabatic Invariants

Adiabatic invariants are quantities whose invariance depends on slow temporal or spatial change of the parameters describing the motion. They have a close relationship with general symmetry principles of physics:

$$\begin{array}{lcl} \textit{complete periodicity} & \leftrightarrow & \textit{conserved quantity} \\ \textit{symmetry} & \leftrightarrow & \textit{conservation law} \end{array}$$

If the motion is nearly-periodic, such as the Larmor rotation in the GC approximation, the associated invariant may not be the same as in the strictly periodic case and its conservation critically depends on the "slowness" of the variation.

In *Hamiltonian mechanics* it is shown that if $q$ and $p$ are the canonical coordinate and momentum of the system and the motion is nearly periodic, then

$$I = \oint p\,dq \tag{3.25}$$

is an adiabatic invariant. This statement requires a proof that we will not discuss here (see, e.g., classical mechanics textbooks by Goldstein or Landau and Lifshitz, or Bellan [2006]).

The momentum of a particle in an electromagnetic field is $\mathbf{p} = m\mathbf{v} + q\mathbf{A}$ and the canonical momentum and the coordinate perpendicular to the magnetic field are $\mathbf{p}_\perp$ and $\mathbf{r}_L$. Thus

$$I = \oint \mathbf{p}_\perp \cdot d\mathbf{r}_L = \frac{2\pi m}{|q|}\,\mu\,, \tag{3.26}$$

which shows that the magnetic moment is an adiabatic invariant.

A classic example of an adiabatic invariant is the *Lorentz–Einstein pendulum* whose length ($l$) changes slowly. This causes a slow change of the frequency $\omega = \sqrt{g/l}$. Changing the length means that work is done on the pendulum and thus the energy of the pendulum per unit mass

$$W = \frac{1}{2} l^2 \dot{\theta}^2 + \frac{1}{2} g l \theta^2 \tag{3.27}$$

is not conserved. A legend tells that Lorentz asked Einstein in 1911, what is the conserved quantity instead. Einstein's reply was: $W/\omega$. This example is closely analogous to the magnetic moment

$$\mu = \frac{W_\perp}{B} = \frac{q}{m}\frac{W_\perp}{\omega_c}\,. \tag{3.28}$$

**Train your brain** by proving that the slowness of the variation is essential in the Lorentz–Einstein pendulum.

### 3.2.1 The first adiabatic invariant

To directly prove that the magnetic moment is an adiabatic invariant is not trivial. Textbooks usually treat some special cases; for a general treatment, see, Goldston and Rutherford [1995]. For our purposes it is instructive to see how the invariance follows from the conservation of the total energy in a static magnetic field in the absence of electric fields:

$$W = W_\parallel + W_\perp = \text{constant} \tag{3.29}$$

$\Rightarrow$

$$\frac{dW_\parallel}{dt} + \frac{dW_\perp}{dt} = 0 \tag{3.30}$$

$W_\perp = \mu B \quad \Rightarrow$

$$\frac{dW_\perp}{dt} = \mu \frac{dB}{dt} + \frac{d\mu}{dt} B\,. \tag{3.31}$$

Now $dB/dt = v_\parallel dB/ds$ is the change of the magnetic field along the GC orbit. The parallel energy is

$$m \frac{dv_\parallel}{dt} = -\mu \nabla_\parallel B = -\mu \frac{dB}{ds}\,. \tag{3.32}$$

Multiplying this by $v_\parallel = ds/dt$ we get

$$\frac{dW_\parallel}{dt} = -\mu \frac{dB}{dt}\,. \tag{3.33}$$

Thus

$$\frac{dW_\parallel}{dt} + \frac{dW_\perp}{dt} = B \frac{d\mu}{dt} = 0\,, \tag{3.34}$$

i.e., $\mu$ is constant if GC approximation is valid and the field is static.

Another case with general interest is when the particle is accelerated by a slow temporal variation of the magnetic field ($\partial/\partial t \ll \omega_c$). Faraday's law implies a presence of an electric field that leads to increase in perpendicular energy

$$\frac{dW_\perp}{dt} = q(\mathbf{E} \cdot \mathbf{v}_\perp)\,. \tag{3.35}$$

During one rotation the particle gains energy

$$\triangle W_\perp = q \int_0^{2\pi/\omega_c} \mathbf{E} \cdot \mathbf{v}_\perp \, dt\,. \tag{3.36}$$

Assuming the slow temporal change we can replace the time integral by a line integral over a closed loop and use Stokes' law

$$\triangle W_\perp = q \oint_C \mathbf{E} \cdot d\mathbf{l} = q \int_S (\nabla \times \mathbf{E}) \cdot d\mathbf{S} = -q \int_S \frac{\partial \mathbf{B}}{\partial t} \cdot d\mathbf{S}\,, \tag{3.37}$$

where $d\mathbf{S} = \mathbf{n}\, dS$, $\mathbf{n}$ is the normal vector of the surface $S$ with the direction defined by the positive circulation of the loop $C$. For small variations of the field $\partial \mathbf{B}/\partial t \to \omega_c \triangle B/2\pi \;\Rightarrow$

$$\triangle W_\perp = \frac{1}{2} |q| \omega_c r_L^2 \triangle B = \mu \triangle B\,. \tag{3.38}$$

On the other hand

$$\Delta W_{\perp} = \mu \Delta B + B \Delta \mu \tag{3.39}$$

and thus $\Delta\mu = 0$. For slow changes $\mu$ is conserved although the inductive electric field accelerates the particle, analogously to the work done on the Lorentz–Einstein pendulum.

### 3.2.2 Magnetic mirror and magnetic bottle

Assume that the total energy $W$ and $\mu = W_{\perp}/B$ are conserved. Let the particle move toward a weak positive gradient of $B$. Now $W_{\perp}$ can increase until $W_{\parallel} \to 0$. The perpendicular velocity is $v_{\perp} = v \sin\alpha$ and

$$\mu = \frac{mv^2 \sin^2\alpha}{2B} \, . \tag{3.40}$$

On the other hand $v^2 \propto W$ is also constant. Thus

$$\frac{\sin^2\alpha_1}{\sin^2\alpha_2} = \frac{B_1}{B_2} \, . \tag{3.41}$$

When $W_{\parallel} \to 0$, $\alpha \to 90^\circ$. The slowing down of the GC motion is due to the *mirror force* $\mathbf{F} = -\mu\nabla_{\parallel}B$. The strength of the *mirror field* $B_m$ depends on the particle's pitch angle at the reference point $B_0$. For the mirror field ($\alpha_m = 90^\circ$) we get

$$\sin^2\alpha_0 = B_0/B_m \, . \tag{3.42}$$

Because $B_m$ is finite, every mirror field is leaky. Particles having a smaller pitch angle than $\alpha_0$ in the field $B_0$ get through the mirror. These particles are said to be in the *loss cone*. Using two opposite mirrors we can build a *magnetic bottle* that confines particles outside the loss cone(s).

The mirror force does not need to be the only force affecting the parallel motion of the GC. The electric field may have a parallel component $\mathbf{E}_{\parallel}$ and the particle may be in a gravitational field. The parallel equation of motion then reads

$$m\frac{d\mathbf{v}_{\parallel}}{dt} = q\mathbf{E}_{\parallel} + m\mathbf{g}_{\parallel} - \mu\nabla_{\parallel}B \, . \tag{3.43}$$

Assuming that the non-magnetic forces can be derived from the potential $U(s)$, we get

$$m\frac{dv_{\parallel}}{dt} = -\frac{\partial}{\partial s}[U(s) + \mu B(s)] \, . \tag{3.44}$$

Thus the GC moves in the effective potential $U(s) + \mu B(s)$. Examples of potentials combined with a mirror force are the gravitational field in the solar atmosphere and parallel electric fields above discrete auroral arcs.

### 3.2.3 The second adiabatic invariant

The *bounce motion* in a magnetic bottle is nearly periodic if the field does not change much during one *bounce period* $\tau_b$

$$\tau_b = 2\int_{s_m}^{s'_m} \frac{ds}{v_\parallel(s)} = \frac{2}{v}\int_{s_m}^{s'_m} \frac{ds}{(1-B(s)/B_m)^{1/2}}\,, \tag{3.45}$$

where $s$ is the arc length along the GC orbit and $s_m$ and $s'_m$ are the coordinates of the mirror points. The bounce period is defined over the whole bounce motion back and forth. This is a sensible approach if $\tau_b \gg \tau_L$. Thus the condition to consider the bounce motion as nearly periodic is more restrictive than in the case of Larmor motion

$$\tau_b \frac{dB/dt}{B} \ll 1\,. \tag{3.46}$$

If this condition is fulfilled, there is an associated adiabatic invariant which turns out to be the *longitudinal invariant*

$$J = \oint p_\parallel\, ds\,. \tag{3.47}$$

To directly prove the invariance of $J$ in a general case is a formidable task. The complete proof is given by Northrop [1963]. The textbook by Goldston and Rutherford [1995] presents the proof for time-independent fields, which is long enough. In space plasmas it is the time-dependence that typically breaks the conservation of $J$.

### 3.2.4 Betatron and Fermi acceleration

Consider the rate of change of the kinetic energy $T$ of a charged particle in a general time-dependent magnetic field $\mathbf{B}$. The time derivative in a moving frame of reference is $d/dt = \partial/\partial t + \mathbf{w}\cdot\nabla$, where $\mathbf{w}$ is the velocity of the frame of reference. In the GCS

$$\frac{dT_{GCS}}{dt} = \mu\frac{dB}{dt} = \mu\left(\frac{\partial B}{\partial t} + \mathbf{w}_\perp\cdot\nabla_\perp B + w_\parallel\frac{\partial B}{\partial s}\right)\,. \tag{3.48}$$

In the frame of reference of the observer (OFR)

$$\frac{dT_{OFR}}{dt} = \frac{dT_{GCS}}{dt} + \frac{d}{dt}\left(\frac{1}{2}mw_\parallel^2\right) + \frac{d}{dt}\left(\frac{1}{2}mw_\perp^2\right)\,. \tag{3.49}$$

With some algebra we get

$$\frac{dT_{OFR}}{dt} = \mu\frac{\partial B}{\partial t} + q\mathbf{w}\cdot\mathbf{E}\,. \tag{3.50}$$

The first term in the right-hand side of (3.50) gives the *betatron acceleration* due to the increasing magnetic flux through the position of the GC. More specifically, we should call this *gyro betatron acceleration*.

The second term contains both magnetic field-aligned acceleration (if $E_\parallel \neq 0$) and another betatron effect, called *drift-betatron acceleration*. When the GC drifts adiabatically across the magnetic field, e.g., due to E×B-drift toward increasing magnetic field ($B_2 > B_1$), the invariance of $\mu$ implies

$$\frac{W_{\perp 2}}{W_{\perp 1}} = \frac{B_2}{B_1} . \tag{3.51}$$

Thus $W_{\perp 2} > W_{\perp 1}$ .

A special case of drift-betatron acceleration is when a particle in a $J$ conserving bounce motion drifts toward a magnetic mirror. This is equivalent to moving the mirror points closer to each other when $\oint ds$ decreases. To compensate this $v_\parallel$ and thus $W_\parallel$ must increase. This mechanism is called *Fermi acceleration*.[1]

Fermi introduced this mechanism to explain the acceleration of *cosmic rays* to very high energies ($10^7 - 10^{10}$ eV) in the magnetic fields of the universe. A typical galactic cosmic ray has wandered around in the galaxy for millions of years. The radius of the Milky Way is of the order of 100 000 light years, and thus the particle has had a lot of time to "collide" with magnetic field structures in the galaxy that have a wide range of velocities. Note that in a given reference frame (e.g., ours) the particle either gains or loses energy when it gets deflected by a magnetic structure (e.g., mirror). As a result, the velocity distribution of the seed population widens and finally some particles end up at very high energies.

The modern version of Fermi acceleration, believed to be responsible for the acceleration of galactic cosmic rays, no longer relies on the conservation of the second adiabatic invariant in a distribution of moving magnetic mirrors. Instead, particles are assumed to be accelerated in shock waves generated in supernova explosions by a mechanism called *diffusive shock acceleration*. In this model, particles gain energy by repeatedly crossing a single shock front from one side to the other (Chap. 11).

The very highest energies of cosmic rays up to about $10^{20}$ eV remain unexplained. It should not even be possible to observe particles with energies higher than this, unless they are created not too far from the observing site. The reason for this is the quantum mechanical interaction of the particles with the blue-shifted cosmic microwave background. Above $6 \times 10^{19}$ eV, known as the *Greisen–Zatsepin–Kuzmin cut-off* this interaction leads to the production of pions that carry the excessive energy away.

### 3.2.5 The third adiabatic invariant

Also the drift across the magnetic field may be nearly-periodic if the field is sufficiently symmetric as, e.g., the quasi-dipolar planetary magnetic fields. The corresponding adiabatic invariant is the magnetic flux through the closed contour defined by the GC drift

$$\Phi = \oint \mathbf{A} \cdot d\mathbf{s} , \tag{3.52}$$

[1] A mechanical analog of Fermi acceleration is hitting a tennis ball with a racket. In the audience's frame the ball is accelerated but in the racket's frame it just mirrors (or actually loses energy due to the elasticity of the ball and racket).

where **A** is the vector potential of the field and $d\mathbf{s}$ is the arc element along the drift path of the GC. The drift period $\tau_d$ has to fulfill $\tau_d \gg \tau_b \gg \tau_L$. The invariant is weaker than $\mu$ and $J$ because much slower changes in the field can break the invariance of $\Phi$.

In the Earth's magnetosphere $\mu$ is often a good invariant. $J$ is invariant for particles that spend at least some time in the magnetic bottle defined by the nearly-dipolar field of the Earth. $\Phi$ is constant for energetic particles in the trapped radiation belts. However, any or all of the invariances can be broken by perturbations to the system.

Let us briefly return to the Hamiltonian mechanics. These three functions $(\mu, J, \Phi)$, whether invariant or not, form a particular set of *canonical action variables* or *action integrals*

$$J_i = \frac{1}{2\pi} \oint_i (\mathbf{p} + q\mathbf{A}) \cdot d\mathbf{s}_i \tag{3.53}$$

with associated *phase angles* $\phi_i$, that in this case are the gyrophase, the bounce phase, and the drift phase. We will return to these in Chap. 10 when we discuss the particle distribution function, or phase-space density, expressed as a function of the action variables.

## 3.3 Motion in the Dipole Field

Charged particle motion in the dipole field is an important application of the orbit theory. Within the distances 2–7 $R_E$ from the Earth's center the dipole is a reasonably good approximation of the geomagnetic field and all particles except high-energy cosmic rays behave adiabatically as long as their orbits are not disturbed by collisions or time-varying electromagnetic fields.

In the following we use "geomagnetically" defined spherical coordinates. The *dipole moment* $\mathbf{M}_E$ is in the origin and points toward the south. Latitude ($\lambda$) is zero at the equator and increases toward the north. Longitude ($\phi$) increases toward the east from a given reference longitude. The SI unit of $M_E$ is $\mathrm{A\,m^2}$. $M_E$ is often replaced by $k_0 = \mu_0 M_E / 4\pi$, which is also called dipole moment. The strength and orientation of the terrestrial dipole moment varies slowly and must be taken into account in time scales of space climate. For our purposes sufficiently accurate approximations are

$$\begin{aligned} M_E &= 8 \times 10^{22}\ \mathrm{A\,m^2} \\ k_0 &= 8 \times 10^{15}\ \mathrm{Wb\,m}\ \ (\mathrm{SI: Wb = T\,m^2}) \\ &= 8 \times 10^{25}\ \mathrm{G\,cm^3}\ \ (\text{Gaussian units},\ \mathrm{G} = 10^{-4}\,\mathrm{T}) \\ &= 0.3\ \mathrm{G}R_E^3 \qquad\quad (R_E \simeq 6370\,\mathrm{km}) \end{aligned}$$

The last (non-SI) expression is useful in practice because the dipole field on the surface of the Earth varies in the range 0.3–0.6 G.

The dipole field is an idealization where the source current is assumed to be shrunk into a point at the origin. The source of a planetary or stellar magnetic field is actually a finite, even large, region within the body giving rise to a whole sequence of higher multipoles. When moving away from the source the non-dipolar (quadrupole, octupole, etc.) contributions vanish faster than the dipole. Outside the source the field is a potential field ($\mathbf{B} = -\nabla\Psi$). The potential for the dipole is

$$\Psi = -\mathbf{k}_0 \cdot \nabla \frac{1}{r} = -k_0 \frac{\sin\lambda}{r^2} . \tag{3.54}$$

It is a standard exercise in elementary electromagnetism to show that

$$\mathbf{B} = \frac{1}{r^3}[3(\mathbf{k}_0 \cdot \mathbf{e}_r)\mathbf{e}_r - \mathbf{k}_0] , \tag{3.55}$$

from which

$$\begin{aligned} B_r &= -\frac{2k_0}{r^3}\sin\lambda \\ B_\lambda &= \frac{k_0}{r^3}\cos\lambda \\ B_\phi &= 0 . \end{aligned} \tag{3.56}$$

The magnitude of the magnetic field is

$$B = \frac{k_0}{r^3}(1+3\sin^2\lambda)^{1/2} \tag{3.57}$$

and the equation for the field line is

$$r = r_0 \cos^2\lambda , \tag{3.58}$$

where $r_0$ is the distance from the dipole to the point where the field line crosses the dipole equator. In dipole calculations we also need the length of the line element

$$ds = (dr^2 + r^2 d\lambda^2)^{1/2} = r_0 \cos\lambda (1+3\sin^2\lambda)^{1/2} d\lambda . \tag{3.59}$$

The geometric factor $(1+3\sin^2\lambda)^{1/2} = (4-3\cos^2\lambda)^{1/2}$ pops up here and there in the dipole expressions.

Every dipole field line is uniquely determined by its (constant) longitude $\phi_0$ and the distance $r_0$. A useful quantity is the *L parameter* $L = r_0/R_E$. For a given $L$ the corresponding field line reaches the surface of the Earth at the latitude

$$\lambda_e = \arccos \frac{1}{\sqrt{L}} . \tag{3.60}$$

The field magnitude along a given field line as a function of latitude is

$$B(\lambda) = [B_r(\lambda)^2 + B_\lambda(\lambda)^2]^{1/2} = \frac{k_0}{r_0^3} \frac{(1+3\sin^2\lambda)^{1/2}}{\cos^6\lambda} . \tag{3.61}$$

For the Earth

$$\frac{k_0}{r_0^3} = \frac{0.3}{L^3}\,\mathrm{G} = \frac{3\times10^{-5}}{L^3}\,\mathrm{T} . \tag{3.62}$$

At the equator on the surface of the Earth the dipole field is 0.3 G, at the poles 0.6 G (i.e., 30 and 60 μT). The observable geomagnetic field has considerable deviations from this because the dipole is not quite in the center of the Earth, the source is not a point, and the conductivity of the Earth is not uniform.

The guiding center approximation can be applied if the particle's Larmor radius is much smaller than the curvature radius of the field defined by $R_C = |d^2\mathbf{r}/ds^2|^{-1}$, which for a static dipole field is

$$R_C(\lambda) = \frac{r_0}{3}\cos\lambda\,\frac{(1+3\sin^2\lambda)^{3/2}}{2-\cos^2\lambda}\,. \tag{3.63}$$

In terms of the particle's *rigidity* $mv_\perp/|q|$ the condition is

$$r_L\left|\frac{\nabla_\perp B}{B}\right| = \frac{mv_\perp}{|q|R_C B} \propto \frac{mv_\perp}{|q|r_0 B}\,, \tag{3.64}$$

i.e., the GC approximation is valid if

$$\frac{mv_\perp}{|q|} \ll r_0 B\,. \tag{3.65}$$

The dipole field is a magnetic bottle and the energetic particles trapped in the bottle around the Earth or magnetized planets are said to form *trapped radiation*. Let $\lambda_m$ be the mirror latitude of a trapped particle and let the subscript 0 refer to the equatorial plane. Then the equatorial pitch angle of the particle is

$$\sin^2\alpha_0 = \frac{B_0}{B(\lambda_m)} = \frac{\cos^6\lambda_m}{(1+3\sin^2\lambda)^{1/2}}\,. \tag{3.66}$$

This shows that the *mirror latitude* does not depend on $L$, but the *mirror altitude* does.

If $\lambda_e$ is the latitude where the field line intersects the surface of the Earth and if $\lambda_e < \lambda_m$, the particle hits the Earth before mirroring and is lost from the bottle. In reality the loss takes place in the upper atmosphere at an altitude that depends on the particle's energy, i.e., on how far it can penetrate before it is lost by collisions. The critical pitch angle in the equatorial plane is

$$\sin^2\alpha_{0l} = L^{-3}(4-3/L)^{-1/2} = (4L^6-3L^5)^{-1/2}\,. \tag{3.67}$$

The particle is in the loss-cone, if $\alpha_0 < \alpha_{0l}$.

The bounce period in a dipolar bottle is

$$\begin{aligned}\tau_b &= 4\int_0^{\lambda_m}\frac{ds}{v_\parallel} = 4\int_0^{\lambda_m}\frac{ds}{d\lambda}\frac{d\lambda}{v_\parallel}\\ &= \frac{4r_0}{v}\int_0^{\lambda_m}\frac{\cos\lambda(1+3\sin^2\lambda)^{1/2}}{1-\sin^2\alpha_0(1+3\sin^2\lambda)^{1/2}/\cos^6\lambda}\,d\lambda\\ &= \frac{4r_0}{v}f(\alpha_0)\,,\end{aligned} \tag{3.68}$$

where

$$v_{\parallel}(\lambda) = v\cos\alpha = v(1-\sin^2\alpha)^{1/2} = v[1-\sin^2\alpha_0 B(\lambda)/B_0]^{1/2} \tag{3.69}$$

and (3.61) has been used. For $30^\circ \leq \alpha_0 \leq 90^\circ$

$$f(\alpha_0) \approx 1.30 - 0.56\sin^2\alpha_0\,. \tag{3.70}$$

The conservation of the second adiabatic invariant requires that the bounce period is much shorter than the variations in the magnetic field. For example, in the inner magnetosphere the bounce times of 1-keV electrons are a few seconds and of 1-keV protons a few minutes. During magnetospheric activity typical time scales of the field changes are minutes. Thus under such conditions $J$ is a good invariant for electrons but not for protons or heavier ions.

Both the gradient and curvature of the dipole field are directed toward the planet. In the dipole field of the Earth positively charged ions drift to the west and electrons to the east.

Because $\nabla \times \mathbf{B} = 0$, we find for $v_{GC}$

$$\begin{aligned} v_{GC} &= \frac{W}{qBR_C}(1+\cos^2\alpha) \\ &= \frac{3mv^2r_0^2}{2qk_0}\frac{\cos^5\lambda(1+\sin^2\lambda)}{(1+3\sin^2\lambda)^2}\left[2-\sin^2\alpha_0\frac{(1+3\sin^2\lambda)^{1/2}}{\cos^6\lambda}\right]. \end{aligned} \tag{3.71}$$

For the drift motion around the Earth, $v_{GC}$ is often less interesting than the angular speed averaged over one bounce period $\langle\dot{\phi}\rangle = \langle v_{GC}/r\cos\lambda\rangle$, which gives the drift rate of the guiding center around the dipole axis. A little exercise gives the result

$$\begin{aligned} \langle\dot{\phi}\rangle &= \frac{4}{v\tau_b}\int_0^{\lambda_m}\frac{v_{GC}(\lambda)(1+3\sin^2\lambda)^{1/2}}{\cos^2\lambda\cos\alpha(\lambda)}\,d\lambda \\ &\equiv \frac{3mv^2r_0}{2qk_0}g(\alpha_0) = \frac{3mv^2R_EL}{2qk_0}g(\alpha_0)\,, \end{aligned} \tag{3.72}$$

where

$$g(\alpha_0) = \frac{1}{f(\alpha_0)}\int_0^{\lambda_m}\frac{\cos^3\lambda(1+\sin^2\lambda)[1+\cos^2\alpha(\lambda)]}{(1+3\sin^2\lambda)^{3/2}\cos\alpha(\lambda)}\,d\lambda\,. \tag{3.73}$$

Within the pitch angle range $30^\circ \leq \alpha_0 \leq 90^\circ$

$$g(\alpha_0) \simeq 0.7 + 0.3\sin(\alpha_0)\,. \tag{3.74}$$

For $\alpha_0 = 90^\circ$

$$\langle\dot{\phi}_0\rangle = \frac{3mv^2R_EL}{2qk_0}\,. \tag{3.75}$$

In the relativistic form this formula is

$$\langle\dot{\phi}_0\rangle = \frac{3mc^2R_EL}{2qk_0}\gamma\beta^2\,. \tag{3.76}$$

The average drift period $\langle \tau_d \rangle$ is

$$\tau_d = \frac{2\pi}{|\langle \dot{\phi} \rangle|} = \frac{4\pi}{3} \frac{|q| k_0}{mc^2 R_E} \frac{1}{L\gamma\beta^2 g(\alpha_0)}$$
$$\approx 1.0 \times 10^4 \frac{m_e}{m} \frac{|q|}{e} \frac{1}{L\gamma\beta^2 g(\alpha_0)} , \qquad (3.77)$$

where the last line gives $\tau_d$ in seconds when the variables are given in SI units. The drift period is inversely proportional to the energy of the particle. In the region where the terrestrial field is most dipolar ($L \simeq 2-7$) the drift periods for 1-keV particles are hundreds of hours whereas those for 1-MeV particles are some tens of minutes, depending on the pitch angles.

In the inner radiation belt ($L \approx 1.5 - 3$) the dominating trapped high-energy population is protons in the energy range 0.1 MeV – 40 MeV, whereas in the outer belt ($L > 4$) the energetic component is mostly electrons in the keV to MeV range. Thus radiation belt protons are mostly non-relativistic whereas a considerable fraction of electrons can be relativistic (Chap. 14).

***Example: Penetration of cosmic rays to the atmosphere***

Most of the galactic cosmic rays are relativistic. In studies of relativistic particles it is common to write $c = 1$. Then energy (eV), momentum (eV $c^{-1}$), and mass (eV $c^{-2}$) are all expressed in units of eV, or actually MeV or GeV. In these units rigidity, whose *physical dimension* is momentum per charge, has the volt as its unit (in the ranges of MV or GV). Rigidity is an important concept for cosmic ray penetration through the geomagnetic field, as it describes which particles can reach the atmosphere.

The relationship between rigidity ($R = |p/q|$) and energy is found by solving the relativistic expression for the total energy $W_T$

$$W_T^2 = p^2c^2 + m_0^2c^4 , \qquad (3.78)$$

where $m_0$ is the (rest) mass of the particle. The result is

$$R = \frac{A}{Z} [(\gamma^2 - 1)^{1/2}] W_{0A} , \qquad (3.79)$$

where $W_{0A}$ is the rest mass energy per nucleon, $A$ is the atomic number and $Z$ the charge state, i.e. $+n$ for $n$ times charged ions. Conversely, if the rigidity is known, the Lorentz factor $\gamma$, and thus the particle speed can be found from

$$\gamma = \left[ \left( \frac{RZ}{A/W_{0A}} \right)^2 + 1 \right]^{1/2} . \qquad (3.80)$$

The ambient magnetic field deviates particles and allows only rigid enough particles to penetrate to a given depth. *Cosmic ray cut-off rigidity* specifies the minimum rigidity that a charged particle must have to be observed at a given position in the geomagnetic field coming from *from a given direction*.

Calculation of the cut-off rigidities is tedious, as the guiding center approximation cannot be used and the incident direction has to be taken into account. In general the cut-off is the higher the more perpendicularly to the magnetic field the particle moves.

In the early days of cosmic ray research Størmer derived the cut-off rigidity formula in the dipole field

$$R_c = M \frac{\cos^4 \lambda}{r^2[1 + (1 - \sin\varepsilon \sin\phi \cos^3 \lambda)^{1/2}]^2} , \tag{3.81}$$

where $M$ is the magnetic moment in the unit system used by Størmer, $\lambda$ is the geomagnetic latitude, $\varepsilon$ the zenith angle, and $\phi$ the azimuthal angle measured from the direction of geomagnetic north with respect to geographic north. Using the terrestrial dipole moment and expressing $r$ in $R_E$ from the dipole center, the numerical terms give the factor 59.6. For *vertical* incidence in the terrestrial dipole field the cut-off rigidity is given by

$$R_c(\mathrm{GV}) = \frac{14.9 \cos^4 \lambda}{r^2} . \tag{3.82}$$

Taking into account the deviations from the dipole field the cut-off rigidity for vertical incidence at sea-level varies between 13 and 17 GV near the equator.

As the $L$-parameter for the dipole field is given by $\cos^2 \lambda = r_0/L$, the cut-off rigidity in the inner magnetosphere (up to $L = 4$) can be estimated by $R_c \approx 16 L^{-2}$ GV. The numerical factor is a little larger than the dipole value due to the external magnetic field contributions. At auroral latitudes the cut-off rigidity is typically less than 1 GV. At the dipole magnetic poles the cut-off rigidity is zero for a particle that has exactly field-aligned direction when entering vertically. In the real magnetosphere the external magnetic field created by magnetospheric currents inhibits the direct entry of low-energy, or small-rigidity, particles.

The final stopping power against the cosmic rays is not the magnetic field but the atmosphere of the Earth. The primary particles collide with atmospheric nuclei cascading first typically to protons, neutrons, and pions, which further decay to photons and muons, etc. Energetic enough photons may form electron-positron pairs. This was the process through which Anderson first identified the positron in 1933. Very high-energy cosmic rays produce large amounts of particles in such cascades. These *air showers* produce Cherenkov radiation in the air, which can be observed by optical means. The neutrons and muons making their way down to the Earth can also be detected directly using ground-based instruments. In fact, neutron and muon fluxes observed on ground are standard means of characterizing the intensity of cosmic ray events.

## 3.4 Motion Near a Current Sheet

The interaction between the terrestrial magnetic field and the solar wind stretches the nightside magnetosphere to a long tail where the field geometry changes from dipolar to that of a thin current sheet. This is just one example of the great variety of current sheets in space plasmas. The current must be there to account for the change of the magnetic field orientation according to Ampère's law $\nabla \times \mathbf{B} = \mu_0 \mathbf{J}$.

### 3.4.1 The Harris model

A two-dimensional current sheet can be described by the *Harris model* whose magnetic field is of the form

$$\mathbf{B} = B_0 \tanh\left(\frac{z}{L}\right) \mathbf{e}_x + B_n \mathbf{e}_z \,, \tag{3.83}$$

where $B_0$ and $B_n$ are constant, $B_n \ll B_0$ and $L$ is the characteristic thickness of the current sheet. If $B_n = 0$, the field is one-dimensional. The electric current points toward the positive $y$-axis and is

$$J_y = \left(\frac{B_0}{\mu_0 L}\right) \mathrm{sech}^2\left(\frac{z}{L}\right) . \tag{3.84}$$

In the one-dimensional case the magnetic field is in magnetohydrostatic equilibrium (Chap. 6) with plasma in the current sheet, whose pressure is

$$P(z) = P_0 \, \mathrm{sech}^2\left(\frac{z}{L}\right) . \tag{3.85}$$

The Harris field can be derived from a vector potential of the form

$$A_y(x,z) = -B_0 F(z) + B_n x \,. \tag{3.86}$$

The particles move in an effective potential of the form

$$U(x,z) = \frac{1}{2m}[p_y - qA_y(x,z)]^2 \,, \tag{3.87}$$

where $p_y$ is the linear momentum in the $y$-direction.

**Train your brain** by calculating $F(z)$ in (3.86).

In simple analytical calculations the Harris model is often approximated as

$$\mathbf{B} = B_0 \left(\frac{z}{L}\right) \mathbf{e}_x + B_n \mathbf{e}_z \,. \tag{3.88}$$

In this approximation the field lines are parabolas

$$x = \frac{B_0}{2B_n L} z^2 + \text{constant} \,. \tag{3.89}$$

In the one-dimensional case ($B_n = 0$) a useful approximation is

$$\begin{aligned} B_x &= B_0 \quad ; \; z \geq L \\ B_x &= \frac{B_0 z}{L} \quad ; \; L \geq z \geq -L \\ B_x &= -B_0 \; ; \; z \leq L\,. \end{aligned} \tag{3.90}$$

In this magnetic field model the components of the equation of motion are

$$\ddot{x} = 0 \tag{3.91}$$

$$\ddot{y} = \left(\frac{qB_0}{mL}\right) z\dot{z} \tag{3.92}$$

$$\ddot{z} = -\left(\frac{qB_0}{mL}\right) z\dot{y}\,. \tag{3.93}$$

The equation of motion perpendicular to the magnetic field can be cast into the form

$$\frac{d}{dt}(\dot{y}^2 + \dot{z}^2) = 0\,, \tag{3.94}$$

which expresses the conservation of energy. After appropriate normalization of $k$ and $z$ the motion in the $z$-direction can be found as a solution of the equation

$$\dot{z}^2 = (1 - k^2 + k^2 z^2)(1 - z^2)\,. \tag{3.95}$$

The general solutions of (3.95) can be expressed in terms of elliptic integrals and Jacobi's elliptic functions. Examples of the orbits are given in Fig. 3.1. Outside the current sheet the motion is normal gyro motion. Within the current sheet the motion is more complicated. The monotonic motion in the $\pm y$-direction is called *Speiser motion.* Particles in the Speiser motion carry most of the current in the current sheet. They do not conserve the magnetic moment but the motion is periodic in the $z$-direction, for which there is another adiabatic invariant [e.g., Büchner and Zelenyi, 1989].

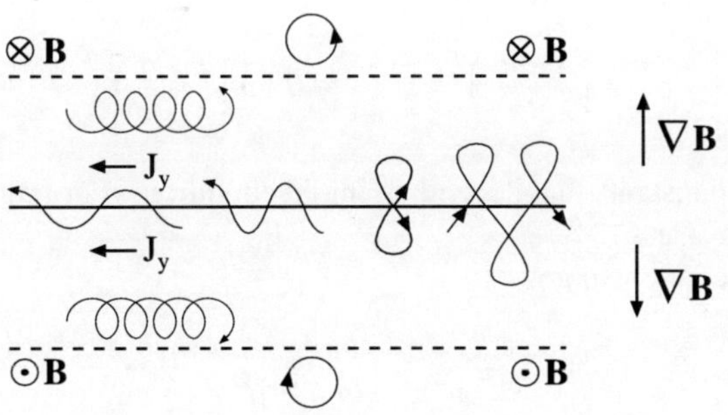


**Fig. 3.1** Orbits of positively charged particles near a one-dimensional current sheet.

### 3.4.2 Neutral sheet with a constant electric field

The earthward plasma convection in the magnetospheric tail induces a dawn-to-dusk directed electric field $\mathbf{E} = E_0\,\mathbf{e}_y$. This electric field has the same direction as the current, i.e., $\mathbf{E}\cdot\mathbf{J} > 0$. Thus Poynting's theorem implies particle energization at the expense of the electromagnetic field. Under these circumstances the equations of motion are

$$\ddot{x} = 0 \tag{3.96}$$

$$\ddot{y} = c_1 z\dot{z} + c_2 \tag{3.97}$$

$$\ddot{z} = -c_1 z\dot{y}\,, \tag{3.98}$$

where $c_1 = qB_0/mL$ and $c_2 = qE_0/m$. Due to the electric field the energy equation is more complicated than above

$$\frac{1}{2}(\dot{y}^2+\dot{z}^2) - c_2 y = \frac{1}{2}(\dot{y}_0^2+\dot{z}_0^2) - c_2 y_0\,, \tag{3.99}$$

where zeros refer to the initial values. The equation in the $z$-direction becomes

$$\ddot{z} = -c_1 z\left[\dot{y}_0 + \left(\frac{c_1}{2}\right)(z^2 - z_0^2) + c_2 t\right]\,. \tag{3.100}$$

This is a nonlinear equation with chaotic solutions.

Assume that a particle remains in this current sheet for a long time (with respect to its gyro period). For large $t$

$$\ddot{z} \approx -c_1 c_2 z t = -\left(\frac{q}{m}\right)^2\left(\frac{E_0 B_0}{L}\right) z t\,. \tag{3.101}$$

This equation has oscillatory solutions in terms of Bessel functions of the first and second kind. For large $t$ the solution can be approximated by

$$z \approx -\frac{t^{-1/4}}{(E_0B_0/L)^{1/12}(q/m)^{1/6}} \times \tag{3.102}$$
$$\left\{A\cos\left[\frac{2}{3}\left(\frac{q}{m}\right)\left(\frac{B_0E_0}{L}\right)^{1/2} t^{3/2}\right] + B\sin\left[\frac{2}{3}\left(\frac{q}{m}\right)\left(\frac{B_0E_0}{L}\right)^{1/2} t^{3/2}\right]\right\}$$

where $A$ and $B$ are constants that depend on initial conditions. For large $t$ the amplitude of the oscillation decays as $t^{-1/4}$.

Now we can integrate $y$ to get

$$y \approx y_0 + \left[\dot{y}_0 - \left(\frac{c_1}{2}\right) z_0^2\right] t + \frac{c_2 t^2}{2}\,. \tag{3.103}$$

Inserting these in the energy equation we find that the kinetic energy increases as $t^2$. Thus all particles execute damped oscillations about $z = 0$, while the positive ions are accelerated in the $+y$-direction and electrons in the $-y$-direction.

### 3.4.3 Current sheet with a small perpendicular magnetic field component

A real current sheet often has a small perpendicular magnetic field component. In the Earth's magnetotail the dipolar field is stretched to form a long current sheet that has a small northward component to a distance of more than $100\,R_E$, except during substorm expansions when the current sheet disrupts closer to the Earth (Chap. 13). Consider the particle dynamics in a 2D model of such a tail

$$\mathbf{B} = B_0 \left(\frac{z}{L}\right) \mathbf{e}_x + B_n \mathbf{e}_z \tag{3.104}$$

with the same electric field as above. Now the equations of motion become

$$\ddot{x} = c_3 \dot{y} \tag{3.105}$$

$$\ddot{y} = -c_3 \dot{x} + c_1 z \dot{z} + c_2 \tag{3.106}$$

$$\ddot{z} = -c_1 z \dot{y} \, , \tag{3.107}$$

where $c_1$ and $c_2$ are the same as previously and $c_3 = qB_0B_n/m$. This leads to equations requiring numerical intergration. The energy integral includes all coordinates

$$\frac{m}{2}(\dot{x}^2 + \dot{y}^2 + \dot{z}^2) + q\varphi = \frac{m}{2}(\dot{x}_0^2 + \dot{y}_0^2 + \dot{z}_0^2) + q\varphi_0 \, , \tag{3.108}$$

where $\varphi$ is the electrostatic potential $\mathbf{E} = -\nabla\varphi$ .

Let us then consider what happens to a proton that approaches the current sheet in Larmor motion. If the energy of the particle is small enough, it starts to execute Speiser motion in the current sheet while simultaneously turning around the weak $B_n$. If there is no electric field, the situation is symmetric and the particle is ejected from the current sheet in a symmetrical position with respect to the axis parallel to the $x$-axis passing through the gyro center of motion in the $xy$-plane. In the presence of $\mathbf{E} = E_0\,\mathbf{e}_y$ the proton is accelerated, which makes it progress farther in the current sheet and being finally ejected with a larger energy.

The capture into Speiser motion and the ejection from the current sheet are very sensitive to the initial conditions, characteristic to *chaotic systems*. Consider a dipole field where the particle motion is adiabatic conserving $\mu$. Stretch the field slowly to a tail-like configuration. When the ratio $R_C/r_L$ becomes smaller than about 10, the invariance of the magnetic moment starts to break and the motion becomes non-adiabatic and the particle loses the guidance of the magnetic field. When the field is stretched further the motion of particles with smaller energies becomes irregular and chaotic. The chaotization changes the pitch angles of the particles which can, for example, fill the loss cone. This is one mechanism to precipitate particles from the magnetosphere to the ionosphere.

## 3.5 Motion in a Time-dependent Electric Field

Understanding charged particles' behavior in time-dependent electric fields is important, as the response of plasma determines the properties of electromagnetic wave propagation (Chap. 4) and time-dependent fields change the energy of the particles through *wave–particle interactions*.

### 3.5.1 Slow time variations

If the magnetic field is static and homogeneous and the time-variation of the electric field slow ($\partial/\partial t \ll \omega_c$), we find the *polarization drift*

$$\mathbf{v}_P = -\frac{m}{qB^2}\frac{d\mathbf{v}_E}{dt}\times\mathbf{B} = \frac{1}{\omega_c B}\frac{d\mathbf{E}_\perp}{dt}\,. \tag{3.109}$$

This drift separates charges and masses which gives a rise to a *polarization current* ($\mathbf{J}_P = nq\mathbf{v}_P$) in the plasma. Due to the large mass ratio between electrons and ions this current is carried mostly by the ion drift.

When $\mathbf{E}$ increases, $\mathbf{J}_P\cdot\mathbf{E} > 0$ for both positive and negative charges, i.e., particles gain energy. This energy gain is the same as the difference in the E×B-drift energy before and after the increase of $\mathbf{E}$. If, on the other hand, $\mathbf{E}$ decreases, the particles lose energy.

Note that $\mathbf{v}_E$ and $\mathbf{v}_P$ are of different order in magnitude

$$\frac{dE}{dt}\sim\omega E \;\Rightarrow\; \frac{v_P}{v_E}\sim\frac{1}{\omega_c B}\omega E\frac{B}{E}\sim\frac{\omega}{\omega_c}\ll 1\,, \tag{3.110}$$

where the last inequality is the basic condition for the existence of the polarization drift.

### 3.5.2 Time variations in resonance with gyro motion

We move now to the case where the rate of change in the electric field is of the same order as the gyro frequency of the particle: $E\propto\exp(-i\omega t)$ and $\omega\approx\omega_c$.

Assuming further a static and homogeneous $\mathbf{B}$ the equation of motion is

$$\frac{d\mathbf{v}}{dt} = \frac{q}{m}(\mathbf{E}\,e^{-i\omega t} + \mathbf{v}\times\mathbf{B})\,. \tag{3.111}$$

Seek a solution of the form $\mathbf{v} = \mathbf{v}_e\exp(-i\omega t) + \mathbf{v}_m$ ($e$ for electric and $m$ for magnetic), where $\mathbf{v}_e$ is time-independent. The equation of motion is then

$$\frac{d\mathbf{v}_m}{dt} - i\omega\mathbf{v}_e\,e^{-i\omega t} = \frac{q}{m}(\mathbf{E}e^{-i\omega t} + \mathbf{v}_m\times\mathbf{B} + \mathbf{v}_e\times\mathbf{B}\,e^{-i\omega t})\,. \tag{3.112}$$

The magnetic part

$$\frac{d\mathbf{v}_m}{dt} = \frac{q}{m}(\mathbf{v}_m\times\mathbf{B}) \tag{3.113}$$

gives the Larmor rotation. The electric part is

$$\frac{q}{m}\mathbf{E} = (-i\omega + \frac{q}{m}\mathbf{B}\times)\mathbf{v}_e \equiv -(i\omega + \boldsymbol{\omega}_c\times)\mathbf{v}_e \,. \tag{3.114}$$

Now $\boldsymbol{\omega}_c$ is the vector $\boldsymbol{\omega}_c = -q\mathbf{B}/m$. Multiplying the above expression from the left by $(i\omega - \boldsymbol{\omega}_c\times)$ we get

$$\frac{q}{m}(i\omega - \boldsymbol{\omega}_c\times)\mathbf{E} = (\omega^2 - \omega_c^2)\mathbf{v}_e + (\boldsymbol{\omega}_c\cdot\mathbf{v}_e)\,\boldsymbol{\omega}_c \,. \tag{3.115}$$

Decompose this into the parallel and perpendicular components

$$\mathbf{v}_{e\parallel} = \frac{i}{\omega}\frac{q\mathbf{E}_\parallel}{m} \tag{3.116}$$

$$\mathbf{v}_{e\perp} = \frac{q}{m}\left(\frac{i\omega - \boldsymbol{\omega}_c\times}{\omega^2 - \omega_c^2}\right)\mathbf{E}_\perp \,. \tag{3.117}$$

We see that $\mathbf{v}_{e\parallel}$ oscillates with the phase lagging 90° behind $\mathbf{E}_\parallel$. The perpendicular velocity can be expressed as

$$\mathbf{v}_{eL} = \frac{q}{m}\frac{i}{\omega - \omega_c}\mathbf{E}_L \tag{3.118}$$

$$\mathbf{v}_{eR} = \frac{q}{m}\frac{i}{\omega + \omega_c}\mathbf{E}_R \,, \tag{3.119}$$

where

$$\mathbf{E}_L = -\frac{1}{2}\left(\mathbf{E}_\perp + i\frac{\boldsymbol{\omega}_c\times\mathbf{E}_\perp}{\omega_c}\right) \tag{3.120}$$

$$\mathbf{E}_R = -\frac{1}{2}\left(\mathbf{E}_\perp - i\frac{\boldsymbol{\omega}_c\times\mathbf{E}_\perp}{\omega_c}\right) \tag{3.121}$$

are the *left-hand* ($\mathbf{E}_L$) and *right-hand* ($\mathbf{E}_R$) *polarized* components of the (wave) electric field. They are in *resonance* with different particle species, the left-hand polarized wave with positive charges, the right-hand polarized wave with negative charges.

NOTE: This is the convention of the sense of circular wave polarization in (modern) plasma physics, i.e., *the electric field of a right-hand polarized wave rotates around the magnetic field in the same sense as an electron.*

### 3.5.3 High-frequency fields

Assume next that $\omega \gg \omega_c$. This allows the use of an approach resembling the GC approximation, called *oscillation center approximation.*

In the zero-order problem we assume that $\mathbf{E}$ is spatially homogeneous with the time dependence of the form $exp(-i\omega t)$. Write the equation of motion in the form

$$\frac{d^2\mathbf{r}}{dt^2} = \frac{q}{m}(\mathbf{E}e^{-i\omega t}) . \tag{3.122}$$

This has the solution

$$\mathbf{r} = \frac{q}{m\omega^2}(\mathbf{E}e^{-i\omega t}) + \mathbf{c}_1 t + \mathbf{c}_2 . \tag{3.123}$$

Include $\mathbf{B}$ and let the fields be weakly inhomogeneous and proportional to $\exp(-i\omega t)$

$$\frac{d^2\mathbf{r}}{dt^2} = \frac{q}{m}[\mathbf{E}(\mathbf{r}) + \frac{d\mathbf{r}}{dt} \times \mathbf{B}(\mathbf{r})]\, e^{-i\omega t} . \tag{3.124}$$

Equation (3.122) is the zero-order approximation of (3.124) if both of the following conditions are valid

1. in the Taylor series

$$\mathbf{E}(\mathbf{r}) = \mathbf{E}(\mathbf{r}_0) + (\mathbf{r}_1 \cdot \nabla_0)\mathbf{E} + \ldots \tag{3.125}$$

- $\mathbf{E}(\mathbf{r}_0)$ dominates
- $\mathbf{r}_0$ is the center of oscillation and $\mathbf{r}_1 = \mathbf{r} - \mathbf{r}_0$ oscillates
- $\mathbf{r}_0$ moves slowly and $\mathbf{E}(\mathbf{r}_0)$ is almost a constant during one oscillation period:

$$\frac{|(\dot{\mathbf{r}}_0 \cdot \nabla)\,\mathbf{E}|}{\omega} \ll E \tag{3.126}$$

2. $(d\mathbf{r}/dt) \times \mathbf{B}$ is small, i.e., $\omega_c \ll \omega$.

Because $d^2r/dt^2 \sim \omega\, dr_1/dt$ and the magnetic term is proportional to $\omega_c dr/dt$, the speed $dr/dt$ must not be much larger than $dr_1/dt$. Under such circumstances we can expand (3.124) as

$$\begin{aligned} &\frac{d^2\mathbf{r}_0}{dt^2} + \frac{d^2\mathbf{r}_1}{dt^2} \\ &\quad = \frac{q}{m}\left[\mathbf{E}(\mathbf{r}_0) + (\mathbf{r}_1 \cdot \nabla_0)\mathbf{E} + \frac{d\mathbf{r}_0}{dt} \times \mathbf{B}(\mathbf{r}_0) + \frac{d\mathbf{r}_1}{dt} \times \mathbf{B}(\mathbf{r}_0)\right] \\ &\quad + \frac{q}{m}\left[\frac{d\mathbf{r}_0}{dt} \times (\mathbf{r}_1 \cdot \nabla_0)\mathbf{B} + \frac{d\mathbf{r}_1}{dt} \times (\mathbf{r} \cdot \nabla_0)\mathbf{B}\right] , \end{aligned} \tag{3.127}$$

where the last line is of the second order.

The second term in the LHS is larger than the first. Thus the zero-order solution is

$$\mathbf{r}_1 = -\frac{q}{m\omega^2}\mathbf{E}_0 . \tag{3.128}$$

For the first-order solution we consider only the time averages of the first-order terms in the same way as in the GC approximation but in this case averaged over the oscillation period. $\langle d\mathbf{r}_0/dt \times \mathbf{B}_0\rangle$ can be neglected because $d\mathbf{r}_0/dt$ is small and $\langle \mathbf{B}_0\rangle = 0$. Now

$$\left\langle \frac{d^2\mathbf{r}_0}{dt^2} \right\rangle = \frac{q}{m}\left\{ \langle (\mathbf{r}_1 \cdot \nabla_0)\mathbf{E}\rangle + \left\langle \frac{d\mathbf{r}_1}{dt} \times \mathbf{B}_0 \right\rangle \right\} \tag{3.129}$$

and $\langle d^2\mathbf{r}_0/dt^2\rangle \approx d^2\mathbf{r}_0/dt^2$. Inserting the zero-order solution for $\mathbf{r}_1$ to the expression above a brief calculation yields

$$\frac{d^2\mathbf{r}_0}{dt^2} = -\frac{q^2}{m^2\omega^2}\left\langle \nabla_0 \frac{E^2}{2} \right\rangle - \frac{q^2}{m^2\omega^2}\left\langle \frac{\partial}{\partial t}(\mathbf{E}_0 \times \mathbf{B}_0) \right\rangle . \tag{3.130}$$

For a standing wave the last term is zero. Thus we have found that the oscillation center is accelerated by the potential

$$\Phi = \frac{q^2}{m^2\omega^2}\left\langle \frac{E^2}{2} \right\rangle . \tag{3.131}$$

This is called *ponderomotive potential*. The oscillation center is accelerated toward smaller $\Phi$. The ponderomotive force $\propto -\nabla\Phi$ is a nonlinear function of the electric field. It can be used to trap particles in the field of a standing wave. This effect appears in various problems of nonlinear plasma physics, e.g., in heating of plasma by intense electromagnetic waves.

# 4. Waves in Cold Plasma Approximation

Plasmas are very rich in wave phenomena. If an equilibrium state of a plasma is perturbed, plasma responds with wave-like behavior. The waves may carry the effects of the perturbation far from their origin, or be damped through interactions with the surrounding plasma. Sometimes the waves may grow to such large amplitudes that the entire plasma configuration is destroyed. Waves are efficient in particle acceleration and plasma heating. Even waves interacting weakly with the plasma can be distorted by the interaction, for example, different frequencies sent at the same time through the plasma arrive at different times and the polarization plane is rotated, as the background magnetic field makes the plasma birefringent.

We start with the traditional introduction to the menagerie of plasma waves discussing them in the *cold plasma approximation*,. The approach is valid when the phase velocities of the waves are larger than the thermal velocity of the background plasma. This is quite sufficient for a wide range of wave phenomena. As we will see, the approach has its natural limitations, e.g., in the context of instabilities and wave–particle interactions, which are discussed in subsequent chapters.

## 4.1 Basic Concepts

An advantage of the cold plasma approach is that it closely resembles the standard treatment of electromagnetic wave propagation in dispersive media and thus the basic concepts of electrodynamics are readily available. In this section we briefly review some of these concepts that are central to plasma wave propagation.

### 4.1.1 Waves in linear media

We start the discussion of wave concepts in linear media, of which the vacuum ($\rho = 0$; $\mathbf{J} = 0$) is the simplest example. From Maxwell's equations we get

H.E.J. Koskinen, *Physics of Space Storms: From the Solar Surface to the Earth*,
Springer Praxis Books, DOI 10.1007/978-3-642-00319-6_4,

$$\nabla^2 \mathbf{H} - \frac{1}{c^2}\frac{\partial^2 \mathbf{H}}{\partial t^2} = 0 \tag{4.1}$$

$$\nabla^2 \mathbf{E} - \frac{1}{c^2}\frac{\partial^2 \mathbf{E}}{\partial t^2} = 0\,, \tag{4.2}$$

where we introduced $\mathbf{H} = \mathbf{B}/\mu_0$ for notational convenience.

The solutions of these equations are waves propagating with the speed of light. Consider a wave that propagates in the $(\pm)z$-direction of a Cartesian coordinate system $(x,y,z)$. The x-component of the wave electric field is

$$E_x(x,y,z,t) = g_1(x,y) f_1(z - ct) + g_2(x,y) f_2(z + ct)\,, \tag{4.3}$$

where $\nabla^2 g_1 = \nabla^2 g_2 = 0$. The most important special cases of these solutions are *plane waves* and *spherical waves*.

For a plane wave propagating in the $z$-direction $\partial/\partial x = \partial/\partial y = 0$ and $g_1$ and $g_2$ are constant. Consequently, there is a plane where $\mathbf{E}$ is constant. A plane wave can be represented by a sinusoidal function

$$E_x(z,t) = E_0 \cos(kz - \omega t)\,, \tag{4.4}$$

where $E_0$ is the *amplitude*, $\omega = 2\pi f$ the *angular frequency*, and $k = 2\pi/\lambda$ the *wave number*. The *phase speed* of the wave is $\omega/k = c$. In vector form we write

$$\mathbf{E}(\mathbf{r},t) = \mathbf{E}_0 \cos(\mathbf{k}\cdot\mathbf{r} - \omega t)\,, \tag{4.5}$$

where $\mathbf{k}$ is the *wave vector*.

Another important class of solutions to the wave equation are spherical waves, for which electric field is constant on the surface of an expanding sphere. For example, the field of a radiating electric dipole antenna far from the source is nearly spherical

$$\mathbf{E}(r,\theta,\phi,t) \approx \frac{a}{r} \sin\theta \cos(kr - \omega t)\,\mathbf{e}_\theta\,. \tag{4.6}$$

In space physics we often, but not always, assume that the source of the wave is so far from the observation site that a plane wave is a good local representation of the wave propagation.

Throughout this book we use the complex notation for plane waves with the following sign convention for the exponentials:

$$\mathbf{E} = \mathbf{E}_0 e^{i(\mathbf{k}\cdot\mathbf{r} - \omega t)}\;;\; \mathbf{B} = \mathbf{B}_0 e^{i(\mathbf{k}\cdot\mathbf{r} - \omega t)}\,. \tag{4.7}$$

If $\mathbf{E}_0$ and $\mathbf{B}_0$ are constant, the temporal and spatial dependencies are said to be *harmonic* and Maxwell's equations can be transformed to an algebraic form

$$\begin{aligned} i\mathbf{k}\cdot\mathbf{D} &= \rho \\ \mathbf{k}\cdot\mathbf{B} &= 0 \\ \mathbf{k}\times\mathbf{E} &= \omega\mathbf{B} \\ i\mathbf{k}\times\mathbf{H} &= \mathbf{J} - i\omega\mathbf{D}\,. \end{aligned} \tag{4.8}$$

Assume that $\rho = 0$, $\mathbf{J} = 0$, $\sigma = 0$ and $\varepsilon$ and $\mu$ are constant but not necessarily equal to $\varepsilon_0$ and $\mu_0$. The solution is modified by $c \to v = 1/\sqrt{\varepsilon\mu}$, i.e., the phase speed becomes different from the speed of light. $\omega$ and $\mathbf{k}$ are related through a *dispersion equation* or *dispersion relation*

$$k = \frac{\omega}{v} = \sqrt{\varepsilon\mu}\,\omega = \frac{n}{c}\,\omega\,, \tag{4.9}$$

where

$$n = \sqrt{\frac{\varepsilon\mu}{\varepsilon_0\mu_0}} \tag{4.10}$$

is the *refractive index* of the medium. The *phase velocity* of the wave is defined by

$$v_p = \frac{\omega}{k} \tag{4.11}$$

and the *group velocity* by

$$v_g = \frac{\partial\omega}{\partial k}\,. \tag{4.12}$$

In this case $v_g = v_p = c/n$ and both are independent of frequency and wave number, i.e., the medium is not dispersive.

As the wave number is the absolute value of the wave vector, the phase and group velocities are also vector quantities. We write the wave vector as $\mathbf{k} = k\mathbf{n}$, where $\mathbf{n}$ is the unit vector defining the *wave normal*. The wave normal is perpendicular to the surface of constant wave phase. The wave normal direction is the *direction of wave propagation* and it thus gives the direction of the phase velocity vector

$$\mathbf{v}_p = \frac{\omega}{k}\,\mathbf{n}\,. \tag{4.13}$$

In isotropic media the direction of wave propagation is the same as the direction of energy flux $\mathbf{S} = \frac{1}{2}\mathbf{E}\times\mathbf{H}^*$. In anisotropic media, e.g., in magnetized plasma, the electric field may have a component $\parallel \mathbf{k}$, implying that $\mathbf{S} \nparallel \mathbf{k}$. The "ray" of the wave may thus propagate in a different direction than $\mathbf{k}$. *Ray-tracing* is a method of following the ray in order to find the direction of energy and information propagation. The propagation velocity of the ray is the group velocity, i.e., the velocity of wave packets

$$\mathbf{v}_g = \frac{\partial\omega}{\partial\mathbf{k}}\,, \tag{4.14}$$

i.e., the gradient of frequency in the $\mathbf{k}$-space.

The angle between the wave and ray propagation can be calculated by letting $\theta$ be the angle between background magnetic field $\mathbf{B}$ and $\mathbf{k}$, and the frequency $\omega$ a function of $k$ and $\theta$. The group velocity is given by

$$\mathbf{v}_g = \frac{\partial\omega}{\partial\mathbf{k}} = \left.\frac{\partial\omega}{\partial k}\right|_\theta \mathbf{e}_k + \left.\frac{1}{k}\frac{\partial\omega}{\partial\theta}\right|_k \mathbf{e}_\theta\,. \tag{4.15}$$

Denoting the angle between $\mathbf{v}_g$ and $\mathbf{v}_p$ by $\delta$ we find that

$$\tan\delta = -\frac{1}{k}\frac{\partial k}{\partial\theta}\bigg|_{\omega} . \tag{4.16}$$

As an example of a dispersive medium, consider a conductive medium whose $\varepsilon$, $\mu$, and $\sigma$ are non-zero constants and $\rho = 0$. Maxwell's equations and Ohm's law ($\mathbf{J} = \sigma\mathbf{E}$) lead to

$$\begin{aligned} \nabla\times\mathbf{E} &= -\frac{\partial\mathbf{B}}{\partial t} \\ \nabla\times\mathbf{B} &= \mu\sigma\mathbf{E} + \mu\varepsilon\frac{\partial\mathbf{E}}{\partial t} \end{aligned} \tag{4.17}$$

$\Rightarrow$

$$\nabla^2\mathbf{E} - \mu\sigma\frac{\partial\mathbf{E}}{\partial t} - \mu\varepsilon\frac{\partial^2\mathbf{E}}{\partial t^2} = 0 . \tag{4.18}$$

This is equation is known as the *telegraph equation*. It is a standard example of how partial differential equations are solved using Fourier transforms. In the plane wave approximation (4.18) is easy to solve in the $(\omega,\mathbf{k})$-space, where Maxwell's equations read

$$\begin{aligned} \mathbf{k}\cdot\mathbf{E} &= 0 \\ \mathbf{k}\cdot\mathbf{H} &= 0 \\ \mathbf{k}\times\mathbf{E} &= \omega\mu\mathbf{H} \\ i\mathbf{k}\times\mathbf{H} &= (\sigma - i\omega\varepsilon)\mathbf{E} . \end{aligned} \tag{4.19}$$

Clearly $\mathbf{k}\perp\mathbf{E}$, $\mathbf{k}\perp\mathbf{H}$, and $\mathbf{E}\perp\mathbf{H}$. Such a wave is called *transverse*. In plasmas also *longitudinal* ($\mathbf{k}\parallel\mathbf{E}$) waves may propagate, e.g, the electrostatic waves discussed in Chap. 5. Selecting the coordinates as $\mathbf{k}\parallel\mathbf{e}_z$, $\mathbf{E}\parallel\mathbf{e}_x$, and $\mathbf{H}\parallel\mathbf{e}_y$, we get

$$\begin{aligned} kE_x &= \omega\mu H_y \\ ikH_y &= -(\sigma - i\omega\varepsilon)E_x . \end{aligned} \tag{4.20}$$

From these we get the dispersion equation

$$k^2 = \varepsilon\mu\omega^2 + i\sigma\mu\omega . \tag{4.21}$$

Denoting $k = |k|\exp(i\alpha)$ we find

$$|k| = \sqrt{\mu\omega\sqrt{\varepsilon^2\omega^2+\sigma^2}} \tag{4.22}$$

$$\alpha = \frac{1}{2}\arctan(\frac{\sigma}{\varepsilon\omega}) . \tag{4.23}$$

Inserting these into the expression for $\mathbf{E}$

$$\mathbf{E} = E_0\mathbf{e}_x\exp[i(|k|(\cos\alpha)z - \omega t)]\exp[-|k|(\sin\alpha)z] \tag{4.24}$$

we have found the plane wave solution for Maxwell's equations in this particular medium. The physical choice of $\alpha$ is given by $\sin\alpha > 0$, i.e., the wave is *damped* when it propagates in the medium, i.e., $e^{-|k|(\sin\alpha)z} \to 0$ with increasing $z$.

Now the phase velocity is

$$v_p = \frac{\omega}{\mathrm{Re}(k)} = \frac{\omega}{|k|\cos\alpha} . \tag{4.25}$$

The distance where the wave is damped by a factor of $e$ is called the *skin depth* of the medium

$$\delta = \frac{1}{\mathrm{Im}(k)} = \frac{1}{|k|\sin\alpha} . \tag{4.26}$$

The *wave impedance* is defined by

$$Z = \frac{E_x}{H_y} = \frac{\mu\omega}{k} = \sqrt{\frac{\mu\omega}{\sqrt{\varepsilon^2\omega^2+\sigma^2}}}\exp\left[-\frac{i}{2}\arctan\left(\frac{\sigma}{\varepsilon\omega}\right)\right] , \tag{4.27}$$

where the argument of the exponential function describes the phase delay between **E** and **H**. The SI unit of impedance is the ohm ($\Omega$).

***Examples***

**Good conductor**: $\sigma >> \varepsilon\omega \Rightarrow \alpha = 45°$; $\delta = \sqrt{\dfrac{2}{\mu\sigma\omega}}$.

$v_p = \delta\omega\tan\alpha = \delta\omega$

For copper (Cu): $\begin{cases} f = 50\ \mathrm{Hz} & \delta = 1\ \mathrm{cm} \quad v_p = 3\ \mathrm{m/s} \\ f = 50\ \mathrm{MHz} & \delta = 10\ \mu\mathrm{m}\ \ v_p = 3\times 10^3\ \mathrm{m/s} \end{cases}$

$Z = \sqrt{\dfrac{\mu\omega}{\sigma}}e^{-i\pi/4} \Rightarrow 45°$ phase shift between **E** and **H**.

**Non-conductive medium**: $\sigma = 0$, $\varepsilon > 0$, $\mu = \mu_0 \Rightarrow \alpha = 0$, i.e., the wave is not damped.

$Z = \sqrt{\dfrac{\mu_0}{\varepsilon}} \equiv Z_0\sqrt{\dfrac{\varepsilon_0}{\varepsilon}}$,

$Z_0$ is called *vacuum impedance*: $\sqrt{\dfrac{\mu_0}{\varepsilon_0}} = 376.73\ \Omega$.

Air is a good vacuum for high-frequency electromagnetic waves; plasma is not when the wave frequency is in the vicinity of plasma or gyro frequencies of the plasma particles.

### 4.1.2 Wave polarization

Polarization is an important property of electromagnetic waves. We use definitions of the right- and left-handedness following the modern plasma literature:

> The wave vector of a right-hand polarized wave, propagating along the magnetic field, rotates in the same sense as an electron.

However, wave polarization must also be defined independently of the background magnetic field. Let a plane wave propagate in the $z$-direction. Consider the plane $z = 0$ and denote $\rho = E_y/E_x = -H_x/H_y$. In general $\rho = |\rho|e^{i\alpha}$ is a complex number.

1. If $\rho$ is a real number, $E_y$ and $E_x$ are in the same phase and the direction of **E** is $(1,\rho,0)$ (if $\rho = \infty$, **E** points along the $y$-axis). The wave is *linearly polarized.*
2. If $\rho = +i$, the phase shift between $E_y$ and $E_x$ is $\alpha = \pi/2$. Looking along the $+z$-axis the vector rotates *clockwise*. This is the *right-hand circularly polarized wave*. In optics this is called the left-hand wave, sometimes it is said to have *positive helicity*. The wave electric field is
$$\mathbf{E} = E_0(\mathbf{e}_x + i\mathbf{e}_y)\, e^{i(kz-\omega t)} . \tag{4.28}$$
3. If $\rho = -i$, $\alpha = -\pi/2$. The wave vector rotates anti-clockwise and the wave is *left-hand circularly polarized* (*negative helicity*). The electric field is
$$\mathbf{E} = E_0(\mathbf{e}_x - i\mathbf{e}_y)\, e^{i(kz-\omega t)} . \tag{4.29}$$
4. The linear and circular polarizations are special cases of *elliptical polarization*, for which $\rho$ ia a complex number.

All polarization states of a plane wave can be constructed as a linear superposition of right-hand and left-hand circularly polarized waves, or of two linearly polarized waves with different planes of polarization, by selecting appropriate amplitudes and phases of the basic polarization components.

### 4.1.3 Reflection and refraction

When waves cross boundaries between different media or propagate in an inhomogeneous medium, they are reflected and refracted. Figure 4.1 defines our notation. The incident wave ($i$) comes from medium 1 and hits the boundary between media 1 and 2.

The properties of the reflected ($r$) and refracted (transmitted, $t$) waves depend on the polarization. For simplicity, consider linear polarization only. Let the electric field $\mathbf{E}_i$ be in the plane of incidence ($xz$-plane) and $\mathbf{H}_i$ perpendicular to this plane. This polarization is called *vertical*. In the opposite case the polarization is *horizontal*. An arbitrary polarization is a linear combination of these two polarization states. Let the medium be such that the polarization state is conserved. If the medium were *birefringent*, the left- and right-hand circularly polarized waves would behave differently. As the linear polarization can be expressed as a sum of left- and right-hand polarized waves, the birefringence results in the rotation of the polarization direction, known as *Faraday rotation*.

Figure 4.1 illustrates the vertical polarization. Now

$$\begin{aligned}
\mathbf{k}_i &= k_i(\sin\theta_i, 0, \cos\theta_i) \\
\mathbf{k}_r &= k_r(\sin\theta_r, 0, -\cos\theta_r)
\end{aligned} \tag{4.30}$$

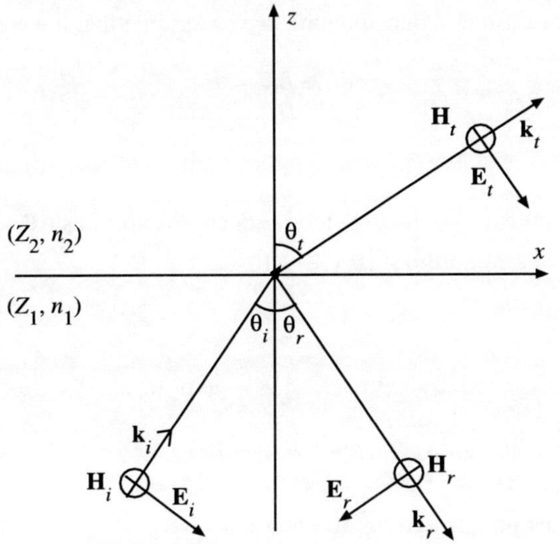


**Fig. 4.1** Reflection and refraction of a vertically polarized wave at the boundary between two different linear media with impedances and refractive indices $(Z_1, n_1)$, $(Z_2, n_2)$.

$$\mathbf{k}_t = k_t(\sin\theta_t, 0, \cos\theta_t)\,.$$

The boundary conditions at the surface $z = 0$ imply that the waves $(i, r, t)$ must be in the same phase at the same time. Thus $\mathbf{k}_i$, $\mathbf{k}_r$, and $\mathbf{k}_t$ are in the same plane. A straightforward calculation gives the fields

$$\begin{aligned}
\mathbf{E}_i &= E_i(\cos\theta_i, 0, -\sin\theta_i)\exp[i(k_i(\sin\theta_i x + \cos\theta_i z) - \omega t)] \\
\mathbf{H}_i &= \frac{E_i}{Z_1}(0,1,0)\exp[i(k_i(\sin\theta_i x + \cos\theta_i z) - \omega t)] \\
\mathbf{E}_r &= E_r(-\cos\theta_r, 0, -\sin\theta_r)\exp[i(k_r(\sin\theta_r x - \cos\theta_r z) - \omega t)] \\
\mathbf{H}_r &= \frac{E_r}{Z_1}(0,1,0)\exp[i(k_r(\sin\theta_r x - \cos\theta_r z) - \omega t)] \\
\mathbf{E}_t &= E_t(\cos\theta_t, 0, -\sin\theta_t)\exp[i(k_t(\sin\theta_t x + \cos\theta_t z) - \omega t)] \\
\mathbf{H}_t &= \frac{E_t}{Z_2}(0,1,0)\exp[i(k_t(\sin\theta_t x + \cos\theta_t z) - \omega t)]\,,
\end{aligned} \tag{4.31}$$

where the direction of the vectors is given by the triplets after the amplitude of the vector.

The boundary conditions derived from Maxwell's equations are:

$$\begin{aligned}
\mathbf{n}_{12} \times (\mathbf{E}_1 - \mathbf{E}_2) &= 0 \\
\mathbf{n}_{12} \times (\mathbf{H}_1 - \mathbf{H}_2) &= \mathbf{K}\,,
\end{aligned}$$

where **K** is a surface current induced by the wave. Assume that the current is zero. Then

$$\begin{aligned} E_{ix} + E_{rx} &= E_{tx} \\ H_{iy} + H_{ry} &= H_{ty} \end{aligned} \tag{4.32}$$

$\Rightarrow$

$$\begin{aligned} &E_i \cos\theta_i \exp[i(k_i \sin\theta_i x - \omega t)] - E_r \cos\theta_r \exp[i(k_r \sin\theta_r x - \omega t)] \\ &= E_t \cos\theta_t \exp[i(k_t \sin\theta_t x - \omega t)] \end{aligned} \tag{4.33}$$

and

$$\begin{aligned} &\frac{E_i}{Z_1} \exp[i(k_i \sin\theta_i x - \omega t)] - \frac{E_r}{Z_1} \exp[i(k_r \sin\theta_r x - \omega t)] \\ &= \frac{E_t}{Z_2} \exp[i(k_t \sin\theta_t x - \omega t)] \,. \end{aligned} \tag{4.34}$$

These equations must be satisfied for all $t$ and $x \Rightarrow$

$$\omega_i = \omega_r = \omega_t = \omega \tag{4.35}$$

$$k_i \sin\theta_i = k_r \sin\theta_r = k_t \sin\theta_t \,. \tag{4.36}$$

The incident and reflected waves propagate in the same medium $(n_1) \Rightarrow$

$$\frac{c}{\omega} k_i = \frac{c}{\omega} k_r \Rightarrow k_i = k_r \Rightarrow \theta_i = \theta_r \tag{4.37}$$

In addition, we find *Snell's law* for the angle of refraction

$$\sin\theta_t = \frac{k_i}{k_t} \sin\theta_i = \frac{n_1}{n_2} \sin\theta_i \,. \tag{4.38}$$

Now we can calculate the *reflection coefficient* for vertical polarization

$$R_\parallel = \frac{E_r}{E_i} = \frac{Z_1 \cos\theta_i - Z_2 \cos\theta_t}{Z_1 \cos\theta_i + Z_2 \cos\theta_t} \,. \tag{4.39}$$

Often $\mu_1 = \mu_2 (= \mu_0)$. Then $Z_1/Z_2 = n_2/n_1$, which leads to *Fresnel's formulas*

$$R_\parallel = \frac{E_r}{E_i} = \frac{n_2 \cos\theta_i - n_1 \cos\theta_t}{n_2 \cos\theta_i + n_1 \cos\theta_t} \tag{4.40}$$

$$T_\parallel = \frac{E_t}{E_i} = \frac{2 n_1 \cos\theta_i}{n_2 \cos\theta_i + n_1 \cos\theta_t} \,. \tag{4.41}$$

$T_\parallel$ is the *transmission coefficient* for vertical polarization. These equations are often given in the form where $\theta_t$ is eliminated using Snell's law. Physically Fresnel's formulas express energy conservation at the reflecting boundary.

In the same way we find Fresnel's formulas for horizontal polarization

$$R_\perp = \frac{n_1 \cos\theta_i - n_2 \cos\theta_t}{n_1 \cos\theta_i + n_2 \cos\theta_t} \tag{4.42}$$

$$T_\perp = \frac{2n_1 \cos\theta_i}{n_1 \cos\theta_i + n_2 \cos\theta_t} \,. \tag{4.43}$$

***Example: Total reflection and transmission***

Consider the Earth's ionosphere as an isotropic non-conductive medium. This is a reasonably good approximation for radio waves with sufficiently high frequency ($\omega \gg \omega_{ce}$ , $\omega \gg \nu_{coll}$ , $\omega \gg \omega_p$).

In the air below the ionosphere: $\sigma = 0,\ \mu = \mu_0,\ n_1 = 1$ .

In the ionosphere, (see 4.48): $n_2 = \dfrac{ck}{\omega} = \sqrt{\dfrac{\varepsilon}{\varepsilon_0}} = \sqrt{1 - \dfrac{\omega_p^2}{\omega^2}} < 1$ .

Now Fresnel's formulas imply: $|R_\perp| \geq |R_\parallel|$, for all $\theta_i$. Thus the horizontal polarization has a larger reflection coefficient and is more efficient for radio wave communication via the ionosphere.

For sufficiently large $\theta_i$ we find the *total reflection*: $\sin\theta_t = (n_1/n_2)\sin\theta_i \geq 1 \Rightarrow |R_\perp| = |R_\parallel| = 1$ .

For a certain angle of incidence, known a the *Brewster angle* ($\theta_B$), the vertically polarized wave is transmitted completely ($R_\parallel = 0$). Note that the horizontally polarized wave is always partially reflected.

## 4.2 Radio Wave Propagation in the Ionosphere

As an example of wave propagation in a dispersive inhomogeneous medium we consider radio wave propagation in the ionosphere. It has considerable practical interest for the physics of space storms: radio waves can be used to probe the state of the ionosphere and, on the other hand, radio communication systems, including satellite navigation, are disturbed by the space storms.

### 4.2.1 Isotropic, lossless ionosphere

We begin with an assumption that the ionosphere is *isotropic* and neglect the Earth's magnetic field. This requires $\omega \gg \omega_{ce} \approx 10^7\,\mathrm{s}^{-1}$. Let the medium be *lossless*, i.e., neglect the effects of collisions; thus $\omega \gg \nu_{coll}$. These requirements are fulfilled at frequencies $f \gg \omega_{ce}/2\pi \approx 1.6\,\mathrm{MHz}$. We consider waves whose frequencies are so high that only electrons respond to the wave electric field, whereas ions form an immobile background. We need to determine the functions $\sigma$ and $\varepsilon$. Here $n$ refers to the refractive index and the electron density is denoted by $n_e$.

Consider the problem again in the plane wave approximation. From the electron equation of motion we find

$$m_e \frac{d\mathbf{v}}{dt} = -i\omega m_e \mathbf{v} = -e\mathbf{E} \tag{4.44}$$

$\Rightarrow$

$$\mathbf{J} = -n_e e \mathbf{v} = \frac{\omega_{pe}^2}{\omega^2} i\omega\varepsilon_0 \mathbf{E} \tag{4.45}$$

$\Rightarrow$

$$\sigma = \frac{\omega_{pe}^2}{\omega^2} i\omega\varepsilon_0 \,. \tag{4.46}$$

Assume that, except for conductivity, the medium has the electromagnetic properties of a vacuum, i.e., $\varepsilon = \varepsilon_0$ and $\mu = \mu_0$. The Ampère–Maxwell law can now be written as

$$i\mathbf{k} \times \mathbf{H} = \frac{\omega_{pe}^2}{\omega^2} i\omega\varepsilon_0 \mathbf{E} - i\omega\varepsilon_0 \mathbf{E} = -i\omega \left(1 - \frac{\omega_{pe}^2}{\omega^2}\right) \varepsilon_0 \mathbf{E}. \tag{4.47}$$

Thus the medium *looks like* a dielectric with permittivity

$$\varepsilon = \left(1 - \frac{\omega_{pe}^2}{\omega^2}\right) \varepsilon_0 \,. \tag{4.48}$$

In plasma physics we often write $\omega_{pe}^2/\omega^2 \equiv X$. Now the refractive index is

$$n = \sqrt{1 - X} \,, \tag{4.49}$$

which is the dispersion equation and can also be written as

$$c = \frac{\omega}{k} \sqrt{1 - X} \,. \tag{4.50}$$

The phase and group velocities are

$$v_p = \frac{\omega}{k} = \frac{c}{\sqrt{1 - X}} \tag{4.51}$$

$$v_g = \frac{\partial \omega}{\partial k} = c\sqrt{1 - X} \,. \tag{4.52}$$

When $k$ increases (short wavelengths), the dispersion equation approaches that of an electromagnetic wave in free space $\omega = ck$. At long wavelengths the wave corresponds to plasma oscillation $\omega = \omega_{pe}$. If the frequency is smaller than the local plasma frequency ($X > 1$), the wave does not propagate. In the ionosphere the maximum electron densities are of the order of $10^{12}\,\mathrm{m}^{-3}$. Because $f_{pe}(\mathrm{Hz}) \equiv \omega_{pe}/2\pi \approx 9\sqrt{n_e(\mathrm{m}^{-3})}$, the maximum plasma frequency in the ionosphere is about 9 MHz.

Let us then then find out what happens to an electromagnetic wave pulse (wave packet) when it propagates vertically toward the ionosphere modeled in this way and becomes reflected. The pulse returns after time $T$. The height $h' = cT/2$ is called the *virtual reflection height*. In reality the wave packet moves with speed $v_g$ and

$$T = 2\int_0^h \frac{dz}{v_g}\,, \tag{4.53}$$

where $h$ is the *real reflection height*. That is the height where the group velocity becomes zero and we get

$$h' = c\int_0^h \frac{dz}{v_g} = \int_0^h \frac{dz}{\sqrt{1-X(z)}}\,. \tag{4.54}$$

If we know the density profile, we can compute the relation between $h'$ and $h$ for different frequencies.

The *ionosonde* is an instrument that is used to study the inverse problem. It transmits radio waves at different frequencies and detects the reflected signal. By measuring $h'$ for different frequencies we can attempt to find the frequency dependence of $h$, which would yield the density profile of the ionosphere. The integral for the virtual height can be solved analytically for sufficiently smooth profiles. The monotonic parts of the profile can be approximated by a piecewise linear function composed of pieces

$$\begin{aligned} n_e &= a(z-z_1) && \text{when } z > z_1 \\ n_e &= 0 && \text{when } z \le z_1\,. \end{aligned} \tag{4.55}$$

The real reflection takes place at the altitude where $\omega^2 = \omega_{pe}^2 \Rightarrow$

$$h = z_1 + \frac{\varepsilon_0 m}{ae^2}\omega^2\,, \tag{4.56}$$

and the virtual reflection at

$$h' = \int_0^h \frac{dz}{\sqrt{1 - \frac{a(z-z_1)e^2}{\varepsilon_0 m\omega^2}}} = z_1 + \frac{2\varepsilon_0 m}{ae^2}\omega^2\,. \tag{4.57}$$

**Train your brain**

Find the expression for the virtual reflection height for a parabolic density profile

$$\begin{aligned} n_e &= n_m\left[1-\left(\frac{z-z_m}{a}\right)^2\right] && \text{when } |z-z_m| < a \\ n_e &= 0 && \text{when } |z-z_m| \ge a\,, \end{aligned}$$

where the subscript $m$ denotes the peak density.

Oblique propagation is important in radio wave communication between two locations. Let $\theta_0$ be the angle between the vertical direction $z$ and $\mathbf{k}$ in the atmosphere and let $y$ denote the horizontal distance. For the wave packet we have $y = ct \sin\theta_0$. Vertical motion is found from the expression for the virtual height replacing $\omega \to \omega\cos\theta_0$

$$h'(t) = ct\cos\theta_0 = \int_0^z \frac{dz'}{\sqrt{1 - \dfrac{\omega_{pe}^2(z')}{\omega^2\cos^2\theta_0}}}\,, \tag{4.58}$$

where $z$ is the real height at time $t$. Eliminating $t$ we get

$$y = \sin\theta_0 \int_0^z \frac{dz'}{\sqrt{\cos^2\theta_0 - \dfrac{\omega_{pe}^2(z')}{\omega^2}}}\,. \tag{4.59}$$

This gives the *ray path*. In an isotropic medium the ray propagates to the direction of the wave normal.

### 4.2.2 Weakly inhomogeneous ionosphere

What happens to the wave when it approaches the reflection point? We assumed above that the reflection takes place when the vertical component of $\mathbf{v}_g$ is zero, i.e., $n = \sin\theta_0$. On the other hand, we know that some reflection always takes place at the interface between media of different refractive indices, except for vertical polarization at the Brewster angle.

Consider a frequency twice the plasma frequency $\omega = 2\,\omega_{pe}$. Now the wave should get through the ionosphere. Let the incident angle be $\theta_i = 0$. The refractive index is

$$n = \sqrt{1 - \omega_{pe}^2/\omega^2}\,.$$

This gives a reflection coefficient $R = 0.07$ and the reflection should be easily observable. However, it is not, and the prediction that the wave gets through the ionosphere is correct.

To solve this apparent paradox construct a simple model for the ionosphere that consists of thin layers of thickness $\triangle z$ (Fig. 4.2). Let $n_e$ increase, and thus $n$ decrease, upwards. Assume, for simplicity, horizontal polarization and $\theta_i = \theta_t = 0$. At each layer

$$R = \frac{n_1 - n_2}{n_1 + n_2} \approx \frac{\triangle n}{2n}\,. \tag{4.60}$$

The *relative phase* of the signals reflected from different layers turns out to be the key to the solution. Let $E_0$ be the electric field of the incident wave and denote the field after reflection by $E = E(z)$. From each layer an amount of $(\triangle n/2n)E(z)$ is reflected and $(1 - (\triangle n/2n))E(z)$ refracted. In this model $\triangle n < 0$ and thus the electric field of the refracted wave increases. We have, however, not found a perpetuum mobile that would create wave energy from nothing. The wave propagates toward an increasing impedance

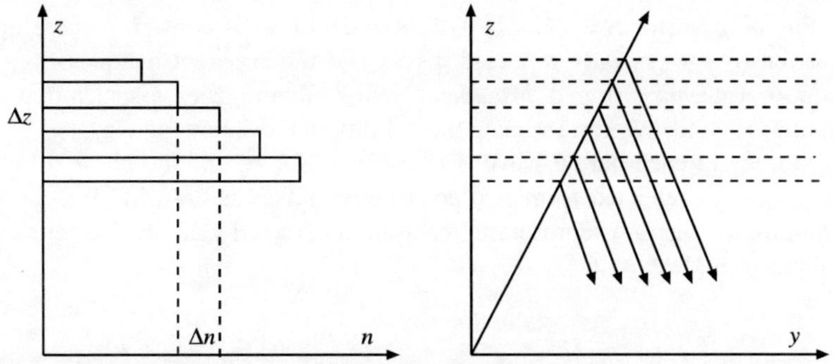


**Fig. 4.2** A model of layered ionosphere. That the electron density increases upward and thus the index of refraction decreases from 1 at the bottom of the ionoshere. The figure is drawn for oblique incidence to be more illustrative, whereas the calculation is simpler for $\theta_i = 0$.

($Z = E/H$) and for the wave magnetic field we find

$$\frac{H_t}{H_i} = T\frac{Z_1}{Z_2} = T\frac{n_2}{n_1} \approx 1 + \frac{\Delta n}{2n} < 1\,. \tag{4.61}$$

At the limit $\triangle z \rightarrow 0$

$$E + dE = \left(1 - \frac{dn}{2n}\right)E \tag{4.62}$$

$\Rightarrow$

$$E = \frac{E_0}{\sqrt{n}}\,;\; H = \sqrt{n}H_0\,. \tag{4.63}$$

At each layer the phase of the wave is shifted by $kdz = nk_0dz$. At the altitude $z$ the accumulated retardation is given by the *phase integral*

$$\int_0^z n(z')k_0\,dz'\,.$$

This method is called the *WKB approximation*. It is best known from quantum mechanics where it was used independently by Wentzel, Kramers, and Brillouin in 1926. The WKB method is useful in studies of wave propagation in (weakly) inhomogeneous media, and not restricted to Schrödinger equation or plasmas. In fact, Jeffreys had already introduced it in his study of linear second order differential equations in 1923 and some authors prefer to call the method the JWKB or WKBJ approximation. The fields in this approximation are given by

$$E_x = \frac{E_0}{\sqrt{n}}\exp[i(k_0\int_0^z n\,dz' - \omega t)] \tag{4.64}$$

$$H_y = \sqrt{n}H_0\exp[i(k_0\int_0^z n\,dz' - \omega t)]\,. \tag{4.65}$$

Now the Poynting vector $\mathbf{S} = (\mathbf{E} \times \mathbf{H}^*)/2 = (E_0 H_0/2)\,\mathbf{e}_z$ is constant. Therefore no energy is carried by the partially reflected waves and the apparent inconsistency with the non-reflection of waves is solved, *provided that* the inhomogeneity is such that the WKB approximation is valid. We can estimate the validity in the following way.

The amplitude of each partial wave is $(\triangle n/2n)(E_0/\sqrt{n})$ and the phase difference between partial waves reflected from two consecutive layers is $2nk_0\triangle z$. We can construct a phase–amplitude diagram representing each partial wave by an arc element $\triangle s$ and an associated phase angle $\triangle \phi$

$$\triangle s = \frac{\triangle n}{2n}\frac{E_0}{\sqrt{n}} \;\; ; \;\; \triangle \phi = 2nk_0 \triangle z \,.$$

Add the arc elements graphically by turning them with respect to each other by the phase angle. Let $\triangle z \to dz$, $\triangle n \to dn$, $\triangle s \to ds$, $\triangle \phi \to d\phi$. If no reflection were to take place, the result would be a circle. However, the resulting curve is a spiral with an increasing radius

$$\triangle r = \lim \frac{\triangle s}{\triangle \phi} = \frac{ds}{d\phi} = \frac{E_0}{4n^{5/2}k_0}\frac{dn}{dz} \,. \tag{4.66}$$

**Train your brain** by drawing the phase–amplitude diagram described above.

The amplitude of the reflected wave is of the order of $\triangle r$ and can thus be neglected if

$$\frac{1}{4n^{5/2}k_0}\left|\frac{dn}{dz}\right| \ll 1 \,. \tag{4.67}$$

Thus the WKB approximation *is not good* if $k_0$ is small, i.e., the wavelength is long compared to the gradient scale length, or if $n \approx 0$, i.e., very close to the point where the wave actually is reflected. Note that local density gradients can also reflect the waves in the case of a smooth background profile.

The WKB approximation can also be used above the reflection region. There $n^2 < 0$, i.e., $n$ is imaginary. The amplitude is

$$E_x = \frac{E_0}{\sqrt{n}} \exp(-i\omega t)\exp(-k|n|z) \; ; \; H_y = i|n|E_x \,. \tag{4.68}$$

This solution is overdamped and the wave is said to be *evanescent*.

When approaching the reflection point, $n \to 0$ and the WKB approximation breaks down. The problem is analytically tractable if the density profile can be assumed to be linear in the vicinity of the reflection point. As the region where the WKB approximation fails is narrow, this is a good assumption if the reflection does not take place very close to a local density maximum. In the latter case a parabolic profile has to be used.

Close to the reflection point the wave is not a plane wave because the spatial dependence is not harmonic. Write Maxwell's equations as

$$\nabla \times \mathbf{E} = -\frac{\partial \mathbf{B}}{\partial t} \Rightarrow \frac{dE_x}{dz} = \mu_0 i\omega H_y$$
$$\nabla \times \mathbf{H} = \mathbf{J} + \frac{\partial \mathbf{D}}{\partial t} \Rightarrow \frac{dH_y}{dz} = n^2 i\omega\varepsilon_0 E_x$$

$\Rightarrow$

$$\frac{d^2E_x}{dz^2} + k_0^2 n^2 E_x = 0$$

$\Rightarrow$

$$\frac{d^2E_x}{dz^2} + k_0^2\left(\frac{z_0 - z}{L}\right) E_x = 0\,, \tag{4.69}$$

where the linear profile was assumed at the last step. With a change of the variable this can be transformed to *Airy's differential equation*

$$\frac{d^2E_x}{d\zeta^2} - \zeta E_x = 0\,, \tag{4.70}$$

whose solutions are expressed in terms of the *Airy integrals* Ai and Bi.

$$E_x(\zeta) = C_1\mathrm{Ai}(\zeta) + C_2\mathrm{Bi}(\zeta)\,. \tag{4.71}$$

Asymptotic expansions for Ai and Bi above the reflection point ($\zeta > 0$) are

$$\mathrm{Ai}(\zeta) \approx \frac{1}{2\sqrt{\pi}}\zeta^{-1/4}\exp(-\frac{2}{3}\zeta^{3/2}) \xrightarrow{\zeta\to\infty} 0$$
$$\mathrm{Bi}(\zeta) \approx \frac{1}{\sqrt{\pi}}\zeta^{-1/4}\exp(\frac{2}{3}\zeta^{3/2}) \xrightarrow{\zeta\to\infty} \infty\,.$$

**Feed your brain**

Look up from some mathematical handbook or from the internet the Airy integrals Ai and Bi and sketch their graphs.

Because the wave must vanish above the reflection point, $C_2$ must be zero. Thus the electric field has the same form as Ai. Approaching the reflection point from the negative side ($\zeta \to 0^-$), its amplitude and period increase and above the reflection point the field rapidly approaches 0. An integral form of Ai is

$$\mathrm{Ai}(\zeta) = \frac{1}{\pi}\int_0^\infty \cos\left(\zeta s + \frac{s^3}{3}\right) ds\,. \tag{4.72}$$

The coefficient $C_1$ is more difficult to determine. For large negative $\zeta$ the solution must join the WKB solution. There are some technical difficulties in finding the asymptotic behavior of Ai for negative argument (roots of negative numbers). A detailed treatment can be found in Budden [1985]. The result is

$$\mathrm{Ai}(\zeta) \approx \frac{1}{2\sqrt{\pi}} \zeta^{-1/4} \left( \exp(-\frac{2}{3}\zeta^{3/2}) + i \exp(\frac{2}{3}\zeta^{3/2}) \right) . \tag{4.73}$$

Matching this with the WKB solution we get

$$C_1 = 2\sqrt{\pi} E_0 (k_0 L)^{1/6} .$$

Finally the electric field is given by

$$\begin{aligned} E_x &= \frac{2E_0}{\sqrt{n}} \cos\left( k_0 \int_z^{z_0} n\,dz' + \frac{\pi}{4} \right) \exp(-i\omega t) \\ &= \frac{E_0}{\sqrt{n}} \left\{ \exp\left[ \frac{i\pi}{4} + i\left( k_0 \int_z^{z_0} n\,dz' - \omega t \right) \right] \right. \\ &\quad \left. + \exp\left[ \frac{-i\pi}{4} + i\left( -k_0 \int_z^{z_0} n\,dz' - \omega t \right) \right] \right\} . \end{aligned} \tag{4.74}$$

This is a sum of upward- and downward-propagating WKB solutions. The phase shift between them ($\pi/2$) comes from the non-WKB region, and it would be quite difficult to guess without doing the actual calculation. This introduces a factor $i$ into the reflection coefficient

$$R = i \exp\left( 2ik_0 \int_z^{z_0} n\,dz' \right) . \tag{4.75}$$

The wave electric field in the reflection region is

$$E_x = 2\sqrt{\pi} E_0 (k_0 L)^{1/6} \mathrm{Ai}(\zeta) \exp(-i\omega t) . \tag{4.76}$$

From this we can estimate how much the field differs from $E_0$. Max[Ai] $\approx 0.55$. Assuming $f = 5\,\mathrm{MHz} \Rightarrow k_0 \approx 0.1\,\mathrm{m}^{-1}$ and let $L \approx 100\,\mathrm{km}$. This gives $E_{x,max} \approx 9\,E_0$. This can be compared with a perfect mirror, for which $E_{x,max} = 2\,E_0$. The wavelength grows in turn by a factor of 14. If the incident wave is sufficiently strong, it can couple to the oscillation modes of the plasma. These may be damped by the plasma particles, resulting in heating of the plasma.

The solution is straightforward to generalize to oblique propagation by substitution $n^2 \rightarrow q^2 = n^2 - \sin^2\theta_i$ .

### 4.2.3 Inclusion of collisions

The interparticle collisions must sometimes be taken into account in radio wave propagation problems, which is very difficult to do analytically. For simplicity, we consider only the average collision frequency $\nu$. This introduces a frictional term to the equation of motion

$$m\frac{d\mathbf{v}}{dt} = -e\mathbf{E} - m\nu\mathbf{v} . \tag{4.77}$$

Assuming again harmonic time dependence we get

$$\mathbf{v} = \frac{e\mathbf{E}_0 \exp(-i\omega t)}{m(i\omega - \nu)} . \quad (4.78)$$

The permittivity and the dispersion equation are modified as

$$\varepsilon = \left(1 - \frac{\omega_{pe}^2}{\omega^2(1 + i\nu/\omega)}\right) \varepsilon_0 \quad (4.79)$$

$$k^2 = \mu_0 \varepsilon_0 \omega^2 \left(1 - \frac{\omega_{pe}^2}{\omega^2(1 + i\nu/\omega)}\right) . \quad (4.80)$$

If we solve $\omega$ from this equation, we see that the collisions have introduced a negative imaginary part to the frequency and the waves are damped. The collision frequency is often denoted by $Z = \nu/\omega$. Now the refractive index is complex

$$n = \sqrt{1 - \frac{X}{1 + iZ}} . \quad (4.81)$$

The WKB solution becomes somewhat different from the non-collisional case. The collisions damp the waves, i.e., energy is lost, and this contributes to the phase shift.

A more complete treatment must start from the Boltzmann equation with an appropriate collision model.

### 4.2.4 Inclusion of the magnetic field

Above the frequency was assumed to be much larger than the electron gyro frequency. In the polar ionosphere $f_{ce} \approx 1.4$ MHz, and in practical applications the unmagnetized theory can be applied only for $f > 5$ MHz. The magnetic field makes the plasma anisotropic and plasma becomes birefringent. We do not have any reason to discuss the details of the rather tedious derivation of the dispersion equation resulting from inclusion of both collisions and the magnetic field but give the basic equations of this *magnetoionic theory* for completeness.

Introduce a new variable $Y = \omega_{ce}/\omega$. Select again $\mathbf{k} \parallel \mathbf{e}_z$ and denote the angle between $\mathbf{B}_0$ and $\mathbf{k}$ by $\psi$. The magnetoionic theory gives the expressions for the polarization $\rho$ and the refractive index $n$

$$\rho = \frac{1}{2}\left(-i\frac{Y \sin^2\psi/\cos\psi}{1 - X - iZ} \pm 2i\sqrt{1 + \frac{Y^2 \sin^4\psi/\cos^2\psi}{4(1 - X + iZ)}}\right) \quad (4.82)$$

$$n^2 = 1 - \frac{X}{1 + iZ - \dfrac{Y^2 \sin^2\psi}{2(1 - X + iZ)} \mp \sqrt{Y^2 \cos^2\psi + \dfrac{Y^4 \sin^4\psi}{4(1 - X + iZ)}}} \quad (4.83)$$

These equations are called *Appleton–Hartree equations*. They have two physically meaningful pairs of solutions $(\rho, n^2)$, corresponding to two selections of signs of the square roots: + and –, or – and +. They are the ordinary (O) and extraordinary (X) modes discussed using a more transparent formalism in the next section.

## 4.3 General Treatment of Cold Plasma Waves

In this section we present the general formalism for waves in magnetized plasma in the cold plasma approximation. Recall that "cold" means here the assumption of the characteristic velocities of the waves being much faster than the thermal velocity of the plasma $\sqrt{2k_BT/m}$. In this approximation thermal effects can be neglected.

### 4.3.1 Dispersion equation for cold plasma waves

To derive the general dispersion equation in a cold plasma we start from Maxwell's equations and Ohm's law where $\sigma$ may be a tensor. In the plane wave approximation we obtain the wave equation

$$\mathbf{k} \times (\mathbf{k} \times \mathbf{E}) + \frac{\omega^2}{c^2} \mathscr{K} \cdot \mathbf{E} = 0\,, \tag{4.84}$$

where

$$\mathscr{K} = \mathscr{I} + \frac{i}{\omega\varepsilon_0}\sigma \tag{4.85}$$

is the *dielectric tensor* and $\mathscr{I}$ the unit tensor. In case of no background fields ($\mathbf{E}_0 = \mathbf{B}_0 = 0$) the dielectric tensor reduces to the already familiar scalar dielectric function

$$K = 1 - \frac{\omega_{pe}^2}{\omega^2} \equiv n^2\,. \tag{4.86}$$

The dielectric tensor $\mathscr{K}$ is a dimensionless quantity expressing the relationship between the displacement and electric fields

$$\mathbf{D} = \varepsilon \cdot \mathbf{E} = \varepsilon_0 \mathscr{K} \cdot \mathbf{E}\,. \tag{4.87}$$

Now the wave equation has particular solutions

$$\begin{array}{ll} \mathbf{k} \parallel \mathbf{E} \Rightarrow \omega^2 = \omega_{pe}^2 & \text{plasma oscillation} \\ \mathbf{k} \perp \mathbf{E} \Rightarrow \omega^2 = k^2c^2 + \omega_{pe}^2 & \text{electromagnetic wave in plasma}\,. \end{array}$$

Include a homogeneous background magnetic field $\mathbf{B}_0$ and consider small perturbations $\mathbf{B}_1$ $(B_1 \ll B_0)$. The total plasma current is

$$\mathbf{J} = \sum_\alpha n_\alpha q_\alpha \mathbf{V}_\alpha\,. \tag{4.88}$$

Note that the assumption of cold plasma means that all particles (of species $\alpha$) are moving at their average velocity $\mathbf{V}_\alpha(\mathbf{r},t)$. Assuming that $\mathbf{V}_\alpha \propto \exp(-i\omega t)$ the first-order equation of motion is

$$-i\omega \mathbf{V}_\alpha = q_\alpha(\mathbf{E} + \mathbf{V}_\alpha \times \mathbf{B}_0)\,. \tag{4.89}$$

Let the background magnetic field be in the direction of the $z$-axis, i.e., $\mathbf{B}_0 \parallel \mathbf{e}_z$, treat the $xy$-plane as a complex plane and use the coordinate system defined by the base $\{\sqrt{1/2}(\mathbf{e}_x + i\mathbf{e}_y), \sqrt{1/2}(\mathbf{e}_x - i\mathbf{e}_y), \mathbf{e}_z\}$. Denote the components in this base by integers $d = \{-1, 1, 0\}$ and express the plasma and gyro frequencies as

$$X_\alpha = \frac{\omega_{p\alpha}^2}{\omega^2}\,,\; Y_\alpha = \frac{s_\alpha \omega_{c\alpha}}{\omega}\,. \tag{4.90}$$

Here $\omega_{c\alpha}$ is a positive quantity and the sign of the charge is given explicitly by $s_\alpha$. Now the components of the current are

$$J_{d,\alpha} = i\varepsilon_0\omega \frac{X_\alpha}{1 - dY_\alpha} E_d\,, \tag{4.91}$$

and the dielectric tensor is diagonal

$$\mathcal{K} = \begin{bmatrix} 1 - \sum_\alpha \frac{X_\alpha}{1 - Y_\alpha} & 0 & 0 \\ 0 & 1 - \sum_\alpha \frac{X_\alpha}{1 + Y_\alpha} & 0 \\ 0 & 0 & 1 - \sum_\alpha X_\alpha \end{bmatrix}\,. \tag{4.92}$$

It is customary to denote the components of the tensor by $R$, $L$, and $P$

$$R = 1 - \sum_\alpha \frac{\omega_{p\alpha}^2}{\omega^2}\left(\frac{\omega}{\omega + s_\alpha \omega_{c\alpha}}\right) \tag{4.93}$$

$$L = 1 - \sum_\alpha \frac{\omega_{p\alpha}^2}{\omega^2}\left(\frac{\omega}{\omega - s_\alpha \omega_{c\alpha}}\right) \tag{4.94}$$

$$P = 1 - \sum_\alpha \frac{\omega_{p\alpha}^2}{\omega^2}\,. \tag{4.95}$$

$R$ has a singularity when $\omega = \omega_{ce}$. The corresponding wave mode, the $R$ *mode*, can be in resonance with electrons. $R$ thus corresponds to the right-hand circularly polarized wave. Similarly the $L$ *mode* can be in resonance with positive ions and corresponds to the left-hand circularly polarized wave. $P$ corresponds to plasma oscillation, which is linearly polarized.

Transforming $\mathcal{K}$ back to the $\{x,y,z\}$-base we get

$$\mathcal{K} = \begin{bmatrix} S & -iD & 0 \\ iD & S & 0 \\ 0 & 0 & P \end{bmatrix}\,, \tag{4.96}$$

where $S = (R+L)/2$ and $D = (R-L)/2$.

The wave equation can be written in terms of the wave normal vector $\mathbf{n} = c\mathbf{k}/\omega$ as

$$\mathbf{n} \times (\mathbf{n} \times \mathbf{E}) + \mathcal{K} \cdot \mathbf{E} = 0 . \tag{4.97}$$

Note that in the following discussion $\mathbf{n}$ consequently refers to the wave normal vector and must not be mixed up with the unit normal vector elsewhere in the text! Recall that $\mathbf{B}_0$ is in the $z$-direction. Select the $x$-coordinate so that $\mathbf{n}$ is in the $xz$-plane and let $\theta$ be the angle between $\mathbf{n}$ and $\mathbf{B}_0$. Now the wave equation is

$$\begin{bmatrix} S - n^2\cos^2\theta & -iD & n^2\cos\theta\sin\theta \\ iD & S - n^2 & 0 \\ n^2\cos\theta\sin\theta & 0 & P - n^2\sin^2\theta \end{bmatrix} \begin{bmatrix} E_x \\ E_y \\ E_z \end{bmatrix} = 0 . \tag{4.98}$$

The non-trivial solutions of the wave equation are found from the dispersion equation

$$An^4 - Bn^2 + C = 0 , \tag{4.99}$$

where

$$\begin{aligned} A &= S\sin^2\theta + P\cos^2\theta \\ B &= RL\sin^2\theta + PS(1+\cos^2\theta) \\ C &= PRL . \end{aligned} \tag{4.100}$$

Solving $n$ would give a generalization of the magnetoionic theory. However, it is more instructive to study the dispersion equation for different angles $\theta$

$$\tan^2\theta = \frac{-P(n^2 - R)(n^2 - L)}{(Sn^2 - RL)(n^2 - P)} . \tag{4.101}$$

Now we can identify the wave modes in various directions. The modes propagating in the direction of the magnetic field ($\theta = 0$) and perpendicular to it ($\theta = \pi/2$) are called the *principal modes*

$$\begin{aligned} &\theta = 0 : \quad P = 0,\ n^2 = R,\ n^2 = L \\ &\theta = \pi/2 : n^2 = RL/S,\ n^2 = P . \end{aligned}$$

These modes have *cut-offs*

$$\begin{aligned} &n^2 \to 0 \ \ (v_p \to \infty,\ k \to 0,\ \lambda \to \infty) \\ &P = 0,\ R = 0,\ \text{or}\ L = 0 \end{aligned}$$

and *resonances*

$$\begin{aligned} &n^2 \to \infty \ \ (v_p \to 0,\ k \to \infty,\ \lambda \to 0) \\ &\tan^2\theta = -P/S\ (\text{provided}\ P \neq 0) . \end{aligned}$$

When the wave approaches a region where it has a cut-off ($n^2 \to 0$), it cannot propagate further and is reflected. At a resonance the wave energy is absorbed by the plasma.

### 4.3.2 Parallel propagation ($\theta = 0$)

The parallel propagating modes are the solutions of $P = 0$, $n^2 = R$, $n^2 = L$. The case $P = 0$ is the trivial plasma oscillation but the right- and left-hand polarized modes are important.

***Right-hand polarized mode (R)***

$$n_R^2 = R = 1 - \frac{\omega_{pi}^2}{\omega(\omega + \omega_{ci})} - \frac{\omega_{pe}^2}{\omega(\omega - \omega_{ce})} . \tag{4.102}$$

The resonance is with the electrons at the electron gyro frequency $\omega = \omega_{ce}$. The cut-off frequency is

$$\omega_R \approx \frac{\omega_{ce}}{2} \left[ 1 + \sqrt{1 + 4\omega_{pe}^2/\omega_{ce}^2} \right] . \tag{4.103}$$

At the limit of low plasma density this reduces to

$$\omega_R \approx \omega_{ce}(1 + \omega_{pe}^2/\omega_{ce}^2) \tag{4.104}$$

and at the limit of high density to

$$\omega_R \approx \omega_{pe} + \omega_{ce}/2 . \tag{4.105}$$

At low frequencies the mode approaches the Alfvén wave to be discussed in Chap. 6. At the limit of high frequency the wave is the electromagnetic wave in an unmagnetized plasma $\omega \to \infty$, $n^2 \to 1 - \omega_{pe}^2/\omega^2$.

***Left-hand polarized mode (L)***

$$n_L^2 = L = 1 - \frac{\omega_{pi}^2}{\omega(\omega - \omega_{ci})} - \frac{\omega_{pe}^2}{\omega(\omega + \omega_{ce})} . \tag{4.106}$$

The resonance is with ions $\omega = \omega_{ci}$. The cut-off frequency is at low density

$$\omega_L = \omega_{ci}(1 + \omega_{pi}^2/\omega_{ci}^2) \tag{4.107}$$

and at high density

$$\omega_L = \omega_{pe} - \omega_{ce}/2 . \tag{4.108}$$

The left-hand mode has a lower cut-off frequency than the right-hand mode. Both modes propagate at all frequencies above their cut-off frequency. At high frequencies both modes approach to the electromagnetic wave in free space ($\omega_{pe}^2/\omega^2 \to 0 \Rightarrow \omega \to ck$) (Fig. 4.3)

***Faraday rotation***

The Faraday rotation is a consequence of the different phase velocities of the left- and right-hand modes. Consider a linearly polarized signal and represent it as a sum of $R$ and

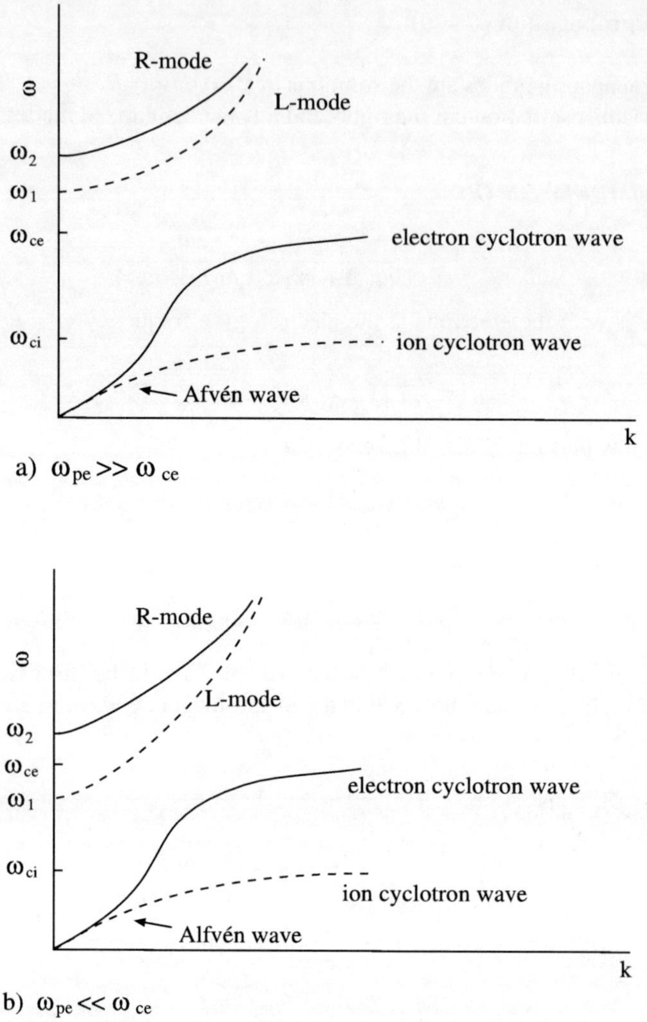


**Fig. 4.3** Parallel propagation for a) high plasma density ($\omega_{pe} \gg \omega_{ce}$) and for b) low plasma density ($\omega_{pe} \ll \omega_{ce}$). The continuous line is the $R$-mode and the dashed line the $L$-mode. Cut-offs are found where the dispersion curve meets the vertical axis ($k = 0$) and resonances are found at large $k$.

$L$ modes

$$\mathbf{E} = [\mathbf{e}_x(E_R e^{ik_R z} + E_L e^{ik_L z}) + i\mathbf{e}_y(E_R e^{ik_R z} - E_L e^{ik_L z})]e^{-i\omega t} \,. \tag{4.109}$$

The anisotropy introduced by the background magnetic field implies that the $R$ and $L$ components of the wave have $k_R \neq k_L$ for the same $\omega$ and

$$\frac{E_x}{E_y} = -i\,\frac{1 + (E_{xL}/E_{xR})\exp[i(k_L - k_R)z]}{1 - (E_{xL}/E_{xR})\exp[i(k_L - k_R)z]} \,. \tag{4.110}$$

Because the sum of the $R$ and $L$ modes is linear, $E_{xL} = E_{xR} \Rightarrow$

$$\frac{E_x}{E_y} = \cot\left(\frac{k_L - k_R}{2} z\right) . \tag{4.111}$$

This means that the plane of polarization rotates when the wave propagates through an anisotropic medium. The degree of rotation $\phi = (k_L - k_R)z/2$ depends on the plasma density and the magnetic field. In astrophysical observations the plasma and gyro frequencies are small compared to the observed electromagnetic signal. Thus the dispersion equations for $L$ and $R$ modes can be approximated as

$$k_{L,R} \approx \frac{\omega}{k}\left[1 - \frac{\omega_{pe}^2}{2\omega^2}\left(1 \pm \frac{\omega_{ce}}{\omega}\right)\right] . \tag{4.112}$$

The differential rotation of the polarization plane is

$$\frac{d\phi}{dz} = \frac{-\omega_{pe}^2 \omega_{ce}}{2c\,\omega^2} = \frac{-e^3}{2m_e^2 \varepsilon_0 c\,\omega^2} n_e B_0 . \tag{4.113}$$

The total rotation from the source to the observer at the distance $d$ is

$$\phi = \frac{-e^3}{2m_e^2 \varepsilon_0 c\,\omega^2} \int_0^d n_e \mathbf{B} \cdot d\mathbf{s} , \tag{4.114}$$

where the integral is taken along the path of the signal. In astrophysics the term *rotation measure* $(RM)$ is introduced by the formula

$$\phi = -RM\, f^{-2} . \tag{4.115}$$

Numerically

$$RM = 23.5 \int_0^d n_e \mathbf{B} \cdot d\mathbf{s} , \tag{4.116}$$

where $f$ is measured in Hz, $n_e$ in $\mathrm{cm}^{-3}$, $B$ in nT and $ds$ in m. Because the direction of rotation is determined modulo $\pi$, it has to be measured at several frequencies in order to resolve how many times the polarization plane has turned during propagation from the source to the receiver.

### *Whistler mode*

In addition to the Alfvén wave there is another important wave mode that propagates only in magnetized plasmas: the *whistler mode*. The $R$ mode has real solutions also in the frequency range between $\omega_{ci}$ and $\omega_{ce}$. If $\omega_{ci} \ll \omega \ll \omega_{ce}$ the dispersion equation can be approximated by

$$k = \frac{\omega_{pe}}{c}\sqrt{\frac{\omega}{\omega_{ce}}} \tag{4.117}$$

$\Rightarrow$

$$v_p = \frac{\omega}{k} = \frac{c\sqrt{\omega_{ce}}}{\omega_{pe}}\sqrt{\omega} \tag{4.118}$$

$$v_g = \frac{\partial \omega}{\partial k} = \frac{2c\sqrt{\omega_{ce}}}{\omega_{pe}}\sqrt{\omega}\,. \tag{4.119}$$

This dispersive mode was found during the First World War as descending, whistling tones heard on communication lines in the frequency band around 10 kHz. The correct explanation for these whistles was not found until 1953 when Storey realized that the waves originated as wide-band electric signals from lightning strokes. Part of the pulse is guided by the magnetic field as a whistler wave to the other hemisphere where it can be detected as a descending tone. The time of arrival depends on the frequency as

$$t(\omega) = \int \frac{ds}{v_g} = \int \frac{\omega_{pe}(s)}{2c\sqrt{\omega\omega_{ce}}}\,ds\,. \tag{4.120}$$

This explanation was not accepted immediately because it requires a higher plasma density in the plasmasphere than was thought to exist at that time.

### 4.3.3 Perpendicular propagation ($\theta = \pi/2$)

Modes propagating perpendicularly to the magnetic field are called, for historical reasons, *ordinary* and *extraordinary* modes. Unfortunately, their definitions are different in different fields of physics. Furthermore, there is nothing really extraordinary about the extraordinary mode.

#### *Ordinary mode (O)*

The ordinary $(O)$ mode is the mode whose index of refraction is

$$n_O^2 = P = 1 - \frac{\omega_{pi}^2}{\omega^2} - \frac{\omega_{pe}^2}{\omega^2} \approx 1 - \frac{\omega_{pe}^2}{\omega^2}\,. \tag{4.121}$$

This corresponds to the "ordinary" electromagnetic wave in isotropic plasma. Its electric field is in the direction of the background magnetic field ($\mathbf{E} \parallel \mathbf{B}_0$). For exactly perpendicular propagation the background magnetic field is not involved in the dispersion equation of the mode. It has a cut-off at $\omega = \omega_{pe}$.

#### *Extraordinary mode (X)*

For the extraordinary $(X)$ mode $n_X^2 = RL/S$. With the obvious approximation $\omega_{ce} \gg \omega_{ci}$ two *hybrid resonances* are found. The upper hybrid resonance is

$$\omega_{UH}^2 \approx \omega_{pe}^2 + \omega_{ce}^2 \tag{4.122}$$

and the lower

$$\omega_{LH}^2 \approx \frac{\omega_{ci}^2 + \omega_{pi}^2}{1 + (\omega_{pe}^2/\omega_{ce}^2)} \approx \omega_{ce}\omega_{ci} \left( \frac{\omega_{pe}^2 + \omega_{ce}\omega_{ci}}{\omega_{pe}^2 + \omega_{pi}^2} \right) . \tag{4.123}$$

The lower hybrid frequency region is particularly important because waves propagating there can be in resonance with both electrons and ions. At the low density limit $\omega_{LH} \rightarrow \omega_{ci}$ and in the high density regime $\omega_{LH} \rightarrow \sqrt{\omega_{ce}\omega_{ci}}$. The cut-offs of the $X$ mode are at low density

$$\omega_X = \begin{cases} \omega_{ce} + \omega_{pe}^2/\omega_{ce} \\ \omega_{ci} + \omega_{pe}^2/\omega_{ce} \end{cases} \tag{4.124}$$

and at high density

$$\omega_X = \omega_{pe} \pm \frac{1}{2}\omega_{ce} . \tag{4.125}$$

At the limit of low frequency

$$n_X^2 \rightarrow 1 + \frac{\omega_{pi}^2}{\omega_{ci}^2} = 1 + \frac{c^2}{v_A^2} , \tag{4.126}$$

where $v_A = B_0/\sqrt{\rho_m \mu_0}$ is the *Alfvén speed*. This is the *magnetosonic wave* in cold plasma approximation. In MHD (Chap. 6) its dispersion equation is found to be

$$\frac{\omega^2}{k^2} = v_s^2 + v_A^2 , \tag{4.127}$$

where $v_s$ is the speed of sound. In cold plasma $v_s$ is small ($\rightarrow 0$), whereas in MHD $c \rightarrow \infty$. In tenuous space plasmas $v_A$ can be a considerable fraction of, or even larger than, $c$. Then the dispersion equation is modified as

$$\frac{\omega^2}{k^2} = \frac{v_s^2 + v_A^2}{1 + v_A^2/c^2} . \tag{4.128}$$

### 4.3.4 Propagation at arbitrary angles

The principal modes $R$, $L$, $O$, $X$ are defined for exactly parallel and perpendicular propagation only, but waves also propagate at other angles. The principal modes are usually illustrated as curves either in the $(\omega, k)$- or $(\omega, n)$-plane. The same can be done for an arbitrary angle $\theta$, or one may select a given mode and follow how it changes as a function of $\theta$.

One way to illustrate wave properties is to use *wave normal surfaces*. Consider the vector $\mathbf{n}/n^2$ whose absolute value is $v_p/c$. This is the phase velocity vector normalized to the speed of light. Draw the tip of the vector as a function of $\theta$ from 0 to $2\pi$ and let the curve rotate around the $z$-axis. The surface of the resulting 3D object is the wave normal surface. In cold homogeneous plasmas there are three topologically different surfaces: spheroid, dumbbell lemniscoid, and wheel lemniscoid. The waves have different wave normal surfaces in different regions of the space parameterized by the plasma frequency

($X$) and the gyro frequency ($Y$). Fig. 4.4 shows the wave normal properties in different regions of this space in the form of a CMA (after Clemmow, Mullaly and Allis) diagram.

Let us look at a couple of examples in the CMA diagram. The frequency is highest in the lower left corner (region 1 in Fig. 4.4) where the wave normal surfaces are spheroids. The wave which is the $R$ mode in the parallel direction goes continuously over to the $X$ mode in the perpendicular direction and the entire surface is often called $RX$ mode. Its phase velocity is in all directions greater than the phase velocity of the $LO$ mode. In region

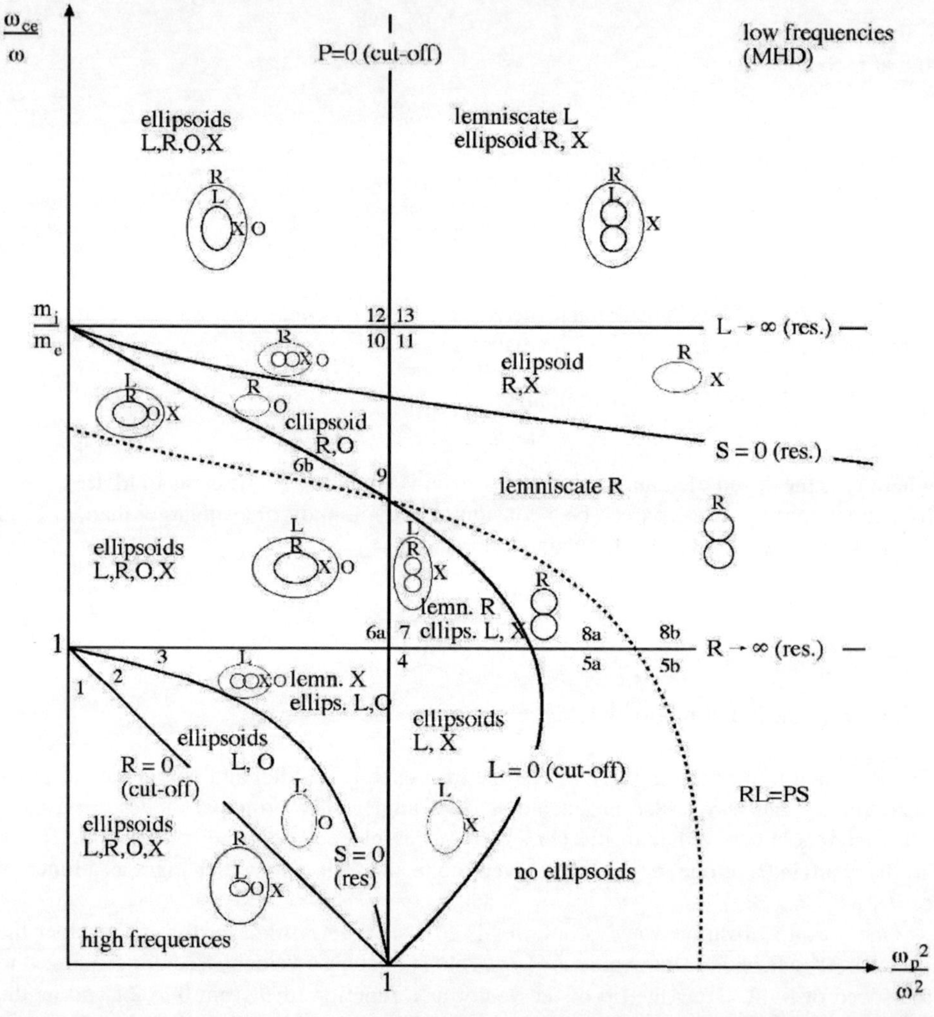


**Fig. 4.4** The CMA diagram. The wave normal surfaces are drawn assuming that the background magnetic field points upward. The principal modes are denoted at the side of each diagram. The variable on the horizontal axis is $X = \omega_{pe}^2/\omega^2 \propto n_e$ and on the vertical axis $Y = \omega_{ce}/\omega \propto B$.

2 there is no *RX* mode, whereas in region 3 the *LO* mode has greater phase velocity than the *X* mode. Note that now the *X* mode is on a wheel lemniscoid, so there is no corresponding parallel propagating mode. In region 7 the faster mode is *LX* and there is also an *R* mode. Now *R* is on a dumbbell lemniscoid, meaning that there is no perpendicular propagating solution. At the lowest frequencies (region 13) we find three MHD solutions, to which we return in Chap. 6.

Another method of presenting the solutions of the dispersion equation is to display them in a 3D $(\omega, k_{\parallel}, k_{\perp})$-space as *dispersion surfaces*. One face of the cube in Fig. 4.5 represents the modes propagating parallel to the magnetic field, another those propagating perpendicular, and the other propagation angles are inside the cube.

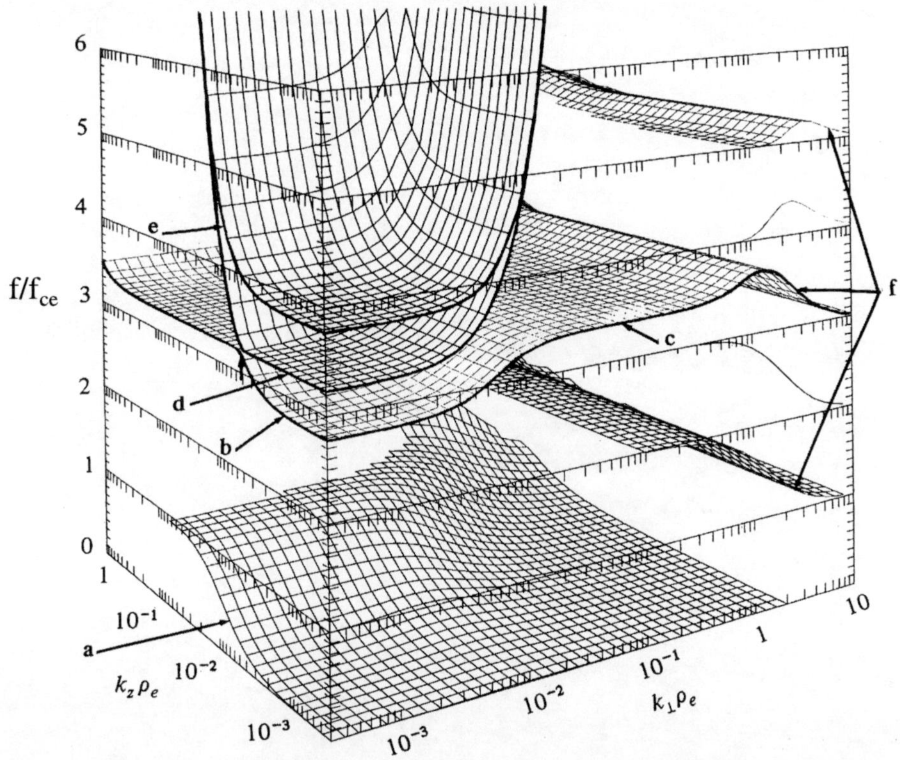


**Fig. 4.5** An example of how to represent the wave modes using dispersion surfaces. The figure illustrates high-frequency waves at the high-density limit ($\omega_{pe} > \omega_{ce}$). The horizontal axes show the perpendicular and parallel wave numbers normalized to the electron Larmor radius and the vertical axis is the frequency normalized to the electron gyro frequency. We identify the following solutions: a) whistler, b) the *L* mode, c) the upper hybrid mode, d) plasma frequency, e) *R* mode and f) Bernstein modes. This figure was calculated using microscopic theory [see André, 1985] that gives more solutions than the cold plasma theory. The Bernstein modes are examples of such. We will return to these modes in Chap. 5. (Figure by courtesy of M. André.)

# 5. Vlasov Theory

The cold plasma approximation of Chap. 4 was based on the assumption that the phase velocities of the waves are much larger than the thermal velocities of the particle populations. This is essentially the same as approximating the particle distribution functions by delta functions, although taking the limit may be tricky and not necessarily mathematically rigorous. The approximation is evidently not valid at resonances and many aspects of wave–particle interactions are lost. In this chapter we introduce the thermal (or kinetic) effects starting from the Vlasov equation

$$\frac{\partial f_\alpha}{\partial t} + \mathbf{v} \cdot \frac{\partial f_\alpha}{\partial \mathbf{r}} + \frac{q_\alpha}{m_\alpha}(\mathbf{E} + \mathbf{v} \times \mathbf{B}) \cdot \frac{\partial f_\alpha}{\partial \mathbf{v}} = 0 \,. \tag{5.1}$$

The *Vlasov theory* is a quite complete description of fully ionized plasmas and it provides a solid foundation for wave–particle interactions. At the same time it often is much too detailed for practical purposes. The theory has its own limitations, in particular in weakly ionized plasmas (e.g., ionosphere, solar photosphere), where plasma–neutral collisions cannot be neglected. The effects arising from the collision term of the Boltzmann equation can be added to the Vlasov treatment, but that reduces the generality of the approach.

## 5.1 Properties of the Vlasov Equation

The Vlasov equation is sometimes regarded as the most important equation of plasma physics. It has several useful properties:

- The Vlasov equation conserves particles. It is straightforward to show that

$$\frac{\partial}{\partial t} \int n_\alpha f_\alpha \, d^3r \, d^3v = 0 \tag{5.2}$$

by integrating (5.1) over the entire $(\mathbf{r}, \mathbf{v})$-space. Here $n_\alpha$ denotes the average density of species $\alpha$ in the volume under consideration.

H.E.J. Koskinen, *Physics of Space Storms: From the Solar Surface to the Earth*,
Springer Praxis Books, DOI 10.1007/978-3-642-00319-6_5,

- Positive probabilities remain positive in the Vlasov description. If $f_\alpha(\mathbf{r},\mathbf{v},t=0) > 0$ for all $(\mathbf{r},\mathbf{v})$, then $f_\alpha(\mathbf{r},\mathbf{v},t) > 0$ for all $t > 0$. This is an important property to be ensured in numerical Vlasov simulations.
- The Vlasov equation conserves entropy. Entropy is defined by

$$S = -\sum_\alpha \int f_\alpha \ln f_\alpha \, d^3r \, d^3v \tag{5.3}$$

$\Rightarrow$

$$\frac{dS}{dt} = -\sum_\alpha \int \left( \frac{df_\alpha}{dt} \ln f_\alpha + \frac{df_\alpha}{dt} \right) d^3r \, d^3v = 0 \,. \tag{5.4}$$

  This is an important issue in the interpretation of Landau's solution of the Vlasov equation to which we turn in the next section.
- The Vlasov equation has many equilibrium solutions. In statistical physics Boltzmann's *H-theorem* states that there is a unique equilibrium in the collisional time scale, the Maxwell distribution. The relevant time scales of the Vlasov theory are much shorter than the collision periods due to the assumption $\partial f/\partial t|_c \to 0$. Let $f_{\alpha 0}$ be any Vlasov equilibrium, then $\partial f_{\alpha 0}/\partial t = 0$ and thus

$$\left[ \mathbf{v} \cdot \frac{\partial}{\partial \mathbf{r}} + \frac{q_\alpha}{m_\alpha} (\mathbf{E} + \mathbf{v} \times \mathbf{B}) \cdot \frac{\partial}{\partial \mathbf{v}} \right] f_{\alpha 0} = 0 \,. \tag{5.5}$$

  In order to generate a general solution to this equation let $(\mathbf{r}'(t'), \mathbf{v}'(t'))$ be the orbit of a particle that intersects the point $(\mathbf{r},\mathbf{v})$ at the time $t' = t$. If functions $a(\mathbf{r}',\mathbf{v}')$, $b(\mathbf{r}',\mathbf{v}')$, ... are constants of motion for particles of species $\alpha$, then any function $f_{\alpha 0}[a(\mathbf{r}',\mathbf{v}'), b(\mathbf{r}',\mathbf{v}'), ...]$ satisfies (5.5) at the time $t' = t$, and thus any function $f_{\alpha 0}[a(\mathbf{r},\mathbf{v}), b(\mathbf{r},\mathbf{v}), ...]$ of the constants of motion is a stationary-state solution of the Vlasov equation.

***Examples of Vlasov equilibria***

1. $\mathbf{E}_0 = \mathbf{B}_0 = 0$. In this case the constants of motion are

$$W = \frac{m_\alpha}{2}(v_x^2 + v_y^2 + v_z^2)$$
$$\mathbf{p} = m_\alpha \mathbf{v} \,.$$

   Examples of equilibrium solutions are now

$$f_{\alpha 0} = \left( \frac{m_\alpha}{2\pi k_B T_\alpha} \right)^{3/2} \exp\left( -\frac{m_\alpha}{2k_B T_\alpha} v^2 \right) \tag{5.6}$$

$$f_{\alpha 0} = C_1 \frac{v_0}{2} \frac{1}{v^4 + v_0^4} \tag{5.7}$$

$$f_{\alpha 0} = C_2 \, v_0 \, \delta(v_x) \, \delta(v_y) \, \delta(v_z^2 - v_0^2) \tag{5.8}$$

$$f_{\alpha 0} = \sqrt{\frac{m_\alpha}{2\pi k_B T_\alpha}} \, \delta(v_x) \, \delta(v_y) \exp\left( -\frac{m_\alpha (v_z^2 - v_{\alpha 0}^2)}{2k_B T_\alpha} \right) , \tag{5.9}$$

where $\delta$'s are the Dirac delta functions and $C_1$ and $C_2$ are appropriate normalization factors. Correct choice of an equilibrium distribution requires physical understanding of the problem under consideration.

2. $\mathbf{E}_0 = 0,\ \mathbf{B}_0 = B_0(\mathbf{r})\,\mathbf{e}_z$. A possible selection of constants of motion is

$$\begin{aligned} W &= \frac{m_\alpha}{2}(v_x^2 + v_y^2 + v_z^2) \\ p_\parallel &= m_\alpha v_z \\ \mathbf{L} &= m_\alpha(xv_y - yv_x)\,\mathbf{e}_z - q_\alpha r A_\phi(r)\,\mathbf{e}_\phi\ , \end{aligned}$$

where $A_\phi$ is the azimuthal component of the vector potential, which is the only non-zero component in this configuration. Another choice of constants of motion could be

$$\begin{aligned} \xi_x &= v_x - \frac{q_\alpha}{m_\alpha}\int B_0(r)\,dy \\ \xi_y &= v_y + \frac{q_\alpha}{m_\alpha}\int B_0(r)\,dx\,. \end{aligned}$$

One of several possible equilibria for a constant $B_0$ is

$$f_{\alpha 0} = F\left(v^2,\ v_y + \frac{q_\alpha B_0}{m_\alpha}x,\ v_x - \frac{q_\alpha B_0}{m_\alpha}y\right)\,. \tag{5.10}$$

## 5.2 Landau's Solution

The Vlasov equation is not easy to solve. It must, of course, be done under the constraint to fulfill Maxwell's equations because the source terms of Maxwell's equations $(\rho, \mathbf{J})$ are determined by the distribution function, which, in turn, evolves according to the Vlasov equation. Furthermore, the force term in the Vlasov equation is nonlinear. Thus the Vlasov equation can be solved analytically only for small perturbations when linearization is possible. We start by writing functions to be solved as sums of equilibirum solutions and small perturbations

$$\begin{aligned} f_\alpha &= f_{\alpha 0} + f_{\alpha 1} \\ \mathbf{E} &= \mathbf{E}_0 + \mathbf{E}_1 \\ \mathbf{B} &= \mathbf{B}_0 + \mathbf{B}_1 \end{aligned}$$

and consider the equations of the first-order terms. However, the problem remains difficult. The general solution for homogeneous plasma in a homogeneous background magnetic field was presented for the first time by Bernstein [1958] and inclusion of inhomogeneities rapidly leads to problems that can be handled by numerical methods only. Landau [1946] solved the field-free case in the following way.

Consider homogeneous plasma free of ambient electromagnetic fields ($\mathbf{E}_0 = \mathbf{B}_0 = 0$) in *electrostatic approximation*: $\mathbf{E}_1 = -\nabla\varphi_1$ ; $\mathbf{B}_1 = 0$. The linearized Vlasov equation is now

$$\frac{\partial f_{\alpha 1}}{\partial t} + \mathbf{v} \cdot \frac{\partial f_{\alpha 1}}{\partial \mathbf{r}} - \frac{q_\alpha}{m_\alpha} \frac{\partial \varphi_1}{\partial \mathbf{r}} \cdot \frac{\partial f_{\alpha 0}}{\partial \mathbf{v}} = 0 , \tag{5.11}$$

where

$$\nabla^2 \varphi_1 = -\frac{1}{\varepsilon_0} \sum_\alpha n_\alpha q_\alpha \int f_{\alpha 1} \, d^3 v . \tag{5.12}$$

Vlasov tried to solve these equations at the end of the 1930s using Fourier transformations in space and time. He ended up with an integral of type

$$\int_{-\infty}^{\infty} \frac{\partial f_{\alpha 0} / \partial v}{\omega - kv} dv ,$$

which has a singularity along the path of integration. Vlasov did not find the correct way of dealing with the singularity.

Landau realized that because the perturbation must begin at some instant, the problem can be treated as an initial value problem and, instead of a Fourier transform, a Laplace transform can be applied in time domain. Once the initial transients of the perturbation have faded away, the *asymptotic solution* gives the intrinsic properties of the plasma, i.e., the dispersion equation.

Thus we write

$$f_{\alpha \mathbf{k}}(\mathbf{v}, t) = \frac{1}{(2\pi)^3} \int f_{\alpha 1}(\mathbf{r}, \mathbf{v}, t) \exp(-i\mathbf{k} \cdot \mathbf{r}) \, d^3 r \tag{5.13}$$

$$\tilde{f}_{\alpha \mathbf{k}}(\mathbf{v}, p) = \int_0^\infty f_{\alpha \mathbf{k}}(\mathbf{v}, t) \exp(-pt) \, dt \, ; \qquad Re(p) \geq p_0 \tag{5.14}$$

and similar transforms for $\varphi(\mathbf{r}, t)$. $p_0$ has to be chosen to ensure the convergence of the integral. After these transforms the equations for $\tilde{f}_{\alpha \mathbf{k}}$ and $\tilde{\varphi}_\mathbf{k}$ become algebraic. After the trivial solution of the algebraic equations the solution in the $(\mathbf{r},\mathbf{v})$-space is found by the inverse transformations

$$f_{\alpha 1}(\mathbf{r}, \mathbf{v}, t) = \int \exp(i\mathbf{k} \cdot \mathbf{r}) d\mathbf{k} \int_{p_0 - i\infty}^{p_0 + i\infty} \exp(pt) \tilde{f}_{\alpha \mathbf{k}}(\mathbf{v}, p) \frac{dp}{2\pi i} \tag{5.15}$$

$$\varphi_1(\mathbf{r}, t) = \int \exp(i\mathbf{k} \cdot \mathbf{r}) d\mathbf{k} \int_{p_0 - i\infty}^{p_0 + i\infty} \exp(pt) \tilde{\varphi}_\mathbf{k}(p) \frac{dp}{2\pi i} . \tag{5.16}$$

The transformed equations (5.11) and (5.12) are

$$(p + i\mathbf{k} \cdot \mathbf{v}) \tilde{f}_{\alpha \mathbf{k}} = f_{\alpha \mathbf{k}}(\mathbf{v}, t = 0) + \frac{q_\alpha}{m_\alpha} \left( i\mathbf{k} \cdot \frac{\partial f_{\alpha 0}}{\partial \mathbf{v}} \right) \tilde{\varphi}_\mathbf{k} \tag{5.17}$$

$$k^2 \tilde{\varphi}_\mathbf{k} = \frac{1}{\varepsilon_0} \sum_\alpha n_\alpha q_\alpha \int \tilde{f}_{\alpha \mathbf{k}} \, d^3 v . \tag{5.18}$$

From these we find the transformed potential

$$k^2 \tilde{\varphi}_{\mathbf{k}} = \frac{\dfrac{1}{\varepsilon_0}\displaystyle\sum_\alpha n_\alpha q_\alpha \int \frac{f_{\alpha\mathbf{k}}(t=0)}{p+i\mathbf{k}\cdot\mathbf{v}}\, d^3v}{1+\dfrac{1}{\varepsilon_0}\displaystyle\sum_\alpha \frac{n_\alpha q_\alpha^2}{m_\alpha}\frac{1}{k^2}\int \frac{\mathbf{k}\cdot\partial f_{\alpha 0}/\partial \mathbf{v}}{ip-\mathbf{k}\cdot\mathbf{v}}\, d^3v} \quad ; \quad Re(p) \geq p_0\,. \tag{5.19}$$

If we now identify $\omega = ip$, the denominator of the RHS corresponds to $K(\mathbf{k},\omega)$ in Chap. 4. Multiplying (5.19) by the denominator and assuming that we would have performed the inverse transform, we can write the equation formally as $\nabla\cdot\mathbf{D}_1 = \rho_1$, where $\rho_1$ is the initial charge density perturbation.

Because $K(\mathbf{k},\omega)$ contains the information we are most interested in, we do not usually need to make the inverse transformation of $\tilde{\varphi}_{\mathbf{k}}$. But we must know how it should be done in order to calculate the integral in

$$K(\mathbf{k},\omega) = 1+\frac{1}{\varepsilon_0}\sum_\alpha \frac{n_\alpha q_\alpha^2}{m_\alpha}\frac{1}{k^2}\int \frac{\mathbf{k}\cdot\partial f_{\alpha 0}/\partial \mathbf{v}}{\omega-\mathbf{k}\cdot\mathbf{v}}\, d^3v\,. \tag{5.20}$$

We can simplify the notation by selecting $\mathbf{k}$ to be in the direction of one coordinate axis and integrating

$$F_{\alpha 0}(u) \equiv \int f_{\alpha 0}(\mathbf{v})\,\delta\left(u-\frac{\mathbf{k}\cdot\mathbf{v}}{|k|}\right) d^3v \tag{5.21}$$

$$\tilde{F}_{\alpha\mathbf{k}}(u) \equiv \int \tilde{f}_{\alpha\mathbf{k}}(\mathbf{v})\,\delta\left(u-\frac{\mathbf{k}\cdot\mathbf{v}}{|k|}\right) d^3v \tag{5.22}$$

$\Rightarrow$

$$K(\mathbf{k},ip) = 1-\sum_\alpha \frac{\omega_{p\alpha}^2}{k^2}\int \frac{\partial F_{\alpha 0}(u)/\partial u}{u-ip/|k|}\, du \quad ; \quad Re(p)\geq p_0\,. \tag{5.23}$$

Taking the inverse Laplace transform we get

$$k^2\varphi_{\mathbf{k}}(t) = \int_{p_0-i\infty}^{p_0+i\infty} \frac{\dfrac{1}{\varepsilon_0}\displaystyle\sum_\alpha n_\alpha q_\alpha \int \frac{F_{\alpha\mathbf{k}}(u,t=0)}{p+i|k|u}\, du}{K(\mathbf{k},ip)} \exp(pt)\,\frac{dp}{2\pi i}\,. \tag{5.24}$$

This integral can be calculated in closed form for some specific equilibrium distributions $F_{\alpha 0}$ and initial perturbations $F_{\alpha\mathbf{k}}(u,t=0)$ only. Landau showed that it is possible to find the *asymptotic* behavior of the potential when $t\to\infty$, i.e., when the transients of the initial perturbation have disappeared and the normal modes of the plasma determine the plasma oscillations.

Before we can integrate (5.24) we need to know the analytic properties of $\tilde{\varphi}_{\mathbf{k}}(p)$. By definition it is analytic when $Re(p)\geq p_0$. In order to make use of residue calculus in the $p$-integration we make an analytic continuation of $\tilde{\varphi}_{\mathbf{k}}(p)$ to the entire complex $p$-plane. The problem is how to continue the integral

$$h(p) = \int_{-\infty}^{+\infty} \frac{g(u)}{u - ip/|k|} du \quad ; \quad Re(p) \geq p_0 \tag{5.25}$$

to $Re(p) < p_0$. Assume that $g(u)$ is analytic when $|u| < \infty$. If $Re(p) > 0$, the pole of the integrand is above the integration path (the real $u$-axis). The analytic continuation requires that the integration contour passes below the pole also in the case $Re(p) \leq 0$

$$h(p) = \begin{cases} \displaystyle\int_{-\infty}^{+\infty} \frac{g(u)du}{u - ip/|k|} & ; Re(p) > 0 \\ \displaystyle P\int_{-\infty}^{+\infty} \frac{g(u)du}{u - ip/|k|} + \pi i g(ip/|k|) & ; Re(p) = 0 \\ \displaystyle\int_{-\infty}^{+\infty} \frac{g(u)du}{u - ip/|k|} + 2\pi i g(ip/|k|) & ; Re(p) \leq 0 \,, \end{cases} \tag{5.26}$$

where $P$ denotes the Cauchy principal value. The integration path is called the *Landau contour* and denoted by $\int_L$. Note that this does not yet define how the contour is to be closed in the upper half plane. It is not always trivial to find a closure whose contribution vanishes at the infinity. Already the Maxwellian distribution is tricky.

**Feed your brain**

Review the basics of analytical continuation from some textbook in complex analysis and show that (5.26) is the correct analytical continuation in the present problem.

In (5.24) the only singularities are the poles at zeros of $K(\mathbf{k}, ip)$. In order to calculate the $p$-integral we move the integration path ($-i\infty \to i\infty$) so far to the negative $Re(p)$ (Fig. 5.1) that the factor $\exp(pt)$ guarantees that the contribution from the vertical parts of the integration contour vanish and the only contributions come from the residues at the poles.

Denoting the residues at $p_j$ by $R_j$ we have

$$\varphi_{\mathbf{k}}(t) = \sum_j R_j \exp(p_j(\mathbf{k})t) + \int_{-i\infty+p_0}^{-i\infty-\alpha} \tilde{\varphi}_{\mathbf{k}}(p)\exp(pt)\frac{dp}{2\pi i} \tag{5.27}$$

$$+ \int_{-i\infty-\alpha}^{i\infty-\alpha} \tilde{\varphi}_{\mathbf{k}}(p)\exp(pt)\frac{dp}{2\pi i} + \int_{i\infty-\alpha}^{i\infty+p_0} \tilde{\varphi}_{\mathbf{k}}(p)\exp(pt)\frac{dp}{2\pi i} \,.$$

The second and fourth terms on the RHS are small because $\tilde{\varphi}_{\mathbf{k}} \to 0$, as $|p| \to \infty$. The third term vanishes exponentially as compared to the residue terms when $t \to \infty$. This yields the asymptotic solution

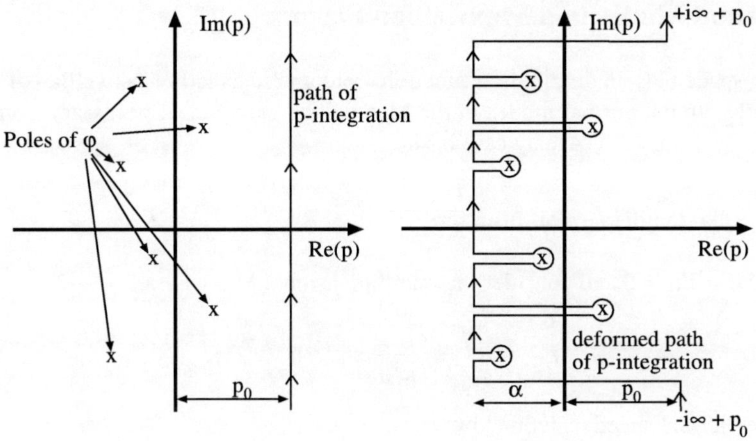


**Fig. 5.1** Integration path in the $p$-plane.

$$\varphi_{\mathbf{k}}(t \to \infty) = \sum_j R_j \exp(p_j(\mathbf{k})t) = \sum_j R_j \exp(-i\omega_j(\mathbf{k})t) , \tag{5.28}$$

where $\omega_j = \omega_r + i\omega_i$ are the solutions of the dispersion equation $K(\mathbf{k}, \omega) = 0$. This means that a long time after the initial perturbation the plasma behavior is determined by the solutions of the dispersion equation, provided that $K$ is calculated along the Landau contour.

$$K(\mathbf{k}, \omega) \equiv 1 - \sum_\alpha \frac{\omega_{p\alpha}^2}{k^2} \int_L \frac{\partial F_{\alpha 0}(u)/\partial u}{u - \omega/|k|} du = 0 . \tag{5.29}$$

Now

$$\begin{array}{ll} Re(p_j) < 0 \Rightarrow \omega_i < 0 & \varphi_{\mathbf{k}} \text{ is damped} \\ Re(p_j) > 0 \Rightarrow \omega_i > 0 & \varphi_{\mathbf{k}} \text{ grows (instability)} . \end{array}$$

For $|\omega_i| \ll |\omega_r|$ the solution is called a *normal mode*. Note that the dispersion equation is calculated only at the time-asymptotic limit.

**Train your brain**

An alternative way to solve the Vlasov equation is to follow Vlasov's approach and end up with the dispersion equation

$$1 + \sum_\alpha \frac{\omega_{p\alpha}^2}{k^2} \int \frac{\mathbf{k} \cdot \partial f_{\alpha 0}/\partial \mathbf{v}}{\omega - \mathbf{k} \cdot \mathbf{v}} d^3 v = 0 .$$

Add weak collisions in the Vlasov equation in the form $\partial f/\partial t|_c = -\nu(f - f_0)$ and show that the Fourier transform method leads to the Landau prescription at the limit $\nu \to 0^+$.

## 5.3 Normal Modes in a Maxwellian Plasma

Although space plasma distribution functions seldom are exactly Maxwellian, it is practical to start with the normal modes in the Maxwellian case and, if necessary, consider the possible deviations on a case by case basis.

### 5.3.1 The plasma dispersion function

Assume $\mathbf{E}_0 = \mathbf{B}_0 = 0$ and consider the one-dimensional Maxwellian

$$F_{\alpha 0} = \sqrt{\frac{m_\alpha}{2\pi k_B T_\alpha}} \exp(-u^2/v_{th\alpha}^2) \,, \tag{5.30}$$

where the thermal speed is defined by

$$v_{th\alpha} = \sqrt{\frac{2k_B T_\alpha}{m_\alpha}} \,.$$

Now the Landau contour is a little problematic because the integrand of

$$\int \frac{\partial F_{\alpha 0}/\partial u}{u - \omega/|k|} \, du \approx \int \frac{u F_{\alpha 0}}{u - \omega/|k|} \, du$$

diverges with $u \longrightarrow \infty$, and the calculation of the closure of the integration path is not trivial. This problem can be solved using methods of complex integration and the result be expressed in terms of the *plasma dispersion function*

$$Z(\zeta) = \frac{1}{\sqrt{\pi}} \int_{-\infty}^{\infty} \frac{\exp(-x^2)}{x - \zeta} \, dx \;\; ; \;\; Im(\zeta) > 0 \tag{5.31}$$

and its derivatives. The plasma dispersion function is related to the error function and numerical routines to evaluate it are available. If we consider electron dynamics only, assuming the ions as a fixed background, the dispersion equation reduces to

$$1 - \frac{\omega_{pe}^2}{k^2 v_{the}^2} Z'\left(\frac{\omega}{k v_{the}}\right) = 0 \,. \tag{5.32}$$

For normal modes ($|\omega_i| \ll |\omega_r|$) the dispersion equation can be expanded around $\omega = \omega_r$

$$1 - \sum_\alpha \frac{\omega_{p\alpha}^2}{k^2} \left(1 + i\omega_i \frac{\partial}{\partial \omega_i}\right) \left[ P \int \frac{\partial F_{\alpha 0}/\partial u}{u - \omega_r/|k|} \, du + \pi i \left(\frac{\partial F_{\alpha 0}}{\partial u}\right)_{u=\omega_r/|k|} \right] = 0 \,. \tag{5.33}$$

From this we can find solutions for the dispersion equation at long and short wavelengths. These correspond to series expansions of $Z$ for large and small arguments, respectively.

### 5.3.2 The Langmuir wave

We start from the long wavelength limit ($\omega/k \gg v_{th}$), which is the same approximation as made in the cold plasma theory (Chap. 4). Now

$$-P\int \frac{\partial F_{\alpha 0}/\partial u}{u-\omega_r/|k|}\,du = \int \frac{\partial F_{\alpha 0}}{\partial u}\left(\frac{1}{\omega/|k|}+\frac{u}{(\omega/|k|)^2}+\frac{u^2}{(\omega/|k|)^3}+\ldots\right)du\,. \tag{5.34}$$

Using this expansion, neglecting the ion dynamics and inserting a Maxwellian distribution for electrons (5.33) yields

$$\omega_r \approx \omega_{pe}(1+3k^2\lambda_{De}^2)^{1/2} \approx \omega_{pe}\left(1+\frac{3}{2}k^2\lambda_{De}^2\right) \tag{5.35}$$

as the real part of the frequency, and the imaginary part is[1]

$$\omega_i = -\sqrt{\frac{\pi}{8}}\frac{\omega_{pe}}{|k^3\lambda_{De}^3|}\exp\left(-\frac{1}{2k^2\lambda_{De}^2}-\frac{3}{2}\right)\,. \tag{5.36}$$

This is the *Langmuir wave*. The finite temperature of the Maxwellian distribution makes the standing cold plasma oscillation to propagate. Furthermore, the negative imaginary part of the frequency indicates that the wave is *damped* at the rate $\omega_i$. This phenomenon is known as the *Landau damping*. The damping is a genuine collective effect characteristic for plasmas. Its interpretation will be discussed in Sect. 5.4.

We can find the same result by expanding $K(\omega,\mathbf{k})$

$$K(\omega,\mathbf{k}) \approx K(\omega_r,\mathbf{k}) + i\omega_i \frac{\partial K(\omega_r,\mathbf{k})}{\partial \omega_r}\,. \tag{5.37}$$

Note that $K(\omega_r,\mathbf{k})$ is a complex function containing an expression of the form

$$\lim_{\varepsilon\to 0^+}\int \frac{\partial F_{\alpha 0}/\partial u}{u-\omega_r/|k|-i\varepsilon}\,du\,.$$

Thus we have

$$K(\omega_r,\mathbf{k}) = K_r(\omega_r,\mathbf{k}) + iK_i(\omega_r,\mathbf{k}) \tag{5.38}$$

$$K_i = -\pi\sum_\alpha \frac{\omega_{p\alpha}^2}{k^2}\left(\frac{\partial F_{\alpha 0}}{\partial u}\right)_{u=\omega_r/|k|} \tag{5.39}$$

$$K_r = 1-\sum_\alpha \frac{\omega_{p\alpha}^2}{k^2}\,P\int \frac{\partial F_{\alpha 0}/\partial u}{u-\omega_r/|k|}\,du\,. \tag{5.40}$$

Note that while the Landau contour is not given explicitly, it is taken care of by the limit $\varepsilon \to 0^+$. Equating the imaginary parts we find

[1] Here and in the following we replace $\approx$ by $=$ once the initial approximation has been introduced.

$$\omega_i = \frac{-K_i(\omega_r, \mathbf{k})}{\partial K_r(\omega_r, \mathbf{k})/\partial \omega_r}, \tag{5.41}$$

where $K_r$ fulfills the dispersion equation

$$K_r(\omega_r, \mathbf{k}) = 0\,. \tag{5.42}$$

### 5.3.3 The ion–acoustic wave

Take then also the ion motion into account. Assume that $T_e \gg T_i$ and look for solutions of the dispersion equation in the phase velocity range

$$\sqrt{\frac{k_B T_i}{m_i}} < \frac{\omega}{k} < \sqrt{\frac{k_B T_e}{m_e}}\,. \tag{5.43}$$

At this limit we can use the same series expansion for the ions as above, but now the cold plasma approximation is no more valid for electrons because $v_p < v_{the}$. The appropriate expansion is

$$P\int \frac{\partial F_{\alpha 0}/\partial u}{u - \omega_r/|k|}\,du \approx 2\int \frac{\partial F_{\alpha 0}}{\partial (u^2)}\,du\,. \tag{5.44}$$

Assuming Maxwellian distributions for both species we get

$$K_r = 1 - \frac{\omega_{pi}^2}{\omega_r^2} + \frac{1}{k^2\lambda_{De}^2} \tag{5.45}$$

$$K_i = \pi \sum_\alpha \frac{\omega_{p\alpha}^2}{k^2}\left(\frac{m_\alpha}{2\pi k_B T_\alpha}\right)^{1/2} \frac{m_\alpha}{k_B T_\alpha}\frac{\omega_r}{|k|}\exp\left(-\frac{\omega_r^2 m_\alpha}{2k^2 k_B T_\alpha}\right). \tag{5.46}$$

By solving the dispersion equation the real part of the frequency is found to be

$$\omega_r^2 = \frac{k^2 c_s^2}{1 + k^2\lambda_{De}^2} \quad ; \quad c_s = \sqrt{\frac{k_B T_e}{m_i}} \tag{5.47}$$

and the damping rate is given by

$$\begin{aligned} \omega_i &= -\frac{K_i}{\partial K_r/\partial \omega_r} \\ &= -\frac{|\omega_r|\sqrt{\pi/8}}{(1+k^2\lambda_{De}^2)^{3/2}}\left[\left(\frac{T_e}{T_i}\right)^{3/2}\exp\left(\frac{-T_e/T_i}{2(1+k^2\lambda_{De}^2)}\right) + \sqrt{\frac{m_e}{m_i}}\right]. \end{aligned} \tag{5.48}$$

This is the *ion–acoustic wave* and $c_s$ is called the *ion–sound speed* or ion–acoustic speed.

Note that the ion–sound speed is determined by the *electron temperature* and the *ion mass*. If the ion temperature is to be taken into account, we should replace $T_e$ by $T_e + T_i$. However, this mode can be treated as a normal mode ($|\omega_i| \ll |\omega_r|$) only if $T_e \gg T_i$, which motivates that the ion temperature was neglected at the beginning. In many practical situations, for example in the auroral ionosphere this condition is not met and the mode is strongly damped. As we will see in Chap. 9, the strongly damped ion–acoustic mode is also important in the scattering of electromagnetic waves.

### 5.3.4 Macroscopic derivation of Langmuir and ion–acoustic modes

Finite temperature effects in plasmas do not always require a Vlasov theory treatment. For example, in MHD (Chap. 6) the temperature is included through the equation of state and energy equation. Thus the normal fluid sound wave, not the ion–acoustic wave, is a part of the dispersion equation for MHD waves.

The Langmuir and ion–acoustic waves can be introduced in a warm unmagnetized plasma description starting from simple electron and ion fluid equations, which is the method applied in many introductory plasma physics textbooks. We sketch the procedure here because the same approach is useful in the discussion of beam–plasma instabilities in Chap. 7.

Assume that the plasma is homogeneous and that there are no background electromagnetic fields. Let the pressure be isotropic, the average velocity zero, and the equation of state or the form $P/\rho_m^{-\gamma}$ = constant. We are looking for plane wave solutions and linearize the continuity equations (2.117) for ions and electrons. The first-order equations are

$$i\omega n_{i1} - in_0 \mathbf{k}\cdot\mathbf{V}_{i1} = 0 \tag{5.49}$$

$$i\omega n_{e1} - in_0 \mathbf{k}\cdot\mathbf{V}_{e1} = 0\,. \tag{5.50}$$

In the momentum equation we retain the electron pressure gradient but neglect the ion pressure effects due to the smaller ion mobility. Considering small mass density perturbations $\rho_{m1} \ll \rho_{m0}$ the equation of state can be written as

$$P_1 = P_0\gamma\frac{\rho_{m1}}{\rho_{m0}}\,. \tag{5.51}$$

Now the momentum equations for ions and electrons are

$$-i\omega\mathbf{V}_{i1} = \frac{e}{m_i}\mathbf{E}_1 \tag{5.52}$$

$$-i\omega\mathbf{V}_{e1} = -\frac{e}{m_i}\mathbf{E}_1 - \frac{i\mathbf{k}\gamma P_0}{n_0 m_e}\frac{n_{e1}}{n_0}\,. \tag{5.53}$$

The first Maxwell equation ties these together

$$i\mathbf{k}\cdot\mathbf{E}_1 = -\frac{e}{\varepsilon_0}(n_{e1}-n_{i1})\,. \tag{5.54}$$

Combining these and writing $P_0 = n_e k_B T_e$ we get

$$\left(1 - \frac{\omega_{pi}^2}{\omega^2} - \frac{\omega_{pe}^2}{\omega^2 - k^2(\gamma k_B T_e/m_e)}\right)\mathbf{k}\cdot\mathbf{E}_1 = 0\,. \tag{5.55}$$

The expression in the parenthesis is now the dielectric function $K(\omega)$, whose zeros yield the dispersion equation $K(\omega) = 0$. This has 4 roots (or 2 roots for $\omega^2$). One pair of solutions yields the dispersion equation

$$\omega^2 = \omega_{pe}^2 + k^2(\gamma k_B T_e/m_e) \,. \tag{5.56}$$

At zero temperature or for infinite wavelength ($k = 0$) this is the standing plasma oscillation. The finite temperature makes the wave propagating and dispersive for finite $k$. The wave is electrostatic (longitudinal, $\mathbf{k} \parallel \mathbf{E}$).

To identify this mode with the Langmuir wave of the Vlasov theory, we must specify the polytropic index $\gamma$, which requires some physical intuition. Let us consider the thermal effect as a small correction to the cold plasma theory or, equivalently, the long wavelength limit. Then we can assume that the thermal effect expands less than a wavelength during one plasma oscillation. During one oscillation period there is thus no heat exchange between the wave and the plasma, and thus the process can be treated as adiabatic. Because the field-free plasma is essentially one-dimensional, we have $\gamma = (d+2)/d = 3$ and

$$\omega^2 = \omega_{pe}^2 \left(1 + 3k^2\lambda_{De}^2\right) \,. \tag{5.57}$$

As long as the thermal correction is small we can approximate the square root as

$$\omega = \omega_{pe}\sqrt{1 + 3k^2\lambda_{De}^2} \approx \omega_{pe}\left(1 + \frac{3}{2}k^2\lambda_{De}^2\right) \,, \tag{5.58}$$

which is the same solution we found in the Vlasov theory.

The second pair of solutions gives the ion–acoustic wave.

$$\omega = \frac{kc_s}{\sqrt{1 + k^2\lambda_{De}^2}} \,, \tag{5.59}$$

where we have introduced the ion–sound speed $c_s = \sqrt{k_B T_e/m_i}$. In this solution we have set $\gamma = 1$, i.e., assumed an isothermal process. Its motivation is the small oscillation frequency of the ions allowing the electrons to thermalize during one oscillation period.

Thus we have found both Langmuir and ion–acoustic modes without needing to invoke the Vlasov theory or the Landau solution. The price to pay was to figure out the appropriate polytropic indices, whereas in Vlasov theory the numerical factors are direct consequences of assumed Maxwellian distributions and the wavelength regimes, where we looked for the solutions. However, by far a more serious deficiency of this macroscopic warm plasma treatment is that it does not give even a hint of the damping of the waves.

## 5.4 Physics of Landau Damping

Landau's original solution was not fully accepted before it was experimentally verified in laboratories in the 1960s. A problem was that the Vlasov equation conserves entropy, whereas the Landau solution does not appear to do so. Consider, e.g., the Langmuir waves. The wave electric field interacts with the Maxwellian electrons accelerating those whose velocity is slightly less than the phase speed of the wave, and decelerating those that move a little faster. Because $\partial f/\partial v < 0$, there are more slower electrons than faster electrons around the phase speed. Thus there is a net energy transfer from the wave to the particles, i.e., the wave is damped and the particle distribution heated, which at the first sight looks like a dissipative process.

To resolve this apparent contradiction consider the perturbed distribution function $f_1(t)$ closer. Recall that

$$\tilde{f}_{\alpha\mathbf{k}}(\mathbf{v},p) = \frac{f_{\alpha\mathbf{k}}(\mathbf{v},t=0)}{(p+i\mathbf{k}\cdot\mathbf{v})} + \frac{q_\alpha}{m_\alpha}\frac{i\tilde{\varphi}_{\mathbf{k}}\mathbf{k}\cdot\partial f_{\alpha 0}/\partial\mathbf{v}}{(p+i\mathbf{k}\cdot\mathbf{v})} \tag{5.60}$$

$$f_{\alpha\mathbf{k}}(\mathbf{v},t) = \frac{1}{2\pi i}\int_{p_0-i\infty}^{p_0+i\infty} \tilde{f}_{\alpha\mathbf{k}}(\mathbf{v},p)\exp(pt)\,dp\,. \tag{5.61}$$

$\tilde{f}_{\alpha\mathbf{k}}(\mathbf{v},p)$ has the same poles as $\tilde{\varphi}_{\mathbf{k}}(p)$, i.e., the solutions of $K=0$. There is an additional pole at $p=-i\mathbf{k}\cdot\mathbf{v}$. At the limit $t\longrightarrow\infty$ we find

$$f_{\alpha\mathbf{k}} = \hat{f}_{\alpha B}\exp(-i\mathbf{k}\cdot\mathbf{v}t) + \sum_{\omega_{\mathbf{k}}}\hat{f}_{\alpha\mathbf{k}}\exp(-i\omega_{\mathbf{k}}t)\,, \tag{5.62}$$

where $\omega_{\mathbf{k}}$ are the solutions of the dispersion equation and the sum is over these solutions. $\hat{f}_{\alpha B}$ and $\hat{f}_{\alpha\mathbf{k}}$ are time-independent amplitudes. The terms in the sum over $\omega_{\mathbf{k}}$ are damped at the same rate as the perturbed field $\varphi_{\mathbf{k}}(t)$. In the first term on the RHS of (5.62) $B$ stands for *ballistic*. The ballistic term is there because the Vlasov equation is formally similar to the Liouville equation and every particle remembers its initial perturbation wherever it goes in the phase space. When $t$ increases, the ballistic term becomes increasingly oscillatory in the $\mathbf{v}$-space (Fig. 5.2) and its contribution to $\varphi_{\mathbf{k}}(t)$ behaves as

$$k^2\varphi_{\mathbf{k}} = \frac{1}{\varepsilon_0}\sum_\alpha q_\alpha n_\alpha\int\hat{f}_{\alpha B}\exp(-i\mathbf{k}\cdot\mathbf{v}t)d^3v \longrightarrow 0\,, \tag{5.63}$$

when $t\longrightarrow\infty$. That is, at the time-asymptotic limit the ballistic terms contain the information of the initial perturbation but they do not contribute to the *observable* electric field.

The existence of ballistic terms leads to a *nonlinear* phenomenon called the *Landau echo*, the laboratory observation of which was an important step towards the acceptance of Landau's solution (Fig. 5.3) as the correct way to deal with the Vlasov equation.

Assume that an intial perturbation took place at time $t_1$ and its spectrum was narrow near $k\approx k_1$. Then

$$f_\alpha = f_{\alpha 0} + f_{\alpha\mathbf{k}1}(u,t=t_1)\exp(ik_1u(t-t_1)) + \ldots\,. \tag{5.64}$$

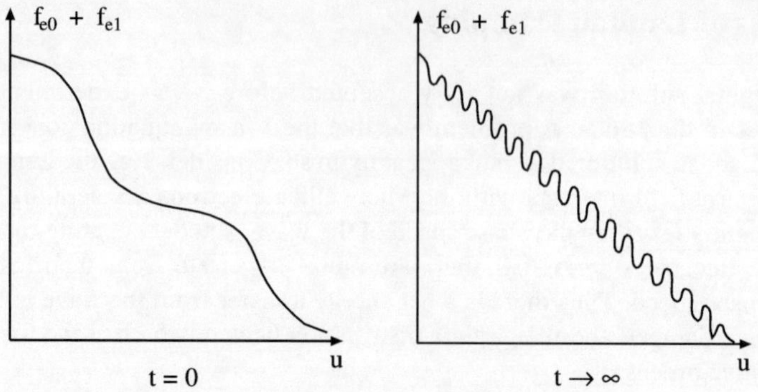


**Fig. 5.2** Evolution of the distribution function when the electrostatic perturbation becomes damped but the ballistic term remains superposed on the equilibrium distribution.

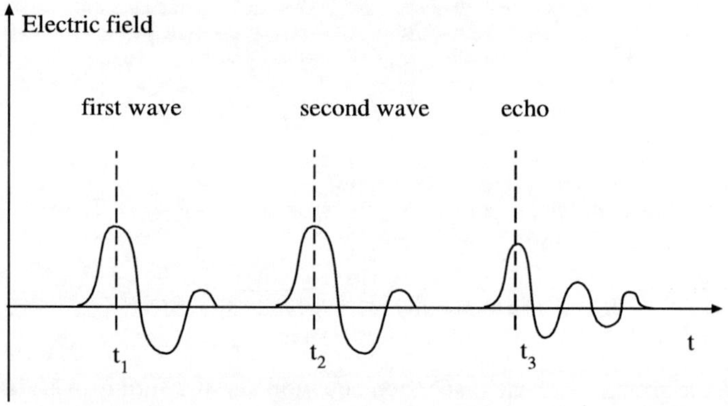


**Fig. 5.3** The Landau echo.

Wait until the perturbation has been damped below the observable limit and only the ballistic term superposed on the equilibrium distribution remains. Then launch another wave ($k \approx k_2$) at time $t_2$ and wait until it also is damped. Add this to $f_\alpha$ and *do not linearize*! Thus

$$f_\alpha = f_{\alpha 0} + f^{(1)}_{\alpha \mathbf{k}1} \exp(ik_1 u(t-t_1)) + f^{(1)}_{\alpha \mathbf{k}2} \exp(ik_2 u(t-t_2)) + f^{(2)}_\alpha + \dots . \tag{5.65}$$

where (1) and (2) indicate the order of the terms. In the second-order term there is a contribution of the form

$$f^{(2)}_\alpha \approx f^{(1)}_{\alpha \mathbf{k}1} f^{(1)}_{\alpha \mathbf{k}2} \exp(ik_1 u(t-t_1)) \exp(-ik_2 u(t-t_2)) \; ; \; t > t_2 . \tag{5.66}$$

At time $t = t_3$ defined by

$$k_1(t_3 - t_1) - k_2(t_3 - t_2) = 0 \tag{5.67}$$

the second-order term is no longer small, and the perturbed charge density

$$\rho_{q2} \approx \int du \exp(ik_1 u(t-t_1) - ik_2 u(t-t_2)) f^{(2)}_{\alpha(\mathbf{k}_2-\mathbf{k}_1)} \tag{5.68}$$

becomes finite and observable. Thus the "beating" of the ballistic terms of the first two perturbations has produced a new observable perturbation, the Landau echo, that is a damped mode of the plasma. It is transient because the beat condition is satisfied only for a short while and the Landau damping acts on this wave as well. The effect has been verified in laboratories and shows that the Landau damping does not need to violate the conservation of entropy in the time scale $\tau \ll \tau_{coll}$.

As collisional time scales in tenuous space plasmas often are very long compared to the relevant time scales of investigated phenomena, the existence of Landau echoes indicates that even in the case of small-amplitude perturbations there can be nonlinear *mixing* of wave modes at the microscopic level. This is one viewpoint to *plasma turbulence*. However, no satisfactory general method of calculating the transport coefficients (resistivity, viscosity, etc.) from plasma kinetic theory has been found.

## 5.5 Vlasov Theory in a General Equilibrium

In space plasmas a background magnetic field is practically always present. Therefore we must look for a more general description including the background fields. The linearized Vlasov equation then reads as

$$\left[\frac{\partial}{\partial t} + \mathbf{v}\cdot\frac{\partial}{\partial \mathbf{r}} + \frac{q_\alpha}{m_\alpha}(\mathbf{E}_0 + \mathbf{v}\times\mathbf{B}_0)\cdot\frac{\partial}{\partial \mathbf{v}}\right] f_{\alpha 1} = -\frac{q_\alpha}{m_\alpha}(\mathbf{E}_1 + \mathbf{v}\times\mathbf{B}_1)\cdot\frac{\partial f_{\alpha 0}}{\partial \mathbf{v}} \,. \tag{5.69}$$

This can be solved employing the *method of characteristics* that can be described as "integration over unperturbed orbits". Define new variables (**r**',**v**',t')

$$\frac{d\mathbf{r}'}{dt'} = \mathbf{v}' \; ; \; \frac{d\mathbf{v}'}{dt'} = \frac{q_\alpha}{m_\alpha}\left[\mathbf{E}_0(\mathbf{r}',t') + \mathbf{v}'\times\mathbf{B}_0(\mathbf{r}',t')\right] \tag{5.70}$$

with boundary conditions

$$\mathbf{r}'(t'=t) = \mathbf{r} \; ; \; \mathbf{v}'(t'=t) = \mathbf{v} \,. \tag{5.71}$$

Consider $f_{\alpha 1}(\mathbf{r}',\mathbf{v}',t')$ and use (5.69) to calculate its total time derivative

$$\begin{aligned} &\frac{df_{\alpha 1}(\mathbf{r}',\mathbf{v}',t')}{dt'} \\ &= \frac{\partial f_{\alpha 1}(\mathbf{r}',\mathbf{v}',t')}{\partial t'} + \frac{d\mathbf{r}'}{dt'}\cdot\frac{\partial f_{\alpha 1}(\mathbf{r}',\mathbf{v}',t')}{\partial \mathbf{r}'} + \frac{d\mathbf{v}'}{dt'}\cdot\frac{\partial f_{\alpha 1}(\mathbf{r}',\mathbf{v}',t')}{\partial \mathbf{v}'} \\ &= -\frac{q_\alpha}{m_\alpha}\left[\mathbf{E}_1(\mathbf{r}',t') + \mathbf{v}'\times\mathbf{B}_1(\mathbf{r}',t')\right]\cdot\frac{\partial f_{\alpha 0}(\mathbf{r}',\mathbf{v}')}{\partial \mathbf{v}'} \,. \end{aligned} \tag{5.72}$$

The boundary conditions imply that $f_{\alpha 1}(\mathbf{r}',\mathbf{v}',t') = f_{\alpha 1}(\mathbf{r},\mathbf{v},t)$ at time $t' = t$. Thus the solution of (5.72) at $t' = t$ is a solution of the Vlasov equation. The point is that (5.72) can be calculated by a direct integration because its LHS is an *exact differential*. The formal solution is

$$f_{\alpha 1}(\mathbf{r},\mathbf{v},t) = -\frac{q_\alpha}{m_\alpha}\int_{-\infty}^{t}\left[\mathbf{E}_1(\mathbf{r}',t') + \mathbf{v}'\times\mathbf{B}_1(\mathbf{r}',t')\right]\cdot\frac{\partial f_{\alpha 0}(\mathbf{r}',\mathbf{v}')}{\partial \mathbf{v}'}\,dt' + f_{\alpha 1}(\mathbf{r}'(-\infty),\mathbf{v}'(-\infty),t'(-\infty))\,. \tag{5.73}$$

This procedure can be interpreted in the following way. $f_{\alpha 1}$ has been found by integrating the Vlasov equation from $-\infty$ to $t$ along the path in the $(\mathbf{r},\mathbf{v})$-space that at each individual point coincides with the orbit of a charged particle in the equilibrium fields $\mathbf{E}_0$ and $\mathbf{B}_0$. From $f_{\alpha 1}$ we can calculate $n_{\alpha 1}(\mathbf{r},t)$ and $\mathbf{V}_{\alpha 1}(\mathbf{r},t)$, which are then inserted in Maxwell's equations

$$\nabla\times\mathbf{E}_1 = -\frac{\partial \mathbf{B}_1}{\partial t} \tag{5.74}$$

$$\nabla\cdot\mathbf{E}_1 = \frac{1}{\varepsilon_0}\sum_\alpha q_\alpha n_{\alpha 1} \tag{5.75}$$

$$\nabla\times\mathbf{B}_1 = \frac{1}{c^2}\frac{\partial \mathbf{E}_1}{\partial t} + \mu_0\sum_\alpha q_\alpha (n_\alpha \mathbf{V}_\alpha)_1\,. \tag{5.76}$$

This set of equations could now (in principle) be solved as an initial value problem in the same way as the Landau solution. However, we can also take a shortcut and accept that the Landau solution is the correct way to deal with the resonant integrals, and assume that the wave fields are of the form $\mathbf{E}_1(\mathbf{r},t) = \mathbf{E}_{\mathbf{k}\omega}\exp(i\mathbf{k}\cdot\mathbf{r} - i\omega t)$ and that $f_{\alpha 1}(\mathbf{r}',\mathbf{v}',t\to -\infty)\to 0$. This yields for $\mathrm{Im}(\omega) > 0$

$$f_{\alpha\mathbf{k}} = -\frac{q_\alpha}{m_\alpha}\int_{-\infty}^{0}(\mathbf{E}_{\mathbf{k}\omega} + \mathbf{v}'\times\mathbf{B}_{\mathbf{k}\omega})\cdot\frac{\partial f_{\alpha 0}(\mathbf{v}')}{\partial \mathbf{v}'}\exp[i(\mathbf{k}\cdot\mathbf{R} - \omega\tau)]\,d\tau\,, \tag{5.77}$$

where $\tau = t' - t$, $\mathbf{R} = \mathbf{r}' - \mathbf{r}$. The solutions for $\mathrm{Im}(\omega) < 0$ are found by analytic continuation of $f_{\alpha\mathbf{k}}$ to the lower half-plane. Inserting this into Maxwell's equations in the $(\omega,\mathbf{k})$ space and eliminating $\mathbf{B}_{\mathbf{k}\omega}$ we get

$$\mathscr{K}\cdot\mathbf{E} = 0\,, \tag{5.78}$$

where $\mathscr{K}$ is the dispersion tensor. The cold plasma theory is, in principle, found at the limit $f(\mathbf{v})\to\delta(\mathbf{v})$, although some care must be exercised with the details of the limit procedure.

***Example***

Assume $\mathbf{E}_0 = \mathbf{B}_0 = 0$ and $f_0 = f_0(v^2)$. Define $F_{\alpha 0}(u) = \int f_{\alpha 0}\delta(u - \mathbf{k}\cdot\mathbf{v}/|k|)\,d^3v$, $E_\mathbf{k} = (\mathbf{k}\cdot\mathbf{E})/|k|$, and $\mathbf{E}_\perp = (\mathbf{k}\times\mathbf{E})/|k| \Rightarrow$

$$\begin{bmatrix} K_\perp & 0 & 0 \\ 0 & K_\perp & 0 \\ 0 & 0 & K_\mathbf{k} \end{bmatrix} \begin{bmatrix} E_{\perp 1} \\ E_{\perp 2} \\ E_\mathbf{k} \end{bmatrix} = 0 , \tag{5.79}$$

where

$$K_\perp = 1 - \frac{k^2c^2}{\omega^2} - \sum_\alpha \frac{\omega_{p\alpha}^2}{\omega} \int \frac{F_{\alpha 0}}{\omega - |k|u} du \tag{5.80}$$

$$K_\mathbf{k} = 1 + \sum_\alpha \frac{\omega_{p\alpha}^2}{\omega} \int_L \frac{F_{\alpha 0}/\partial u}{\omega/|k| - u} du . \tag{5.81}$$

These give
electrostatic modes : $K_\mathbf{k} = 0 \;\; (\mathbf{E}_\perp = 0)$
electromagnetic modes : $K_\perp = 0 \;(\mathbf{E}_\mathbf{k} = 0)$ .

The electrostatic solution is the Landau solution. The dispersion equation for the electromagnetic modes is

$$\omega^2 = k^2c^2 + \sum_\alpha \omega_{p\alpha}^2 \int_{-\infty}^{\infty} \frac{\omega F_{\alpha 0}}{\omega - |k|u} du . \tag{5.82}$$

This has solutions only if $\omega \gg kv_{the}$ and we find the familiar electromagnetic mode in nonmagnetized cold plasma

$$\omega^2 \approx k^2c^2 + \omega_{pe}^2 . \tag{5.83}$$

## 5.6 Uniformly Magnetized Plasma

Assume now that $\mathbf{B}_0 = B_0\mathbf{e}_z$, $\mathbf{E}_0 = 0$, $f_{\alpha 0} = f_{\alpha 0}(v_\perp^2, v_\parallel)$. The derivation of the dielectric tensor is a tedious procedure, which we only outline here. Denote

$$v_x = v_\perp \cos\phi, \; v_y = v_\perp \sin\phi, \; v_z = v_\parallel .$$

Using these variables the particle orbit can be written as

$$\begin{aligned} v_x' &= v_\perp \cos(\phi - \omega_c\tau) \; ; \; x' = x - \frac{v_\perp}{\omega_c}\sin(\phi - \omega_c\tau) + \frac{v_\perp}{\omega_c}\sin\phi \\ v_y' &= v_\perp \sin(\phi - \omega_c\tau) \; ; \; y' = y + \frac{v_\perp}{\omega_c}\cos(\phi - \omega_c\tau) - \frac{v_\perp}{\omega_c}\cos\phi \\ v_z' &= v_\parallel \qquad\qquad\qquad ; \; z' = v_\parallel\tau + z . \end{aligned} \tag{5.84}$$

To integrate $f_{\alpha\mathbf{k}}$ from (5.77), we need the identity

$$\exp\left(i\frac{k_\perp v_\perp}{\omega_c}\sin(\phi - \omega_c\tau)\right) = \sum_{n=-\infty}^{\infty} J_n\left(\frac{k_\perp v_\perp}{\omega_c}\right)\exp[in(\phi - \omega_c\tau)] ,$$

where $J_n$ are the ordinary Bessel functions of the first kind. After a few pages of calculation the dielectric tensor is written in the form

$$\mathscr{K}(\omega,\mathbf{k}) = \left(1-\sum_\alpha \frac{\omega_{p\alpha}^2}{\omega^2}\right)\mathscr{I} - \sum_\alpha \sum_{n=-\infty}^{\infty} \frac{2\pi\omega_{p\alpha}^2}{n_{\alpha 0}\omega^2} \int_0^\infty \int_{-\infty}^{\infty} v_\perp dv_\perp dv_\parallel \left(k_\parallel \frac{\partial f_{\alpha 0}}{\partial v_\parallel} + \frac{n\omega_{c\alpha}}{v_\perp}\frac{\partial f_{\alpha 0}}{\partial v_\perp}\right) \frac{\mathscr{S}_{n\alpha}(v_\parallel, v_\perp)}{k_\parallel v_\parallel + n\omega_{c\alpha} - \omega} . \tag{5.85}$$

The tensor $\mathscr{S}_{n\alpha}$ is of the form

$$\mathscr{S}_{n\alpha}(v_\parallel, v_\perp) = \begin{bmatrix} \frac{n^2\omega_{c\alpha}^2}{k_\perp^2} J_n^2 & \frac{inv_\perp\omega_{c\alpha}}{k_\perp} J_n J_n' & \frac{nv_\parallel\omega_{c\alpha}}{k_\perp} J_n^2 \\ -\frac{inv_\perp\omega_{c\alpha}}{k_\perp} J_n J_n' & v_\perp^2 J_n'^2 & -iv_\parallel v_\perp J_n J_n' \\ \frac{nv_\parallel\omega_{c\alpha}}{k_\perp} J_n^2 & iv_\parallel v_\perp J_n J_n' & v_\parallel^2 J_n^2 \end{bmatrix} \tag{5.86}$$

and $J_n' = dJ_n/d(k_\perp v_\perp/\omega_{c\alpha})$.

$\mathbf{B}_0$ makes the plasma anisotropic. The temperature may now be different in parallel and perpendicular directions as, e.g., in the case of *bi-Maxwellian* distribution

$$f_{\alpha 0} = \frac{m_\alpha}{2\pi k_B T_{\alpha\perp}} \sqrt{\frac{m_\alpha}{2\pi k_B T_{\alpha\parallel}}} \exp\left[-\frac{m_\alpha}{2k_B}\left(\frac{v_\perp^2}{T_{\alpha_\perp}} + \frac{v_\parallel^2}{T_{\alpha_\parallel}}\right)\right] . \tag{5.87}$$

When it is inserted into the elements of $\mathscr{K}$, the resonant integrals in the direction of $v_\parallel$ can be expressed in terms of the plasma dispersion function $Z$. The wave modes are again the non-trivial solutions of $\mathscr{K}\cdot\mathbf{E} = 0$.

The mode structure has grown in complexity from the unmagnetized Landau solution:

- The distinction between electrostatic and electromagnetic modes is no more exact; $\mathbf{E} \parallel \mathbf{k}$ can be satisfied approximately but also the electromagnetic modes may have an electric field component along $\mathbf{k}$.
- The Bessel functions introduce harmonic mode structure organized according to $\omega \approx n\omega_{c\alpha}$ for each species $\alpha$.
- The resonance $\omega = \mathbf{k}\cdot\mathbf{v}$ in the isotropic plasma is replaced by

$$\omega - n\omega_{c\alpha} = k_\parallel v_\parallel . \tag{5.88}$$

Thus only the velocity component $\parallel \mathbf{B}_0$ is associated with Landau damping and only for waves with $k_\parallel \neq 0$.

### 5.6.1 Perpendicular propagation ($\theta = \pi/2$)

For perpendicular propagation $k_\parallel = 0$ and the wave equation reduces to

$$\begin{bmatrix} K_{xx} & K_{xy} & 0 \\ K_{yx} & K_{yy} & 0 \\ 0 & 0 & K_{zz} \end{bmatrix} \cdot \begin{bmatrix} E_x \\ E_y \\ E_z \end{bmatrix} = 0 . \tag{5.89}$$

Assuming an isotropic distribution one of the solutions is

$$K_{zz} = 1 - \frac{k^2c^2}{\omega^2} - \frac{2\pi}{\omega}\sum_\alpha\sum_n \omega_{p\alpha}^2 \int_{-\infty}^{\infty} dv_\parallel \int_0^\infty \frac{J_n^2 f_{\alpha 0} v_\perp}{\omega - n\omega_{c\alpha}} dv_\perp = 0 . \tag{5.90}$$

This is the the Vlasov theory counterpart of the $O$ mode of Chap. 4

$$\omega^2 \approx k^2c^2 + \omega_{pe}^2 . \tag{5.91}$$

An additional series of modes with narrow bands just *above* the harmonics of the cyclotron frequency are found

$$\omega = n\omega_{c\alpha}\left\{1 + O\left[\frac{\omega_{p\alpha}^2}{k^2c^2}(kr_{L\alpha})^{2n}\right]\right\} . \tag{5.92}$$

These modes are *electrostatic cyclotron waves*. Both electrons and all ion species have their own cyclotron mode families.

The remaining solutions are found from the determinant

$$\begin{vmatrix} K_{xx} & K_{xy} \\ -K_{xy} & K_{yy} \end{vmatrix} = 0 . \tag{5.93}$$

These equations express the mode for which $\mathbf{E}\cdot\mathbf{k} \ll \mathbf{E}\times\mathbf{k}$, which is the $X$-mode. It has all branches found in Sect. 4.3.3, including the high-frequency mode, the mode below the upper hybrid resonance, and the mode below the lower hybrid resonance. The lowest frequency mode is called the *magnetosonic mode* that at lowest frequencies (longest wavelengths) is the same as the magnetosonic wave in MHD (Chap. 6).

In addition a new set of electrostatic (i.e., $\mathbf{E}\cdot\mathbf{k} \gg \mathbf{E}\times\mathbf{k}$) modes are found in the Vlasov theory. These modes are called *Bernstein modes* and they exist both for electrons (modes labeled f in Fig. 4.5) and for all ion species (modes labeled C in Fig. 5.4). The exactly perpendicular modes are not Landau damped, but they cannot propagate at the cyclotron frequencies. If the modes have finite $k_\parallel$, they experience damping, which for $n \neq 0$ is called cyclotron damping.

The Bernstein modes and electrostatic cyclotron modes have different characteristics. At frequencies below the hybrid resonance frequencies (both for electron and ion modes) the Bernstein modes can have any frequency within the band $(n\omega_{c\alpha}, (n+1)\omega_{c\alpha})$, whereas above the hybrid frequencies the modes are limited to frequencies above but near each harmonic of the gyro frequency. The electrostatic cyclotron waves, on the other hand, are

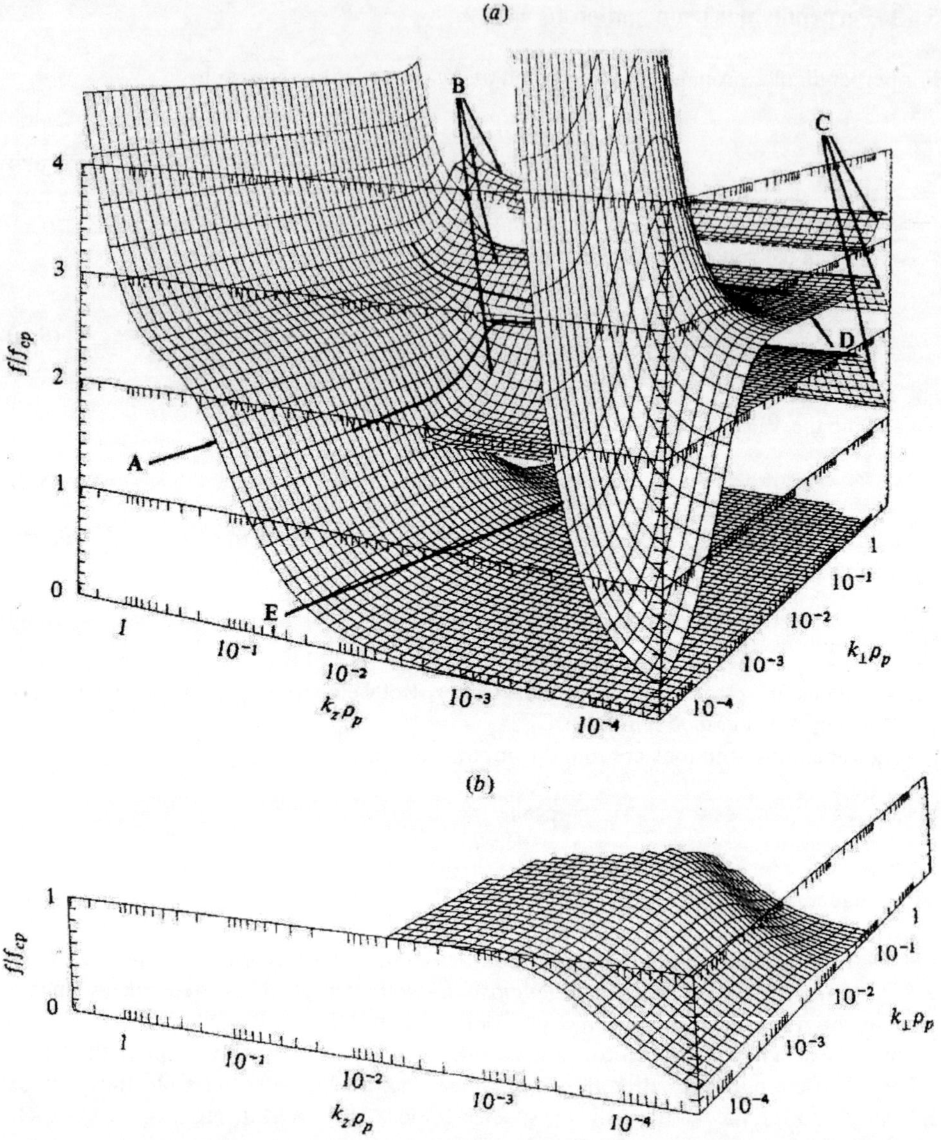


**Fig. 5.4** Dispersion surfaces for low-frequency waves [André, 1985]. Frequencies are normalized to proton gyro frequency and wave numbers to proton gyro radius. A) ion–acoustic waves, B) electrostatic ion cyclotron waves, C) ion–Bernstein modes, D) lower hybrid plateau, E) low-frequency part of the whistler mode. The surface containing the electromagnetic ion cyclotron waves and Alfvén waves is shown separately for the sake of clarity.

always limited to frequencies relatively close to the gyro frequency. This is best illustrated for the ion–Bernstein modes (C) and electrostatic ion cyclotron modes (B) in Fig. 5.4. The figure also illustrates the fact that the electrostatic ion cyclotron modes predominantly have a larger $k_\parallel$ than the Bernstein modes.

### 5.6.2 Parallel propagation ($\theta = 0$)

At the lowest frequencies ($\omega \ll \omega_{ci}$) we find the Alfvén wave

$$\omega_r = \frac{k_\parallel v_A}{\sqrt{1+v_A^2/c^2}} \tag{5.94}$$

that is the same as the MHD mode with the "cold plasma correction" in the denominator arising from the inclusion of the displacement current into Ampère's law. As the Vlasov equation is solved for the full set of Maxwell's equations, the solutions shall have both cold and MHD approximations as limiting cases.

Note that the Vlasov theory introduces damping of Alfvén waves, which will not be found in MHD (Chap. 6)

$$\omega_i = -\frac{\omega_{pi}^2}{|k_\parallel|}\frac{1}{1+c^2/v_A^2}\sqrt{\frac{\pi m_i}{8k_B T_i}}\exp\left(\frac{-B^2}{2\mu_0 n_e k_B T_i}\frac{\omega_{ci}^2}{\omega_r^2}\right). \tag{5.95}$$

The damping rate is very small at low frequencies. When $\omega \to \omega_{ci}$, the mode approaches the ion cyclotron resonance from below the same way as in the cold theory and the damping rate increases. At this limit the mode is called the *electromagnetic ion cyclotron wave*, which is damped by the resonant ions. In Vlasov theory the cyclotron resonance is no more a singularity, but becomes correctly described, including energy transfer from the waves to the particles.

Other parallel modes are, of course, the electromagnetic *R*- and *L*-modes and the whistler mode. Also the whistler mode is damped, although the damping rate is small except at short wavelengths (large $k$). Near the electron gyro frequency the whistler mode goes over to the *electromagnetic electron cyclotron wave*.

The most important differences between electrostatic and electromagnetic cyclotron waves are their polarization and harmonic structures

$$\begin{array}{llll}\text{Electromagnetic}: & \mathbf{k} \parallel \mathbf{B}_0 & \omega \approx \omega_{c\alpha} & \text{no harmonic structure} \\ \text{Electrostatic}: & \mathbf{k} \perp \mathbf{B}_0 & \omega \approx n\omega_{c\alpha} & \text{harmonic structure}\end{array}$$

The electromagnetic cyclotron modes are below the gyro frequencies, whereas the electrostatic modes are above the harmonics of the gyro frequencies.

### 5.6.3 Propagation at arbitrary angles

As in cold plasmas, the waves can propagate at arbitrary angles between 0 and 90°. The dispersion surface description is a convenient way to illustrate the various wave modes

(Fig. 5.4) The figure is similar to Fig. 4.5, but calculated for ion-related modes. Both figures were produced by numerically solving the linearized Vlasov equation for homogeneously magnetized electron–ion plasma where both species have Maxwellian distribution functions.

For example, the whistler mode is on the surface that joins a "plateau" at the lower hybrid frequency in the perpendicular direction. Note that $\omega_i$ varies strongly from one point in the mode structure to another and some parts of the surfaces are strongly damped. For example, the Bernstein modes propagate only close to the perpendicular direction. The electrostatic ion cyclotron waves penetrate somewhat deeper into the cube, especially if electrons are warmer than ions. The ion–acoustic surface is also strongly damped unless $T_e \gg T_i$. The entire mode structure is very sensitive to the actual shape of the velocity distribution function, to the relative temperatures, and also to the ion composition.

# 6. Magnetohydrodynamics

In Chapter 2 we discussed the derivation of MHD equations in the hard way, starting from the Vlasov equation, taking velocity moments and making several approximations. This is not how MHD was first formulated. Instead the starting point was classical hydrodynamics that was reformulated for electrically conductive fluids under the influence of the magnetic field. We begin the discussion with a brief review of this procedure

## 6.1 From Hydrodynamics to Conservative MHD Equations

The equations of ordinary gas dynamics can be written as

$$\frac{\partial \rho}{\partial t} = -\nabla \cdot (\rho \mathbf{V}) \tag{6.1}$$

$$\rho \frac{d\mathbf{V}}{dt} = -\nabla P + \nu \rho \nabla^2 \mathbf{V} \tag{6.2}$$

$$\frac{d}{dt}(P\rho^{-\gamma}) = 0 \,, \tag{6.3}$$

where $d/dt = \partial/\partial t + \mathbf{V} \cdot \nabla$ and $\gamma$ is the polytropic index. This is a set of five equations for five unknowns: density $\rho$, pressure $P$, and three velocity components. Equations (6.2) are known as the *Navier–Stokes equations*, where $\nu$ is the viscosity. If viscosity can be neglected, as we often do in MHD, the equations are called the *Euler equations*.[1]

Of these equations (6.1) is given in the *conservation form* $\partial F/\partial t + \nabla \cdot \mathbf{G} = 0$, where $F$ is the conserved quantity and $G$ the corresponding flux quantity. Often, particularly when doing numerical fluid simulations in conservative systems, it is convenient to write the whole theory in terms of conserved quantities. In hydrodynamics this requires that Euler equations can be used instead of Navier–Stokes equations (i.e., $\nu = 0$) because viscosity causes dissipation, making the system is non-conservative.

[1] In mathematics and physics it is not so easy to keep track of all the equations that have been named in honor of Leonhard Euler!

H.E.J. Koskinen, *Physics of Space Storms: From the Solar Surface to the Earth*, Springer Praxis Books, DOI 10.1007/978-3-642-00319-6_6,

The whole set of equations of conservative hydrodynamics can be formulated as

$$\frac{\partial \rho}{\partial t} = -\nabla \cdot \mathbf{p} \tag{6.4}$$

$$\frac{\partial \mathbf{p}}{\partial t} = -\nabla \cdot \left( \frac{\mathbf{pp}}{\rho} + P\mathscr{I} \right) \tag{6.5}$$

$$\frac{\partial u}{\partial t} = -\nabla \cdot \left[ (u+P)\frac{\mathbf{p}}{\rho} \right] , \tag{6.6}$$

where $\mathbf{p} = \rho \mathbf{V}$ is the momentum density (mass density flux), $\mathbf{pp}$ denotes the direct product (dyad) with components $p_i p_j$, $\mathscr{I}$ is the unit dyad, and $u$ the total energy density related to the pressure as

$$u = \frac{P}{\gamma - 1} + \frac{\mathbf{p}^2}{2\rho} . \tag{6.7}$$

$P/(\gamma-1)$ is the thermal energy density and $\mathbf{p}^2/2\rho$ the kinetic energy density. Variables $(\rho, \mathbf{V}, P)$ are called *primitive variables*, whereas $(\rho, \mathbf{p}, u)$ are *conserved variables.*

**Train your brain** by transforming (6.1)–(6.3) to (6.4)–(6.6)

In MHD we must add Ampère's force $\mathbf{J} \times \mathbf{B}$ to the momentum equation

$$\frac{\partial \mathbf{p}}{\partial t} = -\nabla \cdot \left( \frac{\mathbf{pp}}{\rho} + P\mathscr{I} \right) + \mathbf{J} \times \mathbf{B} . \tag{6.8}$$

This is not yet in the conservation form, but neglecting the displacement current as we normally do in MHD we can express Ampère's force as

$$\mathbf{J} \times \mathbf{B} = -\frac{1}{\mu_0} \mathbf{B} \times (\nabla \times \mathbf{B}) = -\nabla \left( \frac{B^2}{2\mu_0} \right) + \frac{1}{\mu_0} \nabla \cdot (\mathbf{BB}) . \tag{6.9}$$

With this the momentum equation can be written in the conservation form as

$$\frac{\partial \mathbf{p}}{\partial t} = -\nabla \cdot \left[ \frac{\mathbf{pp}}{\rho} + \left( P + \frac{\mathbf{B}^2}{2\mu_0} \right) \mathscr{I} - \frac{1}{\mu_0} \mathbf{BB} \right] . \tag{6.10}$$

Thus the momentum of the magnetic field is taken care in a natural way.

The energy equation (6.3) is automatically conservative. However, it is instructive to write it in the form that explicitly illustrates the conservation of total energy density $u$ in the same way as in gas dynamics. This is straightforward to do because the magnetic energy density is $B^2/(2\mu_0)$. Thus

$$u = \frac{P}{\gamma - 1} + \frac{\mathbf{p}^2}{2\rho} + \frac{\mathbf{B}^2}{2\mu_0} \tag{6.11}$$

and the energy equation can be written as

$$\frac{\partial u}{\partial t} = -\nabla \cdot \left[ \left( u + P - \frac{\mathbf{B}^2}{2\mu_0} \right) \frac{\mathbf{p}}{\rho} + \frac{1}{\mu_0 \rho} \mathbf{B} \times (\mathbf{p} \times \mathbf{B}) \right] . \tag{6.12}$$

We need one more equation to describe the time evolution of the magnetic field. Because we are now interested in conservative MHD, we must limit the discussion to the ideal MHD case, where $\mathbf{E} = -\mathbf{V} \times \mathbf{B}$. Inserting this into Faraday's law we get

$$\frac{\partial \mathbf{B}}{\partial t} = \nabla \times (\mathbf{V} \times \mathbf{B}) , \tag{6.13}$$

which after replacing $\mathbf{V}$ by the conserved quantity $\mathbf{p}$ reads

$$\frac{\partial \mathbf{B}}{\partial t} = \nabla \times \left( \frac{\mathbf{p}}{\rho} \times \mathbf{B} \right) . \tag{6.14}$$

Now we have the complete set of 8 equations (6.4, 6.10, 6.12, 6.14) for 8 conservative variables $(\rho, \mathbf{p}, u, \mathbf{B})$ of the ideal MHD.

**Train your brain**

Use MHD's Ohm's law with Ampère's and Faraday's laws to write the Poynting theorem in the form

$$-\oint_{\partial \mathscr{V}} \mathbf{E} \times \mathbf{H} \cdot d\mathbf{a} = \frac{\partial}{\partial t} \int_{\mathscr{V}} \frac{B^2}{2\mu_0} d^3 r + \int_{\mathscr{V}} \frac{J^2}{\sigma} d^3 r + \int_{\mathscr{V}} \mathbf{V} \cdot \mathbf{J} \times \mathbf{B} d^3 r ,$$

This has already been mentioned in Chap. 1, Eq. (1.12).

Then take the scalar product of $\mathbf{V}$ and the momentum equation

$$\rho_m \left( \frac{\partial}{\partial t} + \mathbf{V} \cdot \nabla \right) \mathbf{V} + \nabla \cdot \mathscr{P} - \mathbf{J} \times \mathbf{B} = 0$$

and derive the energy equation of adiabatic ideal MHD in the form

$$\frac{\partial}{\partial t} \left( \frac{\rho_m V^2}{2} + \frac{P}{\gamma - 1} + \frac{B^2}{2\mu_0} \right) + \nabla \cdot \left( \frac{\rho_m V^2}{2} \mathbf{V} + \frac{\gamma P}{\gamma - 1} \mathbf{V} + \frac{\mathbf{E} \times \mathbf{B}}{\mu_0} \right) = 0 , \tag{6.15}$$

which may be slightly more transparent than (6.12).

## 6.2 Convection and Diffusion

Let us go back to primitive variables without the assumption of ideal MHD. From Ohm's law in resistive MHD and Maxwell's equations it is easy to derive the *induction equation* for the magnetic field

$$\frac{\partial \mathbf{B}}{\partial t} = \nabla \times (\mathbf{V} \times \mathbf{B}) + \frac{1}{\mu_0 \sigma} \nabla^2 \mathbf{B} \, , \tag{6.16}$$

where the *magnetic diffusivity* $\eta = 1/\mu_0 \sigma$ has been assumed to be spatially uniform.

In the frame of reference co-moving with the plasma ($\mathbf{V} = 0$) the induction equation reduces to the *diffusion equation*

$$\frac{\partial \mathbf{B}}{\partial t} = \eta \nabla^2 \mathbf{B} \, . \tag{6.17}$$

Thus if the resistivity is finite, the magnetic field diffuses smoothing out spatial inhomogeneities, local curvature, etc., expressed by the term $\nabla^2 \mathbf{B}$.

The characteristic diffusion time can be found by simple dimensional analysis. Let $L_B$ be the characteristic gradient scale length of the magnetic field. The solution of the diffusion equation is of the form

$$B = B_0 \exp(\pm t/\tau_d) \, , \tag{6.18}$$

where the *magnetic diffusion time* $\tau_d$ is

$$\tau_d = \mu_0 \sigma L_B^2 \, . \tag{6.19}$$

At the limit $\sigma \to \infty$ (ideal MHD), the diffusion term is small and the plasma flow is described by the *convection equation*, which ties the flow and the magnetic field to each other

$$\frac{\partial \mathbf{B}}{\partial t} = \nabla \times (\mathbf{V} \times \mathbf{B}) \, . \tag{6.20}$$

In this case there is no diffusion of the magnetic field across the plasma, and the magnetic field is said to be *frozen-in* to the plasma.

Let us consider the relative strengths of convection and diffusion. Let $\tau$ be the time scale of temporal variations, $V$ the typical velocity, $L_B$ the local gradient scale length, and $\tau_d$ the diffusion time scale. Substituting $\partial/\partial t \to \tau$ and $\nabla \to L_B^{-1}$, and neglecting directions, the induction equation reduces to

$$\frac{B}{\tau} = \frac{VB}{L_B} + \frac{B}{\tau_d} \, . \tag{6.21}$$

The ratio of the terms on the RHS is the dimensionless *magnetic Reynolds number* $R_m$

$$R_m = \mu_0 \sigma L_B V = L_B V/\eta \, . \tag{6.22}$$

$R_m$ has in MHD a role analogous to the Reynolds number in classical hydrodynamics $R = LV/\nu$, where $\nu$ is the viscosity of the fluid.

Although the diffusivity is often small, it is never exactly zero. As a simple example, we can consider the one-dimensional current sheet $B(z,t)\,\mathbf{e}_x$ in the frame of reference co-moving with the plasma ($\mathbf{V} = 0$). Let the initial condition be

$$B(z,0) = \begin{cases} +B_0 \, , \; z > 0 \\ -B_0 \, , \; z < 0 \, . \end{cases} \tag{6.23}$$

In one dimension the diffusion equation is

$$\frac{\partial B}{\partial t} = \eta \frac{\partial^2 B}{\partial z^2} \tag{6.24}$$

with the solution

$$\begin{aligned} B(z,t) &= B_0 \,\mathrm{erf}\left(\frac{z}{\sqrt{4\eta t}}\right) \\ &= \frac{2B_0}{\sqrt{\pi}} \int_0^{z/\sqrt{4\eta t}} \exp(-u^2)\, du \, . \end{aligned} \tag{6.25}$$

The total magnetic flux remains constant (=0) but the energy of the field $\int B^2/2\mu_0 \, dz$ decreases with time. (Strictly speaking, this configuration is infinite, but we can think that there is an outer boundary somewhere.) It is an easy exercise to show that

$$\frac{\partial}{\partial t} \int \frac{B^2}{2\mu_0} dz = -\int \frac{J^2}{\sigma} dz \, . \tag{6.26}$$

Thus the energy is dissipated through *Ohmic heating*, also known as *Joule heating*, in the current sheet.

***Example: Conductivity and diffusivity in the Sun***

Almost everywhere in the Sun the classical resistivity is very small. Important exceptions are the photosphere and lower chromosphere where the degree of ionization is low and collisions with neutrals limit the current flow.

The photospheric conductivity is about $10\,\mathrm{S\,m^{-1}}$ ( = $10\,\Omega^{-1}\,\mathrm{m^{-1}}$ = $10\,\mathrm{mho\,m^{-1}}$). This yields $\eta \approx 10^5\,\mathrm{m^2\,s^{-1}}$. For photospheric granules $L_B \approx 1000$ km and $V \approx 2\,\mathrm{km\,s^{-1}}$). These numbers give $R_m \approx 20\,000 \gg 1$. This predicts very weak diffusion, indeed. This is not consistent with the actually observed magnetic fields, whose evolution implies some 200 times faster diffusivity and correspondingly smaller $R_m$. The explanation is that the turbulence in the upper convection zone introduces *turbulent diffusivity* $\eta_t \approx 2 \times 10^7\,\mathrm{m^2\,s^{-1}}$, but there is no rigorous way to calculate this number.

The solar gas becomes fully ionized above 2000 km. The effective electron collision time can be estimated using *Spitzer's formula*

$$\tau_{ei}(\mathrm{s}) = 0.266 \times 10^6 \, \frac{T^{3/2}(\mathrm{K})}{n_e(\mathrm{m^{-3}}) \ln \Lambda} \, , \tag{6.27}$$

where $\ln \Lambda$ is the Coulomb logarithm (of the order of 20). Now the classical conductivity

$$\sigma = \frac{n_e e^2 \tau_{ei}}{m_e} \tag{6.28}$$

has the numerical value

$$\sigma(\mathrm{S\,m^{-1}}) = 1.53 \times 10^{-2} \frac{T^{3/2}(\mathrm{K})}{\ln \Lambda} \, . \tag{6.29}$$

Using $\ln \Lambda = 20$ the diffusivity is given by

$$\eta(\mathrm{m^2\,s^{-1}}) = 10^9 \times T^{-3/2}(\mathrm{K}) \, . \tag{6.30}$$

For a typical coronal temperature $T = 10^6$ K this yields diffusivity of only $\eta = 1\,\mathrm{m^2\,s^{-1}}$. In the corona the scale lengths and the characteristic speeds also become larger when moving outward. Consequently, the expanding solar wind is an excellent example of ideal MHD plasma.

However, even in the solar wind plasma deviations from the ideal conditions may arise. The reason is that when plasmas originating from different sources convect toward each other, their frozen-in configuration may be very different from each other, e.g., the magnetic field directions may be anti-parallel leading to formation of very thin current sheets. In such cases the collective interactions can give rise to wave–wave and wave–particle interactions resulting, e.g., in turbulent diffusivity or in effective (anomalous) resistivity. Another non-ideal example is the formation of shocks in cases when the relative flow speed is supersonic, or supermagnetosonic. We will return to these effects later.

## 6.3 Frozen-in Field Lines

Hannes Alfvén was the first to realize the importance of the convection of the plasma and the magnetic field together and he introduced the concept of frozen-in field lines to illustrate this. Later he denounced the concept as "pseudopedagogical", which was his expression for something that makes us to think that we have understood a phenomenon, whereas we actually have misunderstood it. Alfvén's criticism was based on the picture of moving magnetic field lines. According to Maxwell's equations the *magnetic field* is a fundamental physical entity that may change both in time and space. The *magnetic field line* is just a mathematical abstraction and there is nothing physical in the motion of magnetic field lines. However, the frozen-in concept is quite useful in plasma physics when interpreted correctly.

The frozen-in concept can be formulated both in differential and integral forms. We start from the differential description assuming ideal MHD. Let two plasma elements move according to Fig. 6.1. Let the elements be on the same field line $\mathbf{B}(t)$ at the time $t$. To be on the same field line means that if we trace the field $\mathbf{B}$ from one plasma element, we end up at the other. In this sense the plasma elements are *magnetically connected* to each other. The (vector) distance between the elements is $\triangle\mathbf{l}$. During the time $dt$ the elements move

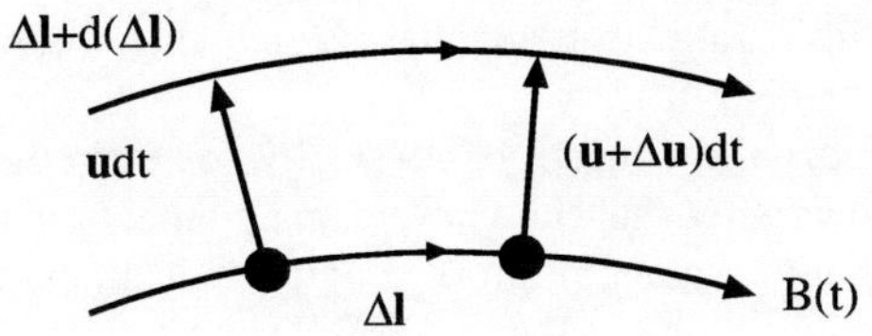


**Fig. 6.1** Illustration of the proof that two plasma elements originally on a common field line are also on a common field line after time $dt$.

the distances $\mathbf{u}\,dt$ and $(\mathbf{u}+\triangle\mathbf{u})\,dt$, where $\mathbf{u}(\mathbf{r},t)$ is the plasma flow velocity. Now we have to show that the elements are on a common field line also at the time $t+dt$, i.e., the path $\triangle\mathbf{l}+d(\triangle\mathbf{l})$ is along the field line of $\mathbf{B}(t+dt)$. Here the spatial differential is denoted by $\triangle$ and the total time differential by $d$.

Write $d(\triangle\mathbf{l})$ in terms of $\mathbf{u}$. The first term in the Taylor series of $\mathbf{u}$ is

$$\triangle\mathbf{u} = (\triangle\mathbf{l}\cdot\nabla)\mathbf{u}\,. \tag{6.31}$$

From Fig. 6.1 we see that

$$\triangle\mathbf{l}+d(\triangle\mathbf{l}) = \triangle\mathbf{l}+(\mathbf{u}+\triangle\mathbf{u})\,dt-\mathbf{u}\,dt\,, \tag{6.32}$$

which leads to

$$\frac{d(\triangle\mathbf{l})}{dt} = \triangle\mathbf{u} = (\triangle\mathbf{l}\cdot\nabla)\mathbf{u}. \tag{6.33}$$

The convection equation gives for the magnetic field

$$\begin{aligned}\frac{\partial\mathbf{B}}{\partial t} &= \nabla\times(\mathbf{u}\times\mathbf{B})\\ &= (\mathbf{B}\cdot\nabla)\mathbf{u}-(\mathbf{u}\cdot\nabla)\mathbf{B}-\mathbf{B}(\nabla\cdot\mathbf{u})\,,\end{aligned} \tag{6.34}$$

where $\nabla\cdot\mathbf{B}=0$ was used. In the frame moving with the plasma

$$\frac{d\mathbf{B}}{dt} = \frac{\partial\mathbf{B}}{\partial t}+(\mathbf{u}\cdot\nabla)\mathbf{B} = (\mathbf{B}\cdot\nabla)\mathbf{u}-\mathbf{B}(\nabla\cdot\mathbf{u})\,. \tag{6.35}$$

Calculate next $d(\triangle\mathbf{l}\times\mathbf{B})/dt$

$$\begin{aligned}\frac{d}{dt}(\triangle\mathbf{l}\times\mathbf{B}) &= \frac{d(\triangle\mathbf{l})}{dt}\times\mathbf{B}+\triangle\mathbf{l}\times\frac{d\mathbf{B}}{dt}\\ &= [(\triangle\mathbf{l}\cdot\nabla)\mathbf{u}]\times\mathbf{B}+\triangle\mathbf{l}\times[(\mathbf{B}\cdot\nabla)\mathbf{u}-\mathbf{B}(\nabla\cdot\mathbf{u})]\,.\end{aligned} \tag{6.36}$$

Because $\triangle\mathbf{l}$ originally was parallel to $\mathbf{B}$, $\triangle\mathbf{l}\times\mathbf{B}=0$, and the third term on the RHS is zero. For the same reason $\triangle\mathbf{l}$ and $\mathbf{B}$ can be interchanged in the first term on the RHS. Thus the first and the second term are the same except for their sign and we have

$$\frac{d}{dt}(\triangle\mathbf{l}\times\mathbf{B}) = 0\,. \tag{6.37}$$

Consequently, $\triangle\mathbf{l}$ remains parallel to $\mathbf{B}$ and plasma elements originally on a common field line remain on a common field line.

This picture of the frozen-in concept is physically sound in the context of Maxwell's equations. We have shown only that the two plasma elements remain magnetically connected at all times without an assumption of moving field lines, although the *picture* of moving field lines is useful as long as the ideal MHD approximation is valid.

We can also analyze the frozen-in concept in the integral formulation by calculating how well the magnetic flux is preserved when the plasma moves. We expect that in ideal MHD the magnetic flux through a closed loop moving with plasma should remain constant

$$\frac{d\Phi}{dt} = \frac{d}{dt}\int \mathbf{B}\cdot d\mathbf{S} = 0\,. \tag{6.38}$$

To prove this we consider a closed contour $C$ at time $t$ moving with the plasma velocity $\mathbf{V}(\mathbf{r},t)$. At time $t+\triangle t$ the loop has moved, and possibly deformed, to $C'$. Let $S$ and $S'$ be the surfaces closed by $C$ and $C'$. Let $d\mathbf{l}$ be an arc element on $C$. It moves in time $\triangle t$ the distance $\mathbf{V}\triangle t$ and sweeps out the area $d\mathbf{l}\times\mathbf{V}\triangle t$. Form now the closed surface consisting of $S$, $S'$ and of the surface swept out by $\mathbf{V}\triangle t$ when $d\mathbf{l}$ is integrated along the closed contour $C$. Because the magnetic field is divergence-free, the total flux through this closed surface at time $t+\triangle t$ must vanish

$$-\int_S \mathbf{B}(t+\triangle t)\cdot d\mathbf{S} + \int_{S'} \mathbf{B}(t+\triangle t)\cdot d\mathbf{S}' + \oint_C \mathbf{B}(t+\triangle t)\cdot d\mathbf{l}\times\mathbf{V}\triangle t = 0\,. \tag{6.39}$$

The positive direction of $d\mathbf{S}$ is outward from the closed volume.

Now we can calculate $d\Phi/dt$ when the contour $C$ moves with the fluid

$$\begin{aligned}
\frac{d\Phi}{dt} &= \lim_{\triangle t\to 0}\frac{\Phi_{C'}(t+\triangle t)-\Phi_C(t)}{\triangle t}\\
&= \lim_{\triangle t\to 0}\frac{\int\mathbf{B}(t+\triangle t)\cdot d\mathbf{S}' - \int\mathbf{B}(t)\cdot d\mathbf{S}}{\triangle t}\\
&= \lim_{\triangle t\to 0}\frac{\int[\mathbf{B}(t+\triangle t)-\mathbf{B}(t)]\cdot d\mathbf{S}}{\triangle t} - \oint_C \mathbf{B}(t+\triangle t)\cdot d\mathbf{l}\times\mathbf{V}\\
&= \int\frac{\partial\mathbf{B}}{\partial t}\cdot d\mathbf{S} - \oint_C(\mathbf{V}\times\mathbf{B})\cdot d\mathbf{l}\\
&= \int\left[\frac{\partial\mathbf{B}}{\partial t} - \nabla\times(\mathbf{V}\times\mathbf{B})\right]\cdot d\mathbf{S}\\
&= -\int\nabla\times(\mathbf{E}+\mathbf{V}\times\mathbf{B})\cdot d\mathbf{S}\,.
\end{aligned} \tag{6.40}$$

Here (6.39) was used to transform $\int\mathbf{B}(t+\triangle t)\cdot d\mathbf{S}'$ to $\int\mathbf{B}(t+\triangle t)\cdot d\mathbf{S}$. The integrand vanishes if

$$\mathbf{E}+\mathbf{V}\times\mathbf{B} = -\nabla\Psi\,, \tag{6.41}$$

where $\Psi$ is a scalar. This is a necessary and sufficient condition to preserve the magnetic flux. Thus clearly in ideal MHD where $\Psi$ is a constant, the magnetic flux and plasma flow together.

In ideal MHD the primary fields are **B** and **V**. The electric current and the electric field are calculated from these using Ampère's law and ideal-MHD Ohm's law. In the Maxwellian sense the current is the source of **B**. Thus the flow of the magnetic field, e.g., with the solar wind, means that plasma particles also carry the current system along the flow.

In space plasmas the first correction to the ideal MHD is often not the resistive term but the Hall term $\mathbf{J} \times \mathbf{B}/(ne)$

$$\mathbf{E} + \mathbf{V} \times \mathbf{B} = \frac{1}{ne} \mathbf{J} \times \mathbf{B} \,. \tag{6.42}$$

This is expected to be the case, e.g., near current sheets separating magnetic fields of different strength and direction. In this *Hall MHD* the magnetic field becomes frozen-in to the electron flow

$$\mathbf{E} = -\mathbf{V}_e \times \mathbf{B} \,. \tag{6.43}$$

Physically this is a consequence of the fact that the electron gyro motion is more strongly tied to the magnetic field than the ion motion. However, with this correction we have lost much of the meaning of the frozen-in concept because the mass flow, determined by the heavier ions, is separated from the evolution of the magnetic field, at least locally.

The breakdown of the frozen-in condition is one of the most important phenomena in space plasmas. The change of interconnection between plasma elements can, in general, be called *reconnection*. Reconnection is one of the most important concepts from the viewpoint of space storms and will be discussed thoroughly in Chap. 8, where we also briefly discuss the other non-ideal contributions to the generalized Ohm's law (Eq. 2.135).

## 6.4 Magnetohydrostatic Equilibrium

Consider next MHD plasma in a time-independent ($d/dt = 0$) equilibrium. Assuming scalar pressure ($\nabla \cdot \mathcal{P} \rightarrow \nabla P$) the momentum equation reduces to

$$\mathbf{J} \times \mathbf{B} = \nabla P \,. \tag{6.44}$$

This gives $\mathbf{B} \cdot \nabla P = 0$ and $\mathbf{J} \cdot \nabla P = 0$. Thus **B** and **J** are vector fields on surfaces of constant pressure.

We have already seen in (6.9) that

$$\mathbf{J} \times \mathbf{B} = -\nabla \left( \frac{B^2}{2\mu_0} \right) + \frac{1}{\mu_0} \nabla \cdot (\mathbf{BB}) \,.$$

The first term on the RHS is the gradient of the magnetic energy density, i.e., of the magnetic pressure $B^2/(2\mu_0)$. The second term is the divergence of the tensor $\mathbf{BB}/\mu_0$, which describes the stress and torsion arising from the inhomogeneities of the magnetic field. By applying Ampère's law we can eliminate the current and write the equation for *magneto-*

*hydrostatic equilibrium* as

$$\nabla \cdot \mathscr{P} = -\frac{1}{\mu_0}\mathbf{B} \times (\nabla \times \mathbf{B}) \,. \tag{6.45}$$

Assuming scalar pressure and negligible $\nabla \cdot (\mathbf{BB})$ the sum of the magnetic and plasma pressures is constant

$$\nabla \left( P + \frac{B^2}{2\mu_0} \right) = 0 \,. \tag{6.46}$$

The *plasma beta*

$$\beta = \frac{2\mu_0 P}{B^2} \tag{6.47}$$

expresses the ratio of the plasma and magnetic pressures.

The current perpendicular to $\mathbf{B}$ is now

$$\mathbf{J}_\perp = \frac{\mathbf{B} \times \nabla P}{B^2} \,. \tag{6.48}$$

This total current is often called *diamagnetic current*. It is the sum of all current elements in the plasma. In addition to gradient and curvature currents an inhomogeneous plasma density may cause net *magnetization current*

$$\mathbf{J}_M = \nabla \times \mathbf{M} \,. \tag{6.49}$$

Here the *magnetization* $\mathbf{M}$ is the density of magnetic moments $\boldsymbol{\mu}$.

The pressure and temperature of the plasma may be anisotropic, so $\beta$ can also be different in the parallel and perpendicular directions ($\beta_\perp \neq \beta_\parallel$). Writing the perpendicular and parallel pressures as $P_\perp = nW_\perp$ and $P_\parallel = 2nW_\parallel$, where $n$ is the number density of the plasma particles, we can express the curvature and gradient currents in terms of the pressure

$$\mathbf{J}_C = \frac{P_\parallel}{B}(\nabla \times \mathbf{b})_\perp \tag{6.50}$$

$$\mathbf{J}_G = P_\perp \nabla \frac{1}{B} \times \mathbf{b} \,, \tag{6.51}$$

where $\mathbf{b} = \mathbf{B}/B$ is the unit vector in the direction of $\mathbf{B}$. The magnetization is $\mathbf{M} = n\boldsymbol{\mu} = -n(W_\perp/B)\mathbf{b}$ and

$$\mathbf{J}_M = \nabla \times \mathbf{M} = -\nabla \left( \frac{P_\perp}{B}\mathbf{b} \right) \,. \tag{6.52}$$

Summing all currents we find

$$\mathbf{J} = \frac{\mathbf{B} \times \nabla P}{B^2} + \frac{P_\parallel - P_\perp}{B}(\nabla \times \mathbf{b})_\perp \,, \tag{6.53}$$

which yields magnetohydrostatic equilibrium equations for anisotropic plasma

$$\mathbf{J} \times \mathbf{B} = \nabla_{\perp} P_{\perp} + (P_{\parallel} - P_{\perp})\mathbf{b} \cdot \nabla \mathbf{b} = (\nabla \cdot \mathscr{P})_{\perp} \quad (6.54)$$

$$(\nabla \cdot \mathscr{P})_{\parallel} = 0 . \quad (6.55)$$

Parker was the first to derive (6.53) from single particle motion and the equation is sometimes named after him.

In time-dependent problems we must include inertial currents, of which the first-order term is the polarization current

$$\mathbf{J}_P = \frac{\rho_m}{B^2} \frac{d\mathbf{E}}{dt} . \quad (6.56)$$

## 6.5 Field-aligned Currents

The Parker equation (6.53) does not say anything of possible currents along the magnetic field. If $\beta \ll 1$ in magnetohydrostatic equilibrium, the pressure gradient is negligible and thus

$$\mathbf{J} \times \mathbf{B} = 0 , \quad (6.57)$$

i.e, the electric current must flow along the magnetic field. Because a current creates a magnetic field around it, the self-consistent *field-aligned current* (FAC) consists of spiraling magnetic field lines and the resulting structure is often characterized as a *flux rope*. Another term is *force-free field* because the magnetic force on the plasma is zero. The force-free equilibrium is an approximation, but often a very good one, to the momentum equation.

### 6.5.1 Force-free fields

The innocent-looking equation $\mathbf{J} \times \mathbf{B} = 0$ is actually pretty hard to solve. The problem lies in its nonlinearity. Using Ampère's law we can write it as

$$(\nabla \times \mathbf{B}) \times \mathbf{B} = 0 . \quad (6.58)$$

That $\mathbf{B}_1$ and $\mathbf{B}_2$ are two solutions of this equation *does not imply* that $\mathbf{B}_1 + \mathbf{B}_2$ would be another solution.

We can express the field-alignment as

$$\nabla \times \mathbf{B} = \mu_0 \mathbf{J} = \alpha(\mathbf{r})\mathbf{B} , \quad (6.59)$$

where $\alpha$ is a function of position. Taking divergence of this we get

$$\mathbf{B} \cdot \nabla \alpha = 0 , \quad (6.60)$$

i.e., $\alpha$ is constant along the magnetic field. If $\alpha$ is constant everywhere, the equation

$$\nabla \times \mathbf{B} = \alpha \mathbf{B} \tag{6.61}$$

is linear. Now the sum $\mathbf{B}_1 + \mathbf{B}_2$ of two solutions is also a solution for the force-free field. Taking a curl of (6.61) we get the Helmholtz equation

$$\nabla^2 \mathbf{B} + \alpha^2 \mathbf{B} = 0 \tag{6.62}$$

that has known solutions. That the field fulfills the Helmholtz equation is a necessary but not sufficient condition for the field to be force-free. Of course, the boundary conditions must also be specified correctly.

A special case of force-free magnetic fields is the current-free configuration $\nabla \times \mathbf{B} = 0$. Then the magnetic field can be expressed as the gradient of a scalar potential $\mathbf{B} = \nabla \Psi$. Because $\nabla \cdot \mathbf{B} = 0$, the magnetic field can be found by solving the Laplace equation

$$\nabla^2 \Psi = 0 \tag{6.63}$$

with appropriate boundary conditions. Thus we can use the well-developed methods of potential theory.

***Example: Linear force-free model of a coronal arcade***

Let us consider a simple model for a coronal arcade above the surface of the Sun (for further discussion of coronal loops, see Chap. 12). Let the configuration look like an arc in the $xz$-plane and extend uniformly in the $y$-direction. Let the structure be sinusoidal in the $x$-direction with wave number $k$. The Helmholtz equation has the second spatial derivative, thus the same $z$-dependence is retained after two derivations for sinusoidal and exponential functions. Because the field should vanish at high altitude, we choose the $z$-dependence as $\exp(-lz)$. These choices fulfill the Helmholtz equation if $\alpha^2 < k^2$. In order to have the structure above the solar surface we consider $z > 0$. Let us then seek solutions of the form

$$\begin{aligned} B_x &= B_{x,0} \sin(kx)\, e^{-lz} \\ B_y &= B_{y,0} \sin(kx)\, e^{-lz} \\ B_z &= B_0 \cos(kx)\, e^{-lz} \, . \end{aligned} \tag{6.64}$$

Now the equation $\nabla \times \mathbf{B} = \alpha \mathbf{B}$ yields

$$\begin{aligned} lB_{y,0} &= \alpha B_{x,0} \\ -lB_{x,0} + kB_0 &= \alpha B_{y,0} \\ kB_{y,0} &= \alpha B_0 \end{aligned} \tag{6.65}$$

and the field can be expressed as

$$\begin{aligned} B_x &= (l/k) B_0 \sin(kx)\, e^{-lz} \\ B_y &= (\alpha/k) B_0 \sin(kx)\, e^{-lz} \\ B_z &= B_0 \cos(kx)\, e^{-lz} \, , \end{aligned} \tag{6.66}$$

where $k, l$, and $\alpha$ must be related by

$$l^2 = k^2 - \alpha^2 . \tag{6.67}$$

The projection of the magnetic field lines on the $xy$-plane are straight lines parallel to each other

$$B_y = \frac{\alpha}{(k^2 - \alpha^2)^{1/2}} B_x , \tag{6.68}$$

whereas the projection to the $xz$-plane are arcs, which we were looking for. Visually the arcade looks like a flux rope, one half of which is above the solar surface (Fig. 6.2).

The arcade is simpler if the current is so weak that we can neglect it and use potential theory. We can look for separable solutions in 2D Cartesian space by writing $\Psi = X(x)Z(z)$. From the Laplace equation

$$\frac{\partial^2 \Psi}{\partial x^2} + \frac{\partial^2 \Psi}{\partial z^2} = 0 \tag{6.69}$$

we find

$$\frac{1}{X}\frac{d^2 X}{dx^2} = -\frac{1}{Z}\frac{d^2 Z}{dz^2} = -k^2 , \tag{6.70}$$

where $k$ is a constant. This is fulfilled, e.g., by $\Psi = (B_0/k)\sin kx\, e^{-kz}$, from which we find the field configuration

$$B_x = \frac{\partial \Psi}{\partial x} = B_0 \cos kx\, e^{-kz} \tag{6.71}$$

$$B_z = \frac{\partial \Psi}{\partial z} = -B_0 \sin kx\, e^{-kz} . \tag{6.72}$$

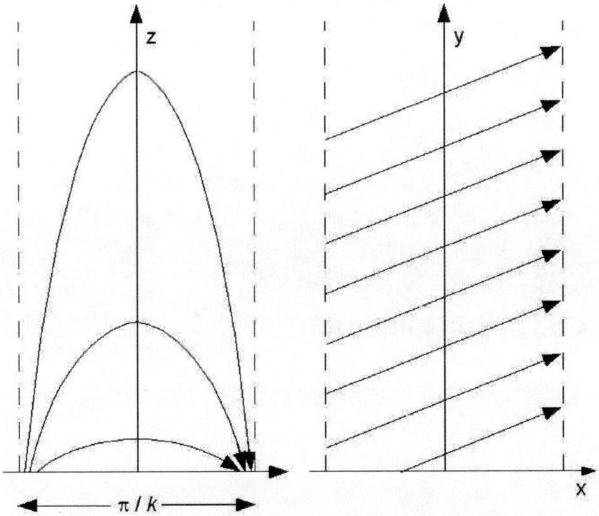


**Fig. 6.2** Sketch of the linear force-free arcade solution.

In the $xz$-plane this looks the same as the force-free solution, but there is no distortion of the arcs in the $y$-direction.

### 6.5.2 Grad–Shafranov equation

The linear force-free arcade magnetic field discussed above is essentially two-dimensional because $B_y$ is the same as $B_x$ multiplied by a constant. Sometimes this kind of translationally symmetric geometry is called $2\frac{1}{2}$-dimensional.

Let us consider the general configuration with translational symmetry, e.g., a large flux rope whose axis can be assumed to be locally straight and $B$ is uniform in the $z$-direction, retaining the scalar plasma pressure $P$ in the calculation. Because $\nabla \cdot \mathbf{B} = 0$, the magnetic field can be written as

$$\mathbf{B} = \left( \frac{\partial A}{\partial y}, -\frac{\partial A}{\partial x}, B_z \right) , \tag{6.73}$$

where $\mathbf{A} = A(x,y)\,\mathbf{e}_z$ is the vector potential. Assume that the magnetic field and the pressure are in force balance, i.e.,

$$\frac{1}{\mu_0}(\nabla \times \mathbf{B}) \times \mathbf{B} - \nabla P = 0 . \tag{6.74}$$

Because none of the functions in (6.74) depends on $z$, its $z$-component reduces to

$$\frac{\partial B_z}{\partial x}\frac{\partial A}{\partial y} - \frac{\partial B_z}{\partial y}\frac{\partial A}{\partial x} = 0 . \tag{6.75}$$

Thus the gradients in the $xy$-plane $\nabla_\perp B_z$ and $\nabla_\perp A$ are parallel to each other and $B_z$ can be expressed as a function of $A$

$$B_z(x,y) = B_z(A(x,y)) . \tag{6.76}$$

Using this we can write the $x$- and $y$-components of (6.74) as

$$\frac{1}{\mu_0}\left(B_z B_z' + \nabla_\perp^2 A\right)\frac{\partial A}{\partial x} + \frac{\partial P}{\partial x} = 0 \tag{6.77}$$

$$\frac{1}{\mu_0}\left(B_z B_z' - \nabla_\perp^2 A\right)\frac{\partial A}{\partial y} + \frac{\partial P}{\partial y} = 0 , \tag{6.78}$$

where the prime indicates the derivative $d/dA$. Also $\nabla_\perp P$ is parallel to $\nabla_\perp A$. Thus

$$P(x,y) = P(A(x,y)) . \tag{6.79}$$

Now (6.77) and (6.78) are both satisfied if

$$\frac{1}{\mu_0}\left(\nabla_\perp^2 A + B_z B_z'\right) + P' = 0 . \tag{6.80}$$

Writing the total pressure as

$$P_t = \frac{B_z^2}{2\mu_0} + P \tag{6.81}$$

we have arrived to the *Grad–Shafranov equation*

$$\frac{1}{\mu_0}\nabla_\perp^2 A + \frac{dP_t}{dA} = 0\,. \tag{6.82}$$

**Train your brain**

The Grad–Shafranov method is not limited to translational symmetry. Find the corresponding equation for azimuthal symmetry ($\partial/\partial\phi = 0$).

Hint: Use cylindrical coordinates, and if you find the problem too hard, consult Boyd and Sanderson [2003].

The Grad–Shafranov equation is a useful tool when looking for ideal MHD solutions under the assumption of translational or rotational symmetries. While it is nonlinear ($B_z^2$), it is a scalar equation and thus much easier to handle than nonlinear vector equations.

In the force-free case the solutions are found by setting $P = 0$. There is no underlying constant-$\alpha$ assumption and thus equation

$$\frac{1}{\mu_0}\nabla_\perp^2 A + \frac{d}{dA}\left(\frac{B_z^2}{2\mu_0}\right) = 0 \tag{6.83}$$

is not limited to the linear force-free configurations.

### 6.5.3 General properties of force-free fields

It is possible to prove a number of useful theorems for force-free fields. For example:

*1. A field with finite magnetic energy cannot be force-free everywhere.*

**Proof**: Because $\mathbf{B}$ falls off faster than $r^{-2}$ at large distances from the origin, the energy can be written as

$$W = \int \frac{B^2}{2\mu_0} d\mathcal{V} = \frac{1}{\mu_0}\int \mathbf{r}\cdot\mathbf{J}\times\mathbf{B}\, d\mathcal{V}\,,$$

which vanishes everywhere if the field is force-free everywhere. Thus a magnetic field that is force-free everywhere must have a singularity. This is trivially true for potential fields, e.g., a dipole has a singularity (the dipole itself) and being current-free it certainly is force-free as well.

*2. If $\mathbf{J}\times\mathbf{B} = 0$ in a finite volume $\mathcal{V}$ and on its boundary S, then $\mathbf{B} = 0$ everywhere.*

**Train your brain** by proving this statement.

This theorem implies that if there is a finite FAC in a finite volume, it must be anchored to the boundary of the volume. This is related to the continuity equation, which states that the sources of parallel currents are sinks of perpendicular currents, and vice versa

$$\nabla \cdot \mathbf{J} = \nabla_{\parallel} \cdot \mathbf{J}_{\parallel} + \nabla_{\perp} \cdot \mathbf{J}_{\perp} = 0 \,. \tag{6.84}$$

*3. An axisymmetric, force-free, poloidal magnetic field must be current-free.*

**Proof**: A poloidal field written in cylindrical coordinates is given by

$$\mathbf{B} = B_r \mathbf{e}_r + B_z \mathbf{e}_z \tag{6.85}$$

without any dependence on $\phi$. Thus the current is according to Ampère's law

$$\mathbf{J} = \frac{1}{\mu_0} \left( \frac{\partial B_r}{\partial z} - \frac{\partial B_z}{\partial r} \right) \mathbf{e}_\phi \,. \tag{6.86}$$

Now the force is

$$\mathbf{J} \times \mathbf{B} = |\mathbf{J}| \left( B_z \mathbf{e}_r - B_r \mathbf{e}_z \right) , \tag{6.87}$$

which is zero only if $\mathbf{J}$ vanishes. This theorem warns us against trying to construct too simple fields in polar coordinates. This is actually one formulation of the famous *Cowling anti-dynamo theorem* that will be proven in Sect. 8.3.2.

### 6.5.4 FACs and the magnetosphere–ionosphere coupling

The continuity equation (6.84) governs the ionosphere–magnetosphere coupling where the FACs above the auroral zone are connected to the horizontal currents in the ionosphere. In the magnetospheric scale the ionosphere can be considered as a thin layer. In the magnetospheric end the transition from field-aligned to perpendicular current flow takes place over a large volume. The current that is field-aligned in the low beta (of the order of $10^{-6}$) plasma above the auroral region becomes more and more perpendicular as increasing $\beta$ allows $\nabla P$ to make $\mathbf{J} \times \mathbf{B}$ non-zero.

Let us discuss the coupling in a quasi-static idealized configuration. We start by calculating the current sources and sinks in the magnetosphere. Let the magnetospheric plasma be anisotropic and use formulas from Sect. 6.4. Because the perpendicular magnetization current ($\nabla \times \mathbf{M}$) is divergence-free, we get

$$\begin{aligned} \nabla_{\perp} \cdot \mathbf{J} &= \nabla_{\perp} \cdot (\mathbf{J}_C + \mathbf{J}_G) \\ &= \nabla_{\perp} \cdot \frac{P_{\parallel} - P_{\perp}}{B} (\nabla \times \mathbf{b})_{\perp} - \nabla_{\perp} P_{\perp} \cdot \nabla_{\perp} \times \frac{\mathbf{b}}{B} \,. \end{aligned} \tag{6.88}$$

Thus the pressure gradient in the direction of the particle drift causes divergence in the perpendicular current and thus acts as a source or sink of field-aligned current. If the pressure is isotropic, this current arises directly from the divergence of the diamagnetic current

$$\nabla_\perp \cdot \mathbf{J} = \nabla_\perp \cdot \left( \frac{-\nabla P \times \mathbf{b}}{B} \right) . \qquad (6.89)$$

In a time-dependent case also the polarization current may have a divergence

$$\begin{aligned} \nabla_\perp \cdot \mathbf{J}_P = \nabla_\perp \cdot \left( \frac{\rho_m}{B^2} \frac{d\mathbf{E}}{dt} \right) &= \frac{\rho_m}{B^2} \frac{d}{dt} (\nabla_\perp \cdot \mathbf{E}) \\ &= \frac{\rho_m}{B} \mathbf{b} \cdot \frac{d}{dt} (\nabla_\perp \times \mathbf{V}) = \frac{\rho_m}{B} \frac{d\Omega}{dt} , \end{aligned} \qquad (6.90)$$

where we have assumed $\mathbf{E} = -\mathbf{V} \times \mathbf{B}$ and introduced the *vorticity* $\Omega = \mathbf{b} \cdot (\nabla \times \mathbf{V})$ in the direction of the magnetic field. Finally, the FAC density is obtained by integrating along the magnetic field

$$J_\parallel = -B \int \nabla_\perp \cdot \mathbf{J} \frac{dl}{B} = -B \int \left[ \nabla_\perp P \cdot \left( \nabla \times \frac{\mathbf{b}}{B} \right) - \frac{\rho_m}{B} \frac{d\Omega}{dt} \right] \frac{dl}{B} . \qquad (6.91)$$

Thus the sources of FACs are pressure gradients and time-dependent vorticity of the plasma flow. Both are thought to be important in the magnetosphere.

The ionospheric end of the current circuit is a non-MHD regime where Ohm's law is given by Eq. (1.62), that is

$$\mathbf{J} = \begin{pmatrix} \sigma_P & \sigma_H & 0 \\ -\sigma_H & \sigma_P & 0 \\ 0 & 0 & \sigma_\parallel \end{pmatrix} \cdot \mathbf{E} .$$

In the following discussion we assume the parallel conductivity to be infinite, although this is not always true above the auroral region (e.g., Eq. (2.138)).

Assume then that the magnetic field is perpendicular to the ionospheric layers, which is, for the present purpose, a good enough approximation in the auroral region where the FACs between the ionosphere and magnetosphere flow. Furthermore, let the magnetic field be constant in the ionosphere. Representing the perpendicular ionospheric current as a sum of Hall and Pedersen currents and integrating along the field line we get

$$J_\parallel = - \int \nabla_\perp \cdot \left( \sigma_P \mathbf{E} - \sigma_H \frac{\mathbf{E} \times \mathbf{B}}{B} \right) dz . \qquad (6.92)$$

Approximating the ionosphere as a thin layer in the magnetospheric scale this equation can be further integrated over the thickness $h$ of the resistive ionosphere

$$J_\parallel \approx -\nabla_\perp \cdot \left( \Sigma_P \mathbf{E} - \Sigma_H \frac{\mathbf{E} \times \mathbf{B}}{B} \right) = -\nabla_\perp \cdot (\Sigma_P \mathbf{E}) + \frac{\mathbf{E} \times \mathbf{B}}{B} \cdot \nabla \Sigma_H . \qquad (6.93)$$

Here the *height-integrated Pedersen and Hall conductivities* are denoted by $\Sigma_P = h\sigma_P$ and $\Sigma_H = h\sigma_H$ (SI unit A m$^{-1}$). Thus the sources and sinks of FACs in the ionosphere are the divergence of the Pedersen current and, in the case of non-uniform Hall conductivity, the gradient of the Hall conductivity.

It is not quite clear how the ionospheric and magnetospheric FACs close to each other in detail. The only region from which we have detailed and statistically representative observations of FACs is above the auroral oval. Regardless of the actual closure mechanisms or the current paths we can estimate the effect of the current system on the electric potential across the polar cap. It, in turn, is a quantity that can be determined by measuring the ionospheric plasma flow using, e.g., ionospheric radars or polar-orbiting satellites.

Let us assume, for simplicity, isotropic magnetosphere and complete north-south symmetry. In that case (6.91) reduces to

$$\frac{J_{I\parallel}}{B_I} \approx \frac{1}{2}\int\left\{\mathbf{b}\cdot\left[\nabla P\times\nabla(1/B^2)\right] - \frac{\rho_m}{B^2}\frac{d\Omega}{dt}\right\}dz\,, \tag{6.94}$$

where $I$ denotes the ionosphere and $J_{I\parallel}$ is thus the ionospheric FAC density caused by the magnetospheric vortices and pressure gradients. The integration extends from the southern auroral ionosphere to the northern.

Because this current must be the same as the current calculated in the ionosphere, we find an equation that ties the auroral and polar region electric field to the plasma flow in the magnetosphere. As the last simplifying assumption let the ionospheric electric field be a potential field ($\mathbf{E} = -\nabla\varphi$). Then the coupling equation between the ionosphere and magnetosphere becomes

$$\begin{aligned}\nabla_\perp(\Sigma_P\nabla\varphi) + \frac{\mathbf{B}_I\times\nabla\varphi}{B_I}\cdot\nabla\Sigma_H \\ = \frac{B_I}{2}\int\left\{\mathbf{b}\cdot\left[\nabla P\times\nabla(1/B^2)\right] - \frac{\rho_m}{B^2}\frac{d\Omega}{dt}\right\}dz\,.\end{aligned} \tag{6.95}$$

The resistive ionosphere continuously dissipates energy from the magnetosphere. To maintain the coupling requires an external source of energy, which is the solar wind flow and its interaction with the terrestrial magnetic field through mechanisms that we shall discuss later.

### 6.5.5 Magnetic helicity

The *magnetic helicity* of a magnetic field configuration is defined by

$$H = \int \mathbf{A}\cdot\mathbf{B}\,d\mathscr{V}\,, \tag{6.96}$$

where $\mathbf{A}$ is the vector potential. Helicity is a measure of the structural complexity of the magnetic field. Because the vector potential is defined only to within a gauge transformation $\mathbf{A}\rightarrow\mathbf{A}' = \mathbf{A}+\nabla\chi$, the helicity is gauge-independent only if the field extends over all space and decreases sufficiently rapidly (and $\chi$ does not increase too rapidly). For magnetic field configurations of finite dimensions the helicity is well defined if and only if $\mathbf{B}\cdot\mathbf{n} = 0$ on the bounding surface.

The helicity of a magnetic field configuration is conserved, if the field is confined within a closed surface $S$, $\mathbf{B}\cdot\mathbf{n} = 0$ on $S$, and the field permeates a perfectly conducting medium that moves in such a way that $\mathbf{B} \cdot \mathbf{V} = 0$ on $S$. To show this we first note that from the convection equation

$$\frac{\partial \mathbf{B}}{\partial t} = \nabla \times (\mathbf{V} \times \mathbf{B})$$

we get, within the given gauge,

$$\frac{\partial \mathbf{A}}{\partial t} = \mathbf{V} \times \mathbf{B} \,. \tag{6.97}$$

Calculate now $dH/dt$

$$\begin{aligned}
\frac{dH}{dt} &= \int \left( \frac{\partial \mathbf{A}}{\partial t} \cdot \mathbf{B} + \mathbf{A} \cdot \frac{\partial \mathbf{B}}{\partial t} \right) d\mathcal{V} \\
&= \int \left[ \frac{\partial \mathbf{A}}{\partial t} \cdot (\nabla \times \mathbf{A}) + \mathbf{A} \cdot \left( \nabla \times \frac{\partial \mathbf{A}}{\partial t} \right) \right] d\mathcal{V} \\
&= \int \nabla \cdot \left( \frac{\partial \mathbf{A}}{\partial t} \times \mathbf{A} \right) d\mathcal{V} \\
&= \int \mathbf{n} \cdot \left( \frac{\partial \mathbf{A}}{\partial t} \times \mathbf{A} \right) dS \\
&= 0 \,,
\end{aligned} \tag{6.98}$$

where we have used the fact that $\partial \mathbf{A}/\partial t \perp \mathbf{B}$ and that both $\mathbf{B}$ and $\mathbf{V}$ are normal to $\mathbf{n}$ on $S$. Thus $\partial \mathbf{A}/\partial t \parallel \mathbf{n}$ on $S$ and the final integral is zero and $H$ is a constant of motion.

***Example: Helicity of two flux tubes linked together***

Consider two flux tubes that have the shapes of tori (doughnuts) and that are linked together through the annuli of each other. The total magnetic helicity is the sum of the contribution from each tube separately $H = H_1 + H_2$. For thin flux tubes $\mathbf{B} = \nabla \times \mathbf{A}$ is approximately normal to the cross-section $S$ of the tube and we can write for tube 1

$$H_1 = \int \mathbf{A} \cdot \mathbf{B} \, d\mathcal{V} = \oint d\mathbf{s} \cdot \mathbf{A} \int dS \, \mathbf{n} \cdot \nabla \times \mathbf{A} \,. \tag{6.99}$$

The surface integral is thus the magnetic flux in tube 1: $\Phi_1$. The line integral goes around tube 2 yielding $\Phi_2$. Thus the helicity contribution from tube 1 is $H_1 = \Phi_1 \Phi_2$. Tube 2 gives the same contribution $H_2 = \Phi_1 \Phi_2$, and the total helicity is

$$H = 2\Phi_1 \Phi_2 \,. \tag{6.100}$$

If two flux tubes are wound around each other $N$ times

$$H = \pm 2N \Phi_1 \Phi_2 \,, \tag{6.101}$$

where the sign depends on the relative orientation of the magnetic field in the flux tubes. If there are more than two interlinked flux tubes, they each contribute by a factor of their respective $\Phi$.

***Woltjer's theorem***

Woltjer showed in 1958 an important property of ideal MHD:

*For a perfectly conducting plasma in a closed volume* $\mathscr{V}_0$ *the integral*

$$\int_{\mathscr{V}_0} \mathbf{A} \cdot \mathbf{B} \, d\mathscr{V} = H_0 \tag{6.102}$$

*is invariant and the state of minimum magnetic energy is a linear, i.e., constant-*$\alpha$*, force-free field*

**Proof**: The invariance was shown above. Consider the magnetic energy

$$W = \int_{\mathscr{V}_0} \frac{B^2}{2\mu_0} \, d\mathscr{V} \tag{6.103}$$

and small perturbations of $\mathbf{A}$ and $\mathbf{B}$ to $\mathbf{A} + \delta\mathbf{A}$ and $\mathbf{B} + \delta\mathbf{B}$ such that $\delta\mathbf{A} = 0$ on $S$ and $\delta\mathbf{B} = \nabla \times \delta\mathbf{A}$. By linearizing and subtracting $\alpha_0 \delta H_0 \equiv 0$, where $\alpha_0$ is constant, we get

$$\begin{aligned} 2\mu_0 \delta W &= \int_{\mathscr{V}_0} [2\mathbf{B} \cdot \delta\mathbf{B} - \alpha_0 (\delta\mathbf{A} \cdot \mathbf{B} + \mathbf{A} \cdot \delta\mathbf{B})] \, d\mathscr{V} \\ &= \int_{V_0} \nabla \cdot (-2\mathbf{B} \times \delta\mathbf{A} + 2\alpha_0 \mathbf{A} \times \delta\mathbf{A}) \, d\mathscr{V} \\ &+ 2 \int_{V_0} (\nabla \times \mathbf{B} - \alpha_0 \mathbf{B}) \cdot \delta\mathbf{A} \, d\mathscr{V} \,. \end{aligned} \tag{6.104}$$

The first integral on the RHS of a divergence can be transformed to a surface integral which vanishes because $\delta\mathbf{B} = \nabla \times \delta\mathbf{A}$, whereas the second integral shows that $\delta W = 0$ for all perturbations if and only if

$$\nabla \times \mathbf{B} = \alpha_0 \mathbf{B} \,. \tag{6.105}$$

This states that if the energy is at minimum, the configuration must be force-free and we have proven Woltjer's theorem.

The converse statement of this result is not necessarily true. If the configuration is force-free, we have shown that the energy has an extremum, but not that it would be minimum.

This result has been postulated to hold also for small but non-zero resistivity (known as Taylor's hypothesis) and thus it is a good starting point to assume that the state of minimum energy in nearly-ideal MHD problems is a force-free configuration.

Note that, e.g., the magnetospheric configuration is determined by perpendicular currents and is thus not force-free although ideal MHD is a reasonable large-scale description of magnetospheric plasma flow. The magnetosphere is not in a minimum energy equilibrium state.

## 6.6 Alfvén Waves

In MHD there are two characteristic speeds: The speed of sound waves

$$v_s = \sqrt{\gamma P/\rho_m} = \sqrt{\gamma k_B T/m} \tag{6.106}$$

and the speed of Alfvén waves in the direction of the magnetic field

$$v_A = \sqrt{\frac{B^2}{\mu_0 \rho_m}}\,. \tag{6.107}$$

A combination of these speeds is the magnetosonic speed, which is the speed of magnetosonic waves perpendicular to the magnetic field

$$v_{ms} = \sqrt{v_s^2 + v_A^2}\,. \tag{6.108}$$

### 6.6.1 Dispersion equation of MHD waves

Elementary plasma physics textbooks often discuss the Alfvén waves starting from the the modes propagating parallel and perpendicular to the ambient magnetic field. However, the linearized MHD equations are straightforward and easy to solve for plane waves propagating at all angles at once. Consider a compressible, non-viscous, perfectly conductive fluid in a magnetic field. This is described by the equations

$$\frac{\partial \rho_m}{\partial t} + \nabla\cdot(\rho_m \mathbf{V}) = 0 \tag{6.109}$$

$$\rho_m \frac{\partial \mathbf{V}}{\partial t} + \rho_m(\mathbf{V}\cdot\nabla)\mathbf{V} = -\nabla P + \mathbf{J}\times\mathbf{B} \tag{6.110}$$

$$\nabla P = v_s^2 \nabla \rho_m \tag{6.111}$$

$$\nabla\times\mathbf{B} = \mu_0 \mathbf{J} \tag{6.112}$$

$$\nabla\times\mathbf{E} = -\frac{\partial \mathbf{B}}{\partial t} \tag{6.113}$$

$$\mathbf{E} + \mathbf{V}\times\mathbf{B} = 0\,. \tag{6.114}$$

From these we can eliminate **J**, **E**, and $P$

$$\frac{\partial \rho_m}{\partial t} + \nabla\cdot(\rho_m \mathbf{V}) = 0 \tag{6.115}$$

$$\rho_m \frac{\partial \mathbf{V}}{\partial t} + \rho_m(\mathbf{V}\cdot\nabla)\mathbf{V} = -v_s^2 \nabla \rho_m + (\nabla\times\mathbf{B})\times\mathbf{B}/\mu_0 \tag{6.116}$$

$$\nabla\times(\mathbf{V}\times\mathbf{B}) = \frac{\partial \mathbf{B}}{\partial t}\,. \tag{6.117}$$

Assume that in equilibrium the density $\rho_{m0}$ is constant and $\mathbf{V} = 0$. Furthermore, let the background magnetic field $\mathbf{B}_0$ be uniform. Considering small perturbations to the variables

$$\mathbf{B}(\mathbf{r},t) = \mathbf{B}_0 + \mathbf{B}_1(\mathbf{r},t) \tag{6.118}$$

$$\rho_m(\mathbf{r},t) = \rho_{m0} + \rho_{m1}(\mathbf{r},t) \tag{6.119}$$

$$\mathbf{V}(\mathbf{r},t) = \mathbf{V}_1(\mathbf{r},t) \tag{6.120}$$

we can linearize the equations by picking up the first-order terms

$$\frac{\partial \rho_{m1}}{\partial t} + \rho_{m0}(\nabla \cdot \mathbf{V}_1) = 0 \tag{6.121}$$

$$\rho_{m0}\frac{\partial \mathbf{V}_1}{\partial t} + v_s^2 \nabla \rho_{m1} + \mathbf{B}_0 \times (\nabla \times \mathbf{B}_1)/\mu_0 = 0 \tag{6.122}$$

$$\frac{\partial \mathbf{B}_1}{\partial t} - \nabla \times (\mathbf{V}_1 \times \mathbf{B}_0) = 0\,. \tag{6.123}$$

From these we find an equation for the velocity perturbation $\mathbf{V}_1$

$$\frac{\partial^2 \mathbf{V}_1}{\partial t^2} - v_s^2 \nabla(\nabla \cdot \mathbf{V}_1) + \mathbf{v}_A \times \{\nabla \times [\nabla \times (\mathbf{V}_1 \times \mathbf{v}_A)]\} = 0\,, \tag{6.124}$$

where we have introduced the Alfvén velocity as a vector

$$\mathbf{v}_A = \frac{\mathbf{B}_0}{\sqrt{\mu_0 \rho_{m0}}}\,. \tag{6.125}$$

Looking for plane wave solutions in the form $\mathbf{V}_1(\mathbf{r},t) = \mathbf{V}_1 \exp[i(\mathbf{k} \cdot \mathbf{r} - \omega t)]$ we get an algebraic equation

$$-\omega^2 \mathbf{V}_1 + v_s^2 (\mathbf{k} \cdot \mathbf{V}_1)\mathbf{k} - \mathbf{v}_A \times \{\mathbf{k} \times [\mathbf{k} \times (\mathbf{V}_1 \times \mathbf{v}_A)]\} = 0. \tag{6.126}$$

It is convenient to expand the vector products. After straightforward vector manipulation this leads to the dispersion equation for ideal MHD waves

$$\begin{aligned} &-\omega^2 \mathbf{V}_1 + (v_s^2 + v_A^2)(\mathbf{k} \cdot \mathbf{V}_1)\mathbf{k} \\ &+(\mathbf{k} \cdot \mathbf{v}_A)[((\mathbf{k} \cdot \mathbf{v}_A)\mathbf{V}_1 - (\mathbf{v}_A \cdot \mathbf{V}_1)\mathbf{k} - (\mathbf{k} \cdot \mathbf{V}_1)\mathbf{v}_A)] = 0\,. \end{aligned} \tag{6.127}$$

### 6.6.2 MHD wave modes

Now it is a good time to look at the limiting cases of perpendicular and parallel propagation.

#### *Perpendicular propagation*

Let $\mathbf{k} \perp \mathbf{B}_0$, which implies $\mathbf{k} \cdot \mathbf{v}_A = 0$, and the dispersion equation reduces to

$$\mathbf{V}_1 = (v_s^2 + v_A^2)(\mathbf{k}\cdot\mathbf{V}_1)\mathbf{k}/\omega^2 \,. \tag{6.128}$$

Clearly $\mathbf{k} \parallel \mathbf{V}_1$, and we have found the *magnetosonic wave*

$$\frac{\omega}{k} = \sqrt{v_s^2 + v_A^2} \,. \tag{6.129}$$

Assuming harmonic behavior also for the magnetic field the convection equation reduces to

$$\omega\mathbf{B}_1 + \mathbf{k}\times(\mathbf{V}_1\times\mathbf{B}_0) = 0 \,, \tag{6.130}$$

which yields the magnetic field of the wave

$$\mathbf{B}_1 = \frac{V_1}{\omega/k}\mathbf{B}_0 \,. \tag{6.131}$$

The electric field can then be computed from the ideal MHD Ohm's law $\mathbf{E} = -\mathbf{V}_1 \times \mathbf{B}_0$. This wave is known as the *compressional (or fast) Alfvén (or MHD) wave.*

***Parallel propagation***

For $\mathbf{k} \parallel \mathbf{B}_0$, the dispersion equation reduces to

$$(k^2 v_A^2 - \omega^2)\mathbf{V}_1 + \left(\frac{v_s^2}{v_A^2} - 1\right) k^2(\mathbf{V}_1\cdot\mathbf{v}_A)\mathbf{v}_A = 0 \,. \tag{6.132}$$

This describes two different wave modes. $\mathbf{V}_1 \parallel \mathbf{B}_0 \parallel \mathbf{k}$ yields the *sound wave*

$$\frac{\omega}{k} = v_s \,. \tag{6.133}$$

The second solution is a transversal wave with $\mathbf{V}_1 \perp \mathbf{B}_0 \parallel \mathbf{k}$. Now $\mathbf{V}_1 \cdot \mathbf{v}_A = 0$ and we find the *Alfvén wave*

$$\frac{\omega}{k} = v_A \,. \tag{6.134}$$

The wave magnetic field is

$$\mathbf{B}_1 = -\frac{\mathbf{V}_1}{\omega/k}B_0 \,. \tag{6.135}$$

The wave magnetic field is perpendicular to the background field. This mode does not perturb the density or pressure. The mode causes shear stress on the magnetic field $(\nabla\cdot(\mathbf{BB})/\mu_0)$ and is often called the *shear Alfvén wave.*

***Propagation at oblique angles***

The Alfvén waves are not limited to parallel and perpendicular propagation. To find the dispersion equation in an arbitrary direction we denote the angle between $\mathbf{k}$ and $\mathbf{B}_0$ by $\theta$ and insert it into the dot products of the dispersion equation. Select the $z$-axis parallel to

$\mathbf{B}_0$ and the $x$-axis so that $\mathbf{k}$ is in the $xz$-plane. Then

$$\begin{aligned}
\mathbf{k} &= k(\mathbf{e}_x \sin\theta + \mathbf{e}_z \cos\theta) \\
\mathbf{v}_A &= v_A \mathbf{e}_z \\
\mathbf{V}_1 &= V_{1x}\mathbf{e}_x + V_{1y}\mathbf{e}_y + V_{1z}\mathbf{e}_z \\
\mathbf{k}\cdot\mathbf{v}_A &= k v_A \cos\theta \\
\mathbf{k}\cdot\mathbf{V}_1 &= k(V_{1x}\sin\theta + V_{1z}\cos\theta) \\
\mathbf{v}_A\cdot\mathbf{V}_1 &= v_A V_{1z}
\end{aligned}$$

and the dispersion equation reads as

$$V_{1x}(-\omega^2 + k^2 v_A^2 + k^2 v_s^2 \sin^2\theta) + V_{1z}(k^2 v_s^2 \sin\theta\cos\theta) = 0 \tag{6.136}$$

$$V_{1y}(-\omega^2 + k^2 v_A^2 \cos^2\theta) = 0 \tag{6.137}$$

$$V_{1x}(k^2 v_s^2 \sin\theta\cos\theta) + V_{1z}(-\omega^2 + k^2 v_s^2 \cos^2\theta) = 0\,. \tag{6.138}$$

The $y$-component yields a linearly polarized mode with the phase velocity

$$\frac{\omega}{k} = v_A \cos\theta\,. \tag{6.139}$$

This is the extension of the shear Alfvén wave to all directions. It does not propagate perpendicularly to the magnetic field because its phase velocity becomes zero when $\theta \to \pi/2$.

The non-trivial solutions of the remaining pair of equations are found setting the determinant of the coefficients of $V_{1x}$ and $V_{1z}$ zero

$$\left(\frac{\omega}{k}\right)^2 = \frac{1}{2}(v_s^2 + v_A^2) \pm \frac{1}{2}[(v_s^2 + v_A^2)^2 - 4 v_s^2 v_A^2 \cos^2\theta]^{1/2}\,. \tag{6.140}$$

The solutions with plus and minus signs are called *fast* and *slow* Alfvén (or MHD) waves. The wave normal surface representation of these modes is given in Fig. 6.3.

## 6.7 Beyond MHD

It is clear that the MHD theory meets its limits when the scales of the investigated physical phenomena become comparable to the scale sizes of individual particle motion, of which the ion gyro radius is usually the first to be encountered in space physics, or when several dynamically important particle species with different particle distribution functions co-exist. In this section we discuss two topics that are closely related to MHD but require different techniques: The hybrid approach to problems where non-fluid aspects of ion dynamics are included and the kinetic effects on Alfvén waves.

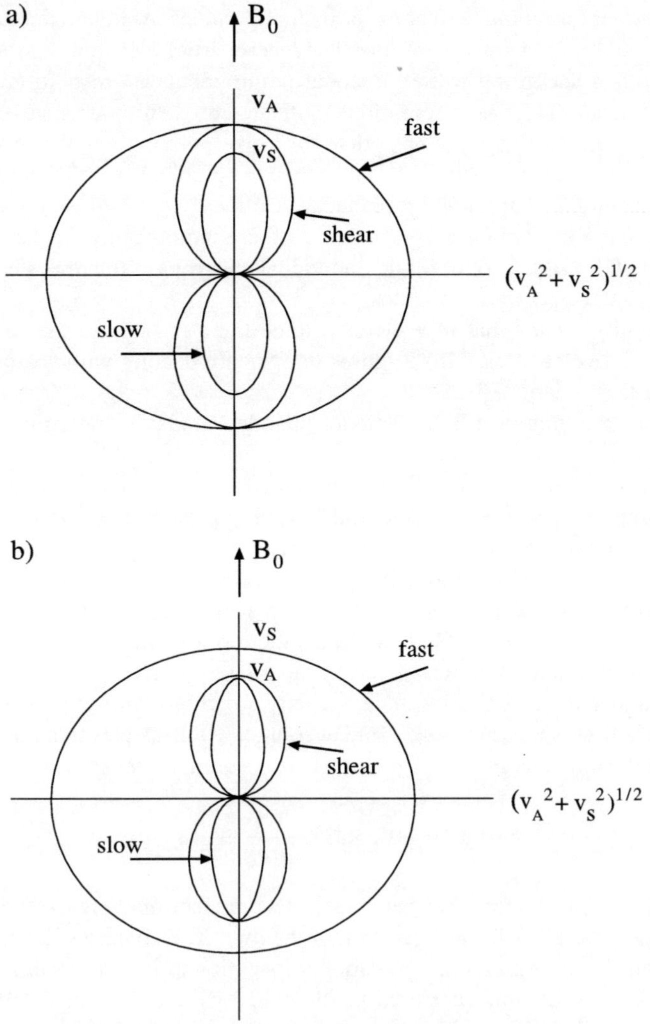


**Fig. 6.3** Wave normal surfaces for slow, fast, and shear Alfvén waves, when a) $v_A > v_s$ and b) $v_s > v_A$.

### 6.7.1 Quasi-neutral hybrid approach

Examples of problems where we need to go beyond the MHD description are shock phenomena to be discussed in Chap. 11 and solar wind interaction with solar system bodies whose size is not large as compared to the gyro radii of plasma particles. While the Vlasov theory could in principle be applied to these problems, it is in turn so detailed that numerical simulations easily require more computing resources than is available today.

A compromise approach utilized in shock and planetary plasma studies is known as the *quasi-neutral hybrid* approach. The underlying idea is to treat the electrons as a macroscopic fluid and represent ions either as individual particles, or in practice, as macroparticles, i.e., reasonable-sized clumps of ions. The macroparticles can during the computation be split and joined according to the practical requirements on resolution or computing time. The treatment of electrons as a fluid is motivated by their much smaller length scales (gyro radius and inertial length $c/\omega_{pe}$) than the corresponding lengths of the ions. Usually the electron scale sizes are much smaller than the gradient scale lengths of the shocks or the plasma configuration around a planet. When these conditions are met, the method brings major saving in computing resources, as there is no need to make the computing grid so small that it could account for the individual electrons or the time step short enough to correspond to the electron time scales.

Quasi-neutrality means that in a given volume there is (nearly) the same amount of positive and negative charges. This requires that we are dealing with spatial scales much larger than the Debye length $\lambda_{De}$.

The plasma equations of a hybrid model include Faraday's and Ampère's laws. The displacement current is not usually included in Ampère's law, although we are interested in faster processes than in MHD. It can be argued that the displacement current is not dynamically important in the quasi-neutral hybrid approach, but this is something that needs to be ensured in the problem being investigated.

Ohm's law is used in the hybrid approach to provide the electric field. The $\mathbf{V}\times\mathbf{B}$ term and the Hall term $\mathbf{J}\times\mathbf{B}/ne$ in the generalized Ohm's law (Eq. 2.135) can be combined into a single term $-\mathbf{V}_e\times\mathbf{B}$ (cf. 6.43). In this respect the formulation resembles the Hall MHD. As the generalized Ohm's law was given in the single-fluid variables, the bulk velocity $\mathbf{V}$ should be understood as the ion bulk velocity weighted with the electric charges of each ion species $i$. This takes correctly into account any ion species that may have charges different from the unit charge $|q_e|$ (e.g. $He^{++}$). Thus Ohm's law can be written in the form

$$\mathbf{E} = -\mathbf{V}_e\times\mathbf{B} + \frac{\nabla P_e}{q_e n_e} + \frac{\mathbf{J}}{\sigma}\,, \tag{6.141}$$

where the electron inertial term has been neglected and isotropic pressure assumed.

The Lorentz force gives the acceleration of the ions. The equation closing the group of hybrid equations is the expression for spatial propagation of ions according to their velocity obtained from the Lorentz acceleration. Now the hybrid equations for the propagation of the field quantities and ions are

$$n_e = |q_e|^{-1}\sum_i q_i n_i \tag{6.142}$$

$$\mathbf{J} = \sum_i q_i n_i \mathbf{v}_i + q_e n_e \mathbf{V}_e \tag{6.143}$$

$$\frac{d\mathbf{r}_i}{dt} = \mathbf{v}_i \tag{6.144}$$

$$\frac{d\mathbf{v}_i}{dt} = \frac{q_i}{m_i}(\mathbf{E}+\mathbf{v}_i\times\mathbf{B}) \tag{6.145}$$

$$\mathbf{E} = -\mathbf{V}_e\times\mathbf{B} + \frac{\nabla P_e}{q_e n_e} + \frac{\mathbf{J}}{\sigma} \tag{6.146}$$

$$\nabla \times \mathbf{E} = -\frac{\partial \mathbf{B}}{\partial t} \tag{6.147}$$

$$\nabla \times \mathbf{B} = \mu_0 \mathbf{J}\,, \tag{6.148}$$

where the quantities indexed by $i$'s refer to individual ion macroparticles and the sums are over all macroparticles.

If other forces acting on the ions, e.g., gravitation, need to be considered, they can be added to the Lorentz force term. For example, this may be required in studies of how planetary low-temperature ions are picked up by the solar wind streaming past the planet.

The treatment of the electron pressure gradient in a hybrid model requires additional assumptions, which may be hard to validate. Such assumptions typically include adiabatic (or isothermal) behavior of the electron fluid and estimates for electron temperature. Also the electron–ion coupling can be difficult to describe in a physically correct way.

The equations of the hybrid approach give a closed set of equations for propagation of particle positions and velocities and the magnetic field from their initial values. This allows ion kinetics to be modeled self-consistently with the dynamical electromagnetic fields. The electrons play a secondary role following tightly the magnetic field lines and obeying the assumed form of equation of state (e.g., adiabatic or isothermal).

Leaving the single-fluid MHD regime introduces some new concerns. The quasi-neutral hybrid equations cannot be cast into a conservative form and one has to carefully monitor, e.g., the conservation of energy during a numerical simulation. Furthermore, because the Lorentz force must be solved for each macroparticle, the numerical simulations become noisy.

An analogous hybrid approach can also be useful in the Vlasov picture. There electrons are again treated as a single fluid, whereas the ion populations are represented by distribution functions. If there is no need to resolve the physics at the electron scales, the *hybrid–Vlasov approach* makes the equations simpler and brings in considerable savings in computer time and memory requirements as compared to fully kinetic computations.

### 6.7.2 Kinetic Alfvén waves

We have so far seen how the Alfvén wave appears in the cold plasma theory, in the Vlasov theory, and in particular in MHD. While the Vlasov theory would, in principle, be sufficient for a kinetic description for these waves, it is sometimes useful to write the dispersion equations in a form that shows the kinetic corrections to the MHD waves.

There are two types of *kinetic Alfvén waves*. For relatively large beta ($\beta > m_e/m_i$), e.g., in the solar wind, at the magnetospheric boundary, or in the magnetotail plasma sheet, the mode is called the *oblique kinetic Alfvén wave* with the phase velocity

$$v_\parallel = v_A \left[1 + k_\perp^2 r_{Li}^2 \left(\frac{3}{4} + \frac{T_e}{T_i}\right)\right]^{1/2} \tag{6.149}$$

$$v_\perp = \frac{k_\parallel v_A}{k_\perp} \left[1 + k_\perp^2 r_{Li}^2 \left(\frac{3}{4} + \frac{T_e}{T_i}\right)\right]^{1/2}\,.$$

Kinetic Alfvén waves are important also in low-beta plasmas, e.g., on magnetic field lines coupling the auroral ionosphere to outer parts of the magnetosphere. The kinetic Alfvén waves are able to carry the field-aligned current and set up small-scale parallel electric fields. For $\beta \ll m_e/m_i$ the electron thermal speed is smaller than the Alfvén speed ($v_{the} < v_A$) and the electron inertia needs to be taken into account. The wave is called the *shear kinetic Alfvén wave, inertial kinetic Alfvén wave*, or just inertial Alfvén wave. Its dispersion equation reads as

$$\omega^2 = k_\parallel^2 v_A^2 \frac{1 + k_\perp^2 r_{Li}^2}{1 + k_\perp^2 c^2/\omega_{pe}^2}\,, \tag{6.150}$$

where the ratio $c/\omega_{pe}$ is the *electron inertial length.*

These two limits ($\beta \ll m_e/m_i$ and $\beta > m_e/m_i$) of the dispersion equation can be found, with quitę some effort, by considering the electron and ion continuation and momentum equations. When moving from the auroral ionosphere out to the magnetosphere along a magnetic field line, the limit $\beta = m_e/m_i$ is crossed somewhere at $4-5\,R_E$ from the center of the Earth, and the inertial wave becomes the oblique kinetic wave. Furthermore, the kinetic Alfvén wave is subject to Landau damping, albeit small, as long as the wavelength is long and $\beta$ is small. All these facts call for kinetic treatment.

Lysak and Lotko [1996] derived such a dispersion equation for the low-frequency long-wavelength modes in uniformly magnetized plasma with $\beta \ll 1$, but reaching to the regime $\beta > m_e/m_i$. The derivation starts writing the determinant of the dielectric tensor (5.85) for a Maxwellian distribution function in a suitable coordinate system and equating it to zero

$$|\mathcal{K}| = 0\,. \tag{6.151}$$

A lengthy calculation finally gives the dispersion equation

$$\left(\frac{\omega}{k_\parallel v_A}\right)^2 = \frac{\mu_i}{1 - \Gamma_0(\mu_i)} + \frac{k_\perp^2 \rho_s^2}{\Gamma_0(\mu_e)[1 + \zeta Z(\zeta)]}\,, \tag{6.152}$$

where $\mu_\alpha = k_\perp^2 r_{L\alpha}^2$, $\rho_s$ is the gyro radius of an ion moving with the ion sound speed, i.e., $\rho_s^2 = c_s^2/\omega_{ci}^2$, $\zeta = \omega/k_\parallel v_{the}$ is the argument of the plasma dispersion function $Z$ (5.31) with $v_{the} = \sqrt{2k_B T_e/m_e}$, $\Gamma_0(\mu) = \exp(-\mu) I_0(\mu)$, and $I_0$ is the modified Bessel function of the first kind.

**Challenge your brain**

Read the article Lysak and Lotko [1996], fill in the steps leading to (6.152), and convince yourself that you get at appropriate limits the expressions (6.149) and (6.150)

# 7. Space Plasma Instabilities

Space storms are extreme manifestations of space plasma instabilities. Onset of a solar flare or a substorm expansion are examples of complex phenomena involving rapid perturbations and system reconfigurations both at macroscopic and microscopic levels. Furthermore, the plasma waves discussed in previous chapters do not appear without reason. They are driven either by an external "antenna" or by local instabilities somewhere in the system. As there are several plasma wave modes, there is also a rich flora of different plasma instabilities. The wave phenomena reach far beyond the family of linear wave modes. A growing instability may develop to a nonlinear regime and cannot any more be described in terms of normal modes. Shocks, to be discussed in Chap. 11, are examples of strongly nonlinear wave phenomena.

Our ability to treat plasma stability analytically is in most cases limited to the linear regime, where we can determine whether plasma is stable or unstable to small perturbations, on not. If plasma is stable, the perturbation will eventually be damped. For a small damping rate ($|\omega_i| \ll \omega_r$) the perturbation is a normal mode of the plasma, but often the damping takes place very quickly and the mode is overdamped. If $\omega_i > 0$, the wave grows and we have an instability. Without doing actual calculations it usually is impossible to say to how large an amplitude a wave can grow. If nothing quenches the growth, the system develops toward a major configurational change. The growth may also lead to a state in which some plasma particles start to interact more strongly with the growing wave, e.g., by heating. This can sometimes be described in terms of *quasi-linear saturation* within the Vlasov theory.

A way of categorizing plasma instabilities is to divide them between microscopic (kinetic) and macroscopic (configurational) instabilities. A *macroinstability* is something that can be described by macroscopic equations in the configuration space. A *microinstability* takes place in the $(\mathbf{r},\mathbf{v})$-space and depends on the actual shape of the distribution function.

Although it may sometimes look as if the plasma would become unstable without any apparent reason, it is not true. The instabilities do not arise without *free energy*. The free energy may be stored in the magnetic or plasma configuration, e.g., as magnetic tension in the Harris current sheet, anisotropic plasma pressure, streaming of plasma particles with respect to each other, etc. Identification of the free energy source is essential to understand

H.E.J. Koskinen, *Physics of Space Storms: From the Solar Surface to the Earth*, Springer Praxis Books, DOI 10.1007/978-3-642-00319-6_7,

a given instability because different forms of free energy can give rise to widely different consequences.

In this chapter we discuss selected space plasma instabilities that are of interest in the context of space storms. Unfortunately, it is impossible to penetrate deeply into the details of all instabilities. The interested reader is encouraged to consult, e.g., the comprehensive discussion of space plasma instabilities in the textbook by Treumann and Baumjohann [1996], which has also inspired the following discussion.

## 7.1 Beam–plasma Modes

Perhaps the simplest electrostatic dispersion equation describing an instability can be constructed by considering an unmagnetized plasma consisting of ions as a non-moving background, a mobile electron background population (density $n_0$, $V_0 = 0$), and a cold electron beam ($n_b$, $\mathbf{V}_b$) streaming through the background. Following the same procedure as in Sect. 5.3.4, but neglecting thermal effects, it is an easy exercise to derive the dispersion equation

$$\varepsilon(\omega,\mathbf{k}) = 1 - \frac{\omega_{p0}^2}{\omega^2} - \frac{\omega_{pb}^2}{(\omega - \mathbf{k}\cdot\mathbf{V}_b)^2} = 0\,. \tag{7.1}$$

This equation describes standing Langmuir oscillations of both background electrons and beam electrons, the latter being Doppler-shifted by the streaming velocity.

If we neglect the background plasma entirely ($\omega_{p0}^2 = 0$), the solutions of the dispersion equation are

$$\omega = \mathbf{k}\cdot\mathbf{V}_b \pm \omega_{pb}\,. \tag{7.2}$$

These solutions are called *beam modes*.

One way to analyze the stability properties of a plasma system is to investigate the energy balance. The energy density of an electromagnetic wave in a plasma can be written as

$$W_w = \varepsilon_0 \delta\mathbf{E}^* \cdot \varepsilon \cdot \delta\mathbf{E} + \frac{|\delta\mathbf{B}|^2}{2\mu_0}\,, \tag{7.3}$$

where $\delta\mathbf{E}, \delta\mathbf{B}$ indicate the wave electric and magnetic fields and $\varepsilon$ is, in general, a tensor. The magnetic permeability of the plasma is assumed to be constant $\mu_0$, which is usually a good approximation in this context and thus the determination of magnetic energy is straightforward. The electric energy is more complicated because it depends on the dielectric properties of the plasma. Transforming into the $(\omega,\mathbf{k})$-space the electric field *spectral energy density*, can be expressed as

$$W_w(\omega,\mathbf{k}) = \frac{\varepsilon_0}{2}\langle|\delta\mathbf{E}(\omega,\mathbf{k})|^2\rangle \frac{\partial[\omega\varepsilon(\omega,\mathbf{k})]}{\partial\omega}\,. \tag{7.4}$$

**Feed your brain** by figuring out how expression (7.4) is derived.

Because the energy density and the spectral energy density are real quantities, (7.4) contains the real part of $\varepsilon(\omega,\mathbf{k})$ only. The formula for $W_w$ expresses both the energy density of the electric field, $W_E = \varepsilon_0|\delta E|^2/2$, and the energy in the wave motion of the particles. This motion provides the polarization field, i.e., the energy that can formally be considered to set up the displacement field **D**.

For the beam-plasma mode, the ratio of the total wave energy and the electric field energy is

$$\frac{W_w}{W_E} = \frac{\partial[\omega\varepsilon(\omega,\mathbf{k})]}{\partial\omega} = \omega\frac{\partial\varepsilon(\omega,\mathbf{k})}{\partial\omega} = \frac{2\omega_{p0}^2}{\omega^2} + \frac{2\omega\omega_{pb}^2}{(\omega-\mathbf{k}\cdot\mathbf{V}_b)^3}\,. \tag{7.5}$$

The first term on the RHS comes from the Langmuir waves and the second is the contribution of the beam mode. If the Doppler-shifted frequency of the beam mode is negative, its energy is *negative*. When the beam moves through the plasma, it must be slowed down by the electromagnetic interaction with the background. Then the negative energy mode loses energy, i.e., its negative amplitude grows. Thus the beam–plasma system can have growing solutions, depending on the actual plasma parameters.

Beam–plasma instabilities arise in many space storm relevant environments from beams propagating into the upstream direction of shock fronts to drifts destabilizing waves in the auroral ionosphere.

### 7.1.1 Two-stream instability

The most fundamental beam–plasma instability is the *two-stream instability*. Figure 7.1 illustrates the coupling between the Langmuir mode and the beam modes in the $(\omega,k)$-space.

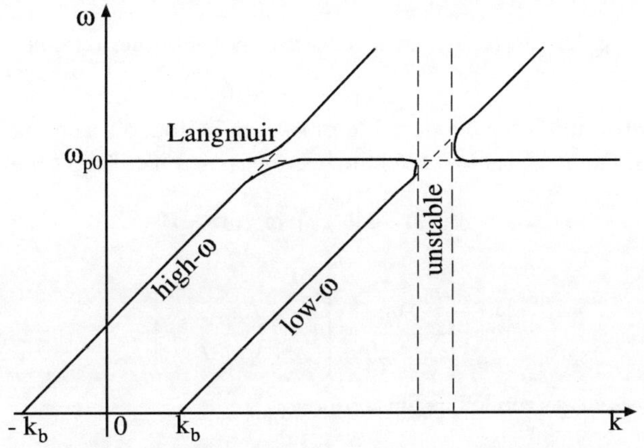


**Fig. 7.1** Coupling of Langmuir and beam–plasma modes.

The dispersion equation

$$1 - \frac{\omega_{p0}^2}{\omega^2} = \frac{\omega_{pb}^2}{(\omega - \mathbf{k} \cdot \mathbf{V}_b)^2} \tag{7.6}$$

is a fourth-order polynomial equation for $\omega$ with four roots in the complex plane. The solutions can be illustrated graphically by plotting both sides of the dispersion equation separately, i.e., the $\varepsilon_l$ (plasma oscillation) and $1 - \varepsilon_b$ (1 – beam modes). Figure 7.2 shows that there are two stable solutions in the real axis, whereas the second pair of solutions are complex numbers. One of these is the unstable solution associated with the negative energy mode.

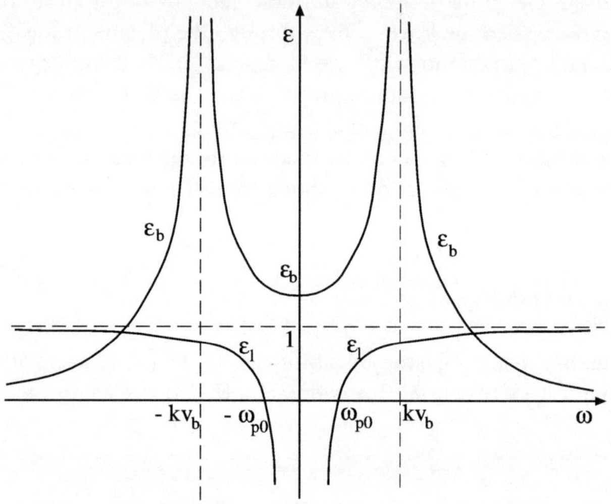


**Fig. 7.2** Graphical solution to illustrate the two-stream instability.

The analytical solution for the unstable mode can be found assuming that close to the phase velocity of the beam ($\omega \approx \pm kV_b$) the beam term is much larger than one. Then

$$\omega_{p0}^2 (\omega - kV_b)^2 + \omega_{pb}^2 \omega^2 = 0 \tag{7.7}$$

$\Rightarrow$

$$\omega = \frac{kV_b}{2 + n_b/n_0} \left[ 1 \pm i \left( \frac{n_b}{n_0} \right)^{1/2} \right] . \tag{7.8}$$

Thus the negative energy mode has the frequency

$$\omega_{ts} = \frac{kV_b}{2 + n_b/n_0} \tag{7.9}$$

and the growth rate ($\gamma = \omega_i$)

$$\gamma = \omega_{ts} \left( \frac{n_b}{n_0} \right)^{1/2} . \tag{7.10}$$

The source of free energy is in the motion of the beam. If the external energy source ceases to feed the system, the instability quenches itself after the beam has slowed down to a certain *threshold*. The threshold is determined by the damping through the background particles, which is a microscopic process and thus beyond the present treatment.

### 7.1.2 Buneman instability

A special case of two-stream instability arises when the entire electron population is streaming with respect to the ions in cold unmagnetized plasma. This is known as the *Buneman instability*. It is an example of *current-driven* instabilities because the relative motion of the particle populations corresponds to a net current in the plasma. If the electrons and/or ions are so warm, that their distribution functions overlap, microscopic treatment becomes necessary.

The instability is easiest to study in the rest frame of the ions. The cold plasma dispersion equation is

$$\varepsilon(\omega, \mathbf{k}) = 1 - \frac{\omega_{pi}^2}{\omega^2} - \frac{\omega_{pe}^2}{(\omega - kV_0)^2} = 0 , \tag{7.11}$$

where $V_0$ is the relative velocity between the populations. Because $\omega_{pe} \gg \omega_{pi}$ , the electron term dominates. The slow negative energy mode $\omega_- \approx kV_0 - \omega_{pe}$ is the unstable mode, whereas the positive energy mode $\omega_+ \approx kV_0 + \omega_{pe}$ is stable. Thus we can write

$$(\omega - \omega_-)\omega^2 = \frac{\omega_{pi}^2 (\omega - kV_0)^2}{\omega - \omega_+} . \tag{7.12}$$

As in the two-stream case the interesting wave number is $k \approx \omega_{pe}/V_0$. Now, however, $\omega \ll \omega_{pe}$. With these approximations

$$\omega^3 \approx -\frac{m_e}{2m_i} \omega_{pe}^3 . \tag{7.13}$$

This has one real root

$$\omega = -\left( \frac{m_e}{2m_i} \right)^{1/3} \omega_{pe} . \tag{7.14}$$

Writing $\omega = \omega_r + i\gamma$ we obtain two equations

$$\omega_r(\omega_r^2 - 3\gamma^2) = -\frac{m_e \omega_{pe}^3}{2m_i} \tag{7.15}$$

$$\gamma^2 = 3\omega_r^2 . \tag{7.16}$$

From these we can solve the frequency for the Buneman mode at the maximum growth rate

$$\omega_{bun} = \left(\frac{m_e}{16m_i}\right)^{1/3} \omega_{pe} \approx 0.03\,\omega_{pe} \tag{7.17}$$

$$\gamma_{bun} = \sqrt{3}\left(\frac{m_e}{16m_i}\right)^{1/3} \omega_{pe} \approx 0.05\,\omega_{pe}\,. \tag{7.18}$$

The growth rate of the mode is of the same order as its frequency. The amplitude grows rapidly and can lead to rapid change of the configuration, e.g., transforming the free energy (current) to heat of the plasma. This is an example of *anomalous resistivity* where the instability takes the role of collisions to resist the current flow.

**Train your brain**

Show that the unstable wave modes of the Buneman instability must have

$$k^2V_0^2 < \omega_{pe}^2\left[1+\left(\frac{m_e}{m_i}\right)^{1/3}\right]^3 . \tag{7.19}$$

Thus the unstable wave must have a minimum wavelength. The frequency has its maximum ($\omega_{bun}$) at this threshold and decreases toward longer waves. The instability quenches itself through a nonlinear process where the growing electric field fluctuations begin to trap electrons slowing them down to a velocity that is below the threshold for the wave growth.

## 7.2 Macroinstabilities

Division between macro- and microinstabilities is mainly a technical matter. There are instabilities that can be treated in macroscopic theory although velocity space effects may become important in some stage of their evolution, in particular, at saturation.

### 7.2.1 Rayleigh–Taylor instability

The *Rayleigh–Taylor* (RT) instability, also known as the Kruskal–Schwarzschild instability, is an example of macroscopic instabilities arising from plasma inhomogeneity. It describes the stability of a system in which heavier fluid is supported above lighter fluid, e.g., by surface tension or magnetic field. The RT instability is also a neutral fluid phenomenon, an example being a carefully prepared colorful cocktail drink, but we are not in that business.

Consider instead a heavy plasma supported against the gravitational force by the magnetic field (Fig. 7.3). Let the boundary between the heavy and light plasmas, as well as the magnetic field, be in the $(x,y)$-plane, $\mathbf{B}_0 = B_0\mathbf{e}_x$. Let the gravitational acceleration $\mathbf{g} = -g\,\mathbf{e}_z$ act downward and the density gradient $\nabla n_0 = [\partial n_0(z)/\partial z]\mathbf{e}_z$ point upward. Such

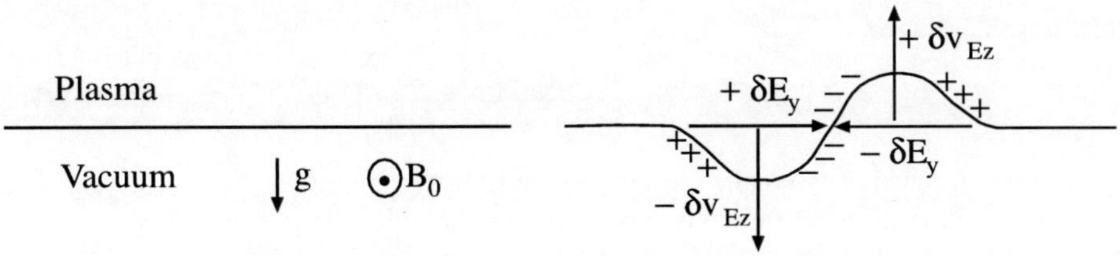


**Fig. 7.3** Principle of the Rayleigh–Taylor instability.

configurations appear, e.g., in the equatorial ionosphere and in the solar atmosphere. Let the plasma be, for simplicity, collisionless and cold.

Consider a small sinusoidal perturbation to the boundary (Fig. 7.3). The gravitational field causes an ion drift in the $-y$-direction. This leads to an electric field perturbation in the $+y$-direction in the region where plasma perturbation is downward, and in the opposite direction in the region where the perturbation is upward. In the downward perturbed region the E×B drift is downward and in the upward perturbed region upward. Thus the E×B drift enhances the perturbation and the system is unstable. The gravitationally supported plasma falls down and the dilute bubbles rise.

The dispersion equation for the RT instability can be found starting from the cold ion equation of motion including the gravitational force. Assuming harmonic perturbations we find

$$\left(\omega + \frac{gk_\perp}{\omega_{ci}}\right)\delta\mathbf{V}_{i\perp} = \frac{e}{m_i}(\mathbf{k}_\perp\delta\varphi - iB_0\mathbf{e}_x \times \delta\mathbf{V}_{i\perp})\,. \tag{7.20}$$

In order to the E×B drift to be effective the frequency of the disturbance must be much smaller than $\omega_{ci}$. Consequently, the expression for the velocity disturbance is

$$\delta\mathbf{V}_{i\perp} = -\delta\varphi\left[i\mathbf{k}_\perp \times \mathbf{e}_x + \frac{\mathbf{k}_\perp}{\omega_{ci}B_0}\left(\omega + \frac{gk_\perp}{\omega_{ci}}\right)\right]. \tag{7.21}$$

The ion continuity equation is now

$$\omega\delta n_i = n_0\mathbf{k}\cdot\delta\mathbf{V}_i - i\delta\mathbf{V}_i\cdot\nabla n_0\,. \tag{7.22}$$

Eliminating $\delta\mathbf{V}_i$ we get an expression for the density disturbance

$$\delta n_i = n_0\delta\varphi\left[\frac{e}{m_i}\left(\frac{k_\parallel^2}{\omega^2} - \frac{k_\perp^2}{\omega_{ci}^2}\right) + \frac{k_\perp}{B_0L_n}\left(\omega + \frac{gk_\perp}{\omega_{ci}}\right)^{-1}\right], \tag{7.23}$$

where $L_n$ is the undisturbed density scale length

$$L_n^{-1} = \frac{d\ln n_0(z)}{dz} > 0\,. \tag{7.24}$$

Assuming that the electrons are cold and do not drift (the gravitational drift of electrons is a factor $m_e/m_i$ slower than the ion drift), we get from the electron continuity and momentum equations the relation

$$\delta n_e = -\delta\varphi \frac{n_0}{B_0}\left(\frac{\omega_{ce}}{\omega}\frac{k_\parallel^2}{\omega} - \frac{k_\perp}{L_n\omega}\right) . \tag{7.25}$$

Because the frequency is small, the charge neutrality is maintained and we can equate $\delta n_e = \delta n_i$. Eliminating the fluctuating potential we finally find the dispersion equation

$$\frac{\omega_{ci}}{\omega}\frac{1}{k_\perp L_n}\left(1 - \frac{\omega}{\omega + gk_\perp/\omega_{ci}}\right) - \left(1 + \frac{m_i}{m_e}\right)\frac{\omega_{ci}^2}{\omega^2}\frac{k_\parallel^2}{k_\perp^2} + 1 = 0 . \tag{7.26}$$

To find exact solutions to this equation is a little tedious. The highest growth rate is found for exactly perpendicular propagation ($k_\parallel = 0$) because in that case the electric field will lead to the largest vertical drift. Assuming further a weak gravitational effect ($\omega \gg k_\perp g/\omega_{ci}$) the first-order solution is

$$\omega^2 = -\frac{g}{L_n} , \tag{7.27}$$

which has a purely growing branch with the growth rate

$$\gamma_{0rt} = \left(\frac{g}{L_n}\right)^{1/2} . \tag{7.28}$$

This is the same growth rate as is found for the RT instability in non-magnetic fluids.

Expanding the dispersion equation to the second order in $k_\perp g/(\omega_{ci}\omega)$ we could find an oscillating solution, but still the growth rate would be much larger than the oscillation frequency. Letting $k_\parallel \neq 0$, solutions remain limited in a narrow cone around the perpendicular direction.

The gravitational acceleration decreases with increasing distance as $r^{-2}$. At the Earth this implies that the RT instability is important only in the ionosphere, and because the magnetic field must be horizontal, only in the equatorial ionosphere. In fact, radar and satellite observations have verified the existence of rising low-density bubbles from the nightside F-region above 200 km within the latitude range from 20°S to 20°N. This effect is known as *equatorial spread-F*. The bubbles can rise up to about 1000 km altitude with upward velocities of about 100 m s$^{-1}$.

Neither the ionosphere nor the partially ionized parts of the solar atmosphere are fully collisionless. Electrons can still be taken as collision-free but the ion–neutral collision rate $\nu_{in}$ and the pressure force must be taken into account. The collisional growth rate is found to be

$$\gamma_{rt} = \gamma_{0rt}\left[1 - \exp\left(-\frac{\gamma_{0rt}}{\nu_{in}}\right)\right] , \tag{7.29}$$

which at the limit of vanishing collisions yields $\gamma_{0rt}$ of (7.28). At the limit of large collision frequency the growth rate becomes

$$\gamma_{rtn} = \frac{g}{\nu_{in}L_n} = \frac{\gamma_{0rt}^2}{\nu_{in}} . \tag{7.30}$$

### 7.2.2 Farley–Buneman instability

The *Farley–Buneman* (FB) instability is somewhat analogous to the RT instability. It is also driven by the horizontal currents, but the currents are not of gravitational origin.

The magnetic field does not need to be horizontal for the FB instability to appear, but it is most transparent to consider it in the equatorial ionosphere, where the magnetic field is horizontal and the electric field points vertically downward $\mathbf{E}_0 = -E_0\,\mathbf{e}_z$. Consequently, the E×B drift is eastward $V_E = -E_0/B_0$ and the linearized electron continuity equation can be written as

$$\delta V_{ey} = \left(\frac{\omega}{k_\perp} - V_E\right)\frac{\delta n}{n_0}\,, \tag{7.31}$$

where quasi-neutrality has been assumed. Neglecting the electron inertia and the gravitation but retaining the electron–neutral collisions, the linearized electron momentum equation has two components

$$\omega_{ce}\delta V_{ey} + \nu_{en}\delta V_{ez} = 0 \tag{7.32}$$

$$\nu_{en}\delta V_{ey} - \omega_{ce}\delta V_{ez} = -ik_\perp\left(\frac{e}{m_e}\delta\varphi - \frac{k_B T_e}{m_e}\frac{\delta n}{n_0}\right)\,. \tag{7.33}$$

Due to high $\nu_{in}$ the ions do not move in the vertical direction and thus the ion continuity and momentum equations are

$$\delta V_{iy} - \frac{\omega}{k_\perp}\frac{\delta n}{n_0} = 0 \tag{7.34}$$

$$(\omega - i\nu_{in})\delta V_{iy} - k_\perp v_{thi}^2\frac{\delta n}{n_0} = \frac{e}{m_i}k_\perp\delta\varphi\,. \tag{7.35}$$

This set of five linear equations has nontrivial solutions when the determinant of the coefficient matrix is zero. This gives us the dispersion equation

$$\omega\left(1 + i\psi_0\frac{\omega - i\nu_{in}}{\nu_{in}}\right) = k_\perp V_E + i\psi_0\frac{k_\perp^2 c_s^2}{\nu_{in}}\,, \tag{7.36}$$

where

$$\psi_0 = \frac{\nu_{en}\nu_{in}}{\omega_{ce}\omega_{ci}}$$

and the ion temperature is retained in the expression for the ion–sound speed $c_s^2 = k_B(T_e + T_i)/m_i$.

For a weakly unstable solution the frequency is

$$\omega_{fb} = \frac{k_\perp V_E}{1 + \psi_0} \tag{7.37}$$

and the growth rate

$$\gamma_{fb} = \frac{\psi_0}{\nu_{in}}\frac{\omega_{fb}^2 - k_\perp^2 c_s^2}{1 + \psi_0}\,. \tag{7.38}$$

Thus the FB instability sets in when the wave phase speed exceeds the ion–sound speed, or equivalently, when the drift speed exceeds the threshold

$$V_E > (1+\psi_0)c_s\,. \tag{7.39}$$

The collision frequencies depend on the neutral density that follows the barometric law $n_n(z) \propto \exp(-z/H)$. In the equatorial ionosphere $\psi_0 \approx 0.22$ at the altitude of 105 km and decreases rapidly upward making the growth rate negligible above altitudes of 130–150 km. Thus the FB instability is limited to the E-region ionosphere.

The FB instability takes place also in the auroral ionosphere, where the geometry is different and there are other mechanisms to make the observed spectra more complicated. The FB fluctuations are useful in diagnostics of ionosphere properties because they scatter electromagnetic waves (Chap. 9). A *coherent ionospheric scatter radar* transmits waves of a few meter's wavelength and receives the backscattered signal. The backscattering occurs when the wave front crosses the background magnetic field at right angles. From the Doppler shift of the backscattered signal it is possible to derive the component of the drift speed in the direction of the wave. Using two such radars pointing to the same scattering volume from to different locations it is possible to determine the two-dimensional velocity field $\mathbf{V}$ and thus the electric field as $\mathbf{E} = -\mathbf{V}\times\mathbf{B}$, assuming that it is perpendicular to the background magnetic field.

### 7.2.3 Ballooning instability

Another analog of the RT instability is the *ballooning instability*, in which the critical forces are the ion pressure gradient and the magnetic tension due to curvature. Ballooning is of particular interest for the theme of this book, as it is one of the instabilities that has been considered to facilitate the current diversion from the Earth's magnetotail through the ionosphere at the time of the substorm onset (Chap. 13). The ballooning instability is also of the interest in the context of prominence eruptions on the Sun (Chap. 12), where it is linked to the traditional RT instability through the gravitational effect.

While the mathematical analysis of the ballooning instability is challenging, its basic idea in the magnetospheric context can be illustrated by Fig. 7.4, which describes an interpretation of observations of the ESA *GEOS*-2 spacecraft. The figure is drawn on the equatorial plane close to midnight near the geostationary distance, which is known to be the interface region where the near-Earth dipole-like magnetic configuration changes to a highly-stretched tail-like configuration prior to the onset substorm expansion.

In this region both $\nabla P$ and $\nabla B$ point earthward. Let the configuration be perturbed by a wave moving in the azimuthal direction. Now electrons and ions located in the earthward side of the wave, i.e., in the region where the pressure is greater will undergo faster drifts to the east and to the west, respectively, than the particles in the tailward side. This leads to a similar polarization field $\delta\mathbf{E}$ as in the case of RT instability and thus to a $\delta\mathrm{E}\times\mathrm{B}$ drift. This drift enhances the initial perturbation and the system is unstable.

The growing positive and negative space charges can act as sinks and sources of field-aligned currents, which tends to stabilize the instability. These FACs have been suggested to build up the substorm current wedge [Roux, 1985], which will be discussed in Chap. 13.

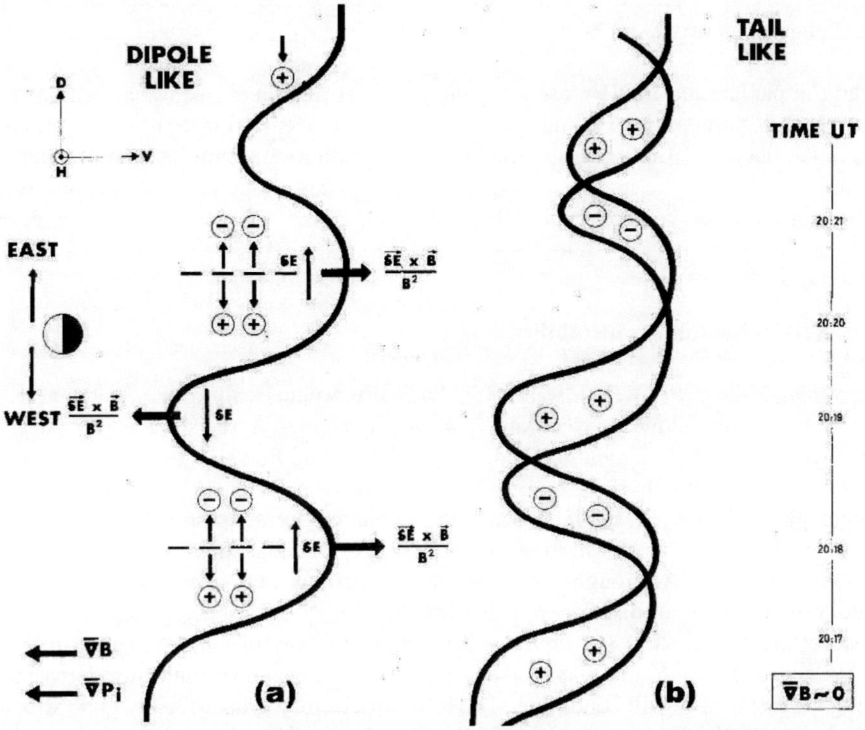


**Fig. 7.4** The idea of the ballooning instability at the interface between the stretched tail-like magnetic field configuration and the near-Earth dipole-like field lines. The structure on the left (a) shows the local growth of the instability analogous to the RT instability. In the structure on the right (b) a westward motion has been added. This leads to a phase shift between the electron and ion dominated regions. The excessive charges are then expected to be neutralized either by sending electrons along the magnetic field line to the ionosphere from the negative charge regions or drawing electrons from the ionosphere to the positive charge regions. The $\{H,V,D\}$ coordinate system indicated in the figure is frequently used to organize data obtained close to the geostationary equatorial orbit. $H$ is northward (the direction of **B** at magnetic equator), $V$ points outward from the Earth, and $D$ to the east. The time line on the right refers to the motion of the *GEOS*-2 satellite, the observations of which were used in the analysis of an event that was simultaneously well covered by ground-based observations in the Scandinavian sector. (Adapted from Roux [1985].)

The actual stability analysis of the ballooning mode in this context has turned out to be a complicated issue. On the tail-like side the magnetic field close to the current sheet is very small and plasma beta larger than 1. Thus the theory must include high-$\beta$ effects and finding out the actual growth rates requires extensive computer simulations.

**Challenge your brain**

Read the paper Liu [1997] presenting an analytical treatment of the ballooning mode associated with the sudden thinning of the cross-tail current prior to the substorm onset. Fill in the details of Liu's analytic treatment. Thereafter search in the scientific journals to find out how far the understanding of the ballooning mode analysis has progressed until the time when you are reading this text.

### 7.2.4 Kelvin–Helmholtz instability

The *Kelvin–Helmholtz* (KH) instability is basically a neutral fluid phenomenon arising, e.g., from the wind blowing over water and causing ripples on the surface. Very beautiful KH vortices can often be seen in clouds due to shear wind flows.

As a space plasma physics example, we consider the solar wind flow along the Earth's magnetopause in the ideal-MHD scale following the presentation by Treumann and Baumjohann [1996]. At the narrow magnetospheric boundary layer kinetic effects lead to anomalous viscosity through wave–particle interactions, but we neglect them as higher-order corrections to our discussion. However, this non-MHD aspect of the KH instability is also of great interest to our topic, as it is one of the key mechanisms how solar wind plasma gets access into the magnetosphere when the dayside reconnection is weak, i.e., during the northward IMF conditions. The KH vortices may grow to really giant structures as demonstrated by Hasegawa et al [2004] using *Cluster* observations on the low-latitude flank of the magnetopause.

Let the magnetic field and the flow be tangential to the boundary and let the velocities be different on each side of the boundary. Assume scalar pressure, linearize around the background $\mathbf{B}_0$ and $n_0$, and consider small displacements $\delta\mathbf{x}$ defined by $\delta\mathbf{V} = d\delta\mathbf{x}/dt$. The strategy is to linearize the induction and momentum equations leading to an expression for $\delta\mathbf{x}$. The linearized equations are

$$\begin{aligned} \delta\mathbf{B} &= \nabla\times(\delta\mathbf{x}\times\mathbf{B}_0) \qquad (7.40)\\ &= \mathbf{B}_0\cdot\nabla\delta\mathbf{x} - \delta\mathbf{x}\cdot\nabla\mathbf{B}_0 - \mathbf{B}_0\nabla\cdot\delta\mathbf{x} \\ \mu_0 m_i n_0 d^2\delta\mathbf{x}/dt^2 &= -\mu_0\nabla\delta P + \qquad (7.41)\\ &\quad -\delta\mathbf{B}\times(\nabla\times\mathbf{B}_0) - \mathbf{B}_0\times(\nabla\times\delta\mathbf{B}) \, , \end{aligned}$$

where the induction equation has been integrated with respect to $t$. Define the first-order perturbation of the total pressure by

$$\mu_0\delta P_{tot} = \mu_0\delta P + \mathbf{B}_0\cdot\delta\mathbf{B} \, . \qquad (7.42)$$

Eliminating the magnetic field perturbation we get

$$m_i n_0\left[(\mathbf{v}_A\cdot\nabla)^2 - \frac{\partial^2}{\partial t^2}\right]\delta\mathbf{x} = \nabla\delta P_{tot} + \mathbf{C} \, . \qquad (7.43)$$

The Alfvén velocity is calculated using the background parameters and the vector **C** contains the remaining terms. This equation illustrates that the Alfvén wave is coupled to pressure fluctuations. Because $\nabla \cdot \mathbf{B} = 0$ and $\nabla \cdot \delta\mathbf{B} = 0$, (7.40) and (7.41) yield another equation for the total pressure perturbation

$$\nabla^2 \delta P_{tot} = -m_i \nabla \cdot \left( n_0 \frac{d^2 \delta\mathbf{x}}{dt^2} \right) + \frac{1}{\mu_0} \nabla \times (\delta\mathbf{B} \cdot \nabla\mathbf{B}_0 + \mathbf{B}_0 \cdot \nabla\delta\mathbf{B}) \,. \tag{7.44}$$

Assume now that the the plasma and the flow are homogeneous on both sides of the boundary. This implies that the plasma perturbation is incompressible ($\nabla \cdot \delta\mathbf{V} = 0$). Thus the RHS of (7.44) vanishes as does **C**, and what remains is a Laplace equation for the pressure perturbation

$$\nabla^2 \delta P_{tot} = 0 \,. \tag{7.45}$$

The pressure disturbance $\delta P_{tot}$ is limited at the thin boundary and fades out with increasing distance from the boundary. Let the boundary be in the $(x,z)$-plane and assume plane wave solutions for both $\delta\mathbf{x}$ and $\delta P_{tot}$ with wave number $\mathbf{k} = k_x \mathbf{e}_x + k_z \mathbf{e}_z$ and frequency $\omega$. Now we can solve the displacement of the boundary

$$\delta\mathbf{x} = \frac{\delta P_{tot}}{m_i n_0 [\omega^2 - (\mathbf{k} \cdot \mathbf{v}_A)^2]} \tag{7.46}$$

and the solution of the Laplace equation for $\delta P_{tot}$ is

$$\delta P_{tot} = P_0 \exp(-k|y|) \exp[-i(\omega t - k_x x - k_z z)] \,, \tag{7.47}$$

where $k^2 = k_x^2 + k_z^2$. The exponential $y$-dependence is introduced to make the wave evanescent outside the boundary because free energy is available only at the boundary.

We consider the boundary as a *tangential discontinuity*, i.e., a boundary through which there is no plasma flow and $B_n = 0$, but where $V_t, B_t, n$, and $P$ may jump. (This and other MHD discontinuities will be discussed more thoroughly in Chap. 11.) We further require that the normal component of the displacement is continuous. Denote the two sides of the boundary by 1 and 2 and let the plasma stream with velocity $\mathbf{V}_0$ in region 1 and the fluid in region 2 be in rest. Because the total pressure $P + B^2/(2\mu_0)$ is continuous, the continuity of the normal component of the displacement yields the dispersion equation for the KH waves

$$\frac{1}{n_{02}[\omega^2 - (\mathbf{k} \cdot \mathbf{v}_{A2})^2]} + \frac{1}{n_{01}[(\omega - \mathbf{k} \cdot \mathbf{V}_0)^2 - (\mathbf{k} \cdot \mathbf{v}_{A1})^2]} = 0 \,. \tag{7.48}$$

This equation has some formal similarity to the equations for streaming instabilities discussed earlier, but now the unstable modes are the Alfvén waves. The dispersion equation has an unstable solution

$$\omega_{kh} = \frac{n_{01} \mathbf{k} \cdot \mathbf{V}_0}{n_{01} + n_{02}} \tag{7.49}$$

corresponding to the complex root for which

$$(\mathbf{k} \cdot \mathbf{V}_0)^2 > \frac{n_{01} + n_{02}}{n_{01} n_{02}} [n_{01} (\mathbf{k} \cdot \mathbf{V}_{A1})^2 + n_{02} (\mathbf{k} \cdot \mathbf{V}_{A2})^2] \,. \tag{7.50}$$

The KH instability occurs thus for sufficiently large $\mathbf{V}_0$. For small $\mathbf{V}_0$ the wave number $k$ would have to be too large, i.e., the wavelength too short, for the MHD description to be valid.

At the limit where the spatial scale becomes comparable to the ion gyro radius the finite gyro radius effects introduce the *kinetic Alfvén* waves discussed in Sect. 6.7.2. For relatively large beta ($\beta > m_e/m_i$), e.g., at the magnetospheric boundary, the relevant mode is the *oblique kinetic Alfvén wave*(6.149) with the phase velocity

$$v_{\parallel} = v_A \left[1 + k_{\perp}^2 r_{Li}^2 \left(\frac{3}{4} + \frac{T_e}{T_i}\right)\right]^{1/2} \tag{7.51}$$
$$v_{\perp} = \frac{k_{\parallel} v_A}{k_{\perp}} \left[1 + k_{\perp}^2 r_{Li}^2 \left(\frac{3}{4} + \frac{T_e}{T_i}\right)\right]^{1/2} .$$

Referring to the KH unstable configuration, we can expect that $\lambda_{\parallel} \gg \lambda_{\perp}$, i.e., $k_{\parallel} \ll k_{\perp}$.

The KH instability is important also in low-$\beta$ plasmas. Above auroral arcs the electric field points toward the arc on both sides. Thus there is a strong shear in the plasma flow. The KH instability arising from this shear is a popular explanation why auroral arcs evolve to folds and spirals.

### 7.2.5 Firehose and mirror instabilities

The *firehose instability* has an analog in a familiar firehose or garden hose with a rapid water flow, in which a small perturbation can cause a violent motion of the loose end of the hose. In ideal anisotropic MHD a magnetic flux tube corresponds to the hose and the parallel pressure to the flowing water.

We can start the analysis from the momentum equation of the CGL theory (2.158)

$$\rho_m \left(\frac{d\mathbf{V}}{dt}\right)_{\perp} + \nabla_{\perp}\left(P_{\perp} + \frac{B^2}{2\mu_0}\right) - \frac{(\mathbf{B}\cdot\nabla)\mathbf{B}}{\mu_0}\left(\frac{P_{\perp} - P_{\parallel}}{B^2/\mu_0} + 1\right) = 0 .$$

Assuming $\mathbf{V}_0 = 0$ and $\mathbf{B} = \mathbf{B}_0 + \mathbf{B}_1$, where $\mathbf{B}_1$ is a small perturbation, and performing the standard linearization procedure with this momentum equation, we get the dispersion equation

$$\omega^2 = \frac{k^2}{2\rho_{m0}} \left\{ \left(\frac{B_0^2}{\mu_0} + P_{\perp} + 2P_{\parallel}\cos^2\theta + P_{\perp}\sin^2\theta\right) + \right. \tag{7.52}$$
$$\left. \pm \sqrt{\left(\frac{B_0^2}{\mu_0} + P_{\perp}(1+\sin^2\theta) - 4P_{\parallel}\cos^2\theta\right)^2 + 4P_{\perp}^2 \sin^2\theta\cos^2\theta} \right\} .$$

**Challenge your brain** by deriving (7.52). It is not the easiest linearization exercise you will encounter in plasma physics.

For perpendicular propagation ($\theta = \pi/2$) (7.52) reduces to

$$\frac{\omega^2}{k_\perp^2} = \frac{2}{\rho_{m0}} \left( \frac{B_0^2}{2\mu_0} + P_\perp \right) . \tag{7.53}$$

This is the stable magnetosonic of MHD mode with the phase velocity $\sqrt{v_A^2 + v_s^2}$.

For parallel propagation there are two solutions. The sound wave

$$\omega^2 = \frac{3k_\parallel^2}{\rho_{m0}} P_\parallel \tag{7.54}$$

and another mode with the dispersion equation

$$\omega^2 = \frac{k_\parallel^2}{\rho_{m0}} \left( \frac{B_0^2}{\mu_0} + P_\perp - P_\parallel \right) . \tag{7.55}$$

At the isotropic limit this is the shear Alfvén wave ($\omega/k = v_A$). If $P_\parallel > P_\perp + B_0^2/\mu_0$, the wave has an unstable solution, which is the firehose instability. The dispersion equation can be written in terms of parallel and perpendicular beta and the growth rate thus becomes

$$\gamma = \frac{k_\parallel v_A}{\sqrt{2}} \left( \beta_{0\parallel} - \beta_{0\perp} - 2 \right)^{1/2} \tag{7.56}$$

and the threshold for the instability can be expressed as

$$\beta_{0\parallel} > \beta_{0\perp} + 2 . \tag{7.57}$$

This implies that $\beta > 2$ and the firehose instability requires very weak magnetic field or strong pressure. This is possible, e.g., in the solar wind and in the magnetotail neutral sheet. Once excited the instability is strong.

The *mirror instability* is complementary to the firehose instability and propagates nearly perpendicular to the magnetic field. Its dispersion equation is straightforward (but not easy) to derive from kinetic theory retaining contributions from all particle species. This procedure yields unstable solutions for both parallel and perpendicular directions. In the parallel direction the firehose threshold is found again, now in the form

$$\sum_\alpha \beta_{\alpha\parallel} > 2 + \sum_\alpha \beta_{\alpha\perp} . \tag{7.58}$$

For perpendicular propagation the threshold for the mirror instability is

$$\sum_\alpha \frac{\beta_{\alpha\perp}^2}{\beta_{\alpha\parallel}} > 1 + \sum_\alpha \beta_{\alpha\perp} . \tag{7.59}$$

Figure 7.5 illustrates how a mirror unstable region looks in satellite data and explains why the mode is called the mirror mode. Part of the plasma is trapped in the local magnetic

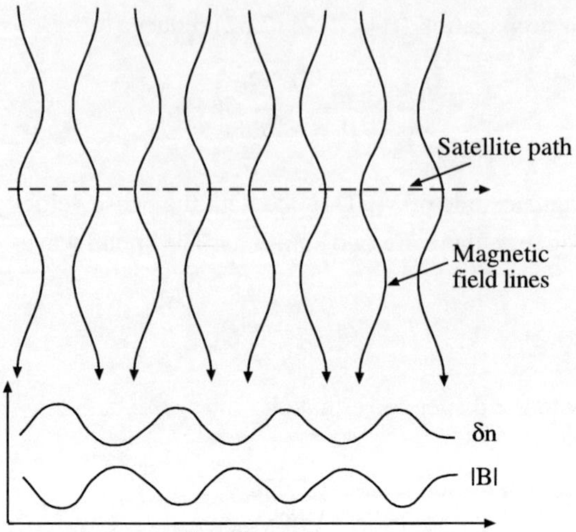


**Fig. 7.5** Sketch of satellite observations of plasma density and magnetic field fluctuations through a mirror unstable region.

bottles of the wave. The mode has been frequently observed in the dayside magnetosheath. The shocked solar wind plasma is adiabatically heated in the perpendicular direction. At the same time the field-aligned flow around the magnetopause lowers the parallel temperature, leading to favorable conditions for the mirror instability to develop. The particle trapping in the mirror structures and the heat flux along the magnetic field are reasons why the mode is difficult to treat at the fluid limit. In particular, the latter violates the assumptions of the CGL theory, which gives a correct description of the firehose branch but not of the mirror branch.

### 7.2.6 Flux tube instabilities

Flux tube instabilities are particularly important in solar physics, as well as in laboratory devices. A powerful method for the stability analysis is based on the *energy principle*. The energy content of the system is calculated in the presence of small perturbations. If the energy variation $\triangle W$ is negative, the system is unstable. The calculations are usually pretty cumbersome.

There are three basic modes of instabilities in flux tubes carrying a longitudinal current. The magnetohydrostatic equilibrium ($\mathbf{J} \times \mathbf{B} = \nabla P$) is in all cases maintained by the azimuthal magnetic field. This arrangement is known as the *linear pinch*.

The *pinch instability* arises from squeezing (pinching) the flux tube. The azimuthal field increases in regions where the tube is pinched and decreases outside. Thus the pinching self-amplifies the instability. This instability is important in certain laboratory settings and it may take place in the active regions of the solar corona.

The *kink instability* resembles the pinch effect. If the tube is kinked, there is an inward pressure gradient in the inner edge of the kink and outward pressure gradient in the outer edge. Again the perturbation is self-amplifying and thus unstable. This process may be excited in the solar corona or in the magnetospheric tail current sheet.

Finally, the *helical instability* is probably very common in strongly twisted, nearly force-free, flux tubes in the solar corona. The instability requires strong enough field-aligned current to flow through the structure. Solar prominences and coronal loops are examples of helical magnetic field structures (Chap. 12).

## 7.3 Microinstabilities

The microinstabilities require a Vlasov theory approach and the practical calculations quickly become intractable with analytical methods and require extensive computer simulations. Here we introduce the topic by looking for growing solutions to the electrostatic dispersion equation.

### 7.3.1 Monotonically decreasing distribution function

Let $f_{\alpha 0}(\mathbf{v})$ decrease monotonically and consider an electrostatic perturbation in the form

$$f_\alpha = f_{\alpha 0} + f_{\alpha 1}(\mathbf{v}) \exp[i(\mathbf{k}\cdot\mathbf{r} - \omega t)] \,, \tag{7.60}$$

where $\omega$ is a solution of the dispersion equation

$$1 - \frac{\omega_{pe}^2}{k^2} \int_{-\infty}^{\infty} \frac{1}{u - \omega/|k|} \frac{\partial}{\partial u} \left[ F_{e0}(u) + \frac{m_e}{m_i} F_{i0}(u) \right] du = 0 \,. \tag{7.61}$$

Here we have assumed two populations (electrons and ions) and $F_{\alpha 0}$ is the one-dimensional distribution function. If the dispersion equation implies $\omega_i > 0$, the distribution function is unstable, otherwise it is stable.

Assume now that there are unstable solutions $\omega_i > 0$. Thus the pole in (7.61) is in the upper half plane and the integral can be taken along the real $u$-axis. Denote $F = F_{e0} + (m_e/m_i)F_{i0}$. The dispersion equation reduces to

$$\begin{aligned} 1 &- \frac{\omega_{pe}^2}{k^2} \int_{-\infty}^{\infty} \frac{u - \omega_r/|k|}{(u - \omega_r/|k|)^2 + \omega_i^2/k^2} \frac{\partial F}{\partial u} du + \\ &- \frac{i\omega_i}{|k|} \frac{\omega_{pe}^2}{k^2} \int_{-\infty}^{\infty} \frac{\partial F/\partial u}{(u - \omega_r/|k|)^2 + \omega_i^2/k^2} du = 0 \,. \end{aligned} \tag{7.62}$$

These integrals do not contain any singularities. Because the real and imaginary parts both must be zero, we have

$$\int_{-\infty}^{\infty} \frac{\partial F/\partial u}{(u-\omega_r/|k|)^2+\omega_i^2/k^2}\,du = 0 \tag{7.63}$$

$$1-\frac{\omega_{pe}^2}{k^2}\int_{-\infty}^{\infty} \frac{u(\partial F/\partial u)}{(u-\omega_r/|k|)^2+\omega_i^2/k^2}\,du = 0\,. \tag{7.64}$$

For a distribution function that decreases monotonically in each direction from the origin $u(\partial F/\partial u) \leq 0$. In that case the integral in (7.64) is negative definite and the equation has no solutions. Thus we have found a contradiction with the assumption $\omega_i > 0$. This applies to all monotonic functions and the result is independent of the frame of reference. The result that a monotonic function is stable is known as *Gardner's theorem*.

### 7.3.2 Multiple-peaked distributions

The cold two-stream instability of Sect. 7.1.1 was produced by a multiple-peaked distribution. To include thermal effects in the analysis we consider the so-called *gentle-bump distribution* for the electrons

$$\begin{aligned} f_{e0} = {} & \frac{n_1}{n_e}\left(\frac{m_e}{2\pi k_B T_1}\right)^{3/2} \exp\left(-\frac{m_e v^2}{2k_B T_1}\right) + \frac{n_2}{n_e}\delta(v_x)\delta(v_y)\left(\frac{m_e}{2\pi k_B T_2}\right)^{1/2} \times \\ & \frac{1}{2}\left\{\exp\left(-\frac{m_e(v_z-V_0)^2}{2k_B T_2}\right) + \exp\left(\frac{m_e(v_z+V_0)^2}{2k_B T_2}\right)\right\}\,. \end{aligned} \tag{7.65}$$

where $n_e = n_1 + n_2 \gg n_2$ , $T_2 \ll T_1$ , $V_0 \gg 2k_B T_1/m_e$. We assume that the ions form a cold background $f_{i0} \sim \delta(v_x)\delta(v_y)\delta(v_z)$. Furthermore, in order to neglect the current driven by the bump we consider an electron distribution that is symmetric about $v_z = 0$ (thus the argument of $f_{e0}$ in Fig. 7.6 is $v_z^2$). This way the problem remains strictly electrostatic.

In the absence of the bump the solution would be the damped Langmuir wave. Now the calculation of $K_r$ and $K_i$ is considerably more tedious than for the Maxwellian distribution. The procedure is, however, straightforward. Start with (5.39) and (5.40). Insert the distribution function (7.65) and consider long wavelengths. With the gentleness assumptions $n_1 \gg n_2$ and $T_1 \gg T_2$ for the bump the solution of the dispersion equation has the real part corresponding to the Langmuir wave

$$\omega_r = \omega_{pe}(1+3k^2\lambda_{De}^2)^{1/2} \approx \omega_{pe}(1+\frac{3}{2}k^2\lambda_{De}^2)\,. \tag{7.66}$$

The imaginary part is modified by a term depending on the relative number densities and temperatures of the bump and the background

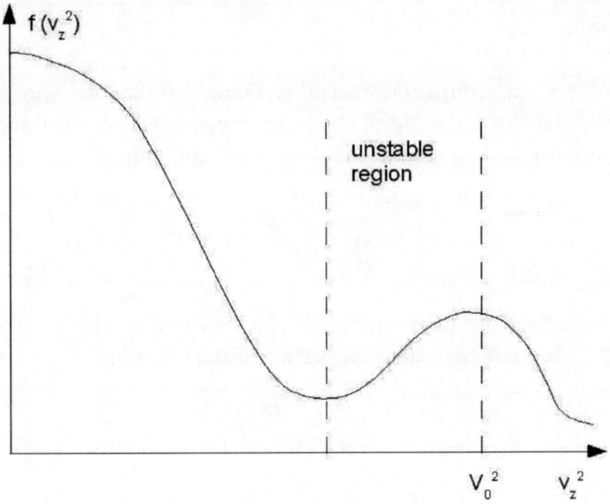


**Fig. 7.6** Gentle-bump distribution.

$$\omega_i = -\sqrt{\frac{\pi}{8}}\frac{\omega_{p1}}{|k^3\lambda_{D1}^3|}\exp\left(-\frac{1}{2k^2\lambda_{D1}^2}-\frac{3}{2}\right)+ \tag{7.67}$$
$$+\frac{n_2}{n_1}\left(\frac{T_1}{T_2}\right)^{3/2}\frac{k^3}{k_z^3}\left(\frac{k_zV_0}{\omega_r}-1\right)\exp\left\{-\frac{T_1/T_2}{2k^2\lambda_{D1}^2}\left(1-\frac{k_zV_0}{\omega_r}\right)^2\right\}.$$

The first term is the Landau damping of the background. The second term is stabilizing to the right from the bump ($v_z > v_0$) where the distribution is decreasing, but it *may* destabilize ($\omega_i > 0$) plasma oscillations to the left from the bump between the two peaks of the distribution function. The essential condition for the instability is whether or not the derivative of the total distribution function is positive $\partial f_{e0}/\partial v > 0$ and *large enough* to overcome the damping by the background. If it is, we have the *gentle-bump instability*. The instability is enhanced if

- the number of particles in the bump is increased,
- the bump becomes sharper (colder),
- the speed of the bump ($V_0$) increases, i.e., the configuration approaches the cold two-stream case.

Note that Gardner's theorem does *not* imply that a non-monotonic distribution would automatically be unstable. If the bump is too gentle, it is not powerful enough to drive an instability. The only way of finding this out is to calculate the imaginary part of the frequency.

**Feed your brain**

There is a more powerful *stability* criterion than Gardner's theorem known as the *Penrose criterion*. It states that for a double-peaked one-dimensional distribution function $F(u)$ with a local minimum at $u_0$, instability is *possible* if and only if

$$\int_{-\infty}^{\infty} \frac{F(u_0) - F(u)}{(u - u_0)^2} du < 0 . \tag{7.68}$$

Using the literature find out how this result can be derived with the so-called *Nyquist method* by considering the analytical properties of the function

$$G = \frac{1}{K(\omega,k)} \frac{dK(\omega,k)}{d\omega} . \tag{7.69}$$

Note that both Gardner's theorem and the Penrose criterion apply to electrostatic problems only.

### 7.3.3 Ion–acoustic instability

In Chapter 5 we found that the damping rate of the ion–acoustic (IAC) wave depends on the ratio $T_e/T_i$. But what happens if the electron and ion distributions are in motion with respect to each other, thus making the total distribution function double-peaked (Fig. 7.7).

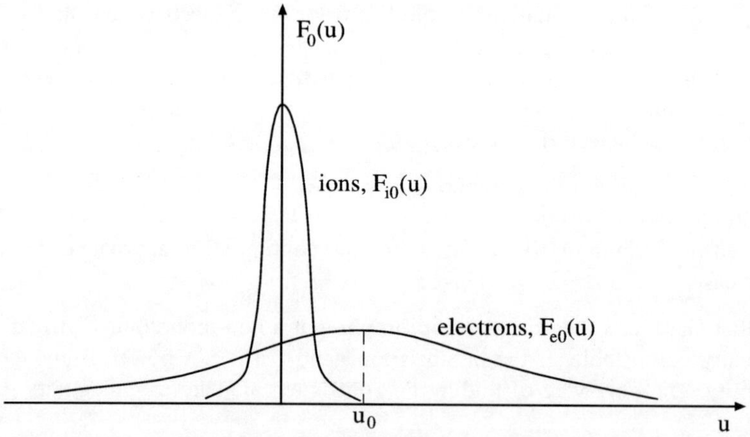


**Fig. 7.7** Maxwellian electron distribution streaming through a Maxwellian ion distribution with velocity $u_0$.

Let the one-dimensional distribution functions be

$$F_{e0} = \sqrt{\frac{m_e}{2\pi k_B T_e}} \exp\left(-\frac{m_e(u-u_0)^2}{2k_B T_e}\right) \tag{7.70}$$

$$F_{i0} = \sqrt{\frac{m_i}{2\pi k_B T_i}} \exp\left(-\frac{m_i u^2}{2k_B T_i}\right). \tag{7.71}$$

Using the Penrose criterion we can show that

- for $T_i = T_e$ the plasma is stable if

$$u_0 < 1.3\sqrt{\frac{k_B T_e}{m_e}}, \tag{7.72}$$

- for $T_i \ll T_e$ the plasma is stable if

$$u_0 < \sqrt{\frac{k_B T_i}{m_i}}, \tag{7.73}$$

otherwise instability is *possible*. Note that in the cold ion case the stability limit is much smaller than in the case of $T_i = T_e$. This appears reasonable for the IAC mode, but the Penrose criterion does not tell anything of the unstable modes. To find the modes the dispersion equation must be solved.

It is physically reasonable to look for solutions in the range

$$\left|\frac{\omega_r}{k}\right| \gg \sqrt{\frac{2k_B T_i}{m_i}}$$

$$\left|\frac{\omega_r}{k} - u_0\right| \ll \sqrt{\frac{2k_B T_e}{m_e}}.$$

A lengthy but straightforward calculation yields

$$\omega_r^2 = \frac{k^2 c_s^2}{1+k^2\lambda_{De}^2} \quad ; \quad c_s = \sqrt{\frac{k_B T_e}{m_i}} \tag{7.74}$$

and

$$\omega_i = -\frac{|\omega_r|\sqrt{\pi/8}}{(1+k^2\lambda_{De}^2)^{3/2}} \times \tag{7.75}$$
$$\left\{\left(\frac{T_e}{T_i}\right)^{3/2} \exp\left(\frac{-T_e/T_i}{2(1+k^2\lambda_{De}^2)}\right) + \sqrt{\frac{m_e}{m_i}}\left(1 - \frac{u_0}{c_s}\sqrt{1+k^2\lambda_{De}^2}\right)\right\}.$$

If $u_0 = 0$, this reduces to the IAC wave introduced in Chap. 5. When $T_e \gg T_i$, the instability condition can be found from the last term in (7.75). Close to the instability threshold the electron streaming and ion damping compete with each other. Again, the positive slope of the distribution must be positive enough to win the damping by the background.

The IAC instability is an example of *current-driven instabilities*. In this setting the current is in the relative drift between electron and ion populations. IAC waves can also be driven by a current carried by an ion beam moving through a warm electron–ion background. In either case, it is not sufficient to consider the net current alone. Both the temperatures and the relative speeds of the populations are critical parameters for the instability to occur.

The Langmuir and the ion–acoustic waves are the most fundamental modes in non-magnetized plasmas. In magnetized plasmas the conditions for the current-driven IAC instability can be met for propagation parallel to the background magnetic field if the field-aligned current is strong enough. The IAC wave is strongly damped for propagation deviating from the direction of the magnetic field.

### 7.3.4 Electrostatic ion cyclotron instability

Strong field-aligned currents can be found, e.g., above the auroral oval, where the ion–acoustic instability competes with other unstable wave modes, the most important of these being the the electrostatic ion cyclotron (EIC) wave (Fig. 5.4). Its dispersion equation can be derived from the general dispersion equation of Chap. 5. We are not going into the details of the analytically complicated calculations, but it is instructive to give a look at the dielectric function in the case of superposition of a Maxwellian electron background and possibly several ion distributions $f_{0i}$ that may be drifting and also have a loss cone around the magnetic field direction

$$K(\omega,\mathbf{k}) = 1 + \frac{\omega_{pe}^2}{\omega_{ce}^2}\frac{k_\perp^2}{k^2} - \frac{1}{k^2\lambda_D^2}Z'(\zeta_e) + \sum_i \frac{\omega_{pi}^2}{k^2}\sum_{n=-\infty}^{n=\infty}\int_{-\infty}^{\infty} dv_\parallel \frac{\hat{G}_{ni} f_{0i}(v_\parallel, v_\perp; \Delta, \Sigma)}{\omega - k_\parallel v_\parallel - n\omega_{ci}} . \tag{7.76}$$

Here the perpendicular integral and the Bessel functions in (5.85) are collected into the operator $\hat{G}_{ni}$

$$\hat{G}_{ni} = 2\pi \int_0^\infty v_\perp dv_\perp J_n^2(k_\perp r_{Li}) \left( k_\perp \frac{\partial}{\partial v_\parallel} + \frac{n\omega_{ci}}{v_\perp}\frac{\partial}{\partial v_\perp} \right) \tag{7.77}$$

and $\Delta$ and $\Sigma$ are parameters to describe the filling ratio of the loss cone and the shape of the distribution within the loss cone.

It is evident that the general solutions of $K(\omega,\mathbf{k}) = 0$ require numerical computations, but there are two motivations for writing these equations down. First, they illustrate that now the combination of parallel and perpendicular derivatives in (7.77) can give rise to a positive growth rate depending on the detailed shape of the ion distribution function. Second, the harmonic structure becomes important and different harmonics of the cyclotron modes have different growth rates.

The classical field-aligned current-driven EIC wave can be found by assuming that the ion distribution function is a Maxwellian drifting along the magnetic field with respect to the electron population. In the first approximation the loss cone is assumed to be filled, thus the parameters $\Delta$ and $\Sigma$ need not to be considered. After a bit of tedious calculation the real part of the frequency turns out to be

$$\omega_r \approx n\,\omega_{ci}\left[1+\frac{T_e}{T_i}\Gamma_n(b_i)\right] \quad ; \; n=1,2,\ldots \tag{7.78}$$

where $b_i = k_\perp^2 r_{Li}^2/2$, $\Gamma_n(b) = I_n(b)\exp(-b)$, and $I_n$ is the modified Bessel function of the first kind.

When $T_e \approx T_i$, the solution for the marginally stable fundamental mode ($n=1$) yields $\omega \approx 1.2\,\omega_{ci}$, $k_\parallel/k_\perp \approx 1/10$, and the critical speed is $u_{0c} \approx 13\,v_{thi}$. For higher harmonics the critical speed is larger and thus the fundamental mode is easiest to destabilize. The EIC mode is particularly important when $T_e \approx T_i$ making the IAC mode strongly damped. However, once the IAC is destabilized, it grows faster than the EIC mode.

**Feed your brain**

Read the classic paper by Kindel and Kennel [1971] on current-driven electrostatic instabilities to learn how these results were obtained for the first time.

A loss cone distribution function may have steep enough perpendicular slope ($\partial f_{0\alpha}/\partial v_\perp$) to lead to positive growth of electrostatic electron and ion cyclotron modes as well as the Bernstein modes at short wavelengths. Loss cone distributions are common in magnetic bottle configurations, e.g., in the quasi-dipolar magnetic field of the inner magnetosphere and thus the loss-cone-related instabilities are of considerable interest to the magnetospheric dynamics during strongly disturbed conditions, i.e., storms.

In a loss cone distribution there is more energy in the perpendicular than in the parallel direction, which provides perpendicular free energy for instability. When a wave mode is excited, particles driving the wave lose part of their perpendicular energy and some of them move into the loss cone. This is process is known as *pitch angle scattering* through wave–particle interaction. Particles scattered to the loss cone are removed from the particle distribution through the end of the bottle. For example, the magnetospheric bottle is leaky and particles precipitate into the upper atmosphere. Another pitch angle scattering mechanism was met in Chap. 3 where we discussed non-conservation of the magnetic moment near a current sheet, which also can move particles into the loss cone.

### 7.3.5 Current-driven instabilities perpendicular to B

Perpendicular currents can also lead to various instabilities, both electromagnetic and electrostatic. In fact, the distinction between electrostatic and electromagnetic becomes less meaningful, in particular if plasma $\beta$ is not small, as is the case near the thin current sheets where reconnection or other mechanisms for current sheet disruption are expected to take place (Chaps. 8, 12, 13). The perpendicular current is often associated with a spatial inhomogeneity, which adds to the difficulties in treating the instabilities. Usually heavy numerical computations with clever physical approximations are needed.

We start this discussion with the *modified two-stream instability* (MTSI), because it is the easiest to discuss in analytical terms resembling the treatment of the unmagnetized two-stream instability (Sect. 7.1.1). MTSI is an electromagnetic instability leading to whistler mode waves propagating oblique to the ambient magnetic field at frequencies above the ion gyro frequency. MTSI is one of the instabilities that have been suggested to be responsible for the disruption of the cross-tail current at the substorm onset (Chap. 13).

The MTSI differs from the unmagnetized two-stream instability by the effect of the ambient magnetic field that constrains more strongly the motion of the electrons than of the ions, e.g., in the stretched magnetic field configuration in the magnetotail. This introduces an "effective" mass to the electrons and the interaction becomes more similar to an ion–ion two-stream instability than the electron–ion Buneman instability.

For strongly magnetized electrons and drifting Maxwellian ions the dispersion equation can be written in a quasi-electrostatic approximation as

$$1+\frac{\omega_{pe}^2}{\omega_{ce}^2}-\frac{k_\parallel^2}{k^2}\frac{\omega_{pe}^2}{\omega^2}+\frac{2\omega_{pi}^2}{k_\perp^2 v_{thi}^2}[1+\zeta_i Z(\zeta_i)]=0\,, \tag{7.79}$$

where $\zeta_i=(\omega-k_\perp v_d)/(k_\perp v_{thi})$, in which $v_d$ and $v_{thi}$ are the ion drift and thermal speeds. Assuming $\zeta_i \gg 1$ the dispersion equation becomes

$$1-\frac{\omega_{lh}^2}{(\omega-k_\perp v_d)^2}-\frac{m_i k_\parallel^2}{m_e k^2}\frac{\omega_{lh}^2}{\omega^2}=0\,, \tag{7.80}$$

where the lower hybrid frequency is given in the approximation

$$\omega_{lh}=\frac{\omega_{pi}}{\sqrt{1+\omega_{pe}^2/\omega_{ce}^2}}\,. \tag{7.81}$$

The dispersion equation is formally the same as (7.11) and can be solved in the same way. Both the frequency and the maximum growth rate turn out to be close to the lower hybrid frequency.

**Train your brain** by calculating the maximum growth rate and the frequency at the maximum growth for the MTSI.

The *ion Weibel instability* (IWI) is another mode that has been studied in the context of the cross-tail current sheet distribution. IWI is more characteristically electromagnetic than MTSI. It is related to the whistler mode propagating along the magnetic field (denoted as the $x$ direction in the coordinates we are using when we describe the current sheet in terms of the Harris model). Again the electrons are considered to be strongly tied to the magnetic field. Assuming that the ions drift perpendicular to $\mathbf{B}$ in the $y$-direction, the wave magnetic field $\delta B_z$ is perpendicular to both the ambient magnetic field and the ion drift direction. Now the ions become bunched between the wave crests. The bunching enhances the the original ion current within the bunches and enhances $\delta B_z$. This feedback leads to an unstable current filamentation and reduces the total current, as free energy is transferred from the current sheet to the growing wave.

Particle drifts perpendicular to the magnetic field can drive drift modes related to practically any plasma oscillations (*drift–cyclotron*, *drift–Alfvén*, etc.). The *lower hybrid drift instability* (LHDI) is a particularly important example. It has been thoroughly investigated in various contexts from fusion to space plasmas, because the lower hybrid waves are ubiquitous in all kinds plasmas due to their capability to interact with both electrons and ions, and even simultaneously. For example, a field-aligned auroral electron beam can have a Landau resonance with a lower hybrid wave, which simultaneously is in gyro resonance with the local ion population. Depending on the details of the actual distribution functions and the magnetic field configuration the energy and momentum transfer may be from the electrons to the ions, or vice versa.

The driver of the LHDI is usually considered to be the diamagnetic drift current $\mathbf{J}_\perp = \mathbf{B} \times \nabla P / B^2$. In space physics the LHDI has been used to explain the broadband electrostatic noise frequently observed in the boundary of the magnetospheric plasma sheet. The mode has also been invoked to explain the current sheet instability at the time of substorm onset (Chap. 13). However, it is unclear whether or not the relatively high plasma $\beta$ of the order of one quenches the instability in the mid-tail current sheet.

### 7.3.6 Electromagnetic cyclotron instabilities

The parallel propagating electromagnetic $R$ and $L$ modes are of special importance for space physics because they can be in *cyclotron resonance* with the charged particles. The real part of the dispersion equation can be taken from the cold plasma theory (Chap. 4) and written in a form that covers both $R$ and $L$ modes

$$\frac{c^2k^2}{\omega^2} = 1 - \frac{\omega_{pe}^2}{\omega(\omega \pm \omega_{ce})} - \sum_i \frac{\omega_{pi}^2}{\omega(\omega \mp \omega_{ci})} . \tag{7.82}$$

We treat the gyro frequencies here as unsigned (positive) quantities. Thus the upper signs correspond to the $L$ mode and the lower signs to the $R$ mode.

However, the cold plasma theory does not give the complete description near the resonances and we need to turn to the tools of Chap. 5, where the resonance condition in a magnetized plasma was found to be

$$k_\parallel v_\parallel = \omega - n\omega_{c\alpha} . \tag{7.83}$$

The case $n = 0$ is the Landau resonance whose contribution to the wave growth or damping is easy to picture in terms of positive or negative gradients of the distribution function in the $v_\parallel$ direction.

For $n \neq 0$ the resonance condition is more difficult to illustrate because it involves both the parallel ($v_\parallel$) and gyro ($\omega_{c\alpha}$) motion of the particles. When $n = 1$ and $k_\parallel = 0$, we have the resonance condition $\omega = \omega_{c\alpha}$ in the rest frame of the wave. Thus a particle sees the wave all the time in the same phase. Depending on the *relative phase* between the wave and a particle the wave electric field either accelerates or decelerates the particle. The net damping or growth, i.e., whether there are more particles to be accelerated or decelerated, thus depends on velocity gradients of the distribution function in both parallel and perpendicular directions. Even a distribution that is monotonic in all directions in the velocity space can still be unstable if the distribution is anisotropic enough. Consequently, neither the Gardner theorem nor the Penrose criterion discussed in the context of electrostatic instabilities are applicable to electromagnetic instabilities.

Exact resonances with the $R$ and $L$ modes always lead to damping of the waves. Consider an electron moving at perfect resonance close to the phase speed of the $R$ mode wave. If it is slower than the wave, it rotates in the same sense as the wave and sees the electric field which accelerates the particle to catch up with the wave and the wave is damped. On the other hand, if the electron moves faster than the wave, the wave seems to move backward in the electron frame and the electron sees the wave vector rotating in the opposite sense and there is no interaction between the wave and the particle and thus no amplification of the wave. The same reasoning applies to ions and the $L$-mode resonance.

Both loss cone distributions and temperature anisotropies with larger perpendicular than parallel temperature ("pancake" distributions) can lead to wave growth, but the required anisotropy threshold for the instability

$$A_\alpha = \frac{T_{\alpha\perp}}{T_{\alpha\parallel}} - 1 > 0 \tag{7.84}$$

must be determined case by case from the microscopic theory. The parallel *resonant energy* of the particles $W_{\alpha\| res}$ can be found directly from (7.82). For the electrons it is

$$\frac{W_{e\| res}}{W_B} = \frac{\omega_{ce}}{\omega}\left(1 - \frac{\omega}{\omega_{ce}}\right)^3 \tag{7.85}$$

and for the ions

$$\frac{W_{i\| res}}{W_B} = \frac{\omega_{ci}^2}{\omega^2}\left(1 - \frac{\omega}{\omega_{ci}}\right)^3 , \tag{7.86}$$

where $W_B = B^2/(2\mu_0 n)$ is the magnetic energy per particle.

Below the electron cyclotron frequency the $R$-mode has the whistler branch that can be driven unstable by anisotropic distribution functions. In terms of the resonant energy and the anisotropy parameter $A_e$ the threshold for the instability can now be shown to be

$$W_{e\| res} > \frac{W_B}{A_e(A_e+1)^2} . \tag{7.87}$$

Similarly, electromagnetic ion cyclotron (EMIC) waves below the ion cyclotron frequency are driven unstable if the threshold of

$$W_{i\| res} > \frac{W_B}{A_i^2(A_i+1)} \tag{7.88}$$

is exceeded

**Feed your brain**

Read the classic paper by Kennel and Engelmann [1966], in which both the whistler mode and EMIC wave growth due to temperature anisotropies were introduced and applied to the loss of particles from the magnetosphere. In particular, derive the threshold expressions (7.87) and (7.88) with the help of this article. Before penetrating to the details of the diffusion theory discussed in the article, it may be useful to read Chap. 10 of this book.

Both the whistler mode and EMIC waves are important in the physics of space storms throughout the sequence from the Sun to the inner magnetosphere. We will meet the whistler waves in Chap. 14 where they are discussed in the context of loss of radiation belt electrons due to pitch angle scattering into the ionospheric loss cones. The EMIC waves, on the other hand, are not limited to the loss of ions from the ring current and inner radiation belt, but they can also scatter relativistic electrons. For large enough energies the increasing Lorentz factor $\gamma$ increases the "effective" mass of the electrons thus lowering their gyro frequency toward the frequency of the EMIC waves. Of course, here the damping is due to the "backward" propagation of the wave with respect to the particles as discussed above.

One reason why the EMIC waves are important in many domains of space plasma physics is that, once generated, they can propagate as Alfvén waves over long distances. For example, ion cyclotron waves generated in the equatorial magnetosphere propagate along the magnetic field down to the Earth where they are observable in form of magnetic pulsations. As they are generated in the region where the ion cyclotron frequency is of the order of 1 Hz, the pulsation periods span from about 1 s to longer periods. In the solar context we already have encountered EMIC waves when discussing the Alfvén waves as one of the possible mechanisms to heat the solar corona in Chap. 1, and we will return to them when discussing the energetic particle events associated with solar storms in Chap. 12.

### 7.3.7 Ion beam instabilities

Finally, distribution functions with negative anisotropy ($A < 0$) can also drive electromagnetic instabilities. Figure 7.8 illustrates an ion distribution that, when superposed with a hot background electron distribution, can be unstable for both $R$ and $L$ modes.

For the $R$ mode the resonance condition is

$$\omega = k_{\|} V_b - \omega_{ci} \, . \tag{7.89}$$

The excited mode is the right-hand polarized component of an Alfvén wave, sometimes called the *Alfvén whistler*. For increasing angle of propagation it goes over to the magnetosonic mode.

The excited $L$ mode is the mode approaching the electromagnetic ion gyro frequency from below. At frequencies well below $\omega_{ci}$ the mode is sometimes called the *ion whistler*. Note that calling these waves Alfvén and ion whistlers is just terminology based on the

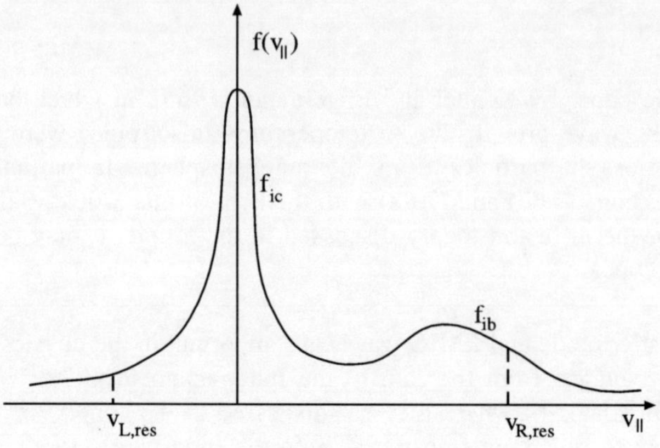


**Fig. 7.8** Ion beam resonances with *R* and *L* modes.

similar whistling characteristics to the mode usually called whistler. Physically they are the same wave modes as discussed in Chaps. 4 and 6.

Ion beams can also drive *non-resonant instabilities* that propagate against the direction of the beam. The non-resonant mode is not a normal mode of a plasma but a purely growing perturbation resembling the firehose instability. These non-resonant modes are supposed to be important in the excitation of the observed turbulent fluctuations of the solar wind upstream of the bow shock.

# 8. Magnetic Reconnection

Magnetic reconnection is a key concept in some of the most important processes associated with space storms, including solar flares, detachment of CMEs from the Sun, interaction between the solar wind and the magnetosphere, and substorm onsets. The idea of reconnection, although not the term, was introduced by Giovanelli [1946] and developed further by Dungey [1953] to explain rapid energy release in solar coronal loops. Later Dungey [1961] applied reconnection to describe magnetospheric convection as a result of interaction between the magnetic field of the solar wind and the magnetosphere.

Over the years a large selection of textbooks and proceedings volumes on magnetic reconnection have been published. Quite interestingly, the first decade of the 21st century has been a period of significant progress in our knowledge of reconnection. Both in situ observations in the magnetosphere and the solar wind together with the remote images of solar coronal processes of unprecedented resolution have strengthened the empirical basis of the concept and encouraged new theoretical and numerical investigations on various scales from the microscopic mechanisms to their macroscopic consequences. A recommendable and modern source for readers wanting to learn more of reconnection is Birn and Priest [2007].

## 8.1 Basics of Reconnection

In collisionless space plasmas described by ideal MHD the magnetic field and plasma flow are frozen-in to each other (Sect. 6.3). This means that plasma elements on the same magnetic field line remain magnetically connected to each other, whereas plasma elements not magnetically connected to each other continue to be so when the system evolves in time. Whenever this connectivity changes, we can literally speak about *reconnection*, which is the most general view on the process. As a local electric field along the magnetic field, arising for any reason, can break the frozen-in flow, this "definition" of reconnection does not require the existence of a current sheet between the reconnecting fields.

Being closely associated with the frozen-in flow the reconnection is usually described in terms of moving magnetic field lines that become cut and reconnected by some, at the

H.E.J. Koskinen, *Physics of Space Storms: From the Solar Surface to the Earth*,
Springer Praxis Books, DOI 10.1007/978-3-642-00319-6_8,

microscopic level unspecified, in fact unknown, physical mechanism. While the picture of spaghetti-like moving field lines often is a powerful picture, it may lead to misunderstanding of the underlying physics. This led Alfvén to denounce his own frozen-in concept as "pseudopedagogical". He noted that there is no reason to assume moving field lines because in the single-particle picture all particles drift across the magnetic field lines. If the magnetic field configuration changes, this should instead be expressed in terms of changing sources of the magnetic field, the currents. This is, of course, a valid but rather extreme view. It is more pragmatic to use the frozen-in picture where it works and interpret reconnection as the change of magnetic connectivity. In the end, what is essential is to understand when, where and how the frozen-in flow is violated.

### 8.1.1 Classical MHD description of reconnection

We start from the induction equation in resistive MHD (Chap. 6)

$$\frac{\partial \mathbf{B}}{\partial t} = \nabla \times (\mathbf{V} \times \mathbf{B}) + \eta \nabla^2 \mathbf{B}\,, \tag{8.1}$$

where the magnetic diffusivity $\eta$ is inversely proportional to conductivity $\eta = 1/(\mu_0 \sigma)$. In collisionless space plasmas the classical diffusivity is extremely small. However, the diffusivity does not need to be determined by classical collisions, as wave–particle interactions or microscopic plasma turbulence can give rise to finite $\eta$ to allow diffusion. How this actually happens is a difficult question, to which we do not always have a good answer.

The induction equation written in the form (8.1) is based on an assumption of the simple MHD form of Ohm's law with uniform conductivity. In case of strong fluctuations the anomalous resistivity may creep into the macroscopic equations, e.g., through the off-diagonal terms of the pressure tensor $\mathcal{P}$. If the non-resistive terms in the generalized Ohm's law

$$\mathbf{E} + \mathbf{V} \times \mathbf{B} = \frac{\mathbf{J}}{\sigma} + \frac{1}{ne}\mathbf{J} \times \mathbf{B} - \frac{1}{ne}\nabla \cdot \mathcal{P}_e + \frac{m_e}{ne^2}\frac{\partial \mathbf{J}}{\partial t} \tag{8.2}$$

are taken into account, the induction equation also becomes more complicated. The Hall term $\mathbf{J} \times \mathbf{B}/(ne)$ de-freezes the ions but not electrons whereas the electron flow may thaw due to electron pressure gradients $\nabla \cdot \mathcal{P}_e$ or inertial effects $\propto \partial \mathbf{J}/\partial t$.

In the induction equation the convective term $\nabla \times (\mathbf{V} \times \mathbf{B})$ describes the ideal frozen-in flow, but the magnetic flux is rearranged by the diffusion process. The diffusion time is given by $\tau_d = L^2/\eta$ , where $L$ is the gradient scale length. In space plasmas $L$ is in general very large and $\eta$ very small making the diffusion a very slow process. However, when two ideal plasma systems flow toward each other with different magnetic field orientations, a thin current sheet develops over which the gradient increases and thus $L$ decreases. Consequently, the diffusion rate increases. If, furthermore, some microscopic process simultaneously enhances the diffusion coefficient $\eta$, the magnetic field can be rearranged very quickly. This is what is usually understood by reconnection. In this sense reconnection is a special diffusion process that can break a thin current sheet separating plasmas of different magnetic connectivity. Of course, this viewpoint on reconnection is more limited than the general concept of any mechanism that breaks the frozen-in flow. However, current sheets are ubiquitous configurations is space plasmas and play a central role in the physics space storms. Thus we limit our discussion to current-sheet-related processes.

Empirically it is clear that reconnection can take place in an explosive manner both in solar eruptions (flares) and in the geomagnetic tail (substorm onsets). The transition from slow diffusion to fast reconnection is among the most challenging problems in theoretical space plasma physics.

### 8.1.2 The Sweet–Parker model

Assuming oppositely ($\pm x$) directed straight magnetic fields on both sides of a current layer it is easy to find the steady-state plasma flow speed toward the boundary for a given diffusivity $\eta$. We use a coordinate system in which the current is directed in the $+y$ direction. In a steady state $\partial \mathbf{B}/\partial t = 0$ and thus

$$\nabla \times \mathbf{E} = \frac{\partial E_y}{\partial z} = 0 , \tag{8.3}$$

i.e., $E_y$ constant. Far from the diffusion region

$$E_y = V B_0 , \tag{8.4}$$

where $B_0$ is the constant magnetic field outside the diffusion region. At the current sheet $B = 0$ and Ohm's law gives

$$E_y = J_y / \sigma . \tag{8.5}$$

Let the thickness of the current sheet be $2l$. Ampère's law yields now

$$J_y = \frac{B_0}{\mu_0 l} \tag{8.6}$$

and thus

$$l = \frac{1}{\mu_0 \sigma V} = \frac{\eta}{V} . \tag{8.7}$$

With increasing inflow speed the current layer finally becomes so thin that the MHD picture is no more valid. However, there is another problem. Even if the diffusion were able to consume the magnetic flux, what happens to the plasma piling up at the current sheet?

The first attempts to solve this question were made, independently, by Sweet [1958] and Parker [1957].[1] They considered the geometry given in Fig. 8.1. The length of the reconnection region $2L$ is assumed to be much longer than its thickness $2l$. Assume, for simplicity, that the inflow (index $i$) and outflow (index $o$) regions are symmetric. This applies, with some reservations that we will discuss later, to the magnetospheric tail current sheet, but, e.g., at the magnetopause the asymmetry is an essential factor and increases the

[1] While Parker's paper was published faster, Sweet was the first to present the model at an IAU Symposium in 1957, which explains the commonly used name of the model.

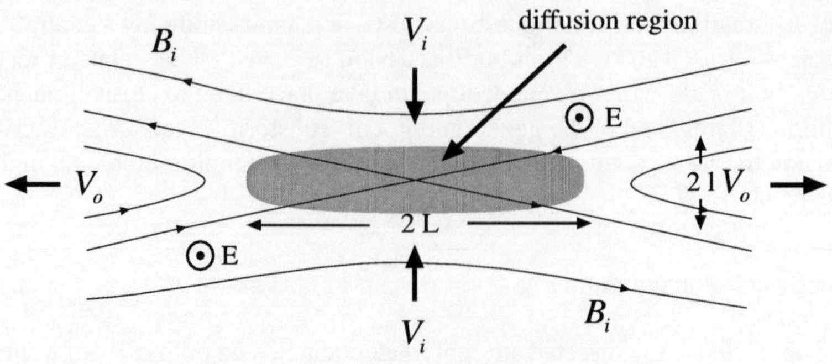


**Fig. 8.1** The Sweet–Parker model of reconnection.

complexity of the problem. In the solar corona the current sheets can be both symmetric and very asymmetric.

In steady state the electric fields in the inflow and outflow regions are the same

$$E = V_i B_i = V_o B_o \,. \tag{8.8}$$

Assume further an incompressible flow $\rho_i = \rho_o = \rho$. Then conservation of mass implies

$$V_i L = V_o l \,. \tag{8.9}$$

Assume next that all inflowing electromagnetic energy is converted to the kinetic energy of the outflow. The inflowing Poynting flux is

$$|\mathbf{S}| = |\mathbf{E}\times\mathbf{H}| = \frac{EB_i}{\mu_0} = \frac{V_i B_i^2}{\mu_0} \,. \tag{8.10}$$

The mass flowing into the diffusion region in unit time $\rho V_i$ is accelerated to the outflow velocity $V_o$. Thus the energy change per unit surface in unit time is

$$\Delta W = \frac{1}{2}\rho V_i (V_o^2 - V_i^2) \,. \tag{8.11}$$

Equating the energy increase and the Poynting flux and noting that $V_o \gg V_i$ we get

$$\frac{V_i B_i^2}{\mu_0} = \frac{1}{2}\rho V_i V_o^2 \tag{8.12}$$

$\Rightarrow$

$$V_o^2 = \frac{2B_i^2}{\mu_0 \rho} = 2v_{Ai}^2 \,. \tag{8.13}$$

Thus the outflow speed is of the order of the Alfvén speed in the inflowing plasma. The factor $\sqrt{2}$ must not be taken literally due to the simplifying assumptions in the derivation of the result. For example, not all electromagnetic energy is converted to kinetic energy in the process.

The inflow speed is found from the width of the diffusion region $2l = 2/\mu_0\sigma V$

$$V_i = v_{Ai}(\sqrt{2}/R_{mA})^{1/2} , \tag{8.14}$$

where $R_{mA} = \mu_0\sigma v_{Ai}L$ is the magnetic Reynolds number calculated for the inflow Alfvén speed, also known as the *Lundquist number*. In space plasmas $R_{mA}$ is very large and thus the inflow speed in the Sweet–Parker model is very slow. For example, in solar flares the energy release through such a slow process would take several days, not a few minutes as is observed.

The amount of magnetic flux reconnected in unit time per unit length along the reconnection line or *X-line* in the $y$-direction is equal to the reconnection electric field $E$ and it is called *reconnection rate*. In this two-dimensional picture the reconnection rate is the same as the electric field at the reconnection point. As the inflow Alfvén Mach number can be written as $M_{Ai} = V_i/v_{Ai} = E/(v_{Ai}B_i)$, it can be used as a measure of the reconnection rate normalized by the *characteristic electric field* $v_{Ai}B_i$. In the Sweet–Parker model the reconnection rate is thus of the order of $(R_{mA})^{-1/2}$.

### 8.1.3 The Petschek model

A few years after the original works by Sweet and Parker, Petschek [1964] improved the reconnection model by noting that all plasma moving from the inflow region to the outflow region does not need to pass through the diffusion region. In his model, illustrated in Fig. 8.2, the flow deviates also outside the diffusion region at slow mode shocks (Chap. 11) connected to the diffusion region.

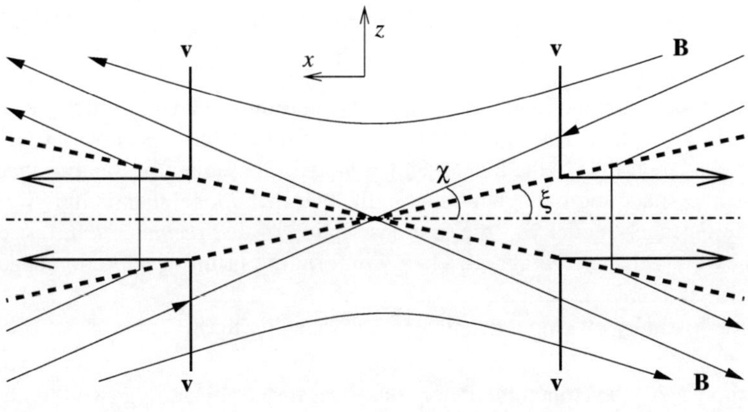


**Fig. 8.2** Petschek model of reconnection. Most of the plasma is accelerated at the slow mode shocks that make an angle $\xi$ with respect to $x$-axis, whereas $\chi$ gives the angle between the $x$-axis and the inflowing magnetic field direction immediately upstream of the shock.

The diffusion region is essential also in Petschek's model because the actual *magnetic* reconnection with the formation of the X-type magnetic neutral line takes place only in that region. However, now the length of the diffusion region ($2L$ in the Sweet–Parker picture above) is microscopic, that is, smaller than the MHD scale. What is "microscopic" is different for ions and electrons and their dynamics play different roles in the actual process breaking the magnetic connection.

As will be discussed in Chap. 11, the slow mode shocks accelerate plasma and the acceleration rate can be calculated from MHD jump conditions at the shocks. As shown in Fig. 8.2 the magnetic field decreases across the shock but the flow field is intensified. Assuming incompressibility this implies the increase of the flow speed in the outflow region. The acceleration depends on the angle $\xi$ between the shock front and the $x$-axis. An alternative viewpoint on acceleration is that since the magnetic field turns at the shock, there is a current and the acceleration is due to the $\mathbf{J} \times \mathbf{B}$ force.

In the coordinate system of Fig. 8.2 the shock is stationary but in the plasma frame the shock propagates at the inflow Alfvén speed. This allows us to calculate the angle $\chi$ between the $x$-axis and the inflowing magnetic field just upstream of the shock. In order to have a standing shock in the coordinates of the figure, the component of the inflow velocity in the direction of the shock normal must be the same as the shock velocity in the direction of its own normal in the plasma frame. This implies

$$V_i \cos\xi = v_{Ai} \sin(\chi - \xi) . \tag{8.15}$$

Assuming a steady state ($E_y$ = constant) a brief calculation gives the outflow speed

$$V_o = v_{Ai} \cos\chi , \tag{8.16}$$

which is again of the order of the inflow Alfvén speed, this time slightly less. A detailed analysis shows that the ratio between the inflow and outflow speeds is

$$\frac{V_i}{V_o} \approx \frac{\pi}{8 \ln R_{mA}} . \tag{8.17}$$

Now the dependence on the Lundquist number is logarithmic and thus much weaker than in the Sweet–Parker model. The estimates for the maximum inflow speed vary $0.01 - 0.1\, v_{Ai}$, which is much larger than in the Sweet–Parker model. Consequently, the reconnection process can handle much more magnetic flux in the Petschek model than in the Sweet–Parker model. The Petschek model was the first description of *fast reconnection*, fast enough to address the rapid release of magnetic energy in form of plasma acceleration and heating.

**Train your brain** by making a detailed analysis leading to (8.17).

Sonnerup [1970] developed the Petschek model further by adding two fast shocks outside the slow shocks. The fast shocks deflect the plasma flow in the same way as the bow shock in front of a magnetosphere (Chap. 11). In this way reconnection could handle even more incoming flux and thus be faster than in the original Petschek model.

There has been some controversy over whether the Petschek model describes the physics correctly. It gives a description of reconnection that is only weakly dependent on the properties of the reconnection region, in particular $\eta$. At the limit of very low $\eta$ the Petschek and Petschek–Sonnerup models predict that if the inflows were pushed more strongly toward each other (larger $E_y$), the outflow cone would widen, not flatten as observations indicate. However, it is possible that in reality microscopic physics enhances the effective $\eta$ enough to make the macroscopic Petschek picture qualitatively correct.

Priest and Forbes [1986] constructed a general mathematical description of MHD reconnection structures of which the Sweet–Parker and Petschek–Sonnerup models are special cases. In their analysis and in various numerical simulations the boundary conditions have been found to play a decisive role. It is possible that the external circumstances dictate if and how the reconnection will take place, whereas the local physics of the diffusion region mostly adjusts to tear the current sheet and dissipate as much magnetic energy as is required.

To produce the Petschek-type reconnection in numerical simulations is not trivial. In practice a numerical MHD code always has some diffusivity due to a finite computing grid and finite time-steps. In order to make, e.g., a realistic magnetotail simulation the resistivity in the inflow region must be as small as possible. However, this may make the current sheet region too ideal and too hard to reconnect, unless the resistivity (or $\eta$) is artificially enhanced in the diffusion region. Thus the developers and users of simulations have somewhat contradictory requirements: in order to describe Petschek-type reconnection, physics that is different from the assumptions of the model needs to be introduced. Even if we found from the observations a shock pattern predicted by the Petschek description, it would not tell us much of the *microphysical mechanism(s)* of reconnection because the Petschek approach is practically independent of the physics of the diffusion region. In fact the importance of determining what causes the enhanced $\eta$ in the diffusion region, or alternatively the role of other terms in the generalized Ohm's law, becomes more urgent.

### 8.1.4 Asymmetric reconnection

The symmetric 2D Sweet–Parker and Petschek cartoons give an oversimplified picture of various relevant realizations of reconnection. Important sites of manifestly *asymmetric reconnection* are the dayside magnetopause of the Earth and the solar coronal configurations where flux tubes of different plasma content and magnetic field magnitudes interact. Also in the geomagnetic tail the reconnecting magnetic field lines may seldom be exactly antiparallel and the plasma density may not be symmetric on each side of the current sheet.

Cassak and Shay [2007] extended the Sweet–Parker scaling laws to asymmetric 2D reconnection allowing both the inflow densities and inflow magnetic field magnitudes to be different on either side of the current sheet. The outflow speed was found to scale as

$$V_o^2 \sim \frac{B_1 B_2}{\mu_0} \frac{B_1 + B_2}{\rho_1 B_2 + \rho_2 B_1}, \tag{8.18}$$

where subscripts 1 and 2 refer to the different inflow regions. At the symmetric limit ($\rho_1 = \rho_2$ and $B_1 = B_2$) this reduces, apart from the undetermined numerical factor, to the Sweet–Parker–Petschek outflow speed.

Without specifying the dissipation mechanism, and thus how fast the reconnection really is, the reconnection rate expressed as the electric field scales as

$$E \sim \left( \frac{B_1 B_2}{B_1 + B_2} \right) V_o \frac{2l}{L} , \tag{8.19}$$

where $l$ and $L$ are the half-thickness and half-length of the diffusion region as in our discussion of the Sweet–Parker reconnection. Thus the *aspect ratio* $l/L$ finally determines how fast reconnection proceeds in the similar manner as it makes the distinction between the reconnection rates in the Sweet–Parker and Petschek models.

A particularly interesting result of this analysis is that the X-line and the plasma stagnation line do not need to be at the same location. Under steady-state conditions Faraday's law ($\nabla \times \mathbf{E} = 0$) implies that the inflow velocities are related as

$$V_1 B_1 = V_2 B_2 , \tag{8.20}$$

where Ohm's law of the ideal MHD is assumed to be valid outside the diffusion region. While according to (8.20) the inflow velocity on the weaker magnetic field side is larger than on the stronger field side, the flux of magnetic energy ($\propto VB^2$) is larger on the stronger field side. By definition there is no flux of magnetic energy across the X-line from one inflow region to the other, and the outflow of kinetic energy is assumed to be relatively evenly distributed across the outflow edge of the diffusion region. Consequently the X-line is shifted toward the weak field side. On the other hand, the plasma stagnation line turns out to be located on that side of the X-line where the Alfvén speed is higher, because there is more mass flux from the side of the lower Alfvén speed. As these results were found independently of the dissipation mechanism, Cassak and Shay [2007] concluded that the separation of the magnetic neutral line and plasma stagnation line is a generic feature of asymmetric reconnection.

At the Earth's dayside magnetopause the inflow from the solar wind side, actually the shocked magnetosheath plasma, has higher density and weaker magnetic field than the inflow from the magnetosphere. Applying the Cassak–Shay reconnection model to the magnetopause, the X-line is shifted toward the solar wind side, whereas the plasma stagnation line is on the magnetosphere side of the X-line, i.e., on closed magnetic field lines. This means that there is plasma flow across the X-line from the open field lines of the magnetosheath to the closed field lines of the magnetosphere.

While it is somewhat outside of the scope of this book, we note that the source mechanisms of the magnetospheric boundary layers illustrated in Fig. 1.18 are among the most complicated and debated issues in magnetospheric physics (e.g., Hultqvist et al [1999] and references therein). Reasons for this complexity are both the highly variable structure of the boundaries, as known from observations, as well as the large variety of physical processes, including reconnection, diffusion through wave–particle interactions, gyroviscous interaction, the direct entry through the cusp regions, etc., that can contribute to the plasma composition of the boundaries. The Cassak–Shay model suggests that even a steady-state dayside reconnection can inject magnetosheath plasma to the magnetospheric boundary layers both on open field lines through the conventional Sweet–Parker outflow region and on the closed field lines made accessible by the separation of the X-line and the plasma stagnation line.

Pritchett [2008] compared the analytical results of Cassak and Shay [2007] with his asymmetric particle-in-cell (PIC) simulations. He noted that for spontaneous reconnection the reconnection rate was considerably smaller than the fast reconnection rates obtained in the symmetric simulations of the so-called GEM Reconnection Challenge to be discussed further in Sect. 8.2.4 below. Introduction of an additional driving electric field on the magnetosheath side, which may be a more realistic assumption for the dayside magnetopause reconnection, led to a sufficiently enhanced reconnection rate consistent with recently estimated reconnection rates from the observations [e.g., Mozer and Retinò, 2007]. Furthermore, the driving electric field strongly changed the structure across the magnetopause. It produced a magnetic field component in the third direction ($B_y$), which led to strong outward Poynting flux ($\propto E_x B_y$) on the magnetosphere side.

On the magnetopause the magnetic fields are antiparallel within limited regions only, and there may be a relatively strong magnetic field component in the current sheet, called the *guide field*. Also the observations in the otherwise rather symmetric magnetotail current sheet indicate that the magnetic field can have a significant $y$ component. Cassak and Shay [2007] claimed that in the case of uniform density the guide field would not considerably change the results of their symmetric analysis, which was also the conclusion by Pritchett [2008] based on the PIC simulations. However, at the magnetopause there is also a strong density gradient, which complicates the issue of the guide field, as it can lead to the lower hybrid drift instability (Sect. 7.3.5) associated with the reconnection process. The role of the guide field is actually an important issue that has turned out to introduce a serious headache for those searching for a theory of collisionless reconnection, to which we turn in the next section.

A more general discussion of 3D magnetic reconnection is beyond the scope of the present book. A good starting point for an interested reader is the book by Birn and Priest [2007] containing an extensive list of references to original works.

## 8.2 Collisionless Reconnection

The previous discussion was implicitly based on resistive diffusion, either collisional or anomalous, in the macroscopic one-fluid MHD picture. However, a thorough understanding of the phenomenon in the context of space storms requires consideration of collisionless microscopic processes. This is presently an active area of research and a satisfactory description of the microscopic aspects of reconnection is yet to come. Various particle and Vlasov simulations yield roughly similar outflow characteristics as the Sweet–Parker or Petschek–Sonnerup models. The inflow speed and the reconnection rate appear to be determined mainly by *ion inertia* and we should like to find an explanation why it is so.

### 8.2.1 The tearing mode

The most intensively studied instability for the formation of the magnetic X-line and reconnection has been the *tearing mode*. There are both collisional and collisionless tearing mode theories. The concept of "tearing" is very suggestive because it literally refers to tearing the current sheet apart. In resistive reconnection the status of tearing as the fundamental concept is strong, although obtaining large enough reconnection rates remains a problem, but the collisionless reconnection is more murky. It is not clear whether the collisionless tearing mode can supply enough (anomalous) resistivity, whether there are some other and faster instabilities involved in the determination of $\eta$ and/or $\nabla \cdot \mathcal{P}_e$, after which the macroscopic process could look like resistive tearing, or whether we should look for something else to solve the problem.

We begin the discussion of the tearing mode from the resistive case. The basic idea is simple. Imagine that the current sheet consists of thin current filaments and perturb their distribution slightly. Because the force between currents flowing in the same direction is attractive and weakens with distance, the force between the filaments that come closer to each other is stronger than the force between those that are moved farther from each other due to the perturbation. This is clearly an unstable configuration and the current filaments tend to bunch, forming magnetic islands as illustrated in Fig. 8.3. The free energy for the instability comes from the tension in the strongly sheared magnetic field. This is one more example of the negative energy modes introduced in Chap. 7: while the perturbation grows, the energy of the magnetic configuration decreases.

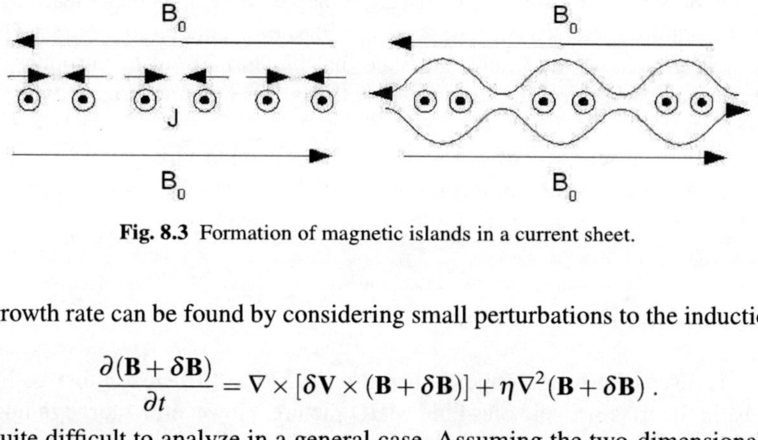


**Fig. 8.3** Formation of magnetic islands in a current sheet.

The growth rate can be found by considering small perturbations to the induction equation

$$\frac{\partial(\mathbf{B}+\delta\mathbf{B})}{\partial t} = \nabla \times [\delta\mathbf{V} \times (\mathbf{B}+\delta\mathbf{B})] + \eta\nabla^2(\mathbf{B}+\delta\mathbf{B}) \,. \tag{8.21}$$

This is quite difficult to analyze in a general case. Assuming the two-dimensional configuration of Fig. 8.3 and linearizing the equations, a lengthy analysis (see, e.g., Treumann and Baumjohann [1996]) leads to the maximum growth rate of the resistive tearing mode

$$\gamma_{tea,max} \approx (2\tau_A \tau_d)^{-1/2} \,, \tag{8.22}$$

where $\tau_A$ is the *Alfvén travel time* across the current layer, i.e., the time during which the Alfvén wave propagates the distance $L = |B/\nabla B|$, and $\tau_d$ is the diffusion time. In the solar wind and in the magnetosphere both Alfvén and diffusion times are very long and the growth rate thus very small. Consequently, the resistive tearing mode cannot explain reconnection under conditions relevant to the physics of space storms without strongly enhanced $\eta$.

### 8.2.2 The collisionless tearing mode

In order to analyze collisionless tearing we consider a 2D current sheet described by a vector potential with only one non-vanishing component $A_y$. This gives the magnetic field

$$\mathbf{B} = \nabla \times \mathbf{A} = \left( -\frac{\partial A_y}{\partial z}, 0, \frac{\partial A_y}{\partial x} \right) . \tag{8.23}$$

Starting from a 1D Harris-type current sheet the tearing mode produces periodic variations along the $x$-axis introducing a finite $B_z$. We represent the perturbed scalar and vector potentials as plane waves

$$\delta A_y(x,z,t) = \delta A(z) \exp(-i\omega t + ikx) \tag{8.24}$$

$$\delta \varphi(x,z,t) = \delta \Phi(z) \exp(-i\omega t + ikx) . \tag{8.25}$$

The stability is analyzed by considering the energy balance between the magnetic field perturbation and the energy dissipated by the current in the current sheet

$$\frac{1}{2\mu_0} \frac{\partial}{\partial t} \int |\delta \mathbf{B}|^2 \, dz = - \int \delta \mathbf{J} \cdot \delta \mathbf{E}^* \, dz . \tag{8.26}$$

If we now find a growing magnetic perturbation when energy is dissipated in the current sheet, we have an instability.

#### *The electron tearing instability*

According to the current filament argumentation the 1D Harris sheet is always unstable. However, the instability may soon saturate because the perturbation introduces a normal component to the magnetic field. As the electrons are magnetized, they introduce stiffness to the magnetic field and thus provide a stabilizing effect against tearing.

In order to understand this effect let us first consider the simple 2D Harris equilibrium with a small normal component $B_{0z}$

$$\mathbf{B}_0 = B_0 \tanh(z/d)\, \mathbf{e}_x + B_{0z}\, \mathbf{e}_z . \tag{8.27}$$

$B_{0z}$ is assumed to be so small that the ions behave unmagnetized but large enough to keep electrons magnetized, i.e.,

$$\frac{r_{Le}}{d} < \left( \frac{B_{0z}}{B_0} \right)^2 < \frac{r_{Li}}{d} . \tag{8.28}$$

In the Harris model the parameter $d$ is the gradient scale length of both the magnetic field and the pressure, whereas in reality the layer carrying most of the current may be much narrower than the plasma sheet.

Now the zero-order undisturbed but inhomogeneous particle distribution functions can be written as

$$f_{\alpha 0}(\mathbf{v},z) = \frac{n_0}{\pi^{3/2} v_{th\alpha}^3} \exp\left( -\frac{v^2 + V_{d\alpha}^2}{v_{th\alpha}^2} + \frac{2m_\alpha V_{d\alpha}(v_y + q_\alpha A_{0y})}{k_B T_\alpha} - \frac{q_\alpha \varphi_0}{k_B T_\alpha} \right), \tag{8.29}$$

where the diamagnetic drift velocity due to the density gradient has been introduced as $V_{d\alpha} = -k_B T_\alpha / q_\alpha B_0 d$ . The solution of the linearized Vlasov equation can be written as

$$\delta f_\alpha(\mathbf{v},z,t) = \frac{q_\alpha f_{0\alpha}}{k_B T_\alpha} \left( v_{d\alpha} \delta A_y - \delta\varphi + i\omega \int\limits_{-\infty}^{t} (v_y \delta A_y - \delta\varphi)\, dt' \right). \tag{8.30}$$

The integral over $t'$ must be calculated along the unperturbed orbits (cf. the general solution of the Vlasov equation in Sect. 5.5). The integral represents the non-adiabatic correction to the first-order distribution function, whereas the terms outside the integral describe the adiabatic particle response. After some non-trivial calculation (8.26) can be rewritten as

$$\begin{aligned} &\frac{\partial}{\partial t}\left\{ \int dz \left[ \left| \frac{d\delta A}{dz} \right|^2 + \left( k^2 - \frac{2}{d^2 \cosh^2(z/d)} \right) |\delta A|^2 \right] \right\} \\ &= -2\mu_0 \,\mathrm{Re} \int dz\, \delta J_{y,ad} \delta E_y^* , \end{aligned} \tag{8.31}$$

where $J_{y,ad}$ is the current carried by the adiabatically moving particles.

For sufficiently narrow current sheets ($k^2 d^2 \ll 1$) the electron tearing mode energy becomes negative and any perturbation leads to instability. Fundamentally the growth of the tearing mode is due to the Landau mechanism. It is sometimes called *inverse Landau damping* but, due to the negative energy of the mode, the word "inverse" may give a false impression. In the "normal" Landau mechanism the particles are energized at the expense of the electromagnetic field of the wave. Here, the particles are also energized by the electromagnetic field, but, due to the fact that the current sheet with tearing islands has lower energy than without the islands, the amplitude of the mode increases until the whole structure goes to the nonlinear regime and something beyond the present description takes place.

The condition $k^2 d^2 \ll 1$ is a long-wavelength approximation for the forming of magnetic islands. This has the advantage that the configuration can be analyzed using the WKB method (Sect. 4.2.2). The flip side of the coin is that the current sheet must be very long in the $x$-direction for the mode to develop.

The growth rate of the electron tearing mode is rather more difficult to calculate (or even to estimate). The result is

$$\gamma_{e,tea} = \sqrt{\pi} \left( 1 + \frac{T_i}{T_e} \right) \left( \frac{r_{Le}}{d} \right)^{5/2} (1 - k^2 d^2)\, \omega_{ce} . \tag{8.32}$$

**Challenge your brain**

Derive the electron tearing mode growth rate (8.32). This is not a task for a beginner. You are recommended to consult the original scientific articles, some of which are referred to in the following discussion.

The original electron tearing mode solution [Coppi et al, 1966] did not take into account the effect of the magnetic normal component. A finite $B_z$ was soon found to exert a strongly stabilizing effect on the electron tearing mode by making the electron orbits adiabatic. Reducing the normal component to zero removes this effect, but then the instability has a very small growth rate. Thus the linear electron tearing mode turned out to be too slow to initiate such explosive events as flares or magnetospheric substorms.

***The ion tearing mode***

While the electrons are magnetized, ions remain unmagnetized for much larger $B_{0z}$. This suggests that the ion inertia might drive the tearing mode [Schindler, 1974]. The ion tearing mode growth rate is found to be of the same form as the electron tearing with the substitution $m_i \rightarrow m_e$ and interchanging the ion and electron temperatures:

$$\gamma_{i,tea} = \sqrt{\pi}\left(1+\frac{T_e}{T_i}\right)\left(\frac{r_{Li}}{d}\right)^{5/2}(1-k^2d^2)\,\omega_{ci}\,. \tag{8.33}$$

Assuming that the ions and electrons are at the same temperature and both modes have the same wavelength, the ion mode grows faster than the electron mode by a factor of $(m_i/m_e)^{1/4}$, which for electron–proton plasma is about 6.5. In the magnetotail current sheet the thermal ions are some 5 times warmer than electrons, which further favors ion tearing. The ion tearing mode can also grow when the electron mode is stable. But again, further analyses showed that the stabilization by adiabatic electrons still is too strong and quenches the growth of the ion mode [Galeev and Zelenyj, 1976].

### 8.2.3 Tearing mode or something else?

After these first failures with the collisionless tearing mode various attempts were made to go around the electron stabilization by looking for mechanisms that would make the electron motion non-adiabatic by microscopic turbulence or wave–particle interactions [e.g., Coroniti, 1980]. In the 1980s chaos theory became popular in many fields of physics, including some problems in space plasma physics. Büchner and Zelenyi [1987] argued that the electron motion in the very stretched tail-like configuration just prior to onset of reconnection would become chaotic and when the electrons lose the guidance of the field, they would no more be able to provide stiffness to the magnetic field configuration. The chaotization was estimated to become significant when the curvature radius of the magnetic field becomes smaller than about $10\,r_{Le}$.

A detailed analysis of microscopic collisionless tearing is a difficult exercise in nonlinear plasma physics. It requires extensive numerical simulations, which have not led to conclusive results. Also here, addition of the guide field introduces complications because it provides another agent to make the electron motion adiabatic. Although the cutting of the plasma sheet looks macroscopically like an evolution of a large tearing island, there is no logical imperative that the microscopic process should be a tearing mode. Even if the scenario assuming anomalous resistivity and thus the growth of a resistive tearing mode were correct, the microscopic instability leading to a finite $\eta$ does not need to be an electron or ion tearing mode.

It is possible that the reconnection process is patchy with numerous overlapping tearing islands *percolating* the whole current sheet (for a review, see Galeev et al [1986]). A magnetic flux tube could migrate through such a percolation, connecting, for example, solar wind and magnetospheric field lines. The percolation may, or may not, lead to a complete collapse of the current sheet by the coalescence of the tearing islands, depending on the external and internal parameters that control the formation of the current sheet.

### 8.2.4 The Hall effect

While the Hall term $\mathbf{J} \times \mathbf{B}/(ne)$ in Ohm's law does not lead to the thawing of the field from the electrons that still remain frozen-in the field, its inclusion in Ohm's law in plasma simulations has turned out to lead into a significant increase of the reconnection rate. In a collaborative study called *GEM Reconnection Challenge* discussed in several articles in the *Journal of Geophysical Research*, vol 106(A3), 2001, different models including the physics of the Hall term (Hall MHD, two-fluid, hybrid models with fluid electrons and kinetic ions, and fully kinetic models) were compared in a 2D Harris current configuration under consistent initial and boundary conditions. All models addressing the Hall physics resulted in practically indistinguishable reconnection rates, with inflow speeds exceeding $0.2\,v_A$ (Fig. 8.4). The resistive models without the Hall effect yield much smaller reconnection rates, unless large localized $\eta$, possibly as a function of current density $\eta = \eta(\mathbf{J})$, is assumed.

While the reconnection rate seems to be of the same order of magnitude in the various models involving the Hall effect, the actual structure of the reconnection region and the resulting outflows are different in different models. This suggests that the Hall effect is critical in determining the rate of the reconnection of the magnetic flux, but it does not describe the details of the magnetic diffusion process. This seems logical, as the electrons remain frozen-in long after the $\mathbf{J} \times \mathbf{B}$ effect has thawed the ions and finally the breakdown of the magnetic field geometry takes place in the electron scale. The ion dynamics determines the reconnection rate, but the electron dynamics is essential to the microscopic process.

It is not quite clear why Hall MHD results in faster reconnection than basic MHD. A suggested explanation is based on the observation that the Hall term $\mathbf{J} \times \mathbf{B}/(ne)$ adds the whistler mode to the slow, intermediate and fast modes of MHD by decoupling the electron and ion motions from each other. The phase speed of the whistler mode (4.118) is inversely proportional to the scale length

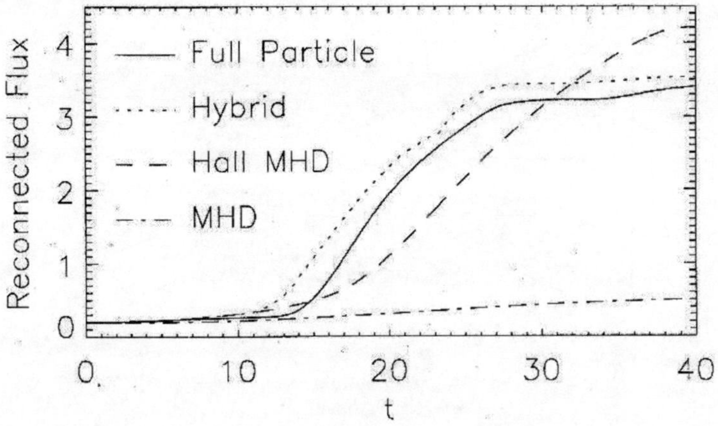


**Fig. 8.4** GEM Reconnection Challenge results. The MHD case was calculated for a small resistivity. The time is normalized to ion gyrofrequency, the length scale to ion inertial length $c/\omega_{pi}$, and the magnetic field and plasma density to unity. (From Birn et al [2001].)

$$v_{ph} \propto k \propto 1/l\,. \tag{8.34}$$

In MHD the outflow speed is of the order of the inflow Alfvén speed, whereas in Hall MHD the electron outflow speed, and thus the removal of magnetic flux from the diffusion region, scales as the whistler mode speed. The electron flux out from the dissipation region in a 2D model is $v_{ph} \cdot l =$ constant, i.e., independent of the width of the current layer. Consequently, the reconnection rate has been argued to become insensitive to the (slow) electron dissipation and thus the ion dynamics would control the reconnection rate. This is somewhat analogous to the above discussion that the ion tearing mode would control the reconnection rate if electrons are chaotic, but this analogy may be just a coincidence.

The GEM Challenge model comparison was done for a 2D configuration only. The reconnection rates for the various models were calculated for a relatively strong initial magnetic perturbation ("tearing island") to a 1D Harris model. This was done to put the system into the nonlinear regime from the beginning because the linear tearing mode is known to lead to different results in different models.

The introduction of the Hall term actually destroys the 2D picture as illustrated in Fig. 8.5. The difference in the electron and ion flows sets up current loops near the edges of the ion dissipation region that give rise to a quadrupolar magnetic field structure out of the plane of the original 2D magnetic field configuration. Multipoint observations with the four Cluster satellites have confirmed the formation of the Hall fields in some fortunate cases when the spacecraft have passed the ion diffusion region in the magnetotail [e.g., Runov et al, 2003].

Adding the Hall term alone to the ideal MHD equations may not be quite sufficient because in reality the electron pressure term $\nabla \cdot \mathcal{P}_e/(ne)$ can be of the same order of magnitude as the Hall term. Malyshkin [2008] presented a Sweet–Parker–Petschek-type analysis of Hall reconnection including both the Hall and electron pressure terms and

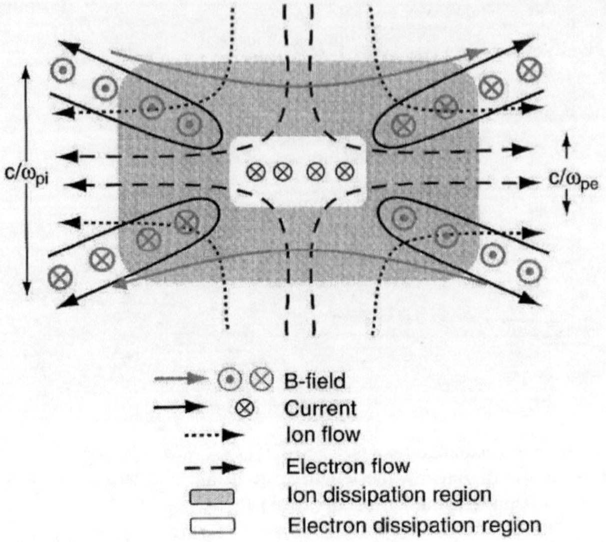


**Fig. 8.5** The quadrupolar magnetic field configuration due to the Hall effect.

considered the generalized Ohm's law in the form

$$\mathbf{E}+\mathbf{V}\times\mathbf{B}=\eta\mathbf{J}+\frac{1}{ne}\mathbf{J}\times\mathbf{B}-\frac{1}{ne}\nabla\cdot\mathscr{P}_e\,. \tag{8.35}$$

The assumptions are similar to the classic reconnection models. Inside the reconnection layer plasma is assumed incompressible, resistivity $\eta$ constant and very small, and the Lundquist number $R_{mA}$ very large. The situation is kept quasi-stationary ($\partial/\partial t = 0$) and quasi-two-dimensional with constant quadrupolar magnetic field components in the third direction ($\pm B_y$) corresponding to Fig. 8.5.

With these assumptions the outflow velocity is again of the order of $v_A$ and the inflow velocity depends on the Lundquist number as

$$V_i \approx \sqrt{3}\,v_A\left(\sqrt{3}\,R_{mA}+\frac{2R_{mA}^2 d_i^2}{L^2}\right)^{-1/2}, \tag{8.36}$$

where $d_i = c/\omega_{pi}$ is the *ion inertial length* and $L$ is the external magnetic field scale length defined by

$$L^2 = -\frac{2B_i}{(\partial^2 B_x/\partial x^2)_i} \tag{8.37}$$

and calculated at the edge of the reconnection layer at the $x$ coordinate of the X-line.

At the first sight (8.36) appears to indicate that just a slight modification of the slow Sweet–Parker reconnection has been found and the inflow remains very small for very large $R_{mA}$. In fact, at the limit $d_i \ll L/\sqrt{R_{mA}}$ the solution is the Sweet–Parker solution. However, Malyshkin [2008] pointed out that the calculation has been made in the inflow region within an infinitesimal environment of a line cutting through the current sheet and crossing the X-line, where $L$ and $B_i$ are calculated. If one calculates the reconnection rate, i.e., the electric field, at the opposite limit $d_i \gg L/\sqrt{R_{mA}}$ it turns out to be independent of the resistivity and thus of $R_{mA}$, i.e.,

$$E_y = v_A B_i \frac{d_i}{L} \,. \tag{8.38}$$

Thus we can think that the velocity with which the magnetic field lines are drawn to the process is

$$V_R \approx \frac{E_y}{B_i} \approx \frac{d_i}{L} v_A \,. \tag{8.39}$$

In Malyshkin's model $L$ is a given parameter and its value needs to be determined either from observations or by numerical simulations. However, various Hall MHD simulations suggest that the ratio $d_i/L$ can be of the order of 0.1, i.e., a fast reconnection rate $E_y \approx 0.1\, v_A B_i$ would be obtained in this model.

As the electric field has become independent of collisions, something else than $\eta \mathbf{J}$ must balance it at the X-line, where the Hall term is zero by definition. It is here the electron pressure term, and possibly the electron inertial term $\propto \partial \mathbf{J}/\partial t$ in the case of time-dependent reconnection, become important. While the $\mathbf{J} \times \mathbf{B}/(ne)$ term decouples the electron and ion motion near the current layer in a rather straightforward way, the $\nabla \cdot \mathcal{P}_e/(ne)$ term appears to act as the agent to let the electrons to diffuse. The off-diagonal elements of $\mathcal{P}_e$, rising from non-gyrotropic particle distributions are of particular interest (e.g., Kuznetsova et al [2007] and references therein).

Both the GEM Reconnection Challenge and Malyshkin's theoretical analysis were two-dimensional and it is not fully clear how much of the conclusions can be carried over to a 3D geometry, in particular when the deviations from two-dimensionality become large. Some 3D PIC simulations [e.g., Pritchett and Coroniti, 2004] suggest that a strong enough guide field, i.e., a pre-existing magnetic field component out of the plane of Fig. 8.5 would strongly reduce the Hall effect. Thus the applicability of Hall reconnection may be limited to rather symmetric current sheets only.

It is also possible that electromagnetic fluctuations may provide sufficient anomalous resistivity that could take the role of $\eta \mathbf{J}$ and the generalized Ohm's law should be written in the form

$$\mathbf{E} + \mathbf{V} \times \mathbf{B} = \frac{1}{ne}\mathbf{J} \times \mathbf{B} - \frac{1}{ne}\nabla \cdot \mathcal{P}_e + \frac{m_e}{ne^2}\frac{\partial \mathbf{J}}{\partial t} - \frac{1}{n}[\langle \delta \mathbf{E} \delta n \rangle + \langle \delta (n\mathbf{V}_e) \times \delta \mathbf{B} \rangle] \,, \tag{8.40}$$

where the last term describing the anomalous resistivity was derived by Yoon and Lui [2006].

**Challenge your brain**

Study carefully the paper by Yoon and Lui [2006] and fill the gaps in the derivation of the anomalous resistivity term in (8.40).

## 8.3 Reconnection and Dynamo

Reconnection annihilates magnetic flux converting magnetic energy to kinetic energy of the plasma and thus causes decay of the current systems. On the other hand, plasma motion associated with reconnection can also lead to creation of magnetic flux through a *dynamo action*, either in the vicinity of reconnection or further away from it. Let us return to the cartoon of the reconnecting magnetosphere (Fig. 8.6). For the southward directed IMF the electric field $\mathbf{E} = -\mathbf{V} \times \mathbf{B}$ points from the dawn to the dusk and consequently the dayside and tail reconnection regions are loads in the electromagnetic system ($\mathbf{J} \cdot \mathbf{E} > 0$), whereas the tail magnetopause is a dynamo ($\mathbf{J} \cdot \mathbf{E} < 0$). But how does this dynamo work?

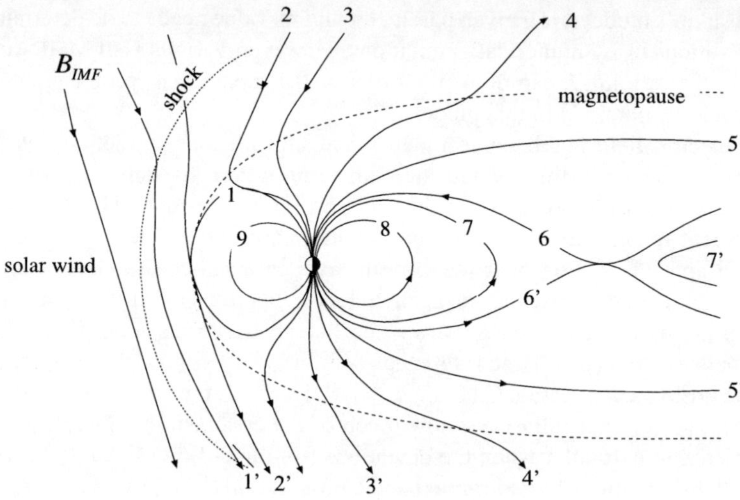


**Fig. 8.6** Reconnecting magnetosphere.

### 8.3.1 Current generation at the magnetospheric boundary

A simple mechanical analogue for the boundary layer current generation is the *MHD generator* of Fig. 8.7. Let plasma flow across the magnetic field (**B** and **u** in the figure). The Lorentz force $q\mathbf{u} \times \mathbf{B}$ deflects positive charges toward the upper electrode and negative toward the lower. If the electrodes are connected through an external load, the plasma current flows upward. Now Ampère's force $\mathbf{J} \times \mathbf{B}$ decelerates the plasma flow **u**. Thus the external current and the magnetic field associated with it are driven at the expense of the kinetic energy of the plasma.

Consider next plasma that has penetrated to the magnetospheric LLBL, e.g., by dayside reconnection or diffusion through the magnetopause (Fig. 8.8). The flow is decelerated in the same way as in the toy model of Fig. 8.7 and the LLBL feeds magnetospheric current systems. The outer "electrode" is now coupled to the magnetopause current and the inner to the Region 1 field-aligned current.

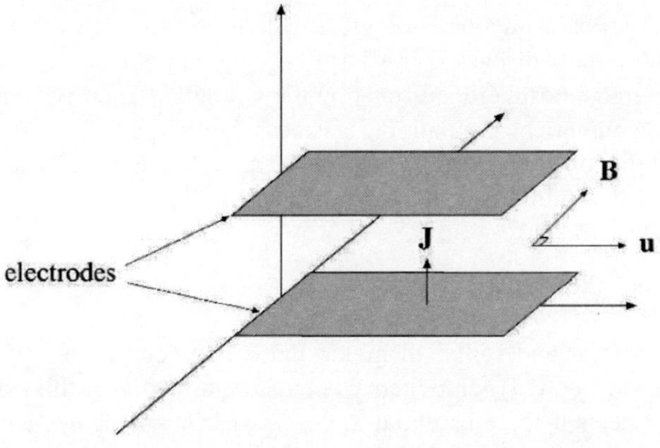


**Fig. 8.7** The basic principle of an MHD generator. In reality MHD generators are more complicated structures and the connection of the external load can be arranged in many different ways.

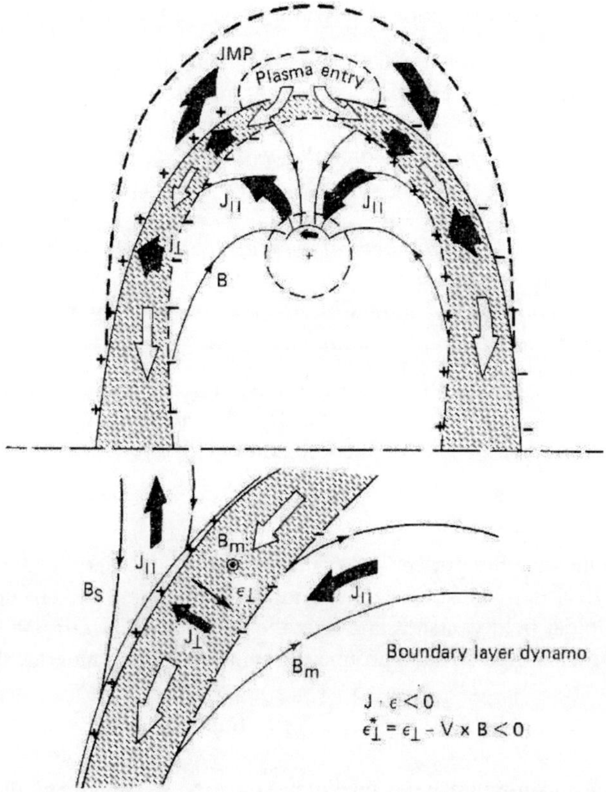


**Fig. 8.8** A model for a dynamo in the LLBL. The open arrows describe the plasma flow and the black arrows the electric currents. Note that the magnetopause current (JMP) and the FAC $J_{\|}$ are not in the plane of the figure, which should be imagined as a projection seen from the tail. (From Lundin and Evans [1985].)

Note that this is just a phenomenological description that is consistent with magnetospheric current systems discussed in Chap. 1. To obtain quantitative description of the generation of the magnetospheric currents by the boundary layer dynamo, more rigorous analysis, requiring numerical simulations, is necessary. It is actually quite remarkable how little is known of the local physics of this dynamo and of its efficiency under various solar wind conditions and within different parts of the magnetopause.

### 8.3.2 Elements of solar dynamo theory

The generation of solar and stellar magnetic fields is a central and difficult problem in MHD. While the hard-core dynamo theory is outside the scope of this book, it is possible to look at some aspects that are essential to the 22-year magnetic cycle of the Sun, which determines the basic climatic period of space storms.

The existence of MHD dynamos is not trivial. A special class of MHD theorems, some of which were discussed in Chap. 6, are the *anti-dynamo theorems* that constrain the configurations where the dynamo action is possible. Some of the most famous are

- *Cowling's theorem: A steady axisymmetric magnetic field cannot be maintained.* This means that a dynamo must produce a more complicated configuration than the simple dipole.
- *A two-dimensional magnetic field cannot be maintained by dynamo action.* This means that in any 3D coordinate system $(x,y,z)$ $\mathbf{B}$ cannot be independent of any of the coordinates.
- *An incompressible motion in a spherical volume having a zero radial component everywhere cannot maintain a magnetic field.*
- *Zeldovich's theorem: An incompressible motion in which $V_z \equiv 0$ in some Cartesian coordinate system cannot maintain a magnetic field.*

**Proof** of Cowling's theorem: Write a steady axisymmetric magnetic field as a sum of a toroidal component ($B_\phi$) and a poloidal vector $\mathbf{B}_p$, the latter of which is a sum of the radial and axial components in the cylindrical coordinates $(r,\theta,\phi)$

$$\mathbf{B} = B_\phi\, \mathbf{e}_\phi + \mathbf{B}_p \,. \tag{8.41}$$

Due to the axisymmetry the projections of field lines to any meridional plane must look the same and form closed loops. On each meridional plane there must be an O-type neutral point, where poloidal field vanishes and thus the field is purely azimuthal (or zero). The neutral points form a closed circle $C$ around the symmetry axis. Integrate the MHD Ohm's law along this circle

$$\oint_C \frac{\mathbf{J}}{\sigma} \cdot d\mathbf{l} = \oint_C \mathbf{E} \cdot d\mathbf{l} + \oint_C \mathbf{V} \times \mathbf{B} \cdot d\mathbf{l} \,. \tag{8.42}$$

Using Stoke's law and noting that the current has only the $\phi$ component, this can be written as

$$\oint_C \frac{J_\phi}{\sigma}\, dl = \int_S \nabla \times \mathbf{E} \cdot d\mathbf{S} + \oint_C \mathbf{V} \times \mathbf{B} \cdot d\mathbf{l} \,. \tag{8.43}$$

For a steady magnetic field $\nabla \times \mathbf{E} = 0$. Along $C$ the magnetic field is in the direction of the line element $\mathbf{B} \parallel d\mathbf{l}$, which also makes the last term in (8.43) vanish. However, $J_\phi$ does not vanish on $C$ and thus the LHS of (8.43) cannot be zero. We have found a contradiction that proves Cowling's theorem.

We leave the proof of the other theorems as exercises for the interested reader. The main point of these theorems is that it is useless to look for oversimple dynamo solutions. As we will see, e.g., the $\alpha\omega$ dynamo discussed below is far from the symmetries of these theorems.

To construct a model for the solar dynamo let us again start from the induction equation

$$\frac{\partial \mathbf{B}}{\partial t} = \nabla \times (\mathbf{V} \times \mathbf{B}) + \eta \nabla^2 \mathbf{B} . \tag{8.44}$$

The plasma dynamo is easiest to describe at the *kinematic* level where the velocity field $\mathbf{V}$ is assumed to be given and not affected by the evolution of the magnetic field. This is a reasonable starting point in a hydrostatic object like the Sun where the pressure and gravitation balance each other and the Lorentz force $\mathbf{J} \times \mathbf{B}$ is negligible. In this case the induction equation is *linear*, which makes an analytical approach feasible.

In reality the magnetic force may not be negligible and $\mathbf{V}$ may be a function of $\mathbf{B}$ making the induction equation *nonlinear*. In that case the analysis requires a simultaneous solution of the momentum equation and we have to solve a *dynamic* problem, which in the case of the Sun means a combined solution of the convective motion and magnetic field generation. This is evidently possible only through numerical simulations. Due to the wide range of scales such computations easily reach beyond the limits of present-day computers.

Figure 8.9 illustrates our understanding of the differential rotation before and after the analysis of solar oscillations. In the older picture the rotation was assumed to resemble concentric cylinders, the outer of which would rotate faster. In this case there would be considerable velocity shear throughout the convection zone. The analysis of rotational modes of solar oscillations has, however, shown that the isocontours of the rotation speed are almost radial in the convection zone and the main shear takes place close to the bottom of the convection zone. Unfortunately, the kinematic approach does not seem be a quite sufficient description of the actual situation, where the field generation is likely to take place within a narrow region at the bottom of the convection zone.

Regardless of these reservations, we discuss the dynamo mechanism at the kinematic level within the *mean-field electrodynamics* approach. While this does not provide a complete description of the solar dynamo, it illustrates some of the basic principles and introduces the alpha effect that belongs to the basic jargon of the physics of MHD dynamos.

We assume the velocity field given and write the magnetic and velocity fields as sums of the average fields $(\langle \mathbf{B} \rangle, \langle \mathbf{V} \rangle)$ and the fluctuating parts $(\mathbf{b}, \mathbf{u})$

$$\mathbf{B} = \langle \mathbf{B} \rangle + \mathbf{b} \tag{8.45}$$

$$\mathbf{V} = \langle \mathbf{V} \rangle + \mathbf{u} . \tag{8.46}$$

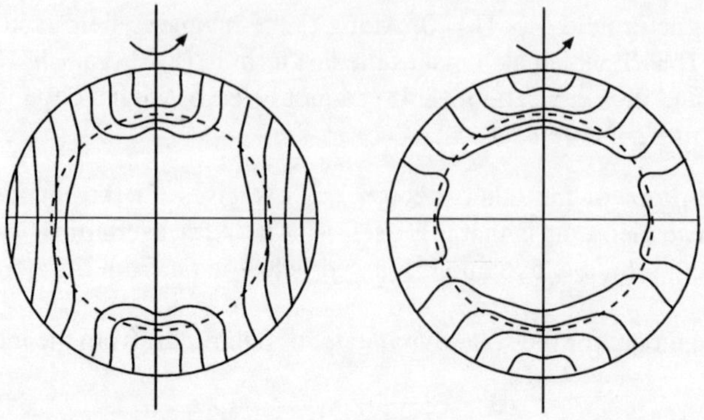

**Fig. 8.9** Isocontours of differential rotation as assumed before helioseismological results (left) and the present view (right). The dashed circle illustrates the boundary between the inner radiative zone and the convective zone at about $0.72 R_\odot$.

Now the field **u** represents the turbulent motion. In the context of turbulent motion "average" often means an ensemble average. Here we can think that the mean values are taken over longitudes ($\phi$) but not over latitudes ($\theta$), because the differential rotation is an essential part of the problem.

If the total velocity field were known, we could solve the kinematic problem directly. However, we cannot express the turbulent motion in an analytical form. Instead we have to be satisfied with knowing $\langle \mathbf{V} \rangle$ and *assuming* reasonable statistical properties of **u**. Substituting the above expressions into the induction equation and separating the mean and fluctuating parts we obtain

$$\frac{\partial \langle \mathbf{B} \rangle}{\partial t} = \nabla \times (\langle \mathbf{V} \rangle \times \langle \mathbf{B} \rangle + \mathscr{E} - \eta \nabla \times \langle \mathbf{B} \rangle) \tag{8.47}$$

$$\frac{\partial \mathbf{b}}{\partial t} = \nabla \times (\langle \mathbf{V} \rangle \times \mathbf{b} + \mathbf{u} \times \langle \mathbf{B} \rangle + \mathbf{G} - \eta \nabla \times \mathbf{b}) , \tag{8.48}$$

where

$$\mathscr{E} = \langle \mathbf{u} \times \mathbf{b} \rangle \tag{8.49}$$

$$\mathbf{G} = \mathbf{u} \times \mathbf{b} - \langle \mathbf{u} \times \mathbf{b} \rangle . \tag{8.50}$$

$\mathscr{E}$ is the mean electric field induced by the fluctuating motion. If we could calculate it, we would have a solution for the mean magnetic field $\langle \mathbf{B} \rangle$, but the calculation is, in general, too difficult. However, we assume a linear relationship between **b** and $\langle \mathbf{B} \rangle$ and, thus, also between $\mathscr{E}$ and $\langle \mathbf{B} \rangle$. Thus we can expand this relationship to the first order as

$$\mathscr{E}_i = \alpha_{ij} \langle B_j \rangle + \beta_{ijk} \partial_k \langle B_j \rangle + \dots , \tag{8.51}$$

where the summation over the repeated indices is assumed. In mathematical language the coefficients $\alpha_{ij}$ and $\beta_{ijk}$ are pseudotensors, which relate an axial vector $\langle \mathbf{B} \rangle$ to a polar vector $\mathscr{E}$. In the kinematic approach the coefficients represent the statistical properties of $\mathbf{u}$ and are independent of $\mathbf{B}$.

The mean electric field is possible to calculate explicitly if the vector $\mathbf{G}$ defined by (8.50) can be neglected. Unfortunately, this requires that either the magnetic Reynolds number must be small or $u\tau \ll l$, where $(u, \tau, l)$ are the characteristic scales of $\mathbf{u}$ and $\mathbf{b}$. In the Sun, $R_m$ is large and $u\tau \sim l$. In this respect the geodynamo is easier, because in the Earth's liquid core $R_m$ is relatively small. This does not mean that the geodynamo would in any absolute terms be an easier problem. Both are hard and far from being fully understood.

In order to proceed, we make a *first-order smoothing approximation* and neglect $\mathbf{G}$. Assume further that the turbulence is isotropic, which reduces the coefficients to the form $\alpha_{ij} = \alpha\,\delta_{ij}$ and $\beta_{ijk} = \beta\,\varepsilon_{ijk}$, where $\varepsilon_{ijk}$ is the antisymmetric permutation symbol. Under these approximations the mean electric field is

$$\mathscr{E} = \alpha\langle\mathbf{B}\rangle - \beta\,\nabla\times\langle\mathbf{B}\rangle + \dots \,. \tag{8.52}$$

$\alpha$ and $\beta$ are determined by the statistical properties (correlations) of the field $\mathbf{u}$ as

$$\alpha = -\frac{1}{3}\int_0^\infty \langle\mathbf{u}(t)\cdot\nabla\times\mathbf{u}(t-t')\rangle\, dt' \tag{8.53}$$

$$\beta = \frac{1}{3}\int_0^\infty \langle\mathbf{u}(t)\cdot\mathbf{u}(t-t')\rangle\, dt' \,. \tag{8.54}$$

Thus $\alpha$ describes the correlation of $\mathbf{u}$ to its own vorticity and $\beta$ its autocorrelation. Substituting these into (8.47) we find

$$\frac{\partial\langle\mathbf{B}\rangle}{\partial t} = \nabla\times\left(\langle\mathbf{V}\rangle\times\langle\mathbf{B}\rangle + \alpha\langle\mathbf{B}\rangle - \eta_t\nabla\times\langle\mathbf{B}\rangle\right), \tag{8.55}$$

where the total diffusivity is $\eta_t = \eta + \beta$. The turbulent contribution

$$\beta \approx \frac{1}{3}u^2\tau \approx \frac{1}{3}ul \gg \eta \tag{8.56}$$

dominates over the classical diffusion. In the solar convective zone $\eta_t \approx \beta \approx 10^8 - 10^9\ \mathrm{m^2\,s^{-1}}$. This reduces the global time scale of diffusive decay to the order of 10–100 years. This is a quite reasonable number considering that the entire solar cycle is 22 years!

An essential feature of this description is that the rate of change of the mean magnetic field is related to the field itself through the coefficient $\alpha$. This is the *alpha effect*.

### 8.3.3 The kinematic $\alpha\omega$ dynamo

The correlation of the turbulent motion to its own vorticity (8.53) implies that the motion is helical. The helical motion can sustain a dynamo alone; such processes are called $\alpha^2$

dynamos. When the helical motion is combined with rotation, as in the Sun, the dynamo is known as $\alpha\omega$ dynamo. The $\alpha\omega$ dynamo can be described as follows.

Use spherical polar coordinates and assume known $\alpha(r,\theta)$ and angular speed of the differential rotation $\Omega(r,\theta)$. Let $\alpha$ be asymmetric with respect to the equatorial plane

$$\alpha(r,\pi-\theta) = -\alpha(r,\theta) \tag{8.57}$$

and the angular velocity symmetric

$$\Omega(r,\pi-\theta) = \Omega(r,\theta)\,. \tag{8.58}$$

Furthermore, assume that besides rotation there is no other mean motion

$$\langle\mathbf{V}\rangle = (0,0,\Omega r\sin\theta)\,. \tag{8.59}$$

Separate the mean field into poloidal and toroidal parts $\langle\mathbf{B}\rangle = \mathbf{B}_p + \mathbf{B}_t$ where

$$\mathbf{B}_p = \nabla\times(0,0,A(r,\theta,t)) \tag{8.60}$$

$$\mathbf{B}_t = (0,0,B(r,\theta,t))\,. \tag{8.61}$$

The mean-field induction equation can also be separated into its poloidal and toroidal parts. Assuming, for simplicity, constant $\eta_t$ we get

$$\frac{\partial A}{\partial t} = \alpha B + \eta_t \nabla_1^2 A \tag{8.62}$$

$$\frac{\partial B}{\partial t} = \frac{\partial\Omega}{\partial r}\frac{\partial}{\partial\theta}(A\sin\theta) - \frac{1}{r}\frac{\partial\Omega}{\partial\theta}\frac{\partial}{\partial r}(rA\sin\theta) - \frac{1}{r}\frac{\partial}{\partial r}\left[\alpha\frac{\partial}{\partial r}(rA)\right] - \frac{1}{r^2}\frac{\partial}{\partial\theta}\left[\frac{\alpha}{\sin\theta}\frac{\partial}{\partial\theta}(A\sin\theta)\right] + \eta_t\nabla_1^2 B\,, \tag{8.63}$$

where $\nabla_1^2 = \nabla^2 - (r\sin\theta)^{-2}$.

Now the role of the alpha effect becomes clear. With $\alpha = 0$ the vector potential determining the poloidal field would decay exponentially and with the disappearance of $A$, the same would happen to $B$. The alpha effect generates a poloidal field from the toroidal field, whereas the differential rotation ($\nabla\Omega$, only the derivatives of $\Omega$ are involved) produces a toroidal field from the poloidal field, etc.:

$$\ldots\ \mathbf{B}_t \overset{\alpha}{\rightarrow} \mathbf{B}_p \overset{\nabla\Omega}{\rightarrow} \mathbf{B}_t\ \ldots\,.$$

This is the $\alpha\omega$ dynamo.

The $\alpha\omega$ cycle qualitatively corresponds to the solar cycle. At the solar minimum there are no sunspots and the large-scale magnetic field is as poloidal as possible. The differential rotation destroys this by winding the magnetic field lines around the Sun thus creating a toroidal component of the magnetic flux in the convection zone. Some of the toroidal field lines penetrate through the solar surface and produce sunspot pairs. As a consequence of this the magnetic polarity of the spots follows a consequent pattern. Because the wound-up magnetic field lines point in opposite azimuthal directions on the opposite hemispheres, the leading spots on one hemisphere have positive polarity and the following spots negative polarity, whereas on the other hemisphere the polarities are reversed.

After the maximum epoch the alpha effect becomes more efficient than the omega effect and starts to reorganize the poloidal field. During this process upward convection, rotation and meridional circulation together lead to a helical twisting of the magnetic field (the alpha effect). After a twist of about 180° the magnetic loops become detached through local reconnection and form new poloidal loops with a magnetic field orientation opposite to the poloidal field of the past minimum. These loops merge (again reconnection!) and produce a new minimum configuration of the poloidal field opposite to the previous minimum.

After the new minimum differential rotation starts again to wind the toroidal component, but the now-emerging sunspot pairs have reversed polarities as compared to the previous cycle. A new cycle is determined by the appearance of sunspot pairs with the new polarity and has a length of about 11 years, whereas the complete magnetic cycle is 22 years. Thus we can understand the Hale's polarity law discussed in Chap. 1.

**Train your brain**

Sketch the evolution of the mean magnetic field in the convective zone following to the description given above. If you have difficulties with this, you can certainly find both good and bad pictures of this on the internet!

Although the kinematic approach does not provide a complete description of the solar dynamo, it is possible to adjust the parameters so that the oscillatory behavior of the solar cycle can be reproduced. For a more complete, or correct, description of solar magnetic field generation and the solar cycle, more complicated nonlinear methods are needed.

An alert reader may wonder if the $\alpha\omega$ description violates the anti-dynamo theorem according to which the magnetic field should depend on all three coordinates. While the average magnetic field is two-dimensional, the underlying turbulence giving rise to the alpha effect is assumed to be three-dimensional. Note further that the alpha effect is a result of the phenomenological mean-field description of the turbulent convection. It is not a property of the exact equations.

# 9. Plasma Radiation and Scattering

Understanding of radiation and scattering processes in space plasmas is essential to correctly interpret the storm signatures in the solar spectrum. Radar scattering is also a powerful tool to probe the properties of the ionosphere.

## 9.1 Simple Antennas

In Chap. 2 we found that the expressions for radiation electric and magnetic fields are proportional to the inverse distance from the source $R^{-1}$. Because the field of a static electric charge decays as $R^{-2}$ and the fields of static electric and magnetic dipoles as $R^{-3}$, the radiation fields far from the source are much stronger than the fields of static charge and current configurations.

Assume that the radiation source is much smaller than the wavelength of the radiation ($d \ll \lambda$) and consider the region far from the source. In the Lorenz gauge the potentials are

$$\varphi = \frac{1}{4\pi\varepsilon_0}\int \frac{[\rho]}{R}\, d^3r' \; ; \; \mathbf{A} = \frac{\mu_0}{4\pi}\int \frac{[\mathbf{J}]}{R}\, d^3r' \,,$$

where the brackets indicate that the functions are evaluated at the retarded time. For radiation is enough to consider $\mathbf{A}$ alone, because the electrostatic field vanishes as $R^{-2}$. In the far region $\mathbf{J} = 0$ but $[\mathbf{J}(r')] \neq 0$ because the integral is taken over the volume including the source. Assuming harmonic time dependence we get

$$\mathbf{A}(\mathbf{r},\omega) = \frac{\mu_0}{4\pi}\int \mathbf{J}(\mathbf{r}',\omega)\frac{\exp(ik|\mathbf{r}-\mathbf{r}'|)}{|\mathbf{r}-\mathbf{r}'|}\, d^3r' \,. \tag{9.1}$$

Far from the source $|\mathbf{r}-\mathbf{r}'| \approx r - \mathbf{e}_r \cdot \mathbf{r}' \;\Rightarrow$

$$\mathbf{A}(\mathbf{r},\omega) = \frac{\mu_0}{4\pi}\frac{\exp(ikr)}{r}\int \mathbf{J}(\mathbf{r}',\omega)\exp(-ik\mathbf{e}_r\cdot\mathbf{r}')\, d^3r' \,. \tag{9.2}$$

H.E.J. Koskinen, *Physics of Space Storms: From the Solar Surface to the Earth*,
Springer Praxis Books, DOI 10.1007/978-3-642-00319-6_9,

The exponential in the integral is convenient to expand as a power series

$$\mathbf{A}(\mathbf{r},\omega) = \frac{\mu_0}{4\pi}\frac{\exp(ikr)}{r}\sum_n \frac{(-ik)^n}{n!}\int \mathbf{J}(\mathbf{r}',\omega)(\mathbf{e}_r\cdot\mathbf{r}')^n\, d^3r' \,. \tag{9.3}$$

Under the assumptions $kr \to \infty$ and $kd \ll 1$ the series converges rapidly and the first non-zero term dominates far from the source.

### *Electric dipole*

For an electric dipole $n = 0$ in (9.3). The current in the source region is found from the continuity equation $i\omega\rho = \nabla\cdot\mathbf{J} \Rightarrow$

$$\int \mathbf{J} d^3r' = -\int \mathbf{r}'(\nabla'\cdot\mathbf{J})\, d^3r' = -i\omega\int \mathbf{r}'\rho(\mathbf{r}')\, d^3r' \,. \tag{9.4}$$

The dipole moment is $\mathbf{p} = \int \mathbf{r}'\rho(\mathbf{r}')\, d^3r' \;\Rightarrow$

$$\mathbf{A}(\mathbf{r},\omega) = -i\frac{\mu_0\omega}{4\pi}\mathbf{p}(\omega)\frac{\exp(ikr)}{r} \,. \tag{9.5}$$

The fields are now easy to calculate

$$\begin{aligned}\mathbf{B} = \nabla\times\mathbf{A} &= \frac{k^2}{4\pi\varepsilon_0 c}\mathbf{e}_r\times\mathbf{p}\frac{\exp(ikr)}{r}\left(1-\frac{1}{ikr}\right) \\ &\xrightarrow{kr\gg1} \frac{k^2}{4\pi\varepsilon_0 c}\mathbf{e}_r\times\mathbf{p}\frac{\exp(ikr)}{r}\end{aligned} \tag{9.6}$$

$$\mathbf{E} = \frac{ic}{k}\nabla\times\mathbf{B} \xrightarrow{kr\gg1} c\mathbf{B}\times\mathbf{e}_r \,. \tag{9.7}$$

To estimate the average radiated power we use the definition of the Poynting vector $\mathbf{S}$. The power $dP$ radiated into the solid angle $d\Omega$ is $dP = \mathbf{S}\cdot d\mathbf{a} = \mathbf{S}\cdot\mathbf{e}_r r^2\, d\Omega \;\Rightarrow$

$$\frac{dP}{d\Omega} = \frac{1}{2\mu_0}\mathrm{Re}\{r^2\mathbf{e}_r\cdot\mathbf{E}\times\mathbf{B}^*\} = \frac{ck^4}{2(4\pi)^2\varepsilon_0}|(\mathbf{e}_r\times\mathbf{p})\times\mathbf{e}_r|^2 \,. \tag{9.8}$$

The cross product contains phase information. If all Fourier components of $\mathbf{p}$ are in the same phase, the intensity as a function of angle $\theta$ measured from the direction of the dipole axis is

$$\frac{dP}{d\Omega} = \frac{ck^4}{32\pi^2\varepsilon_0}|\mathbf{p}|^2\sin^2\theta \,. \tag{9.9}$$

The total radiated power is now

$$P = \frac{ck^4}{12\pi\varepsilon_0}|\mathbf{p}|^2 \,. \tag{9.10}$$

### *Magnetic dipole and electric quadrupole*

These multipoles correspond to the $n = 1$ term in (9.3). Now

$$\mathbf{A}(\mathbf{r},\omega) = \frac{\mu_0}{4\pi}(-ik)\frac{\exp(ikr)}{r}\int \mathbf{J}(\mathbf{r}',\omega)(\mathbf{e}_r\cdot\mathbf{r}')\,d^3r' \,. \tag{9.11}$$

Write the integrand as a sum of symmetric and asymmetric terms

$$(\mathbf{e}_r\cdot\mathbf{r}')\mathbf{J} = \frac{1}{2}[(\mathbf{e}_r\cdot\mathbf{r}')\mathbf{J} + (\mathbf{e}_r\cdot\mathbf{J})\mathbf{r}'] + \frac{1}{2}(\mathbf{r}'\times\mathbf{J})\times\mathbf{e}_r \,.$$

The asymmetric part corresponds to the magnetization due to $\mathbf{J}$

$$\mathbf{M} = \frac{1}{2}(\mathbf{r}\times\mathbf{J}) \tag{9.12}$$

and the magnetic dipole moment is

$$\mathbf{m} = \int \mathbf{M}\,d^3r \,. \tag{9.13}$$

The asymmetric part of the vector potential is

$$\mathbf{A}_A(\mathbf{r},\omega) = \frac{\mu_0}{4\pi}ik\frac{\exp(ikr)}{r}\mathbf{e}_r\times\mathbf{m} \,. \tag{9.14}$$

This gives the radiation field of a magnetic dipole. It is of the same form as the field of the electric dipole if we substitute

$$\mathbf{B}\longrightarrow\mathbf{E}/c,\ \mathbf{E}\longrightarrow -c\mathbf{B},\ \mathbf{m}\longrightarrow\mathbf{p} \,.$$

The difference between the electric and magnetic dipole radiation is the different polarization. The electric dipole radiation is called *TM mode* (transverse magnetic). Its electric field vector is in the plane defined by $\mathbf{p}$ and $\mathbf{e}_r$. The magnetic dipole radiates in the *TE mode* (transverse electric). Its electric field is perpendicular to the plane defined by $\mathbf{m}$ and $\mathbf{e}_r$.

**Train your brain**

Sketch the radiation patterns of electric and magnetic dipole and indicate the electric and magnetic field polarization in both cases.

The symmetric part of the vector potential is found to be

$$\mathbf{A}_S(\mathbf{r},\omega) = -\frac{\mu_0 c}{4\pi}\frac{k^2}{2}\frac{\exp(ikr)}{r}\int \mathbf{r}'(\mathbf{e}_r\cdot\mathbf{r}')\rho(\mathbf{r}')\,d^3r' \tag{9.15}$$

$\Rightarrow$

$$\mathbf{B} = \nabla\times\mathbf{A} \longrightarrow \frac{-i\mu_0 c}{8\pi}k^3\frac{\exp(ikr)}{r}\int(\mathbf{e}_r\times\mathbf{r}')(\mathbf{e}_r\cdot\mathbf{r}')\rho(\mathbf{r}')\,d^3r' \,. \tag{9.16}$$

This magnetic field can be expressed in terms of the *quadrupole moment tensor*

$$\mathscr{Q} = \int (3\mathbf{r}'\mathbf{r}' - r'^2 \mathscr{I})\rho(\mathbf{r}')\, d^3 r' \tag{9.17}$$

⇒

$$\mathbf{B} = \frac{\mu_0 c}{4\pi} \frac{-ik^3}{6} \frac{\exp(ikr)}{r} \mathbf{e}_r \times (\mathscr{Q} \cdot \mathbf{e}_r) \, . \tag{9.18}$$

The power radiated to the angle $d\Omega$ is found to be

$$\frac{dP}{d\Omega} = \frac{ck^6}{1152\pi^2 \varepsilon_0} \frac{4\pi}{5} \sum_{i,j} |Q_{ij}|^2 \, , \tag{9.19}$$

where $Q_{ij}$'s are the elements of tensor $\mathscr{Q}$.

## 9.2 Radiation of a Moving Charge

In the radiation by a moving charge relativistic effects need to be taken into account. We use the standard notation

$$\boldsymbol{\beta} = \mathbf{v}/c \, ; \; \gamma = 1/\sqrt{1-\beta^2} \, .$$

Let $\mathbf{r}' = \mathbf{r}_0(t)$ denote the orbit of a point-like charge. The sources of the electromagnetic fields are

$$\rho = q\delta(\mathbf{r} - \mathbf{r}_0(t)) \tag{9.20}$$

$$\mathbf{J} = q\dot{\mathbf{r}}_0 \delta(\mathbf{r} - \mathbf{r}_0(t)) \, . \tag{9.21}$$

*Green's function* for the retarded potential is

$$G(\mathbf{r}, \mathbf{r}'; t, t') = \frac{1}{4\pi} \frac{\delta(t' + R/c - t)}{R} \, , \tag{9.22}$$

where $R = |\mathbf{R}| = |\mathbf{r} - \mathbf{r}'|$. $G$ fulfills the wave equation

$$\left( \frac{\partial^2}{\partial t^2} \right) G = \delta^{(3)}(\mathbf{r} - \mathbf{r}')\delta(t - t') \, . \tag{9.23}$$

Green's function can be used to integrate over $\mathbf{r}'$, and in the integration over $t$ the identity

$$\int f(x)\delta(g(x))\, dx = \sum_i \frac{f(x_i)}{|g'(x_i)|} \, , \tag{9.24}$$

where $g(x_i) = 0$, is useful. Denoting $\mathbf{n} = \mathbf{R}/R$ this procedure leads to the *Liénard–Wiechert potentials*:

$$\varphi(\mathbf{r},t) = \frac{q}{4\pi\varepsilon_0}\left[\frac{1}{(1-\mathbf{n}\cdot\boldsymbol{\beta})R}\right]_{ret} \tag{9.25}$$

$$\mathbf{A}(\mathbf{r},t) = \frac{q}{4\pi\varepsilon_0 c}\left[\frac{\boldsymbol{\beta}}{(1-\mathbf{n}\cdot\boldsymbol{\beta})R}\right]_{ret} . \tag{9.26}$$

The radiation fields are found by straightforward derivation. They are, of course, the same as found in Chap. 2. Using the notation of the present section we write the fields as

$$\mathbf{E}(\mathbf{r},t) = \frac{q}{4\pi\varepsilon_0}\left[\frac{(\mathbf{n}-\boldsymbol{\beta})(1-\beta^2)}{(1-\mathbf{n}\cdot\boldsymbol{\beta})^3R^2} + \frac{1}{c}\frac{\mathbf{n}\times((\mathbf{n}-\boldsymbol{\beta})\times\dot{\boldsymbol{\beta}})}{(1-\mathbf{n}\cdot\boldsymbol{\beta})^3R}\right]_{ret} \tag{9.27}$$

$$\mathbf{B}(\mathbf{r},t) = \frac{\mathbf{n}\times\mathbf{E}}{c} . \tag{9.28}$$

At the non-relativistic limit ($\beta \ll 1$) the radiation fields are

$$\mathbf{E}(\mathbf{r},t) = \frac{q}{4\pi\varepsilon_0 c^2}\frac{\mathbf{n}\times(\mathbf{n}\times\dot{\mathbf{v}})}{R} \tag{9.29}$$

$$\mathbf{B}(\mathbf{r},t) = \frac{q}{4\pi\varepsilon_0 c^3}\frac{\dot{\mathbf{v}}\times\mathbf{n}}{R} , \tag{9.30}$$

from which we get the Poynting vector

$$\mathbf{S} = \frac{1}{\mu_0 c}\mathbf{E}\times(\mathbf{n}\times\mathbf{E}) = \varepsilon_0 c|\mathbf{E}|^2\mathbf{n} = \frac{1}{Z_0}|\mathbf{E}|^2\mathbf{n} . \tag{9.31}$$

Here $Z_0$ is the vacuum impedance $\approx 120\,\pi$ ohm.

The power radiated to the angle $d\Omega$ is

$$\frac{dP}{d\Omega} = \frac{q^2}{16\pi^2\varepsilon_0 c}\left|\mathbf{n}\times(\mathbf{n}\times\dot{\boldsymbol{\beta}})\right|^2 = \frac{q^2|\dot{\mathbf{v}}|^2}{16\pi^2\varepsilon_0 c^3}\sin^2\theta , \tag{9.32}$$

where $\theta$ is the angle between $\dot{\mathbf{v}}$ and $\mathbf{n}$. The electric field is in the plane of $\dot{\mathbf{v}}$ and $\mathbf{n}$. Integrating over $d\Omega$ we get the *Larmor formula* for the total power

$$P = \frac{q^2\dot{v}^2}{6\pi\varepsilon_0 c^3} . \tag{9.33}$$

For relativistic particles the distinction between $t$ and $t'$ is essential. Define the radiated power as the power radiated in the particle's *own time* ($t'$) and *own position* ($\mathbf{r}'$)

$$\begin{aligned}\frac{dP}{d\Omega} &= \varepsilon_0 c|R\mathbf{E}|^2\frac{dt}{dt'} = \frac{1-\mathbf{n}\cdot\boldsymbol{\beta}}{Z_0}|R\mathbf{E}|^2 \\ &= \frac{q^2}{16\pi^2\varepsilon_0 c}\frac{\left|\mathbf{n}\times((\mathbf{n}-\boldsymbol{\beta})\times\dot{\boldsymbol{\beta}})\right|^2}{(1-\mathbf{n}\cdot\boldsymbol{\beta})^5} .\end{aligned} \tag{9.34}$$

The total power is found either by integrating this expression or making a Lorentz transformation of the Larmor formula. The result is

$$P = \frac{q^2}{6\pi\varepsilon_0 c}\gamma^6(\dot{\beta}^2 - (\boldsymbol{\beta}\times\dot{\boldsymbol{\beta}})^2)\,. \tag{9.35}$$

When $\beta \to 1$, the significance of the denominator of $dP/d\Omega$ increases and the radiation lobes start to stretch into the direction of the particle's motion. The maximum intensity is obtained when $\theta \to 1/(2\gamma)$ and the width of the lobe is $\approx 1/\gamma$. These formulas are applicable both to the bremsstrahlung and to the cyclotron and synchrotron radiation to be discussed shortly.

In addition to the radiated power we often want to know the *spectrum* of the radiation. We discuss this using the observer's time $t$. Denote

$$\frac{dP(t)}{d\Omega} = |\mathbf{G}(t)|^2\,. \tag{9.36}$$

The total *energy* radiated into the angle $d\Omega$ is

$$\frac{dW}{d\Omega} = \int_{-\infty}^{\infty} |\mathbf{G}(t)|^2\,dt\,. \tag{9.37}$$

Define the Fourier transform of $\mathbf{G}$ as

$$\widehat{\mathbf{G}}(\omega) = \frac{1}{\sqrt{2\pi}}\int_{-\infty}^{\infty} \mathbf{G}(t)\exp(i\omega t)\,dt\,. \tag{9.38}$$

When $t$ and $\mathbf{G}(t)$ are real we can apply *Parseval's formula* to write

$$\frac{dW}{d\Omega} = \int_{-\infty}^{\infty} |\widehat{\mathbf{G}}(\omega)|^2\,d\omega\,. \tag{9.39}$$

The negative frequencies can be eliminated by using the identity $\widehat{\mathbf{G}}(-\omega) = \widehat{\mathbf{G}}^*(\omega)$. Define now the *energy spectrum per solid angle* as $d^2W/(d\Omega\,d\omega)$. This tells how much energy is radiated into the angle element $d\Omega$ within the frequency interval $d\omega$. Writing

$$\frac{dW}{d\Omega} = \int_0^{\infty} \frac{d^2W}{d\Omega\,d\omega}\,d\omega \tag{9.40}$$

we can identify

$$\frac{d^2W}{d\Omega\,d\omega} = |\widehat{\mathbf{G}}(\omega)|^2 + |\widehat{\mathbf{G}}(-\omega)|^2 = 2|\widehat{\mathbf{G}}(\omega)|^2\,. \tag{9.41}$$

To evaluate this for a point charge is straightforward but tedious (see, e.g., the classic textbook on electrodynamics by Jackson [1999]). The result is

$$\frac{d^2W}{d\Omega\, d\omega} = \frac{q^2}{16\pi^3\varepsilon_0 c}\left|\int_{-\infty}^{\infty} \frac{\mathbf{n}\times((\mathbf{n}-\boldsymbol{\beta})\times\dot{\boldsymbol{\beta}})}{(1-\mathbf{n}\cdot\boldsymbol{\beta})^2}\exp\left[i\omega\left(t'-\frac{\mathbf{n}\cdot\mathbf{r}_0(t')}{c}\right)\right]dt'\right|^2 \tag{9.42}$$

or, after a partial integration,

$$\frac{d^2W}{d\Omega\, d\omega} = \frac{q^2\omega^2}{16\pi^3\varepsilon_0 c}\left|\int_{-\infty}^{\infty} \mathbf{n}\times(\mathbf{n}\times\boldsymbol{\beta})\exp\left[i\omega\left(t'-\frac{\mathbf{n}\cdot\mathbf{r}_0(t')}{c}\right)\right]dt'\right|^2 . \tag{9.43}$$

At the non-relativistic limit

$$\frac{d^2W}{d\Omega\, d\omega} = \frac{q^2\omega^2}{16\pi^3\varepsilon_0 c^3}\left|\int_{-\infty}^{\infty} \mathbf{n}\times(\mathbf{n}\times\mathbf{v})\exp(i\omega t)\,dt\right|^2 . \tag{9.44}$$

Integrated over all angles this yields the Larmor formula in the form

$$\frac{dW}{d\omega} = \frac{q^2}{6\pi^2\varepsilon_0 c^3}\left|\int_{-\infty}^{\infty} \dot{\mathbf{v}}\exp(i\omega t)\,dt\right|^2 . \tag{9.45}$$

Thus we can calculate the radiation of the particle, once we know its orbit.

## 9.3 Bremsstrahlung

Let us apply the above results to the *bremsstrahlung* of electrons moving in a plasma. Space storm related examples of this are flare-accelerated electrons being decelerated in the solar atmosphere or energetic electrons precipitating into the ionosphere. For simplicity, we neglect the background magnetic field. This is actually a good approximation except close to the multiples of the electron gyro frequency.

Assume that the plasma is so tenuous that the electron's motion at each moment of time can be regarded to take place in the Coulomb field of a single stationary ion

$$|\dot{\mathbf{v}}| = \frac{Ze^2}{4\pi\varepsilon_0 m_e r^2} . \tag{9.46}$$

Now the Larmor formula yields the power radiated by *one* electron

$$P_e = \frac{e^2}{6\pi\varepsilon_0 c^3}\left(\frac{Ze^2}{4\pi\varepsilon_0 m_e r^2}\right)^2 . \tag{9.47}$$

Assume that the electrons arrive the plasma as a beam with number density $n^-$. Calculate first he total radiation in the field of one ion as

$$P = \frac{2}{3} Z^2 \left( \frac{e^2}{4\pi\varepsilon_0} \right)^3 \frac{n^-}{m_e^2 c^3} \int_{r_{min}}^{\infty} \frac{4\pi r^2}{r^4} \, dr$$

$$= \frac{8\pi}{3} Z^2 \left( \frac{e^2}{4\pi\varepsilon_0} \right)^3 \frac{n^-}{m_e^2 c^3 r_{min}} . \tag{9.48}$$

Based on quantum mechanical reasoning we set the lower limit of the integral to the electron's *de Broglie wavelength*

$$r_{min} \cong \frac{\hbar}{\langle p \rangle} = \frac{\hbar}{\sqrt{m_e k_B T}} . \tag{9.49}$$

Introducing the *fine structure constant* $\alpha = e^2/(4\pi\varepsilon_0 \hbar c) \approx 1/137$ and the *electron's classical radius* $r_0 = e^2/(4\pi\varepsilon_0 m_e c^2) \approx 2.82 \times 10^{-15}$ m, we can write the radiated power as

$$P = \frac{8\pi}{3} Z^2 \alpha r_0^2 m_e c^2 \sqrt{\frac{k_B T}{m_e}} n^- . \tag{9.50}$$

Finally, to find the radiated power per unit volume, we multiply this by the ion number density $n^+$

$$P_{vol} = \frac{8\pi}{3} Z^2 \alpha r_0^2 m_e c^2 \sqrt{\frac{k_B T}{m_e}} n^- n^+ . \tag{9.51}$$

For a fundamentally quantum mechanical phenomenon, this quasi-classical analysis gives a remarkably good result. A more rigorous analysis introduces a correction factor of about 1.1 to (9.51).

To find out the energy spectrum we must know the orbit of the particle. Because in plasma small angle collisions dominate, we may approximate the orbit as a straight line and calculate the acceleration at each point on this line due to the forces acting on the particle (cf. the *Born approximation* in quantum mechanics). The closest distance of this line to the scattering center is called the *impact parameter* and we denote it by $b$. Let $L$ be the effective range of the interaction. The collision time is $\tau \sim L/v \Rightarrow \omega \sim v/L$. Let $L \gg b_{min}$, where $b_{min}$ is the smallest value of the impact parameter, for which the rectilinear motion is a good approximation. Now $\omega \ll v/b_{min}$, which limits the applicability of our analysis to radio and microwave frequencies (kHz–GHz).

The equation for the orbit is

$$r_0^2(t) = b^2 + v^2 t^2 \tag{9.52}$$

and the acceleration is given by

$$\dot{v}_\perp(t) = \frac{Ze^2}{4\pi\varepsilon_0 m_e} \frac{b}{(b^2 + v^2 t^2)^{3/2}} \tag{9.53}$$

$$\dot{v}_\parallel(t) = \frac{Ze^2}{4\pi\varepsilon_0 m_e} \frac{vt}{(b^2 + v^2 t^2)^{3/2}} . \tag{9.54}$$

Substitution of these into (9.45) gives

$$\frac{dW}{d\omega} = \frac{e^2}{6\pi^2\varepsilon_0 c^3}\left(\frac{Ze^2}{4\pi\varepsilon_0 m_e}\right)^2 \left|\int_{-\infty}^{\infty} \frac{b\mathbf{e}_\perp + vt\mathbf{e}_\parallel}{(b^2+v^2t^2)^{3/2}} \exp(i\omega t)\, dt\right|^2 . \tag{9.55}$$

The integration can be performed with the help of modified Bessel functions of the second kind $K_0$ and $K_1$

$$\int_{-\infty}^{\infty} \frac{x\sin(ax)}{(b^2+x^2)^{3/2}}\, dx = aK_0(ab)$$

$$\int_{-\infty}^{\infty} \frac{\cos(ax)}{(b^2+x^2)^{3/2}}\, dx = \frac{a}{b}K_1(ab)$$

$$K_0'(x) = -K_1(x)$$

$\Rightarrow$

$$\begin{aligned}\frac{dW}{d\omega} &= \frac{e^2}{6\pi^2\varepsilon_0 c^3}\left(\frac{Ze^2}{4\pi\varepsilon_0 m_e}\right)^2 \left|\frac{2}{v}\left(b\mathbf{e}_\perp \frac{\omega}{vb}K_1(\omega b/v) + \mathbf{e}_\parallel \frac{\omega}{v}K_0(\omega b/v)\right)\right|^2 \\ &= \frac{2}{3}\frac{(Ze)^2\omega^2 c\, r_0^2}{\pi^2\varepsilon_0}\frac{1}{v^4}\left[K_0^2(\omega b/v) + K_1^2(\omega b/v)\right] . \end{aligned} \tag{9.56}$$

The energy spectrum of a single collision is sketched in Fig. 9.1.

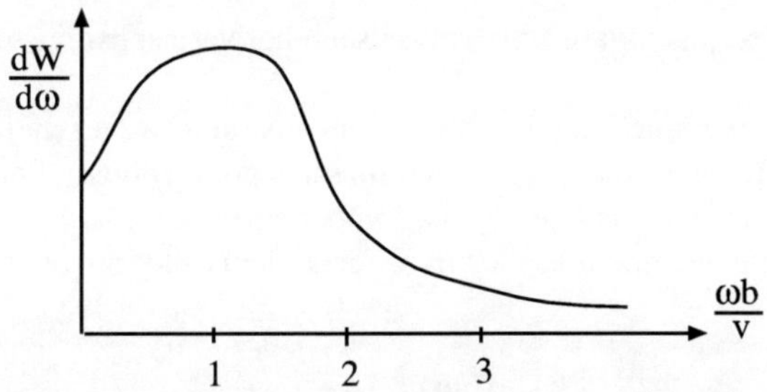


**Fig. 9.1** The shape of the energy spectrum of a single collision.

In the parameter range $b \to b+db$ there are $2\pi b\, db\, n^+ v$ collisions per unit time. Thus a single electron radiates in the frequency range $d\omega$ the power

$$P_e = 2\pi \int_{b_{min}}^{\infty} \frac{dW}{d\omega} n^+ v b\, db . \tag{9.57}$$

Next we must sum over all electrons. Assume an isotropic distribution $f = f(v)$ which implies that in the velocity interval $v \to v + dv$ there are $4\pi v^2 f(v)\,dv$ electrons per unit volume. This gives the *power spectrum*

$$\frac{dP^{tot}}{d\omega} = 8\pi^2 \int_0^\infty f(v) \left( \int_{b_{min}}^\infty \frac{dW}{d\omega} n^+ v b\, db \right) v^2\, dv \tag{9.58}$$

$$= \frac{16}{3} n^+ \frac{(Ze)^2 c r_0^2}{\varepsilon_0} \int_0^\infty f(v)\, \omega b_{min} \mathrm{K}_0 \left( \frac{\omega b_{min}}{v} \right) \mathrm{K}_1 \left( \frac{\omega b_{min}}{v} \right) dv\,,$$

where the identity $d(x\mathrm{K}_0\mathrm{K}_1)/dx = x(\mathrm{K}_0^2 + \mathrm{K}_1^2)$ was used. At small frequencies $\mathrm{K}_0 \sim -\ln x$, $\mathrm{K}_1 \sim 1/x \Rightarrow$

$$\frac{dP^{tot}}{d\omega} = \frac{16}{3} n^+ \frac{(Ze)^2 c r_0^2}{\varepsilon_0} \int_0^\infty f(v)\, v \ln \left( \frac{v}{\omega b_{min}} \right) dv\,. \tag{9.59}$$

For a Maxwellian electron distribution the integral is possible to calculate in a closed form using the formula

$$\int_0^\infty \exp(-\mu x^2) x \ln x\, dx = -\frac{1}{4\mu}(\gamma + \ln \mu)\,,$$

where $\gamma \approx 0.577...$ is known as the *Euler constant*. The total power spectrum is thus

$$\frac{dP^{tot}}{d\omega} = \frac{4}{3} n^+ n^- \frac{(Ze)^2 c r_0^2}{\varepsilon_0 \pi} \left( \frac{m_e}{2\pi k_B T} \right)^{1/2} \left[ \ln \left( \frac{2 k_B T}{m_e \omega^2 b_{min}^2} \right) - \gamma \right]. \tag{9.60}$$

The dependence on $b_{min}$ is logarithmic and as such not very sensitive to the actual value of $b_{min}$.

In this particular example "low frequency" means radio waves and microwaves (kHz–GHz). For these frequencies the classical treatment is good enough. For higher frequencies (X- or $\gamma$-rays) quantum mechanical calculation becomes necessary.

The derivation did not include plasma effects. Thus the above result is valid in the frequency range

$$\omega_p \ll \omega \ll \sqrt{\frac{2 k_B T}{m_e b_{min}^2}}\,. \tag{9.61}$$

Due to the Debye shielding the upper limit of the $b$ integration is not really infinity and we have not included multiple scattering, nor large-angle collisions. However, (9.60) is good enough for many practical purposes.

## 9.4 Cyclotron and Synchrotron Radiation

In magnetized space plasmas the radiation due to the curved path of Larmor motion is important. In the non-relativistic case this is called *cyclotron radiation* and the relativistic version is called *synchrotron radiation*. In solar physics and astrophysics the synchrotron radiation from electrons with a relatively small Lorentz factor ($\gamma \sim$2–3) is sometimes called *gyrosynchrotron radiation* and the term synchrotron is reserved for radiation by ultrarelativistic electrons.

Consider first the Larmor motion of an electron at the non-relativistic limit in the coordinate system where $\mathbf{B}$ is along the $z$ axis, $\mathbf{n}$ is the direction toward the observer in the $xz$ plane and $\theta$ is the angle between $\mathbf{B}$ and $\mathbf{n}$. Denote the gyro frequency by $\omega_0$. Now

$$
\begin{aligned}
\mathbf{n} &= \mathbf{e}_x \sin\theta + \mathbf{e}_z \cos\theta \\
\mathbf{r} &= r_L(\mathbf{e}_x \sin\omega_0 t + \mathbf{e}_y \cos\omega_0 t) \\
\mathbf{v} &= v_\perp(\mathbf{e}_x \cos\omega_0 t - \mathbf{e}_y \sin\omega_0 t) \\
v_\parallel &= 0\,.
\end{aligned}
$$

For a non-relativistic particle the loss of energy during one Larmor period is negligible. To find the radiated energy we note first that

$$
\mathbf{n}\times(\mathbf{n}\times\mathbf{v}) = v_\perp(-\mathbf{e}_x \cos\omega_0 t \cos^2\theta + \mathbf{e}_y \sin\omega_0 t + \mathbf{e}_z \cos\omega_0 t \sin\theta\cos\theta)
$$

and

$$
\int_{-\infty}^{\infty} \begin{pmatrix} \sin\omega_0 t \\ \cos\omega_0 t \end{pmatrix} \exp(-i\omega t)\, dt = \pi \times \begin{cases} -i\delta(\omega-\omega_0) + i\delta(\omega+\omega_0) \\ \delta(\omega-\omega_0) + \delta(\omega+\omega_0)\,. \end{cases}
$$

The energy spectrum radiated into the angle $d\Omega$ is then

$$
\begin{aligned}
\frac{d^2W}{d\Omega\, d\omega} &= \frac{e^2\omega_0^2 v_\perp^2}{16\pi\varepsilon_0 c^3} |\mathbf{e}_x \cos^2\theta + \mathbf{e}_y i + \mathbf{e}_z \sin\theta\cos\theta|^2 [\delta(\omega-\omega_0)]^2 \\
&= \frac{e^2\omega_0^2 v_\perp^2}{16\pi\varepsilon_0 c^3}(1+\cos^2\theta)[\delta(\omega-\omega_0)]^2\,. \qquad (9.62)
\end{aligned}
$$

The square of the delta function is a nasty singularity. We can handle it using the same trick as in quantum mechanics by introducing a finite radiation time $T$ and writing $\delta^2$ as

$$
\begin{aligned}
[\delta(\omega-\omega_0)]^2 &= \delta(\omega-\omega_0)\frac{1}{2\pi}\lim_{T\to\infty}\int_{-T/2}^{T/2} \exp(-i(\omega-\omega_0)t)\, dt \\
&= \lim_{T\to\infty}\frac{T}{2\pi}\delta(\omega-\omega_0)\,. \qquad (9.63)
\end{aligned}
$$

Dividing the energy spectrum by $T$ we get the power spectrum per unit solid angle $d\Omega$

$$\frac{d^2P}{d\Omega\, d\omega} = \frac{e^2\omega_0^2 v_\perp^2}{32\pi^2\varepsilon_0 c^3}(1+\cos^2\theta)\,\delta(\omega-\omega_0)\,. \tag{9.64}$$

The remaining delta function is important because it tells that there is exactly one spectral line at the gyro frequency. The total power is now

$$P = \frac{e^2\omega_0^2 v_\perp^2}{6\pi\varepsilon_0 c^3}\,, \tag{9.65}$$

which is again the Larmor formula (replace $\omega_0 v_\perp \to dv/dt$). Because $P \propto \omega_0^2 \propto 1/m^2$, electrons radiate much more efficiently than ions.

Next we take into account the relativistic corrections and include $v_\parallel$, but still neglect energy losses. Replace $\omega_0/\gamma \to \omega_0$. The energy spectrum must be computed using either (9.42) or (9.43). The integrand contains the factor

$$\begin{aligned} &\exp\left[i\omega\left(t-\frac{\mathbf{n}\cdot\mathbf{r}_0(t)}{c}\right)\right] = \\ &\exp\left[i\omega\left(t-\frac{\beta_\perp}{\omega_0/\gamma}\sin\theta\sin\left(\frac{\omega_0}{\gamma}t\right)-\beta_\parallel t\cos\theta\right)\right]. \end{aligned} \tag{9.66}$$

Using the well-known property of the Bessel functions

$$\exp(ix\sin y) = \sum_{l=-\infty}^{\infty} \mathrm{J}_l(x)\exp(ily)$$

we get

$$\begin{aligned} &\exp\left[i\omega\left(t-\frac{\mathbf{n}\cdot\mathbf{r}_0(t)}{c}\right)\right] = \\ &\sum_{l=-\infty}^{\infty} \mathrm{J}_l\left(\frac{\omega\beta_\perp}{\omega_0/\gamma}\sin\theta\right)\exp\left[i(\omega-\frac{l\omega_0}{\gamma}-\omega\beta_\parallel\cos\theta)t\right]. \end{aligned} \tag{9.67}$$

Next we expand the product $\mathbf{n}\times(\mathbf{n}\times\boldsymbol{\beta})$. We denote $x = (\omega\beta_\perp)/(\omega_0/\gamma)\sin\theta$ and make use of formulas

$$\begin{aligned} \mathrm{J}_{l-1}(x)-\mathrm{J}_{l+1}(x) &= 2\mathrm{J}_l'(x) \\ \mathrm{J}_{l-1}(x)+\mathrm{J}_{l+1}(x) &= \frac{2l}{x}\mathrm{J}_l(x)\,. \end{aligned}$$

After integration we again encounter the $\delta^2$ singularity, of which we get rid with the same $T$-trick as before. The result is

$$\frac{d^2P}{d\Omega\, d\omega} = \frac{e^2\omega^2}{8\pi^2\varepsilon_0 c}\,\delta\left(\frac{l\omega_0}{\gamma}-\omega(1-\beta_\parallel\cos\theta)\right)\times \tag{9.68}$$

$$\sum_{l=1}^{\infty}\left[\left(\frac{\cos\theta-\beta_{\parallel}}{\sin\theta}\right)^2 J_l^2\left(\frac{\omega\beta_{\perp}}{\omega_0/\gamma}\sin\theta\right)+\beta_{\perp}^2 J_l'^2\left(\frac{\omega\beta_{\perp}}{\omega_0/\gamma}\sin\theta\right)\right].$$

The spectrum is composed of peaks at frequencies

$$\omega_l=\frac{l\omega_0\sqrt{1-\beta^2}}{1-\beta_{\parallel}\cos\theta}\ ;\ l=1,2,\ldots\ . \tag{9.69}$$

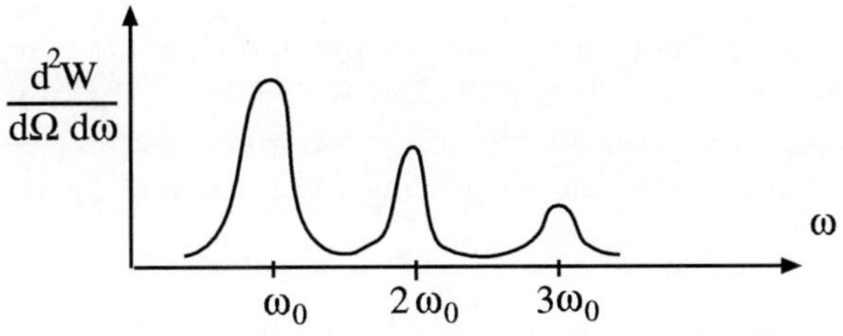


**Fig. 9.2** Spectral lines of cyclotron (or gyrosynchrotron) radiation.

Thus the peaks are shifted from the harmonics of $\omega_0$ due to relativistic mass increase ($\gamma$) and Doppler shift ($1-\beta_{\parallel}\cos\theta$). Integrations over $d\omega$ and $d\Omega$ and sum over $l$ finally yield the total power

$$\sum_{l=1}^{\infty}P_l=P^{tot}=\frac{e^2\omega_0^2}{6\pi\varepsilon_0 c}\left(\frac{\beta_{\perp}^2}{1-\beta^2}\right). \tag{9.70}$$

**Train your brain**

1. The spectrum (9.69) consists of discrete peaks at single frequencies. What physical reasons lie behind the broadening of the spectral lines of Fig. 9.2
2. Show that the energy radiated during one Larmor period is vanishingly small compared to the total energy of the electron.

At the non-relativistic limit ($\beta \ll 1$) but still retaining $v_{\parallel}\neq 0$ it is possible to show that $P_{l+1}/P_l \sim \beta^2$ for large $l$. Thus it is sufficient to consider a few of the first peaks in Fig. 9.2. Most of the power is emitted at the fundamental frequency giving rise to the *cyclotron emission line*

$$\frac{dP}{d\Omega}\approx\frac{e^2\omega_0^2}{32\pi^2\varepsilon_0 c}\beta_{\perp}^2(1+\cos^2\theta). \tag{9.71}$$

This result is calculated for a single electron. Multiplying by the electron number density and writing $v_{\perp}^2=k_BT/m$ we find that $dP/d\Omega \propto P_e$, i.e., the intensity of the cyclotron line is proportional to the electron pressure. The cosine factor tells that the intensity in the direction of **B** is twice the intensity in the perpendicular direction. At the limit $\beta\to 1$ the line separation $\omega_0(1-\beta^2)\to 0$ and highly relativistic electrons radiate a continuous spectrum (Fig. 9.3).

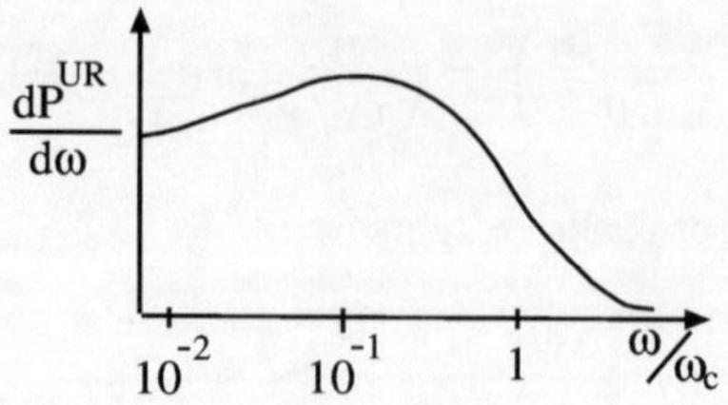


**Fig. 9.3** Continuous synchrotron spectrum of ultrarelativistic electrons.

For a long time it was thought that the cosmic radio emissions were mostly broad-banded bremsstrahlung. In the mid-1950s Ginzburg argued that the strong radio emissions from, e.g., the *Crab nebula* must actually be synchrotron emission [see Ginzburg, 1959]. This was an important milestone in the growing appreciation of the role of the magnetic fields in cosmic plasma systems.

## 9.5 Scattering from Plasma Fluctuations

The basic principles of reflection and refraction of electromagnetic waves at the interface between two macroscopic media with different refractive indices were introduced in Chap. 4. These were applied to the ionosphere where the refractive index was allowed to change smoothly. The problem of partial reflection was swept under the rug at the WKB limit, which was found to be valid when

$$\frac{1}{4n^{5/2}k_0}\left|\frac{dn}{dz}\right| \ll 1\,. \tag{9.72}$$

Thus the wavelength ($\lambda = 2\pi/k_0$) must be shorter than the density scale length. However, at the microscopic limit the electromagnetic wave (i.e., the photons) may scatter both from individual electrons and from collective fluctuations. Thus even a high-frequency wave, which can penetrate through the plasma, is partially scattered and this scattering is observable using sufficiently powerful tools, e.g., ionospheric *scatter radars*. Our discussion follows closely the presentation by Nygrén [1996].

We start the discussion from the macroscopic picture assuming that there are spatial and/or temporal fluctuations in the refractive index ($n$) and thus in the permittivity ($\varepsilon$). The fluctuations may be thermal (damped Langmuir or ion–acoustic modes), they may be characterized as turbulence, or they may be waves driven by plasma instabilities (e.g., ion–acoustic, electrostatic ion cyclotron, two-stream, Rayleigh–Taylor, or Farley–Buneman modes).

Let us consider, for simplicity, an isotropic medium with scalar $\varepsilon$. For high enough frequencies this is a good approximation also in magnetized plasmas because the terms proportional to $Y = \omega_{ca}/\omega$ in the dispersion equation are small enough to be neglected. Formally, we can relate the permittivity and the polarization

$$\mathbf{P} = (\varepsilon - \varepsilon_0)\mathbf{E}\,. \tag{9.73}$$

Denote the spatial and temporal average of $\varepsilon$ by $\langle\varepsilon\rangle$ and write

$$\varepsilon(\mathbf{r},t) = \langle\varepsilon\rangle + \triangle\varepsilon(\mathbf{r},t)\,, \tag{9.74}$$

where $\triangle\varepsilon$ is the fluctuation of the permittivity. Now

$$\mathbf{P} = (\langle\varepsilon\rangle - \varepsilon_0)\mathbf{E} + \triangle\varepsilon\mathbf{E} = \mathbf{P}_{\langle\varepsilon\rangle} + \triangle\mathbf{P}\,, \tag{9.75}$$

where $\triangle\mathbf{P}$ is the fluctuation of the polarization. When the electromagnetic wave propagates in ideal homogeneous medium, there is no reflection and the displacement current is $\dot{\mathbf{D}} = \langle\varepsilon\rangle\dot{\mathbf{E}}$, where the dot denotes the time derivative. The fluctuations introduce another displacement current contribution

$$\triangle\dot{\mathbf{D}} = \triangle\dot{\mathbf{P}} = \triangle\varepsilon\dot{\mathbf{E}}. \tag{9.76}$$

Thus the fluctuations emit radiation through this current *in the presence* of the electromagnetic wave. This emission is *scattered radiation.*

Consider again the plane waves $\mathbf{E} = \mathbf{E}_0\exp[i(\mathbf{k}\cdot\mathbf{r} - \omega_0 t)]$. The displacement current is given by

$$\triangle\mathbf{J} = \triangle\dot{\mathbf{P}} = -i\omega_0\triangle\varepsilon(\mathbf{r},t)\mathbf{E}_0\exp[i(\mathbf{k}\cdot\mathbf{r} - \omega_0 t)]\,. \tag{9.77}$$

Let $\triangle\mathbf{J}\,d^3r'$ be a current element in the volume element $d^3r'$. The retarded vector potential element is now

$$d\mathbf{A}(\mathbf{r},t) = \tag{9.78}$$
$$\frac{-i\omega_0\mu_0\mathbf{E}_0}{4\pi}\frac{\triangle\varepsilon(\mathbf{r}',t-|\mathbf{r}-\mathbf{r}'|/c)}{|\mathbf{r}-\mathbf{r}'|}\exp[i\mathbf{k}\cdot\mathbf{r}' - i\omega_0(t-|\mathbf{r}-\mathbf{r}'|/c)]\,d^3r'\,.$$

Assume that the scattering volume $\mathscr{V}$ is small compared to the distance $R$ from the observer and that the fluctuations are so slow that $\varepsilon$ changes only little during the time the wave propagates through the scattering volume. Then the retarded time from all points within $\mathscr{V}$ can be replaced by a single time $t' \approx t - R/c$, and $|\mathbf{r}-\mathbf{r}'|$ in the denominator by $R$. Now

$$\omega_0|\mathbf{r}-\mathbf{r}'|/c = k|\mathbf{r}-\mathbf{r}'| = \mathbf{k}_s\cdot(\mathbf{r}-\mathbf{r}')\,, \tag{9.79}$$

where $\mathbf{k}_s$ is the wave vector of radiation scattered into the direction of $\mathbf{r}-\mathbf{r}'$. Inserting this into the expression for $\mathbf{A}$ and integrating over $\mathscr{V}$ we get

$$\mathbf{A}(\mathbf{r},t) = \tag{9.80}$$
$$\frac{-i\omega_0\mu_0\mathbf{E}_0}{4\pi R}\exp[i\mathbf{k}_s\cdot\mathbf{r} - i\omega_0(t)]\int_{\mathscr{V}}\triangle\varepsilon(\mathbf{r}',t')\exp[i(\mathbf{k}-\mathbf{k}_s)\cdot\mathbf{r}']\,d^3r'\,.$$

This is a plane wave propagating into the direction of $\mathbf{r}-\mathbf{r}'$. Its amplitude is proportional to the three-dimensional spatial Fourier transform of $\triangle\varepsilon$, i.e., $\triangle\varepsilon(\mathbf{K})$, where $\mathbf{K} = \mathbf{k} - \mathbf{k}_s$. Thus we can write

$$\mathbf{A}(\mathbf{r},t) = \frac{-i\omega_0\mu_0\mathbf{E}_0}{4\pi R}\triangle\varepsilon(\mathbf{K})\exp[i\mathbf{k}_s\cdot\mathbf{r} - i\omega_0(t)]\,. \tag{9.81}$$

From this it is straightforward to calculate the electric and magnetic fields of the scattered wave.

Let $S$ be the intensity (i.e., the absolute value of the Poynting vector) of the incident wave. Then the intensity of the scattered wave is

$$S_s \propto |\triangle\varepsilon(\mathbf{K})|^2 S. \tag{9.82}$$

In isotropic plasma the refractive index is $n = \sqrt{1-\omega_p^2/\omega_0^2}$. This allows us to interpret the above result in terms of density fluctuations $\triangle n_e(\mathbf{K})$

$$S_s \propto |\triangle n_e(\mathbf{K})|^2 S\,. \tag{9.83}$$

Thus the wave vector spectrum of density fluctuations determines the intensity of scattered radiation. Now a little exercise in geometry tells that $|\mathbf{k}| = |\mathbf{k}_s| = 2\pi/\lambda \;\Rightarrow\; |\mathbf{K}| = 2k\cos\phi$, where $2\phi$ is the angle between the incident and scattered radiation, divided to equal halves by the normal direction. The corresponding wavelength is $\Lambda = \lambda/(2\cos\phi)$. If the wave scattered from parallel planes separated by a distance $d$, the difference in the path lengths would be $\delta = 2d\cos\phi$. Thus there will be *constructive interference* of waves scattered from two consecutive planes if

$$d = \frac{\lambda}{2\cos\phi} = \Lambda\,. \tag{9.84}$$

**Train your brain** by sketching the geometric construction that proves (9.84)

Now we can give a physical interpretation to the dependence of scattered intensity on the density fluctuations. The fluctuation is composed of plane waves propagating in all directions as determined by its spatial Fourier transform. The scattering process selects from this wave spectrum the component that gives constructive interference in the direction of the observer, i.e., the scattering of exactly those waves whose wave vector $\mathbf{K}$ is enhanced. The constructive interference requires that the direction of the wave normal ($\mathbf{n} = \mathbf{K}/K$) divides the angle between the incident and scattered waves to equal halves and the wavelength is $\Lambda = \lambda/(2\cos\phi)$. This is analogous to the *Bragg scattering* in a crystal lattice. The case $\phi = 0$ (i.e., $\Lambda = \lambda/2$) is called *backscattering*.

Typical ionospheric backscatter radars transmit waves at frequencies from 10 MHz to 200 MHz. Frequencies around 150 MHz (2 m) are particularly important because they backscatter from 1-m density fluctuations, which happens to be a typical length of Farley–Buneman waves in the auroral E-region ionosphere.

## 9.6 Thomson Scattering

The scattering of an electromagnetic wave off an electron is known as *Thomson scattering*. Thomson scattering is of particular interest in physics of space storms because it is responsible for the white light that we see in the coronagraph images of CMEs. Also the *incoherent scatter radars* utilize Thomson scattering.

Let an electron oscillate in the field of an electromagnetic wave with the frequency $\omega_0$: $\mathbf{E}_i = \mathbf{E}_0 \exp(-i\omega_0 t)$. The acceleration of the electron is $d^2\mathbf{r}_e/d^2t = -(e/m)\mathbf{E}_i$, from which we obtain the velocity

$$\mathbf{v}_e = -\frac{e}{m}\int \mathbf{E}_i\, dt = \frac{-ie\mathbf{E}_0}{m\omega_0}\exp(-i\omega_0 t)\,. \tag{9.85}$$

The current of a single electron is

$$\mathbf{J}(\mathbf{r},t) = -e\mathbf{v}_e(t)\delta[\mathbf{r}-\mathbf{r}_e(t)] = \frac{ie^2\mathbf{E}_0}{m\omega_0}\exp(-i\omega_0 t)\delta[\mathbf{r}-\mathbf{r}_e(t)]\,. \tag{9.86}$$

Thus the vector potential can be written as

$$\begin{aligned}\mathbf{A}(\mathbf{r},t) &= \frac{\mu_0}{4\pi}\int \frac{\mathbf{J}(\mathbf{r}',t')}{|\mathbf{r}-\mathbf{r}'|}\,{}^3r' \\ &= \frac{i\mu_0 e^2\mathbf{E}_0}{4\pi m\omega_0}\int \frac{\exp(-i\omega_0 t')}{|\mathbf{r}-\mathbf{r}'|}\delta[\mathbf{r}'-\mathbf{r}_e(t')]\,d^3r' \\ &= \frac{i\mu_0 e^2\mathbf{E}_0}{4\pi m\omega_0}\frac{\exp[-i\omega_0(t-|\mathbf{r}-\mathbf{r}_e|/c)}{|\mathbf{r}-\mathbf{r}'|}\,.\end{aligned} \tag{9.87}$$

This can be simplified by letting the electron oscillate about the origin of the frame of reference

$$\mathbf{A}(\mathbf{r},t) = \frac{i\mu_0 e^2\mathbf{E}_0}{4\pi m\omega_0}\frac{\exp[-i(\omega_0 t-\mathbf{k}_s\cdot\mathbf{r})]}{|\mathbf{r}|}\,. \tag{9.88}$$

Thus the oscillating electron radiates a spherical wave whose amplitude decreases as $1/r$. Again the radiation is due to the incident wave, i.e, this is another example of scattering of radiation. The scattered magnetic field is

$$\mathbf{B}_s = \nabla\times\mathbf{A} = -\frac{\mu_0 e^2(\mathbf{k}_s\times\mathbf{E}_0)}{4\pi m\omega_0}\frac{\exp[-i(\omega_0 t-\mathbf{k}_s\cdot\mathbf{r})]}{|\mathbf{r}|} \tag{9.89}$$

and the scattered electric field is found using Faraday's law, which gives

$$|\mathbf{E}_s| = c|\mathbf{B}_s|\,. \tag{9.90}$$

The average of the scattered Poynting flux is

$$\langle|\mathbf{S}_s|\rangle = \frac{\varepsilon_0 c}{2}r_0^2\frac{E_0^2\sin^2\chi}{r^2}\,, \tag{9.91}$$

where $r_0 = e^2/(4\pi\varepsilon_0 mc^2)$ is the electron classical radius and $\chi$ the angle between $\mathbf{k}_s$ and $\mathbf{E}_0$. The relation between the incident and scattered Poynting vectors is

$$\langle|\mathbf{S}_s|\rangle = \frac{r_0^2}{r^2}\sin^2\chi\,\langle|\mathbf{S}_i|\rangle\,. \tag{9.92}$$

**Train your brain**

Calculate the average of scattered Poynting flux (9.91).

The total scattered power is

$$P_T = \int \langle|\mathbf{S}_s|\rangle r^2 d\Omega = 2\pi\int_0^\pi \langle|\mathbf{S}_s|\rangle r^2 \sin\chi\, d\chi$$

$$= 2\pi r_0^2\langle|\mathbf{S}_i|\rangle \int_0^\pi \sin^3\chi\, d\chi = \frac{8\pi}{3} r_0^2\langle|\mathbf{S}_i|\rangle\,, \tag{9.93}$$

where $(8/3)\pi r_0^2 \approx 6.65\times 10^{-29}\,\mathrm{m}^2$ is the *Thomson cross-section*. It is quite a small area, and one may wonder if it really is possible to detect Thomson scattering, e.g., from ionospheric electrons.

Let us consider an experimental setting where a powerful transmitter sends electromagnetic radiation into the ionosphere and another antenna listens to the scattered signal. Let the frequency of the transmitted signal be 1 GHz ($\lambda = 30$ cm). The transmitted signal is amplified by an antenna consisting of a large parabolic dish (say, with a diameter of 32 m). The *antenna gain* $G$ is an important factor. In our case it is $G \approx 4\pi A/\lambda^2$, where $A$ is the cross-section area of the paraboloid. Let the distance from the transmitter to the scattering electron be $r_1$ and the distance from the electron to the receiver be $r_2$. The incident signal at the electron is

$$S_i = \frac{P_t G}{4\pi r_1^2}\,, \tag{9.94}$$

where $P_t$ is the transmitted power. The signal at the receiver with effective aperture $A_r$ is

$$S_r = 4\pi r_0^2 \sin^2\chi \frac{P_t G}{4\pi r_1^2}\frac{A_r}{4\pi r_2^2}\,. \tag{9.95}$$

This relation is known as the *radar equation*. The quantity

$$\sigma_0 = 4\pi r_0^2 \sin^2\chi \approx 10^{-28}\sin^2\chi\ \mathrm{m}^2 \tag{9.96}$$

is the *electron's radar cross-section*.

***Example: EISCAT***

In the northern auroral zone there is a powerful radar system EISCAT (European Incoherent Scatter Radar Facility) located in Tromsø, Kiruna, and Sodankylä with an extension to Longyearbyen in Svalbard. One of the radars, transmitting from Tromsø and receiving in Tromsø, Kiruna, and Sodankylä, operates at the frequency of 930 MHz. Assume that the transmitter power is 1 MW and the cross-section of the radiated beam at the location where the scattering takes place at the distance of 300 km is $10^3\,\mathrm{m} \times 10^3\,\mathrm{m} = 10^6\,\mathrm{m}^2$. Thus the incident intensity is 1 W/m$^2$. A typical electron density in the F-layer is $n_e \approx 10^{12}\,\mathrm{m}^{-3}$ and the scattering volume $\mathscr{V} \approx 10^3 \times 10^3 \times 10^4\,\mathrm{m}^3 = 10^{10}\,\mathrm{m}^3$ (the width of the beam multiplied by a height of 10 km). The total scattering cross-section from volume $\mathscr{V}$ is thus $\sigma_{tot} = n_e \mathscr{V} \sigma_0 \approx 10^{-6}\,\mathrm{m}^2$. Assume that the receiver is also at the distance of 300 km from the scattering volume and its effective aperture is $100\,\mathrm{m}^2$. Using the equations above we find that the received power is only of the order of $10^{-16}$ W. That is a factor of $10^{22}$ less than the transmitted power.

Gordon [1958] proposed that Thomson scattering from the ionosphere should be possible to detect with such large antennas. He suggested also that the thermal motion of electrons would broaden the scattered spectrum, which would yield a measurement of electron temperature.

Let us assume that the frequency of the incident radiation is $f_0$. According to the relativistic formula for the Doppler effect the electron moving at speed $v$ relative to the radar "sees" the radiation at frequency

$$f' = f_0 \sqrt{\frac{c+v}{c-v}} \,. \tag{9.97}$$

The electron emits this frequency in its own frame of reference and finally the receiver detects the frequency

$$f = f' \sqrt{\frac{c+v}{c-v}} = f_0 \frac{c+v}{c-v} \approx f_0 \left(1 + \frac{2v}{c}\right) . \tag{9.98}$$

Thus the radar measures the *Doppler shift* $\delta f = 2v/\lambda_0$.

If we assume that the electron velocity distribution is Maxwellian

$$\frac{dn_e}{dv} \propto \exp\left(-\frac{v^2}{v_{th}^2}\right) \; ; \; v_{th} = \sqrt{\frac{2k_B T}{m_e}} \,, \tag{9.99}$$

the *half-width* of the spectrum is approximately

$$\Delta f \approx \frac{4}{\lambda_0} \sqrt{\frac{k_B T}{m_e}} \,. \tag{9.100}$$

Inserting values corresponding to a typical EISCAT measurement at 930 MHz and assuming $T = 1000\,\mathrm{K}$ we get a half-width of about 1.5 MHz. However, already the first incoherent scatter radar observations in Long Branch, Illinois, indicated that the actually scattered radiation had a *much narrower* spectrum, the observed Doppler shifts being in the kilohertz range only. What was wrong in Gordon's suggestion?

The explanation lies in the plasma physics of the scattering volume. As long as the radar wavelength is longer than the Debye length of the plasma, as is the case in the ionosphere with 30-cm waves, the observed signal actually rises from the density fluctuations of the plasma. While the basic scattering process is the *incoherent Thomson scattering* off electrons, the Doppler-broadened shape of the scattered signal is determined by collective fluctuations in the electron density. The density fluctuations have the phase speed determined by the dispersion equations for electron plasma waves and ion–acoustic waves. The scattering process picks up the wavelength according to formula (9.84). In the case of backscattering this is $\lambda_0/2$. Assuming that there are waves propagating in all directions the radar sees both an upshifted and a downshifted Doppler-broadened signal. For the ion waves the phase speed is smaller, thus the peaks are closer to $f_0$ whereas the electron (Langmuir) lines are further away from $f_0$.

The larger mass of the ions dominates the density fluctuations making the ion line much stronger than the electron line. It is actually very difficult to observe the electron plasma oscillation at all, except when there is another, more energetic, electron population present, which can enhance the electron oscillation by the Landau mechanism. These waves are known as the *electron–acoustic waves* and are analogous to the ion–acoustic waves. In the dispersion equation the cool electron background replaces the ions and the electron population of the ion–acoustic case is replaced by the hot electron population.

From the ion lines it is possible to derive a surprising amount of information. The total electron density determines how much of the radiation is scattered, thus the total power yields an estimate for $n_e$. The ion acoustic oscillations are strongly damped by the Landau mechanism. This deviates the waveform from its ideal sinusoidal form and broadens the spectrum. The upward and downward Doppler-shifted ion lines merge to a double-humped spectrum (Fig. 9.4). The total width from one hump to the other is relative to the ion thermal speed ($\propto 4v_{th,i}/\lambda_0$). If we know the ion species, we can determine the ion temperature. The width of the spectrum together with the depth of the minimum between the humps is determined by the Landau damping and thus depends on $T_e/T_i$. If the entire ion line (both humps) is Doppler-shifted, the plasma is in motion with a speed corresponding to the velocity component in the direction of the scattering **k** vector.

Furthermore, the sharpness of the edges of the ion line is a measure of the ion–neutral collision frequency. Careful fitting of the observed spectra into the models of different relative ion abundances gives information of the relative ion concentration. In fact, extracting physical parameters from a backscattered signal of $10^{-16}$ W is a pretty challenging task of scientific data analysis, which can be performed with astonishing success.

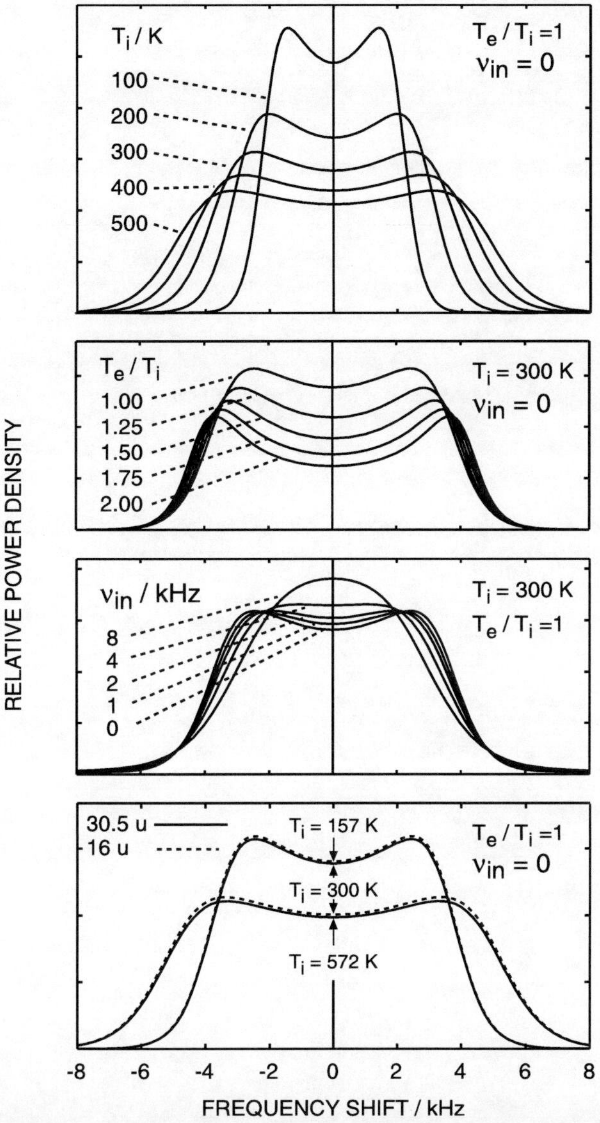


**Fig. 9.4** Dependence of radar power spectrum on various parameters. The ion mass in the top three panels is 30.5 amu, corresponding to an ionospheric mixture of $O_2^+$ and $NO^+$ ions. The top panel is calculated for $T_e = T_i$ and illustrates the effect of the variation of changing ion temperature. The second panel is calculated for $T_i = 300\,\mathrm{K}$ and illustrates the effect of varying electron to ion temperature ratio. The third panel, in turn, shows the effect of ion–neutral collisions. The increasing collision frequency tends to fill the gap between the Doppler-shifted humps. Finally, the bottom panel demonstrates the effect of ion mass in the spectrum. (Figure by courtesy of T. Nygrén.)

# 10. Transport and Diffusion in Space Plasmas

We have already encountered diffusion several times when applying the induction equation of MHD. In this chapter we introduce the Fokker–Planck equation, which is a general kinetic equation to deal with diffusion due to collisions or wave–particle interactions at the level of plasma distribution function. We start by redefining the phase space density and discuss the appropriate coordinates for particle flux calculations.

## 10.1 Particle Flux and Phase Space Density

In Chaps. 2 and 5 we studied "the plasma theorist's" distribution function $f(\mathbf{r},\mathbf{p},t)$ and defined the particle flux as the first-order velocity moment of $f$ by (2.108). Here we consider a more practical approach to the particle flux, namely, how it is determined from observations and how it is related to the distribution function.

We start by defining the *differential unidirectional flux* $j$ as the number of particles $dN$ coming from a given incident direction (unit vector $\mathbf{i}$) that hit a surface of unit area $dA$, oriented perpendicular to the particles' direction of incidence, per unit time $dt$, unit solid angle $d\Omega$ and unit kinetic energy $dW$. Hence we can write

$$dN = j\, dA\, dt\, d\Omega\, dW\ . \tag{10.1}$$

In general

$$j = j(\mathbf{r},\mathbf{i},W,t) \tag{10.2}$$

contains full information on the particles' spatial ($\mathbf{r}$), angular ($\mathbf{i}$) and energy ($W$) distribution at a given time. The flux $j$ is a quantity measured by an ideal directional instrument. It is traditionally given in units $\mathrm{cm^{-2}\,s^{-1}\,ster^{-1}\,keV^{-1}}$, even in literature otherwise using SI units.

In reality particle detectors are only seldom simple surface plates. Instead they may consist of a complicated network of time-of-flight measuring arrangements, electric and magnetic deflectors, stacks of detector plates, etc. Furthermore, real detectors do not sample infinitesimal solid angles or energy intervals. Thus the conversion from the *detector*

H.E.J. Koskinen, *Physics of Space Storms: From the Solar Surface to the Earth*,
Springer Praxis Books, DOI 10.1007/978-3-642-00319-6_10,

*counting rate* to $j$ requires careful consideration of sensitivity, resolution and configuration of the individual instrument. A real detector has a low-energy cut-off and the flux is often convenient to represent as an *integral directional flux* as

$$j_{>E} = \int_E^\infty j \, dW \,. \tag{10.3}$$

Other important concepts are the *omnidirectional flux* $J$ defined by

$$J = \int_{4\pi} j \, d\Omega \tag{10.4}$$

and the corresponding integral flux

$$J_{>E} = \int_E^\infty J \, dW \,. \tag{10.5}$$

Assume next that the particle distribution function is smooth and free of interactions in a locally homogeneous magnetic field $\mathbf{B}$. The magnetic field gives a natural axis for the frame of reference. The direction of incidence $\mathbf{i}$ can be given by the pitch angle $\alpha$ and the azimuthal angle $\phi$ around $\mathbf{B}$. If particles are uniformly distributed in the gyro phase, the angle of incidence, and thus $j$, will depend only on $\alpha$. Thus the number of particles, whose pitch angles lie within the interval from $\alpha$ to $\alpha + d\alpha$ crossing a given point per second from all azimuthal directions $\phi$ per unit perpendicular area and energy, can be expressed as

$$\frac{dN}{dA \, dW \, dt} = 2\pi j \sin\alpha \, d\alpha = -2\pi j \, d(\cos\alpha) \,. \tag{10.6}$$

The flux is called *isotropic* if the number of incoming particles depends only on the size of the solid angle of acceptance and is independent of the direction of incidence, i.e., $j$ is constant with respect to $\alpha$

$$\frac{dN}{d(\cos\alpha)} = \text{const} \,. \tag{10.7}$$

Consequently, in an isotropic distribution equal numbers of particles arrive at the detector from equal intervals of pitch angle cosines. For this reason the flux is often written as a function of $\mu = \cos\alpha$[1]

$$j = j(\mathbf{r}, \mu, W, t) \,. \tag{10.8}$$

For an isotropic flux

$$J = 4\pi \int_0^1 j \, d\mu = 4\pi j \,. \tag{10.9}$$

In the absence of sources and losses the Liouville equation states that the density of particles in the phase space $f_p(\mathbf{r}, \mathbf{p})$, the *phase space density*, is constant along the particle trajectory, i.e.,

$$f_p = \frac{dN}{dx \, dy \, dz \, dp_x \, dp_y \, dp_z} = \text{const} \,. \tag{10.10}$$

[1] This is an example of the unfortunate overloading of the symbol $\mu$. In this context, the magnetic moment is usually given by some other symbol, e.g., $M$.

Let the $z$ axis be along the velocity vector. Then $dx\,dy = dA$, $dz = v\,dt$, and $dp_x\,dp_y\,dp_z = p^2\,dp\,\sin\theta\,d\theta\,d\phi = p^2\,dp\,d\Omega$. Now $v\,dp = dW$ and

$$f_p = \frac{dN}{p^2\,dA\,dt\,d\Omega\,dW} = \frac{j}{p^2}\,. \tag{10.11}$$

Thus we have found the relationship between the differential unidirectional flux and the phase space density. For nonrelativistic particles $f_p \approx j/2mW$. Note that the SI units of $f_p$ are $\mathrm{kg}^{-3}\,\mathrm{s}^3\,\mathrm{m}^{-6}$. We retain the familiar velocity space distribution function by writing $f = m_0^3 f_p$, where $m_0$ is the particle's rest mass.

## 10.2 Coordinates for Particle Flux Description

The action integrals in the electromagnetic field were defined in Chap. 3 by (3.53) as

$$J_i = \frac{1}{2\pi}\oint(\mathbf{p}_i + q\mathbf{A})\cdot d\mathbf{s}_i$$

with associated phase angles $\phi_i$. We already are familiar with one set of these from Chap. 3, namely $\{\mu, J, \Phi\}$, the phase angles of which are the gyro phase, the bounce phase, and the drift phase. From here on $\mu$ again denotes the magnetic moment.

If all action variables $\mu$, $J$ and $\Phi$ are conserved, i.e., adiabatic invariants, we can average over the corresponding phase angles and reduce the 6-dimensional phase space to 3-dimensional space with coordinates $\{\mu, J, \Phi\}$. Let us denote the averaged phase space density, for the time being, by

$$\overline{f} = \overline{f}(\mu, J, \Phi; t)\,. \tag{10.12}$$

Note that this function does not in general satisfy the Liouville equation because it represents an average over particles that have followed different dynamical trajectories before reaching the point of observation.

While the triplet $\{\mu, J, \Phi\}$ can be seen as the most natural set of coordinates in the nearly dipolar magnetic field of the inner magnetosphere, it is not always the most convenient. Both $\mu$ and $J$ depend on particle momentum, which is not quite efficient for computational purposes, as in a general time-dependent field all three coordinates of $\mathbf{p}$ must be given.

$J$ is often convenient to replace by the purely field-geometrical quantities $K$ or $I$ defined by

$$K = \frac{J}{\sqrt{8m_0\mu}} = I\sqrt{B_m} = \int_{s_m}^{s'_m}[B_m - B(s)]^{1/2}\,ds\,, \tag{10.13}$$

where $m_0$ is the rest mass of the particle and $B_m$ the mirror field. The integral $I$ is thus

$$I = \int_{s_m}^{s'_m}\left[1 - \frac{B(s)}{B_m}\right]^{1/2} ds \tag{10.14}$$

and $J = 2pI$.

Note that for a static field in the absence of external forces the drift shell of the particle is completely specified by the pair $\{I, B_m\}$. Now the differential directional flux for a given energy can be expressed as a function of these shell parameters only

$$j = j(I, B_m, W, t) . \tag{10.15}$$

However, in an azimuthally asymmetric field particles on a joint drift shell at some longitude are not on the same drift shell elsewhere as they mirror at different field strengths $B_m$ and, consequently, their $I$ integrals are different. This is known as *drift shell splitting*.

In a symmetric or nearly symmetric field such as the Earth's quasi-dipolar field in the inner magnetosphere, the shell splitting disappears, which is known as *shell degeneracy*. For a pure dipole field $I$ can be replaced by the $L$ parameter and we can map the omnidirectional flux $J$ (do not mix up with the longitudinal invariant $J$) everywhere writing

$$J = J(L, B_m, W, t) . \tag{10.16}$$

These $(B_m, L)$ coordinates are frequently used in studies of particle fluxes in the inner magnetosphere. The point of this description is that if we can determine the omnidirectional flux at a point in space, we can map it everywhere with the same $L$ shell and same mirror magnetic field $B_m$ as long as the adiabatic invariants are conserved in the mapping process.

**Train your brain**

Show that the relationship $L = L(I, B_m)$ in the dipole field is of the form

$$\frac{L^3 R_E^3 B_m}{k_0} = F\left(\frac{I^3 B_m}{k_0}\right) , \tag{10.17}$$

where the function $F$ must be integrated numerically.

Further out in the magnetosphere mapping of particle fluxes from one place to another becomes increasingly complicated, as the field begins to deviate from the dipolar configuration. It is possible to generalize [cf., Roederer, 1970] the $L$ parameter defining

$$L^* = \frac{2\pi k_0}{\Phi R_E} . \tag{10.18}$$

For the dipole field $L^* = L$. Physically $L^*$ is the radial distance to the equatorial points of the symmetric $L$ shell on which the particles would be found if all nondipolar perturbations of the magnetic field were turned off adiabatically. This method can be applied also to the internal field perturbations close to the Earth. To make practical use of $L^*$ is computationally much more demanding than just to trace one field line for a given $L$, but with modern computers this is no more such a problem as it was when Roederer introduced the concept.

However, as Schulz [1996] has pointed out, $L^*$ depends on the Earth's dipole moment and thus it is not invariant over long time periods due to the secular variation of the geomagnetic field. This is an effect that can, in fact, already be seen in radiation belt data from the life-time of an individual satellite, e.g., *SAMPEX*, that was launched in 1992 and returned data for longer than one full solar cycle. Consequently, Schulz [1996] gives a clear preference to the triplet $\{\mu, K, \Phi\}$ as the basic coordinate system for radiation belt models.

## 10.3 Elements of Fokker–Planck Theory

In practical space physics problems we are often interested in the temporal evolution of the particle distribution at a given location, or several locations, in the phase space, i.e., $\partial f/\partial t$. If the distribution function is not far from thermal equilibrium, it may be sufficient to use the Krook model (2.104)

$$\left(\frac{\partial f}{\partial t}\right)_c = -\nu_c(f - f_0)\,.$$

While this is a computationally simple approach, it has some serious drawbacks. First of all, the Krook model does not conserve particles, momentum or energy. Furthermore, there usually is more than one collision frequency that affects the distribution function. As we will discuss in Chap. 14, even in the almost collisionless inner magnetosphere Coulomb collisions, charge exchange collisions and various wave–particle "collisions" need to be included *at the same time* in computations of the ring current and radiation belt dynamics.

Neither is the classical Boltzmann collision integral for $(\partial f/\partial t)_c$ satisfactory. In a plasma the collisions are not binary short-range interactions, but a plasma particle interacts simultaneously with all particles within its Debye sphere through the long-range Coulomb force. For Coulomb interactions the *Fokker–Planck approach* is more suitable. In simple terms the Fokker–Planck theory is a method to include frictional and diffusion effects in the RHS of the Boltzmann equation.

Before going into the details it may be instructive to recall the diffusion equation for the magnetic field introduced in Chap. 6

$$\partial\mathbf{B}/\partial t = \eta\nabla^2\mathbf{B}\,.$$

This is the simplest form of diffusion equations in physics. Here the magnetic diffusivity $\eta$ is the *diffusion coefficient D*, which in this case was assumed spatially homogeneous. Here the diffusion takes place in the configuration space and its SI units are $\mathrm{m^2\,s^{-1}}$. In general the diffusion coefficient has the form

$$D_{xx} \propto \frac{\langle\delta x\,\delta x\rangle}{\tau}\,, \tag{10.19}$$

where $x$ is the coordinate in which the diffusion takes place, e.g, $\alpha$, $L$, $W$, etc., and $\delta x$ gives its deviation during the *diffusion time* $\tau$.

To formally derive the Fokker–Planck equation consider the function $\psi(\mathbf{v}, \triangle\mathbf{v})$ that gives the probability that a particle's velocity $\mathbf{v}$ is deflected by a small increment $\triangle\mathbf{v}$ in time $\triangle t$. Integrating over all possible deflections likely to occur during $\triangle t$ before the time $t$ gives

$$f(\mathbf{r},\mathbf{v},t) = \int f(\mathbf{r},\mathbf{v}-\triangle\mathbf{v},t-\triangle t)\,\psi(\mathbf{v}-\triangle\mathbf{v},\triangle\mathbf{v})\,d(\triangle\mathbf{v})\,. \tag{10.20}$$

As $\psi$ is independent of $t$, the collisional process has no memory of earlier collisions. Thus the collisions are treated as a *Markovian random walk.*

Next we Taylor expand the integral in (10.20) in powers of $\triangle\mathbf{v}$

$$\begin{aligned} f(\mathbf{r},\mathbf{v},t) = \\ \int d(\triangle\mathbf{v}) \Big[ f(\mathbf{r},\mathbf{v},t-\triangle t)\,\psi(\mathbf{v},\triangle\mathbf{v}) - \triangle\mathbf{v}\cdot\frac{\partial}{\partial\mathbf{v}}(f(\mathbf{r},\mathbf{v},t-\triangle t)\,\psi(\mathbf{v},\triangle\mathbf{v})) \\ + \frac{1}{2}\triangle\mathbf{v}\triangle\mathbf{v} : \frac{\partial^2}{\partial\mathbf{v}\partial\mathbf{v}}(f(\mathbf{r},\mathbf{v},t-\triangle t)\,\psi(\mathbf{v},\triangle\mathbf{v})) + \ldots \Big]\,, \end{aligned} \tag{10.21}$$

where : indicates tensor product (summing over both indices). The total probability of all deflections is unity $\int \psi(\triangle\mathbf{v}) = 1$ and we can calculate the rate of change due to collisions to be

$$\begin{aligned} \left(\frac{\partial f}{\partial t}\right)_c &\equiv \frac{f(\mathbf{r},\mathbf{v},t) - f(\mathbf{r},\mathbf{v},t-\triangle t)}{\triangle t} \\ &= -\frac{\partial}{\partial\mathbf{v}}\cdot\left(\frac{f\langle\triangle\mathbf{v}\rangle}{\triangle t}\right) + \frac{1}{2}\frac{\partial^2}{\partial\mathbf{v}\partial\mathbf{v}} : \left(\frac{f\langle\triangle\mathbf{v}\triangle\mathbf{v}\rangle}{\triangle t}\right)\,, \end{aligned} \tag{10.22}$$

where the averages $\langle\triangle\mathbf{v}\rangle$ and $\langle\triangle\mathbf{v}\triangle\mathbf{v}\rangle$ are defined as

$$\langle\ldots\rangle = \int \psi(\mathbf{v},\triangle\mathbf{v})(\ldots)d(\triangle\mathbf{v}) \tag{10.23}$$

and the terms of second or higher order in $\triangle t$ have been dropped. Note that both of the retained averages really are proportional to $\triangle t$ because in a random walk process the mean square displacements increase linearly with time.

Using (10.22) as the collision term in the Boltzmann equation we have the *Fokker–Planck equation*. Thus far we have nothing more than a formal equation and the hard task is to determine the correct form of the probability function $\psi$. The diffusion through Coulomb collisions is treated in many advanced plasma physics textbooks [e.g., Boyd and Sanderson, 2003]. We skip the technical derivation here but note that the first term in (10.22) describes the deceleration ($\propto \triangle\langle\mathbf{v}\rangle/\triangle t$) of a test particle due to collisions, i.e., dynamical friction. The second term is the diffusion term. This time the diffusion coefficient is $D_{\mathbf{vv}} \propto \langle\triangle\mathbf{v}\triangle\mathbf{v}\rangle/\triangle t$ because the diffusion takes place in the velocity space. Note further that the diffusion can take place both in the direction of the velocity, which in magnetized plasmas corresponds to *pitch angle diffusion* and in the absolute value of $v$ corresponding to *energy diffusion*.

## 10.4 Quasi-Linear Diffusion Through Wave–Particle Interaction

The Fokker–Planck theory is fundamentally a collisional theory but also the wave–particle interactions can be casted to the same formulation in the framework of *quasi-linear theory*. Quasi-linear theory is an intermediate stage between the linear kinetic theory discussed in Chap. 5 and a fully nonlinear plasma physics (e.g., shocks discussed in Chap. 11). In this context also the term *weak turbulence* is often used. The basic idea of the quasi-linear theory is to separate the wave growth or damping and the particle diffusion form each other. This is facilitated by considering the space-independent and fluctuating parts of the distribution function separately.

This separation is easiest to illustrate within the same model as used in the derivation of the Landau solution in Chap. 5, i.e., by considering electrostatic waves in unmagnetized plasma. The critical assumption is that evolution of the (electron) distribution function $f(\mathbf{r},\mathbf{v},t)$ takes place much more slowly than the oscillations of the growing waves. Thus we can separate $f$ to a slowly varying part $f_0$, which is the average of $f$ over the fluctuations, and to the fluctuating part $f_1$. We further assume that $f_0$ is spatially uniform. Thus we write

$$f(\mathbf{r},\mathbf{v},t) = f_0(\mathbf{v},t) + f_1(\mathbf{r},\mathbf{v},t)\,. \tag{10.24}$$

Now the Vlasov equation is

$$\frac{\partial f_0}{\partial t} + \frac{\partial f_1}{\partial t} + \mathbf{v}\cdot\frac{\partial f_1}{\partial \mathbf{r}} - \frac{e}{m}\mathbf{E}\cdot\frac{\partial f_0}{\partial \mathbf{v}} - \frac{e}{m}\mathbf{E}\cdot\frac{\partial f_1}{\partial \mathbf{v}} = 0 \tag{10.25}$$

and we need also the first Maxwell equation

$$\nabla\cdot\mathbf{E} = -\frac{e}{\varepsilon_0}\int f_1\, d\mathbf{v}\,. \tag{10.26}$$

The electron charge due to the slow variation of $f_0$ is assumed to be neutralized by the given background ion population.

The average of (10.25) over the rapid fluctuations, denoted by $\langle\ldots\rangle$ is

$$\frac{\partial f_0}{\partial t} = \frac{e}{m}\left\langle \mathbf{E}\cdot\frac{\partial f_1}{\partial \mathbf{v}}\right\rangle\,. \tag{10.27}$$

Only the nonlinear term has been retained because the averages of functions linear in $f_1$, including $\mathbf{E}$, are assumed to vanish. Thus (10.27) is the equation describing the evolution of $f_0$.

By subtracting (10.27) from (10.25) we get an equation for the rapid variations of $f_1$

$$\frac{\partial f_1}{\partial t} + \mathbf{v}\cdot\frac{\partial f_1}{\partial \mathbf{r}} - \frac{e}{m}\mathbf{E}\cdot\frac{\partial f_0}{\partial \mathbf{v}} = \frac{e}{m}\left(\mathbf{E}\cdot\frac{\partial f_1}{\partial \mathbf{v}} - \left\langle \mathbf{E}\cdot\frac{\partial f_1}{\partial \mathbf{v}}\right\rangle\right)\,. \tag{10.28}$$

In this equation we neglect the second-order nonlinear terms on the RHS as smaller than the linear terms on the LHS, which leads to

$$\frac{\partial f_1}{\partial t} + \mathbf{v} \cdot \frac{\partial f_1}{\partial \mathbf{r}} - \frac{e}{m} \mathbf{E} \cdot \frac{\partial f_0}{\partial \mathbf{v}} = 0 . \tag{10.29}$$

This is formally the same equation as the linearized Vlasov equation (5.11) with the exception that now $f_0$ is time-dependent according to (10.27).

From here on we continue in the same way as in the derivation of the Landau solution. Assuming, for simplicity, that there is only one pole in the complex $p$-plane corresponding to the (complex) frequency $\omega_0$ we find the fluctuating part of the distribution function in the $\mathbf{k}$-space

$$f_1(\mathbf{k}, \mathbf{v}, t) = \frac{ie\mathbf{E}(\mathbf{k}, t)}{m(\omega_0 - \mathbf{k} \cdot \mathbf{v})} \cdot \frac{\partial f_0}{\partial \mathbf{v}} , \tag{10.30}$$

where

$$\mathbf{E}(\mathbf{k}, t) = \frac{ie\mathbf{k} \exp(-i\omega_0 t)}{\varepsilon_0 k^2 (\partial K(\mathbf{k}, \omega)/\partial \omega)_{\omega_0}} \int \frac{f_1(\mathbf{k}, \mathbf{v}, 0)}{(\omega_0 - \mathbf{k} \cdot \mathbf{v})} d\mathbf{v} . \tag{10.31}$$

Finally, by substituting (10.30) and (10.31) to (10.27) and making the inverse Fourier transformation back to the $\mathbf{r}$-space the evolution of $f_0$ is obtained from the diffusion equation

$$\frac{\partial f_0}{\partial t} = \frac{\partial}{\partial v_i} D_{ij} \frac{\partial f_0}{\partial v_j} . \tag{10.32}$$

The components of the diffusion matrix $D_{ij}$ are given by

$$D_{ij} = \frac{ie^2}{m^2 \mathscr{V}} \int \frac{\langle E_i(-\mathbf{k}, t) E_j(\mathbf{k}, t) \rangle}{(\omega_0 - \mathbf{k} \cdot \mathbf{v})} d\mathbf{k} , \tag{10.33}$$

where $\mathscr{V}$ is the volume of the plasma.

Thus we have found how to calculate the diffusion of the distribution function in the velocity space if we can determine the spectrum of electric field fluctuations for a given wave mode $(\omega_0, \mathbf{k})$. In practical diffusion problems the calculations must be done numerically with realistic estimates for the wave amplitudes, preferably based on direct observations.

**Train your brain** by filling in the details of the derivation of (10.32).

For the physics of space storms the virtue of this academic electrostatic example is in its transparency. Space plasmas are magnetized, which makes the analytical treatment considerably more complicated. The general quasi-linear theory of velocity space diffusion due to small-amplitude waves in a magnetized plasma was presented by Kennel and Engelmann [1966]. A somewhat more reader-friendly discussion is given in the textbook by Lyons and Williams [1984].

After some rather tedious calculations Kennel and Engelmann [1966] ended up with the diffusion equation for $f_0$

$$\frac{\partial f_0}{\partial t} = \frac{\partial}{\partial \mathbf{v}} \cdot \left( \mathscr{D} \cdot \frac{\partial f_0}{\partial \mathbf{v}} \right) , \tag{10.34}$$

where the *diffusion tensor* $\mathscr{D}$ is defined by

$$\mathscr{D} = \lim_{\mathscr{V}\to\infty} \frac{1}{(2\pi)^3 \mathscr{V}} \sum_n \frac{q^2}{m^2} \int d^3k \frac{i}{\omega_{\mathbf{k}} - k_{\|} v_{\|} - n\omega_c} (\mathbf{a}_{n,\mathbf{k}})^* (\mathbf{a}_{n,\mathbf{k}}) \, . \tag{10.35}$$

Here the vectors $\mathbf{a}_{n,\mathbf{k}}$ contain information on the amplitude of the wave electric field and the polarization, the asterisk indicates the complex conjugate, $\omega_{\mathbf{k}}$ is the complex frequency corresponding to the wave vector $\mathbf{k}$, and $\|$ refers to the direction of the background magnetic field. Of course, finding out the polarizations and the frequencies $\omega_{\mathbf{k}}$ requires a numerical solution of the dispersion equation.

**Challenge your brain**

Perform the calculations in Kennel and Engelmann [1966] and derive (10.35). If you find it too challenging, at least write out the vectors $\mathbf{a}_{n,\mathbf{k}}$ using the expressions given in the paper but transforming them to the notations and units of this book.

Kennel and Engelmann [1966] went further to define a positive definite functional

$$H = \frac{1}{2} \sum_\beta \int d^3v \, f_{0\beta}^2 \, , \tag{10.36}$$

where $\beta$ is over the particle species, and proved that $dH/dt \leq 0$, which indicates that the diffusion brings the system to a marginally stable state for all wave modes. There was no assumption of a small growth rate and thus the same formalism describes both the non-resonant adiabatic diffusion and the *resonant diffusion* at the limit where the imaginary part of the frequency $\omega_{\mathbf{k}i} \to 0$. At this limit the singularity in the denominator of (10.35) is replaced by a delta function that picks up the waves for which

$$\omega_{\mathbf{k}r} - k_{\|} v_{\|} + n\omega_{c\alpha} = 0 \tag{10.37}$$

for some integer $n$, i.e., particles that are either in Landau resonance ($n = 0$) or in gyro harmonic resonance ($n \neq 0$) with the waves. All naturally occurring nearly linear waves in the magnetosphere can in practice be treated under the assumption of resonant diffusion.

A common assumption is that the particle distributions are gyrotropic and thus the problem can be formulated in the two-dimensional $(v_\perp, v_\|)$-space. As discussed, e.g., in Chap. 5 of Lyons and Williams [1984], it is a straightforward exercise to transform the diffusion equation to the $(v, \alpha)$-space, where it reads

$$\begin{aligned}\frac{\partial f}{\partial t} &= \nabla \cdot (\mathscr{D} \cdot \nabla f) \\ &= \frac{1}{v \sin\alpha} \frac{\partial}{\partial \alpha} \sin\alpha \left( D_{\alpha\alpha} \frac{1}{v} \frac{\partial f}{\partial \alpha} + D_{\alpha v} \frac{\partial f}{\partial v} \right) \\ &\quad + \frac{1}{v^2} \frac{\partial}{\partial v} v^2 \left( D_{v\alpha} \frac{1}{v} \frac{\partial f}{\partial \alpha} + D_{vv} \frac{\partial f}{\partial v} \right) ,\end{aligned} \tag{10.38}$$

where we have dropped the subscript 0 and denote the velocity distribution function by $f$. Thus the wave–particle interactions can cause diffusion both in the absolute value of the velocity, i.e., in energy ($W = mv^2/2$) and in pitch angle similar to the diffusion by Coulomb collisions. The details of the diffusion depend on the characteristics of the waves and the velocities of the particles. Generally the pitch angle scattering is the dominant effect.

## 10.5 Kinetic Equation with Fokker–Planck Terms

If all action integrals $\{J_i\}$ are adiabatic invariants, the kinetic equation for the phase space density averaged over the phase angles $\overline{f}(\{J_i\})$ (10.12) reduces to

$$\frac{\partial \overline{f}}{\partial t} + \sum_i \frac{\partial}{\partial J_i}\left[\left\langle \frac{dJ_i}{dt}\right\rangle_{\mathrm{v}} \overline{f}\right] = \sum_{ij} \frac{\partial}{\partial J_i}\left[D_{ij}\frac{\partial \overline{f}}{\partial J_i}\right] - \frac{\overline{f}}{\tau_q} + \overline{S}\,, \tag{10.39}$$

where $\langle dJ_i/dt\rangle_{\mathrm{v}}$ are the frictional transport coefficients and $D_{ij}$ the elements of the diffusion tensor, and we have added terms describing source and loss processes. $\tau_q$ is the lifetime of immediate loss processes (e.g., charge exchange) and $\overline{S}$ represents the drift-averaged sources of $\overline{f}$, e.g., beta decay or the process called *CRAND* (*cosmic ray albedo neutron decay*, see Sect. 14.2.1).

From here on we simplify the notation by dropping the bar above $f$ and $S$. Sometimes it is convenient to write the kinetic equation in some other coordinates $\{Q_i\}$ than the action integrals $\{J_i\}$. For example, we may want to use a coordinate system where the $D_{ij}$ becomes diagonal. A straightforward coordinate transformation results in

$$\frac{\partial f}{\partial t} + \frac{1}{\mathscr{J}}\sum_i \frac{\partial}{\partial Q_i}\left[\mathscr{J}\left\langle \frac{dQ_i}{dt}\right\rangle_{\mathrm{v}} f\right] = \frac{1}{\mathscr{J}}\sum_{ij} \frac{\partial}{\partial Q_i}\left[\mathscr{J}\tilde{D}_{ij}\frac{\partial f}{\partial Q_j}\right] - \frac{f}{\tau_q} + S\,, \tag{10.40}$$

where $\mathscr{J} = \det\{\partial J_k/\partial Q_l\}$ is the *Jacobian* of the transformation from the coordinates $\{J_k\}$ to coordinates $\{Q_l\}$ and $\tilde{D}_{ij}$ denotes the transformed diffusion coefficients.

**Train your brain**

Show that the Jacobian for the transformation from $\{J_i\} = \{\mu, J, \Phi\}$ to $\{Q_i\} = \{\mu, K, \Phi\}$ is $\mathscr{J} = 4\pi(2m_0^3\mu)^{1/2} \propto \mu^{1/2}$. Thus, if the magnetic moment is conserved, $\mathscr{J}$ is constant and can be canceled from (10.40).

In practice the first two action integrals are often adiabatic invariants but the third is not. In such cases the kinetic equation can still be averaged over the angular variables $\phi_1$ and $\phi_2$ but the convective derivative with respect to $(J_3, \phi_3)$ must be retained

$$\begin{aligned}\frac{\partial f}{\partial t} + \frac{dJ_3}{dt}\frac{\partial f}{\partial J_3} + \frac{d\phi_3}{dt}\frac{\partial f}{\partial \phi_3} \\ = -\sum_i \frac{\partial}{\partial J_i}(D_i f) + \sum_{i,j}\frac{\partial}{\partial J_i}\left(D_{ij}\frac{\partial f}{\partial J_i}\right) + S - Lo\,,\end{aligned} \tag{10.41}$$

where the friction coefficients $D_i$, diffusion coefficients $D_{ij}$, radial transport $\partial J_3/\partial t$ and azimuthal transport $\partial \varphi_3/\partial t$ are averaged over the gyration and bounce motion and $Lo$ indicates the loss processes.

In problems associated with space storms the kinetic equation appears in different disguises. For example the evolution of the proton distribution function in the inner radiation belt is sometimes given in the form

$$\frac{\partial f}{\partial t} = L^2 \frac{\partial}{\partial L}\left(\frac{D_{LL}}{L^2}\frac{\partial f}{\partial L}\right) + \frac{G(L)}{\mu^{1/2}}\frac{\partial f}{\partial \mu} - \Lambda f + S, \tag{10.42}$$

where terms on the RHS are: radial diffusion, Coulomb collisions, charge exchange and CRAND. Thus effect of Coulomb collisions is treated as friction, whereas the wave–particle interactions are embedded in $D_{LL}$.

### *Example: The RAM model*

One of the most advanced kinetic models of plasma transport at the time of writing this book is the ring current–atmosphere interactions model (RAM) (see Jordanova et al [2008] and references therein). The model solves the kinetic equation for relativistic electrons and major ion species ($H^+$, $He^+$, $O^+$) as a function of radial distance, magnetic local time, energy and pitch angle. It can handle time-dependent convective transport, radial diffusion and all major loss processes. As we will see in Chap. 14, among the most important wave modes for the dynamics of the inner magnetosphere are the plasmaspheric hiss and whistler mode chorus waves, which both depend on the plasma parameters in the plasmasphere. To address this RAM is coupled to a dynamical plasmasphere model.

RAM is a four-dimensional model the coordinates of which are the radial distance in the equatorial plane ($R_0$), geomagnetic east longitude ($\phi$), energy ($W$), and pitch angle at the equatorial plane ($\alpha_0$) represented in the following by its cosine ($\mu_0 = \cos\alpha_0$). The model is bounce-averaged ($R$ and $\alpha$ given in the equatorial plane) but not drift-averaged ($\phi$-dependence). The kinetic equation for the distribution function $f$ in these variables taking into account the relativistic effects is

$$\begin{aligned}\frac{\partial f}{\partial t} &+ \frac{1}{R_0^2}\frac{\partial}{\partial R_0}\left(R_0^2\left\langle\frac{dR_0}{dt}\right\rangle f\right) + \frac{\partial}{\partial \phi}\left(\left\langle\frac{d\phi}{dt}\right\rangle f\right) \\ &+ \frac{1}{\gamma p}\frac{\partial}{\partial W}\left(\gamma p\left\langle\frac{dW}{dt}\right\rangle f\right) + \frac{1}{h(\mu_0)\mu_0}\frac{\partial}{\partial \mu_0}\left(h(\mu_0)\mu_0\left\langle\frac{d\mu_0}{dt}\right\rangle f\right) \\ &= \left\langle\left(\frac{\partial f}{\partial t}\right)_{rd}\right\rangle + \left\langle\left(\frac{\partial f}{\partial t}\right)_{loss}\right\rangle .\end{aligned} \tag{10.43}$$

Here $p$ is the relativistic momentum of the particle and $\gamma$ the Lorentz factor. The averages $\langle\ldots\rangle$ are taken between the mirror points and $h(\mu_0) = l_b/(2R_0)$, where $l_b$ is the half-bounce path length.

The LHS of (10.43) describes the adiabatic drift of the charged particles and the RHS the diffusive transport $\langle(\partial f/\partial t)_{rd}\rangle$ and the loss processes. In practice the determination of the wave–particle diffusion coefficients is a major task when the model is applied, e.g., to the radiation belt or ring current problems. This is similarly true with other models attempting to model the wave–particle interactions properly.

# 11. Shocks and Shock Acceleration

Shock waves are common phenomena in fluid dynamics. When an obstacle moves faster than the velocity of the wave mode that transfers information in the medium, a shock is formed ahead of the obstacle. An example known to everyone is the sonic shock wave in the air caused by an aircraft moving faster than the speed of sound. This example examples is from the domain of collision-dominated neutral fluids, where the shocks are very thin, only a few collisional mean-free paths, and thus can be described as infinitesimal discontinuities in the mathematical description of the fluid flow.

In the collisionless space plasmas the question of shocks is much more subtle. For example, the collisional mean-free path in the solar wind at $1\,AU$ is of the order of $1\,AU$. A bow shock forming in front of a planetary magnetosphere cannot be that thick. Thus, while the solar wind was expected to be supersonic and super-Alfvénic, it remained unclear whether a bow shock would form, or not, until the first spacecraft observations. Today we know that there is a bow shock in front of all solar system bodies with either a magnetosphere or atmosphere exposed to the solar wind flow. Furthermore, there are shock structures within the solar wind itself when a fast enough interplanetary coronal mass ejection (ICME) propagates through the background wind, or when fast solar wind catches up slower solar wind and the corotating interaction region (CIR) steepens.

The existence of collisionless shocks is an example of collective electromagnetic behavior of the plasma. The microphysical description of shocks is difficult because they are inherently nonlinear phenomena. It is not evident what physical processes take the role of collisions in collisionless plasmas. If the ambient magnetic field is strong enough, the Larmor radii of the particles give characteristic length scales, but this works only perpendicular to the magnetic field. Along the magnetic field and in weakly magnetized plasmas the electron and ion *inertial lengths* ($c/\omega_{p\alpha}$) are natural scale lengths. As these are different for different particle species, charge separation electric fields and electric currents arise and further complicate the physics of collisionless shocks.

While difficult to treat theoretically and extremely challenging for numerical simulations, shocks are of utmost importance for the physics of space storms. They are responsible for the most effective particle acceleration in the solar corona and solar wind, they are essential to the interaction of the solar wind with the magnetosphere, and they also have an important role in macroscopic reconnection models as we saw in Chap. 8.

H.E.J. Koskinen, *Physics of Space Storms: From the Solar Surface to the Earth*,
Springer Praxis Books, DOI 10.1007/978-3-642-00319-6_11,

## 11.1 Basic Shock Formation

Before going to shock formation in magnetized plasmas it is useful introduce some of the basic concepts within the framework of neutral fluids.

### 11.1.1 Steepening of continuous structures

The shock waves are generated by steepening of large-amplitude compressive disturbances. We start by studying the steepening in a simple model describing a neutral gas. Assume that the system has a one-dimensional geometry, i.e., the spatial changes are in the $x$-direction and the velocity of the gas is also directed along the $x$-axis. Thus, we can write the continuity and momentum equations as

$$\frac{\partial \rho}{\partial t} + \frac{\partial}{\partial x}(\rho V) = 0 \tag{11.1}$$

$$\rho \left( \frac{\partial V}{\partial t} + V \frac{\partial V}{\partial x} \right) + \frac{\partial P}{\partial x} = 0\,. \tag{11.2}$$

Let us close this set of equations by considering a cold gas ($P = 0$). The effect we are going to demonstrate exists also in a finite temperature, but the treatment is mathematically a bit more cumbersome. For isothermal gas the assumption of a negligible pressure gradient is equivalent to the assumption of high sonic Mach number, $M_s = V/v_s$, of the flow. The assumption of negligible $\mathbf{J} \times \mathbf{B}$ force, on the other hand, is realized in many magnetized plasma configurations as well. Both assumptions are reasonable, e.g., in the solar wind, so this simple example is of direct relevance to our topic.

Under these assumptions the flow speed fulfills the equation

$$\frac{\partial V}{\partial t} + V \frac{\partial V}{\partial x} = 0\,. \tag{11.3}$$

This equation can be solved in the closed form. Consider the curves

$$\frac{dx}{dt} = V \tag{11.4}$$

in the $(x,t)$-plane. The total time derivative of $V$ along these curves is

$$\frac{dV}{dt} = \frac{\partial V}{\partial t} + \frac{dx}{dt}\frac{\partial V}{\partial x} = \frac{\partial V}{\partial t} + V \frac{\partial V}{\partial x}\,. \tag{11.5}$$

Thus, if $V$ is a solution of (11.3), we know that $V$ is a constant along the *characteristic curves* given by (11.4). Because $V$ is a constant along the curves, the integral of the equation is trivially $x = x_0 + Vt$, where $x_0$ is the position of the fluid element at time $t = 0$. $x_0$ labels the characteristic curves. The solution of (11.3) can now be written as

$$V = V_0(x_0) = V_0(x - Vt)\,, \tag{11.6}$$

where $V_0(x) = V(x,0)$ is the initial velocity profile as a function of $x$.

Consider, as an example, an initial velocity profile (Fig. 11.1)

$$V_0(x) = \begin{cases} V_1, & x < -L \\ \langle V \rangle - \Delta V \frac{x}{2L}, & -L \leq x \leq L \\ V_2, & x > L, \end{cases} \tag{11.7}$$

where $V_1 > V_2$, $\Delta V = V_1 - V_2$, and $\langle V \rangle = (V_1 + V_2)/2$.

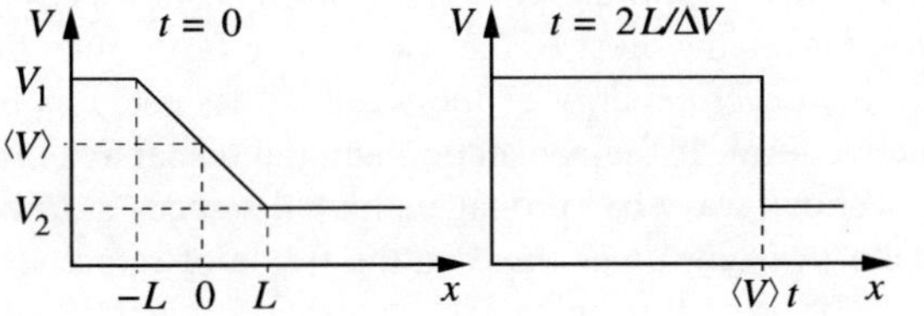


**Fig. 11.1** Steepening of an initially continuous velocity profile to a discontinuity.

The originally continuous profile describes two homogeneous parts of the fluid in relative motion combined with a linearly decreasing part of the velocity profile. The profile evolves as

$$V(x,t) = \begin{cases} V_1, & x < V_1 t - L \\ \langle V \rangle - \Delta V \dfrac{x - \langle V \rangle t}{2L - \Delta V t}, & V_1 t - L \leq x \leq V_2 t + L \\ V_2, & x > V_2 t + L. \end{cases} \tag{11.8}$$

The wave structure separating the two parts of the fluid moves at speed $\langle V \rangle$, while its width $2L - \Delta V t$ decreases linearly with time. Thus the wave steepens until at time $t^* = 2L/\Delta V$ a discontinuity forms at

$$x = V_2 t^* + L = V_1 t^* - L = L \frac{V_1 + V_2}{V_1 - V_2}. \tag{11.9}$$

The discontinuity separates two parts of the fluid with speeds $V_1$ and $V_2$. From that point on our toy model is no more adequate to describe the wave because effects neglected at the beginning (pressure gradient, heat conduction, and viscosity) become important at large gradients.

Note that the initial velocity profile is not critical for the result. If we had used, e.g., a sinusoidal wave, we would have found that after some time the profile would have an infinite $x$ derivative at its steepest points and from that point on discontinuities would develop in the flow. The crucial point is that fluid elements retain their velocities during the propagation. Thus, if the initial velocity profile has high-speed elements following elements of lower speeds, the faster are bound to overtake the slower after long enough time. Consequently, hydrodynamic equations have a built-in tendency for large-amplitude perturbations to develop shock waves due to the nonlinear character of the Euler equation (11.2).

### 11.1.2 Hydrodynamic shocks

In shock studies the choice of an appropriate frame of reference is critical. A shock in the solar wind may propagate either backward or forward in the solar wind frame. Because the solar wind is supersonic, the shock in both cases most likely passes an observer into the downwind direction. On the other hand, a planetary bow shock is stationary in the rest frame of the planet and thus propagates fast against the solar wind flow.

In hydrodynamics the basic mode of propagation is the sound wave. As seen in the previous section, if the wave amplitude for some reason becomes large, the nonlinear term in the Euler equation makes the crest of the wave move faster than the trough. The wave steepens and finally the excess energy of the wave is dissipated as heat. The steepening is due to the convective term. If the convection and the dispersive properties of the wave balance each other, a shock wave can propagate long distances in form of a *soliton*.

Let us consider the *hydrodynamic shock* in the frame of reference of the shock itself. Assume the shock to be very thin in the relevant hydrodynamical scales. The "ahead" or "upstream" region is denoted by subscript 1 and the "behind" or "downstream" by 2. The thermal energy per unit mass is denoted by $U = P/[(\gamma-1)\rho]$. Conservation of mass, momentum, and energy gives the relationships

$$\rho_2 V_2 = \rho_1 V_1 \tag{11.10}$$

$$P_2 + \rho_2 V_2^2 = P_1 + \rho_1 V_1^2 \tag{11.11}$$

$$P_2 V_2 + \left(\rho_2 U_2 + \frac{1}{2}\rho_2 V_2^2\right) V_2 = P_1 V_1 + \left(\rho_1 U_1 + \frac{1}{2}\rho_1 V_1^2\right) V_1 \,. \tag{11.12}$$

These equations are often written using the notation $[f] = f_1 - f_2$, e.g., $[\rho V] = 0$. They are known as *Rankine–Hugoniot relations* and they can be expressed as jumps of the parameters over the shock layer, such as

$$\frac{\rho_2}{\rho_1} = \frac{(\gamma+1)M_1^2}{2+(\gamma-1)M_1^2} \tag{11.13}$$

$$\frac{V_2}{V_1} = \frac{2+(\gamma-1)M_1^2}{(\gamma+1)M_1^2} \tag{11.14}$$

$$\frac{P_2}{P_1} = \frac{2\gamma M_1^2 - (\gamma-1)}{\gamma+1} \,, \tag{11.15}$$

where $M_1 = V_1/v_{s1}$ is the *sonic Mach number* on the upstream side $v_{s1} = \sqrt{\gamma P_1/\rho_1}$ and $\gamma$ is the polytropic index. Thermodynamics tells us that the entropy $S = c_V \log(P/\rho^\gamma)$ cannot decrease, $S_2 \geq S_1$. The equality holds for same conditions on both sides, i.e., when there actually is no shock. From these conditions we can infer the following properties of hydrodynamic shocks

1. $M_1 \geq 1$, i.e., $V_1 \geq v_{s1}$ ahead of the shock
2. $V_2 \leq v_{s2}$, flow is subsonic behind the shock
3. $P_2 \geq P_1$ and $\rho_2 \geq \rho_1$, the shock is compressive

4. $V_2 \leq V_1$ and $T_2 \geq T_1$, the flow is slowed down and the gas heated up
5. $1 \leq \rho_2/\rho_1 < (\gamma+1)/(\gamma-1)$, the maximum compression ratio is $(\gamma+1)/(\gamma-1)$, but the pressure increases $\propto M_1^2$ due to heating.

## 11.2 Shocks in MHD

In a collisional fluid the steepening of the wave front continues until *dissipation* balances the convection. The waves in collisionless plasmas are not dissipative and the balance must be obtained between the convective and *dispersive* properties of the waves. As we shall see below different dispersion properties of different wave modes lead to the escape of fluctuations from the shock front into both the upstream and the downstream direction, which makes the shock structures both complicated and spatially extended. And in fact, the individual wave modes with associated particle dynamics also limit the applicability of the MHD theory in detailed shock studies.

As we have seen in Chap. 6, there are three different MHD wave modes: the slow, the intermediate (shear Alfvén), and the fast mode. The shear Alfvén mode is not compressive and can thus have a large amplitude without steepening. Consequently, it does not form shocks, whereas the compressive slow and fast modes do.

### 11.2.1 Perpendicular shocks

The angle $\theta$ between the shock normal and the magnetic field is important. The simplest case is the perpendicular shock ($\theta = \pi/2$). In that case the magnetic field lines are in a plane parallel to the shock (Fig. 11.2) and the flow arrives along the direction of the shock normal. This shock resembles the hydrodynamic case.

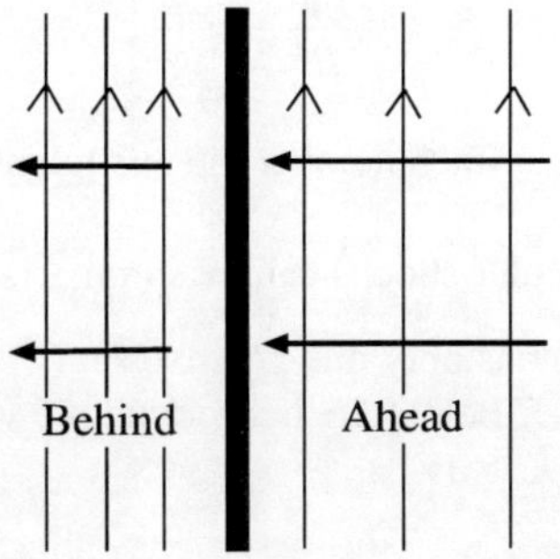


**Fig. 11.2** Perpendicular shock. The thick arrows indicate the flow direction and the thinner lines the magnetic field direction.

Now the Rankine–Hugoniot relations describing the mass, momentum, energy, and magnetic flux conservation are

$$\rho_2 V_2 = \rho_1 V_1 \tag{11.16}$$

$$P_2 + \rho_2 V_2^2 + \frac{B_2^2}{2\mu_0} = P_1 + \rho_1 V_1^2 + \frac{B_1^2}{2\mu_0} \tag{11.17}$$

$$\left(P_2 + \frac{B_2^2}{2\mu_0}\right) V_2 + \left(\rho_2 U_2 + \frac{1}{2}\rho_2 V_2^2 + \frac{B_2^2}{2\mu_0}\right) V_2 =$$

$$\left(P_1 + \frac{B_1^2}{2\mu_0}\right) V_1 + \left(\rho_1 U_1 + \frac{1}{2}\rho_1 V_1^2 + \frac{B_1^2}{2\mu_0}\right) V_1 \tag{11.18}$$

$$B_2 V_2 = B_1 V_1 \,. \tag{11.19}$$

From these we find the jumps

$$\frac{V_2}{V_1} = \frac{\rho_1}{\rho_2} = \frac{1}{X} \tag{11.20}$$

$$\frac{B_2}{B_1} = X \tag{11.21}$$

$$\frac{P_2}{P_1} = \gamma M_1^2 \left(1 - \frac{1}{X}\right) - \frac{1 - X^2}{\beta_1} \,, \tag{11.22}$$

where the *compression ratio* $X = \rho_2/\rho_1$ is the positive root of

$$2(2-X)X^2 + [2\beta_1 + (\gamma - 1)\beta_1 M_1^2 + 2]\gamma X - \gamma(\gamma+1)\beta_1 M_1^2 = 0 \tag{11.23}$$

and the upstream plasma beta $\beta_1$ is given by

$$\beta_1 = \frac{2\mu_0 P_1}{B_1^2} = \frac{2v_{s1}^2}{\gamma v_{A1}^2} \,. \tag{11.24}$$

Thus in addition to the upstream Mach number the upstream $\beta$ is a characteristic parameter of a shock.

The properties of perpendicular shocks can be summarized as

1. Because $1 < \gamma < 2$, (11.23) has only one positive root.
2. The magnetic field reduces $X$ below the hydrodynamic value.
3. The shock is compressive ($X \geq 1$).
4. $V_1 \geq v_{ms} \equiv \sqrt{v_{s1}^2 + v_{A1}^2}\,.$
5. The magnetic compression is limited to $1 < B_2/B_1 < (\gamma+1)/(\gamma-1)\,.$

### 11.2.2 Oblique shocks

In case of oblique propagation the upstream **V** and **B** can be at any angle with respect to each other. Thus it is convenient to transform to a coordinate system known as the *de Hoffmann–Teller (dHT) frame* [de Hoffmann and Teller, 1950]. It is a frame moving on the shock plane with such a velocity that the upstream convective electric field disappears, i.e., $\mathbf{V}_1 \times \mathbf{B}_1 = 0$. Such a coordinate transformation generally exists, except in the case of the exactly perpendicular shock discussed above. In reality, space plasma shocks are seldom, if ever, exactly perpendicular.

In the dHT frame the problem is two-dimensional and we denote the component normal to the the shock front by $x$ and the component on the shock plane by $y$ (Fig. 11.3).

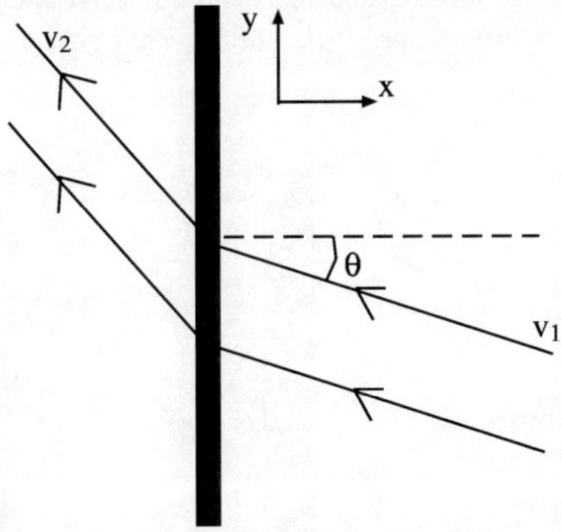


**Fig. 11.3** The shock geometry for arbitrary orientation.

The jump conditions for the oblique shock become

$$\frac{V_{2x}}{V_{1x}} = \frac{\rho_1}{\rho_2} = \frac{1}{X} \tag{11.25}$$

$$\frac{V_{2y}}{V_{1y}} = \frac{V_1^2 - v_{A1}^2}{V_1^2 - Xv_{A1}^2} \tag{11.26}$$

$$\frac{B_{2x}}{B_{1x}} = 1 \tag{11.27}$$

$$\frac{B_{2y}}{B_{1y}} = \frac{(V_1^2 - v_{A1}^2)X}{V_1^2 - Xv_{A1}^2} \tag{11.28}$$

$$\frac{P_2}{P_1} = X + \frac{(\gamma-1)XV_1^2}{2v_{s1}^2}\left(1 - \frac{V_2^2}{V_1^2}\right) . \tag{11.29}$$

Now the compression ratio $X = \rho_2/\rho_1$ is found as a solution of

$$(V_1^2 - Xv_{A1}^2)^2 \left[ Xv_{s1}^2 + \frac{1}{2} V_1^2 \cos^2\theta (X(\gamma-1) - (\gamma+1)) \right] \tag{11.30}$$
$$+\frac{1}{2} v_{A1}^2 V_1^2 \sin^2\theta X \left[ (\gamma + X(2-\gamma)) V_1^2 - X v_{A1}^2 ((\gamma+1) - X(\gamma-1)) \right] = 0 \,.$$

**Train your brain** by deriving the jump conditions for an oblique shock.

The shocks associated with slow, Alfvén, and fast modes look different (Fig. 11.4). It is important to note that the parallel (to the shock normal) component of the magnetic field ($B_x$) does not change over the shock. As already noted above the shear Alfvén wave is not compressive and thus does not steepen to a shock. This case is known as the *rotational discontinuity*.

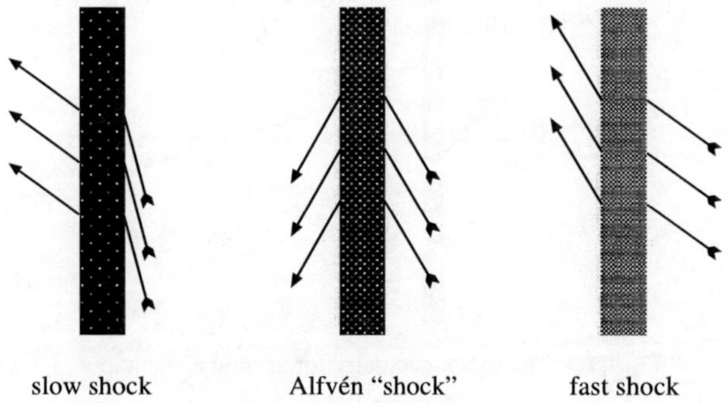


**Fig. 11.4** The magnetic field lines through MHD shocks.

The *slow* and *fast shocks* have following properties

1. They are compressive.
2. $B_x$ remains unchanged over the shock.
3. They conserve the sign of $B_y$ .
4. At the slow shock $B_2 < B_1$ .
5. At the fast shock $B_2 > B_1$ .
6. $V_{1x}$ exceeds the slow/fast speed ahead the shock while $V_{2x}$ is smaller than the slow/fast speed behind the shock.
7. $V_{2x} < V_{1x}$
8. At the limit $B_x \to 0$, the fast shock becomes a perpendicular shock whereas the slow shock becomes a tangential discontinuity ($V_x \to 0$) with arbitrary jumps in $V_y$ and $B_y$ subject to total pressure balance over the shock (see discussion below).

The shock at the limit $\theta \to 0$ is called the *parallel shock*. The arbitrary directions are often described as *quasi-parallel* and *quasi-perpendicular* depending on whether they are closer to parallel or perpendicular. The quasi-parallel shocks are more complicated than the quasi-perpendicular shocks because individual particles can be reflected from the shock and, in the quasi-parallel case, move long distances upstream leading to instabilities beyond the MHD description.

There are two special cases of the shocks.

1. If, in the case of the slow shock, the upstream speed is equal to Alfvén speed ($V_1^2 = v_{A1}^2$) but $X \neq 1$, the tangential component ($B_y$) of the downstream magnetic field vanishes. Such a shock is called a *switch-off shock* as it "switches off" the tangential component.

2. As for the fast parallel shock $X = V_1^2/v_{A1}^2 > 1$, the magnetic field becomes compressed behind the shock and, in order to keep the field divergence-free, it must bend. A parallel fast shock "switches on" the tangential component and is called a *switch-on shock*.

### 11.2.3 Rotational and tangential discontinuities

Not all MHD discontinuities satisfying the jump conditions are shocks. What is characteristic for a shock, is the mass flux and compression across the discontinuity.

At the rotational discontinuity (Fig. 11.4) there is no jump in the mass density ($[\rho_m] = 0$), nor in the velocity normal to the shock front ($[V_n] = 0$). Because $V_n = \pm B_n/\sqrt{\mu_0 \rho_m} \neq 0$, there is, however, mass flux through the discontinuity. The tangential component of the magnetic field changes its sign $\mathbf{B}_{t1} = -\mathbf{B}_{t2}$ and thus the magnetic field rotates across the discontinuity. Mass flux, magnetic rotation, and propagation at the speed $v_A$ are characteristic to reconnection when it tears the current sheet (Chap. 8).

A non-reconnecting current sheet is also a discontinuity. In such a case $\mathbf{B}_t$, $\mathbf{V}_t$, $\rho_m$ and $P$ can all be discontinuous, but there is no mass flux across the boundary, which now forms a *tangential discontinuity*. If the magnetosphere were completely closed, the magnetopause would be an ideal tangential discontinuity. This is, however, topologically impossible, and the formation of the polar cusps (Fig. 1.14) is unavoidable. When reconnection opens the magnetopause, it becomes a rotational discontinuity, also when reconnection takes place tailward of the cusps in the case of northward IMF. Recall, however, that the concepts of tangential and rotational discontinuities are introduced within ideal MHD and, as we have seen in Chap. 8, the actual structure of the reconnection region is already more complicated in the fluid picture.

For completeness, there is an even more simple MHD discontinuity called a *contact discontinuity*, where only the density of the plasmas on each side is different but the plasma flows on on both sides are identical. Contact discontinuities are of little interest to our study.

### 11.2.4 Thickness of the shock front

The thickness of a shock front $\delta x$ in a neutral gas scales as the collisional particle mean free path $\lambda$. Because this is far too large to account for the shock thickness in the interplanetary medium, there must be other means besides binary collisions to dissipate the ordered kinetic energy as heat at the solar system shocks and in astrophysical shocks in general.

In plasma dissipation is also provided by ohmic heating, in which the electromagnetic field does work on the particles and heats the population. The heating rate (i.e., heat input per unit time and unit volume) can be approximated according to (6.26) as

$$\frac{\delta W}{\delta t} \approx \frac{J^2}{\sigma}, \tag{11.31}$$

where $\delta$'s denote small increments. The current density in a shock wave is according to Ampère's law ($\mu_0 \mathbf{J} = \nabla \times \mathbf{B}$)

$$J \approx \frac{B_{2t} - B_{1t}}{\mu_0\, \delta x} \approx \frac{3B_{1t}}{\mu_0\, \delta x}. \tag{11.32}$$

The last approximation holds for a *strong shock* ($M_A \gg 1$) in a non-relativistic monoatomic gas, where the compression ratio is $X \approx 4$. For a strong shock we can estimate the dissipated energy density to be $\delta W \approx \rho_1 V_{1x}^2/2$, and the time available to dissipate it $\delta t \approx \delta x / V_{1x}$. Thus

$$\frac{\frac{1}{2}\rho_1 V_{1x}^2}{\delta x / V_{1x}} \approx \frac{9B_1^2 \sin^2\theta}{\sigma \mu_0^2 (\delta x)^2} \tag{11.33}$$

$\Rightarrow$

$$\delta x \approx \frac{18 B_1^2 \sin^2\theta}{\mu_0^2 \sigma \rho_1 V_{1x}^3} = \frac{18\tan^2\theta}{{M_A}^2} \frac{\eta}{V_{1x}}, \tag{11.34}$$

where $\eta = 1/\mu_0\sigma$ is the magnetic diffusivity of the plasma. Classical resistivity ($1/\sigma$) can be related to the electron mean free path $\lambda_e$ as

$$\frac{1}{\sigma} = \frac{m_e v_{the}}{n_e e^2 \lambda_e}. \tag{11.35}$$

Thus

$$\delta x \approx \frac{18\tan^2\theta}{{M_A}^2} \frac{\lambda_{De}}{\lambda_e} \frac{c}{V_{1x}} \frac{c}{\omega_{pe}}. \tag{11.36}$$

Because the mean free path in a collisionless plasma is always several orders of magnitude larger than the Debye length, this equation predicts a shock thickness that is much smaller than the electron inertial length (skin depth) $c/\omega_{pe}$ of the plasma. It is thus *much too small* to describe real shocks. Ohmic heating can therefore be considered as a relevant dissipation mechanism only if resistivity is substantially larger than its classical value. Such anomalous resistivity may be provided by wave–particle interactions or turbulence.

A simple way of obtaining a minimum thickness of a strong collisionless quasi-perpendicular shock is to consider particle orbits. We may argue that the shock front cannot be much thinner than the Larmor radius of the downstream ions, which have a thermal speed of about

$$v_{thi} \sim \sqrt{\frac{P_2}{\rho_2}} = \sqrt{\frac{\gamma-1}{\gamma}\frac{V_1^2-V_2^2}{2}} \approx \frac{V_{1x}}{\sqrt{5}}\sqrt{1-\frac{1}{X^2}} \approx \frac{\sqrt{3}}{4}V_{1x} \tag{11.37}$$

giving

$$\delta x \gtrsim \frac{\sqrt{3}V_{1x}}{4\,\omega_{ci}}, \tag{11.38}$$

where $\omega_{ci}$ is the ion cyclotron frequency in the downstream field. Note that this argument only applies to the thickness of the density structure and, in fact, assumes that the magnetic field structure will be thinner.

A model for weak collisionless shocks is obtained by considering nonlinear steepening of normal wave modes of the collisionless plasma. Relevant low-frequency wave modes propagating parallel to the magnetic field are the ion whistler waves, i.e., $L$ mode waves below the ion cyclotron frequency. The dispersion equation can be written at the limit $\omega \ll \omega_{ce}$ as

$$k^2 v_A^2 = \omega^2 \frac{\omega_{ci}}{\omega_{ci}+\omega}. \tag{11.39}$$

These waves are dispersionless at small wavenumbers, but become strongly dispersive as the wave frequency approaches $\omega_{ci}$, i.e., at $kv_A \sim \omega_{ci}$. The waves compress the magnetic field and are subject to nonlinear steepening.

**Train your brain**

Derive the dispersion equation (11.39) and show that it becomes dispersive at the limit $kv_A \sim \omega_{ci}$.

For wave numbers $k > \omega_{ci}/v_A$ the phase and group speeds of the waves are increasing functions of the wave number. This implies that the steepening of a large-amplitude (small $k$) wave stops at $k \sim \omega_{ci}/v_A$ because the Fourier components with larger $k$ propagate faster than the rest of the structure and escape from it. This means that the main shock transition is *preceded* by small-amplitude small-wavelength fluctuations in the flow. This predicts the shock thickness in the direction parallel to the magnetic field to be approximately

$$\delta x \sim \frac{v_A}{\omega_{ci}}, \tag{11.40}$$

if it results from steepening of an ion whistler wave (Fig. 11.5).

In the perpendicular direction, the relevant wave mode is the magnetosonic wave (fast Alfvén wave). In that case, dispersion becomes significant at $kv_A \sim \sqrt{\omega_{ci}\omega_{ce}}$, i.e., the magnetic field profile of the shock has the thickness of

$$\delta x \sim \frac{v_A}{\sqrt{\omega_{ci}\omega_{ce}}} \sim \frac{c}{\omega_{pe}}. \tag{11.41}$$

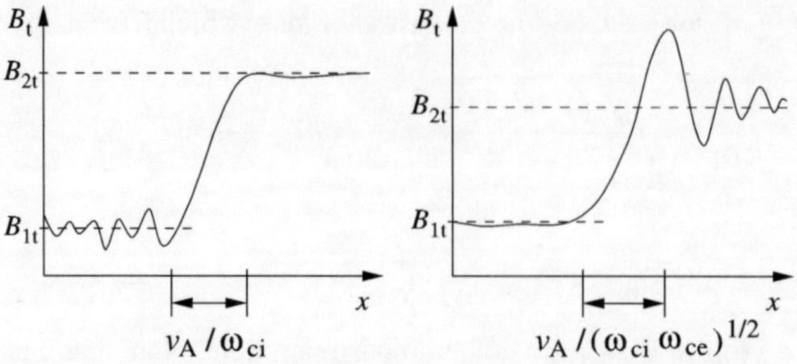


**Fig. 11.5** The transverse magnetic field profile across a quasi-parallel (left) and quasi-perpendicular (right) collisionless shock wave according to our qualitative model. (Figure by courtesy of R. Vainio.)

Thus, magnetic shock structures are predicted to be thinner in the perpendicular than in the parallel direction and, in fact, much thinner than the ion Larmor radius, as was required in the discussion of the density structure above. Another difference between the shock structures obtained this way is that the large-$k$ magnetosonic waves propagate slower than the small-$k$ waves, which means that steepening results in the large-$k$ small-amplitude waves *trailing the perpendicular shock* (Fig. 11.5).

Finally, we note that in order for a collisionless shock wave to represent a super- to sub-magnetosonic transition between two uniform states, there has to be a collisionless (microscopic) dissipation process operating at the shock front because entropy has to increase and this is not possible without dissipation. Dispersion and steepening together without dissipation lead typically to soliton-like waves.

### 11.2.5 Collisionless shock wave structure

Collisionless shocks have many features that are beyond the MHD description. In the absence of collisions the ion and electron populations may have different temperatures downstream of the shock. They may equilibrate, at least partially, via collisionless interactions with fluctuating electromagnetic fields (plasma waves or turbulence) that are always present behind strong shocks. Strong collisionless shocks can also be very efficient particle accelerators, indicating that a major fraction of particle pressure can be carried by non-thermal particles that do not obey fluid equations.

Intuitively, a proton incident on a shock front can penetrate much deeper into the shock structure than an electron due to its much larger inertia. This leads to charge separation close to the thin magnetic front, called the *shock ramp*. The charge separation corresponds to an electric field in the ramp. It decelerates the ions and accelerates the electrons to counteract the effect of inertia. The charge separation field is electrostatic

$$E_x = -\frac{d\varphi}{dx}, \tag{11.42}$$

where the potential $\varphi$ is constant outside the shock front. The potential difference $\Delta\varphi = \varphi_2 - \varphi_1$ across the shock structure can be expressed as a fraction $a$ of the incident ion kinetic energy,

$$e\Delta\varphi = \frac{a}{2} m_p V_{1x}^2 \,. \tag{11.43}$$

If the effect of the electric field is to slow down the ions, the fraction $a$ is notable.

When such an electric field is present in the ramp, some of the protons incident on the shock from the upstream region may be reflected by the electric field, i.e., those whose velocity component along the shock normal is less than $\sqrt{a}V_{1x}$. A quasi-perpendicular shock front can reflect ions at large numbers if its Mach number exceeds a critical value at which the downstream plasma becomes subsonic. Thus, if $V_{1x}/v_A > M_c$, where the *critical Mach number* $M_c$ is between 1.1 and 2.2, depending on the upstream plasma parameters, the shock front reflects ions back to the upstream medium. Such a shock is said to be *supercritical*. The reflected ions drag electrons along with them and create a *shock foot* ahead of the ramp, where the magnetic field already starts to increase. The thickness of the foot is typically of the order of the ion Larmor radius as discussed above. On the downstream side the collisionless dissipation process operates on the large-$k$ fluctuations trailing the shock wave (Fig. 11.5) and the amplitude of these fluctuations decreases. Thus, the magnetic field typically *overshoots* just behind the shock front, i.e., it exceeds the downstream $B_2$ far from the shock. This is also the prediction of the fully nonlinear calculation of the shock structure.

***Example: The Earth's bow shock***

When the supermagnetosonic solar wind hits the magnetosphere of the Earth, a collisionless shock front is formed in the solar wind ahead of the magnetopause (Fig. 11.6). All magnetized planets have essentially similar bow shocks. A corresponding structure can also be found in interplanetary shocks, but not all features listed below accompany all interplanetary shocks. During the early years of the 21st century the *Cluster* mission has produced a large amount of very detailed observations of the bow shock. We have to skip any systematic discussion of these, but the interested reader is referred to the reviews by Bale et al [2005] on the quasi-perpendicular and Burgess et al [2005] on the quasi-parallel shocks.

The shock wave has a large overall structure. In most regions the shock is supercritical. About 1% of the incident solar wind energy flux on the shock is transferred to suprathermal particles, and the reflected ions take most of this energy. The reflected ions are observed as beams propagating sunward along the magnetic field lines at a velocity of about

$$|V_{b\parallel}| \gtrsim \frac{V_{1x}}{\cos\theta} = V_{sw}\frac{\cos(\theta-\psi)}{\cos\theta} \,. \tag{11.44}$$

Here $V_{sw}$ is the radial solar wind speed, $\psi$ is the angle between the radial direction and the direction of the magnetic field and $\theta$ is the angle between the shock normal and the flow direction, and we have assumed, for simplicity, that the shock normal is in the plane defined by the radial direction and the magnetic field (region F in Fig. 11.6). Note that the beam velocity is measured in the upstream plasma frame. The ions escape from the shock approximately tangentially in the shock frame. Ion reflection from the bow shock seems to switch on at regions, where the shock normal angle falls below ~70°. Thus, the *ion*

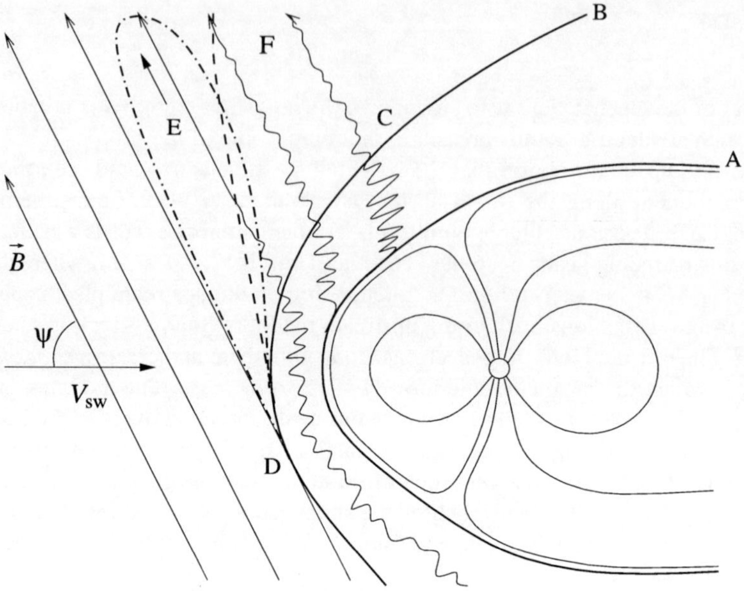


**Fig. 11.6** Structure of a planetary bow shock. Thin solid curves depict magnetic field lines. The thick curve labeled 'A' is the magnetopause. The thick curve labeled 'B' is the bow shock ahead of the magnetopause in the solar wind, which flows with velocity $V_{sw}$ at an angle of $\psi$ relative to the IMF. In the region near point 'C' the shock is quasi-parallel, and in the region near point 'D' it is quasi-perpendicular. Upstream of the bow shock there are regions 'E' and 'F' bounded by the bow shock and the dashed–dotted and the dashed curves. These are the electron and the ion foreshock regions, respectively. The arrow inside region E indicates that the electron foreshock can extend to very large distances from the bow shock. (Adapted from Benz [2002] by R. Vainio.)

*foreshock* extends from this point of the bow shock tangentially outward to the solar wind. The ion beam velocities correspond to proton energies of a few keV.

Also a diffuse ion population can be observed in the foreshock with a broad energy spectrum extending beyond 100 keV. These ions either are accelerated at the shock wave or they may be originally foreshock ions that have undergone stochastic acceleration in the upstream fluctuations.

Electron beams are also observed upstream of the bow shock. They originate from the quasi-perpendicular region of the shock and have plasma frame velocities much larger than $V_{sw}$, corresponding to proton energies of 1–2 keV. As a 1-keV electron is 43 times faster than a 1-keV proton, the electrons move upstream in an almost field-aligned direction, and the boundary of the *electron foreshock* is almost tangential to the upstream field lines. The region is limited to the field lines connecting to quasi-perpendicular regions of the shock (region E in Fig. 11.6). These electrons are probably accelerated at the shock wave by the shock drift mechanism, discussed in the next section.

The supra-thermal particle populations upstream the bow shock are a source of free energy and can drive a variety of plasma waves unstable. The exact relationship between the different plasma wave and particle populations is, however, far from clear. Low-frequency

MHD waves (magnetosonic and Alfvén waves) observed in the foreshock region are probably produced by the diffuse ion population through streaming instabilities. Whistler waves are produced by the ion beams and/or electron populations having large temperature anisotropies ($T_\perp > T_\parallel$). Also various electrostatic waves, e.g., Langmuir waves generated by electron beams and ion–acoustic waves caused by the diffuse ions, are observed in the foreshock region.

## 11.3 Particle Acceleration in Shock Waves

Particle acceleration in shock waves is the most widely accepted model to account for cosmic rays, i.e., relativistic charged particles bombarding the Earth's atmosphere from space. The cosmic ray spectrum at energies below $\sim 10^{15}$ eV in the near-Earth space has three main components, all of which are believed to be accelerated in shock waves: (1) *galactic cosmic rays* (GCR) are accelerated most likely in supernova remnant shock waves in our galaxy; (2) *anomalous cosmic rays* (ACR) are accelerated in the heliospheric termination shock; and (3) *solar cosmic rays* (SCR) are accelerated in coronal and interplanetary shocks related to solar eruptions.

Below $\sim$10 GeV GCR and ACR fluxes are modulated by the 11- and 22-year solar cycles, so they are quasi-stationary cosmic-ray components in the time scales of space storms. SCRs, in contrast, are observed in transient events related to solar flares and coronal mass ejections. GCRs and SCRs are mainly protons whereas ACRs mainly consist of heavier nuclei, such as helium and oxygen.

**Feed your brain** by finding out what literature or internet sources tell about the origin of ACRs.

There are also three types of cosmic-ray electrons: GCR and SCR electrons, and the *Jovian electrons* that originate from the magnetosphere of Jupiter and can be observed near the Earth at intervals of about 13 months when the Earth and Jupiter are magnetically connected by the Parker spiral structure. Supernova shock waves are most probably the source of the accelerated GCR electrons, whereas in the acceleration of SCR and Jovian electrons other mechanisms than shocks are also important, in particular inductive electric fields associated with solar flares and reconnection in the Jovian magnetosphere.

In addition to in situ observations, energetic electrons can be observed remotely through the radiation they produce. Radiating accelerated electron populations can be found in almost all astrophysical objects where we expect violent processes to occur. For example, solar flares, supernova remnants, and astrophysical jets are strong sources of non-thermal radiation generated by accelerated electrons.

Particle acceleration in shock waves takes places through the Fermi mechanism (Sect. 3.2.4). In this process, particles gain energy by reflecting off magnetic irregularities carried by the plasma flow that converges at the shock. There are two main forms of this acceleration process, the *shock drift acceleration*, known also a *fast Fermi accelera-*

*tion* and *diffusive shock acceleration.* In addition, there is a rapid acceleration mechanism that relies on the cross-shock electric potential called *shock surfing mechanism.*

### 11.3.1 Shock drift acceleration

Shock drift acceleration occurs, when a particle interacts once with a quasi-perpendicular shock front (Fig. 11.7). The particle drifts due to the electric field $\mathbf{E} = -\mathbf{V}_1 \times \mathbf{B}_1$ with the upstream speed $V_{1x}$ toward the shock. When the ion or electron hits the shock front, it starts to gyrate in the stronger downstream magnetic field. This means that its Larmor radius is smaller in the downstream side than in the upstream side and its guiding center drifts parallel (ion) or anti-parallel (electron) to the electric field, and the particle gains energy. When interacting with the shock wave, the particle conserves its first adiabatic invariant[1] $\propto p_\perp^2/B$. The maximum gain in energy is obtained when $p_\parallel$ vanishes. For a perpendicular shock

$$p_2 = p_1\sqrt{\frac{B_2}{B_1}} = p_1\sqrt{X}\,, \tag{11.45}$$

where $X$ is the shock compression ratio. So the particle momentum increases by a factor of approximately $\sqrt{X}$ across the shock.

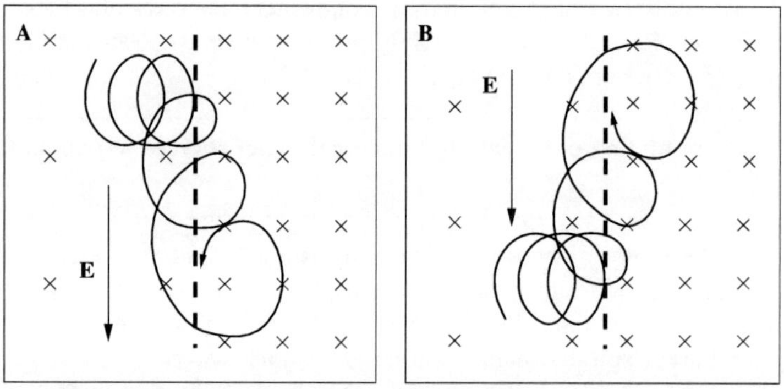


**Fig. 11.7** Shock drift acceleration. An energetic charged particle is convected to a quasi-perpendicular shock from upstream by the electric-field drift. In the shock front ions drift parallel (A) and electrons drift anti-parallel (B) to the electric field and both gain energy. (Figure by courtesy of R. Vainio.)

In a parallel shock the magnetic field is not compressed, so there is no shock drift acceleration. In oblique shocks, particle interaction with the shock wave is most easily treated in the dHT frame, where the electric field vanishes. In that frame the particle energy is constant, but changes in the direction of the particle motion can lead to substantial energy gain when viewed in the upstream rest frame (recall that the dHT frame is a moving frame). Conservation of the first adiabatic invariant requires $p_\perp^2/B$ to be constant. If the

[1] As cosmic rays often are relativistic particles, it is more appropriate to use momentum than velocity in the mathematical expressions.

total momentum $p$ is also constant, $\sin^2\alpha/B$ is constant. Clearly, if $\sin^2\alpha > B_1/B_2$, the particle cannot get into the downstream region, and such particles are reflected back to the upstream region. Thus the particle momentum parallel to the magnetic field changes by

$$|\Delta p_{\parallel}| = 2p|\cos\alpha| \leq 2p\sqrt{\frac{B_2-B_1}{B_2}}\,. \tag{11.46}$$

In the upstream rest frame the change in the parallel momentum is

$$\begin{aligned} |\Delta p'_{\parallel}| &= |\Delta p_{\parallel}|/\sqrt{1-(V_{1x}/\cos\theta_1)^2/c^2} \\ &\leq 2p\sqrt{\frac{B_2-B_1}{B_2(1-(V_{1x}/\cos\theta_1)^2/c^2)}}\,. \end{aligned} \tag{11.47}$$

The momentum gain in this case can be shown to be at most of the same order as in the perpendicular shock. Thus shock drift acceleration produces $\Delta p' \lesssim p'$. From (11.47) it is evident that slow shocks with $B_1 > B_2$ do not accelerate particles by the shock drift acceleration mechanism.

### 11.3.2 Diffusive shock acceleration

One encounter with the shock does not lead to a substantial particle acceleration. If, however, particles can interact with the shock many times, acceleration becomes more efficient. The particle's interaction with magnetic irregularities in the plasma flow can change its propagation direction relative to the shock front enabling several encounters with the shock. As particle propagation in this case resembles diffusion relative to the local plasma flow, the mechanism is called *diffusive shock acceleration.*

Diffusive shock acceleration is easiest to understand by considering parallel shock waves because then the particle's velocity vector does not change when it crosses the shock front. When the particle is moving relative to the plasma under the influence of frozen-in magnetic scattering centers, its energy is conserved in the *local plasma frame* while simultaneously its pitch angle changes. Upstream (downstream) particles are thus confined in the velocity space on semicircles centered at $(v_{\parallel}, v_{\perp}) = (V_{1(2)}, 0)$ (Fig. 11.8) Due to pitch angle scattering energetic ($v' > V$) particles can propagate in either direction relative to the shock. When the flow speed at the shock decreases ($V_2 < V_1$), particles crossing the shock many times systematically gain speed as shown in Fig. 11.8. The figure is drawn for a non-relativistic particle, but the mechanism is valid also for relativistic particles.

At large energies ($v \gg V$) the scattering leads to almost isotropic distribution. This enables us to obtain the energy spectrum of accelerated particles resulting from diffusive shock acceleration by calculating the mean particle momentum

$$\langle p_n \rangle = p_0 \exp\left(\frac{4}{3}\sum_{j=1}^{n}\frac{\Delta V}{v_j}\right) \tag{11.48}$$

after $n$ shock crossings and the probability

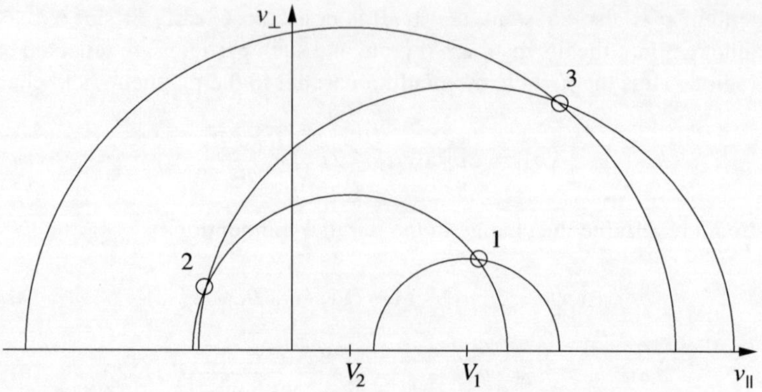


**Fig. 11.8** Diffusive shock acceleration. An energetic charged particle scatters off magnetic irregularities frozen-in to the local plasma flow. The numbered points depict successive crossings of the shock front, where the speed of the scattering centers changes. Because points with odd numbers must have $v_\parallel > 0$ and points with even numbers must have $v_\parallel < 0$, the shock crossings lead to a systematic gain of energy $W \propto v^2$. (Figure by courtesy of R. Vainio)

$$P_n = \exp\left(-4\sum_{j=1}^{n}\frac{V_2}{v_j}\right) \tag{11.49}$$

of a particle performing at least $n$ crossings of the shock (for details, see Drury [1983]). By combining these, the differential momentum spectrum can be given as

$$\frac{dN}{dp} = \frac{3N_0}{(X-1)p_0}\left(\frac{p_0}{p}\right)^{(X+2)/(X-1)}, \tag{11.50}$$

where $N_0$ is the total number of particles injected to the acceleration process and $p_0 \gg mV_1$ is the injection momentum. Thus, shock-accelerated particles have a power law spectrum (in momentum) with the spectral index

$$\sigma = \frac{d\ln N}{d\ln p} = \frac{(X+2)}{(X-1)}, \tag{11.51}$$

which is determined by the compression ratio only. Note that this result applies to oblique shocks as well [Drury, 1983]. The energy spectrum of accelerated particles behind the shock wave is

$$\frac{dN}{d\varepsilon} = \frac{1}{v}\frac{dN}{dp} = \frac{3N_0}{(X-1)p_0c}\left(\frac{p_0}{mc}\right)^{\sigma}\frac{\Gamma}{(\Gamma^2-1)^{(\sigma+1)/2}}, \tag{11.52}$$

where $\varepsilon = \Gamma mc^2$ is the total energy of the particle. We denote here the Lorentz factor by $\Gamma$ in order not to mix up with the polytropic index. At relativistic energies, therefore,

$$\frac{dN}{d\varepsilon} \propto \varepsilon^{-\sigma} , \tag{11.53}$$

which is the result often applied to electron spectra, when calculating their emission.

The spectral index $\sigma$ is actually determined by the shock's compression ratio only if $M \gg 1$. If the Mach number of the shock is of the order of unity, the magnetic scattering centers in the flow (MHD waves) can no more be seen as static magnetic fluctuations, but have substantial phase speeds $v_p \sim v_A$ relative to the flow. Thus these phase speeds need to be taken into account when determining the compression ratio of the actual scattering centers

$$\sigma = \frac{X_{sc} + 2}{X_{sc} - 1} , \tag{11.54}$$

where

$$X_{sc} = \frac{V_{1x} + v_{p1}}{V_{2x} + v_{p2}} \overset{M\to\infty}{\longrightarrow} \frac{V_{1x}}{V_{2x}} = X . \tag{11.55}$$

In parallel fast-mode shocks, for which $1 < M_A < 2$, this may lead to extremely large compression ratios and flat ($\sigma \approx 1$) particle spectra. In slow-mode shocks, the scattering centers always have larger phase speeds than the fluid speeds. Thus the scattering centers do not converge in slow shocks under many circumstances and then the Fermi mechanism is not effective.

Finally, the power-law spectrum does not extend to infinite energies, but is cut-off at some energy determined by the age and the size of the system. Obviously, if the time $\tau$ to accelerate the particles is limited, they cannot be accelerated beyond energies determined by $\dot{p} \sim p/\tau$, where $\dot{p}$ is the rate of momentum gain related to the scattering rates and flow velocities in the system. Similarly, when the particle's Larmor radius ($v_\perp/\omega_c$) becomes of the order of the system size, the particle cannot be accelerated any further.

### 11.3.3 Shock surfing acceleration

In addition to the Fermi-acceleration models discussed above we mention, for completeness, the shock surfing acceleration as another mechanism that has been proposed to account for the acceleration of ions in quasi-perpendicular collisionless shocks. This relies on the existence of a cross-shock potential, which tends to decelerate the incident ions in the shock normal direction ($x$ axis). If an ion has a small velocity component along the shock normal $0 < v_x \ll V_{1x}$, it will be reflected by the cross-shock electric field. But once it is moving back into the upstream magnetic field, the Lorentz force will turn it around as in the shock drift acceleration. The ion will be trapped between the upstream magnetic field and the cross-shock electric field, drift along the upstream convective electric field and gain energy. The ions appear to surf along the shock wave.

Let the $y$-axis point to the direction of the upstream electric field $\mathbf{E}_1 = -\mathbf{V}_1 \times \mathbf{B}_1$. Assume, that $v_z = 0$ and $B_{1x} = B_{1y} = 0$. The equation of motion in the $y$-direction is

$$\dot{v}_y = q(E_y - v_x B_z) , \tag{11.56}$$

where the right-hand side is equal to $qV_{1x}B_{1z}$ for small $v_x$. Thus, the particle experiences linear acceleration along the $y$-direction as long as it stays trapped between the cross-shock potential and the upstream magnetic field. On the other hand, the equation of motion for $v_x$ in the shock ramp is

$$\dot{v}_x = q(E_x + v_y B_z) = q(-\delta\varphi/\delta x + v_y B_z) \,. \tag{11.57}$$

Obviously the particle will no longer stay trapped, once $v_y \approx -E_x/B_z = \delta\varphi/B_z\delta x$, where $\delta x$ is the thickness of the shock. Assuming $\delta x \sim c/\omega_{pe}$ we get

$$v_y \approx \frac{a m_p V_{1x}^2 \omega_{pe}}{2 e B_z c} \approx \frac{a V_{1x}^2 \omega_{pe}}{2 \omega_{ci} c} \approx \frac{a V_{1x}^2}{2 V_A} \sqrt{\frac{m_i}{m_e}} \,, \tag{11.58}$$

amounting to a velocity gain by a factor of the order of 100 for typical perpendicular shocks in the interplanetary medium. While potentially efficient, this mechanism is very sensitive to the thickness of the shock and thus limited to magnetic fields oriented very closely perpendicular to the shock normal.

Acceleration of electrons is not possible with the simple shock surfing mechanism, but particle simulations have revealed that the electric potential in the shock ramp is not monotonic. There appear to be structures inside the ramp, where the electric field points towards downstream, thus being capable of trapping electrons and rapidly accelerating them to relativistic energies. Whether shock surfing is a relevant acceleration mechanism in the solar system shocks is unclear.

# 12. Storms on the Sun

Solar flares and coronal mass ejections are the most important storm phenomena in the atmosphere of the Sun. The observation that a geomagnetic storm commenced only some 17 hours after the flare observed by Carrington and Hodgson in 1859, and many subsequent events suggesting a similar flare–storm relationship, led to the hypothesis that the flares were the drivers of the *nonrecurrent* magnetic storms at the Earth. The evidence for a causal connection from flares to storms was, however, not particularly good. Large flares can be observed without ensuing magnetic storms and storms, also nonrecurrent ones, often occur without any notable preceding flare activity on the Sun. But if it is not a flare, what would be the driver? The answer came with the first CME-observations using a space-borne coronagraph [Tousey, 1973]. The misconception of addressing the flares as primary storm drivers prevailed in some parts of the solar–terrestrial physics community for a long time, even after subsequent spacecraft observations of CMEs and their in situ characteristics in the solar wind had convincingly shown that CMEs are the main drivers of nonrecurrent magnetic storms. (For a discussion of this "solar flare myth", see Gosling [1993].) Finally, the excellent *SOHO* coronagraph images of CMEs during solar cycle 23 brought the real storm drivers to the attention of the entire community concerned with severe space weather.

Although solar flares have lost some of their status in the studies of magnetospheric storms, they are the most dramatic storm phenomena on the Sun. The flares also accelerate charged particles that arrive to the Earth much faster than the CMEs and thus affect the near-Earth environment practically immediately after the storm onset on the Sun. Whatever the association is, flares are often, but not always, associated with CMEs and this relationship needs to be understood better than is the case today. New high-resolution images of flares and eruptive prominences provide fresh views of the reconnection process, making it almost visual. As the strength of the solar flares can be routinely monitored using space-borne X-ray detectors, they provide a useful, though an incomplete, warning method of approaching stormy weather in the geospace.

H.E.J. Koskinen, *Physics of Space Storms: From the Solar Surface to the Earth*,
Springer Praxis Books, DOI 10.1007/978-3-642-00319-6_12,

## 12.1 Prominences and Coronal Loops

Before starting discussion of the solar eruptions we briefly discuss *solar prominences* and *coronal loops*. While these structures may look similar and be related to each other, the prominences are relatively cold, whereas the word "loop" usually refers to hot structures of a variety of shapes seen typically in EUV or X-ray pictures of the Sun.

The prominences are giant gas clouds traditionally observed on the limb of the Sun. They consist of plasma that is much more dense ($n \approx 0.5 - 1.0 \times 10^{17}\,\mathrm{m}^{-3}$, i.e., by a factor of 500) than the ambient coronal density and much cooler (5000–10 000 K) than the surrounding corona. Consequently, they can be seen on the limb as bright arcs, or arcades, against the tenuous background. Due to the lower temperature than in the chromosphere, the prominences look like dark *filaments* in H$\alpha$ images of the chromosphere.

The prominence clouds typically have the form of vertical sheets. The sheets are remarkably stable, surviving up to 300 days, and thus they can be used in studies of solar rotation. There are two basic types of prominences, *quiescent* and *active-region prominences*. The quiescent prominences are typically larger and extend higher up (above 30 000 km). The magnetic field of a quiescent prominence is 0.5–1 mT. In the active regions the prominences are smaller and mostly below 30 000 km, but their density is somewhat higher and the magnetic field is stronger (2–20 mT). The prominences often fade away, but in active regions they can erupt in association with solar flares and/or CME releases. In fact, a rapid disappearance of a filament is a frequently used method of determining the location where a CME originated and about 70% of CMEs originating from the visible disk have been associated with prominence/filament disappearance.

The line-of-sight (i.e., the nearly vertical) component of the magnetic field reverses over a filament but the direction in which the field passes through the prominence may be the same as we would expect for a simple arcade (normal polarity, about 25% of all prominences) or opposite to it (inverse polarity, about 75% of all prominences) as indicated in Fig. 12.1. The stability of the prominences is not directly related to the polarity. The high-latitude quiescent prominences mostly have the inverse polarity, whereas the active-region prominences can have either of the polarities. The filament itself is a plasma sheet carrying a sheet current ($\mathrm{A\,m}^{-1}$) directed either away from the page (normal polarity) or into the page (inverse polarity) in Fig. 12.1.

The formation of prominences is still an active area of research. The process starts with a magnetic flux tube rising through the solar surface as an arc. Priest et al [1989] suggested that when the distance between the footpoints of the flux tube becomes sufficiently long, a radiative instability sets in and the cool plasma starts to accumulate in the central part of the arc. Also twisting of the flux tube plays a role in the process by forming a magnetic dip in the lower part of the horizontal portion of the flux-tube where cool plasma starts to accumulate (panes (b) and (c) in Fig. 12.2). If the flux tube becomes too long or the twist too strong, the prominence becomes unstable and erupts.

The coronal loops are magnetic flux tubes filled with hot plasma reaching out to the corona. They are much hotter than prototypical prominences, which makes them observable at the wavelengths from EUV to soft X-rays. The loops are not in pressure balance with the surrounding corona, but they are confined by the strong magnetic field in the flux

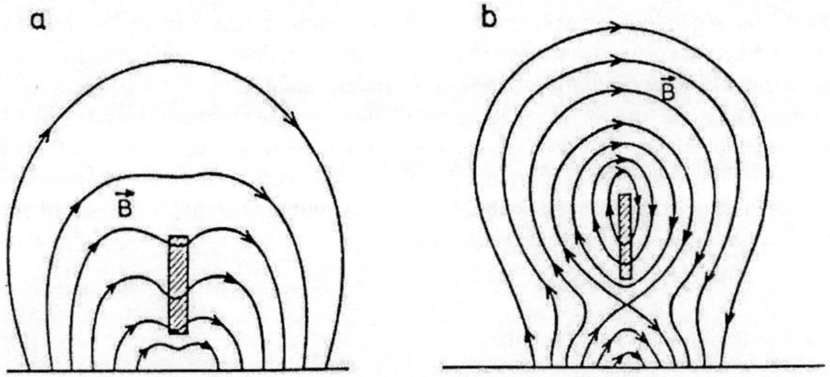


**Fig. 12.1** Magnetic field configuration in association with prominences. On the left the polarity of the prominence is normal, on the right it is inverse. (From Anzer and Priest [1985].)

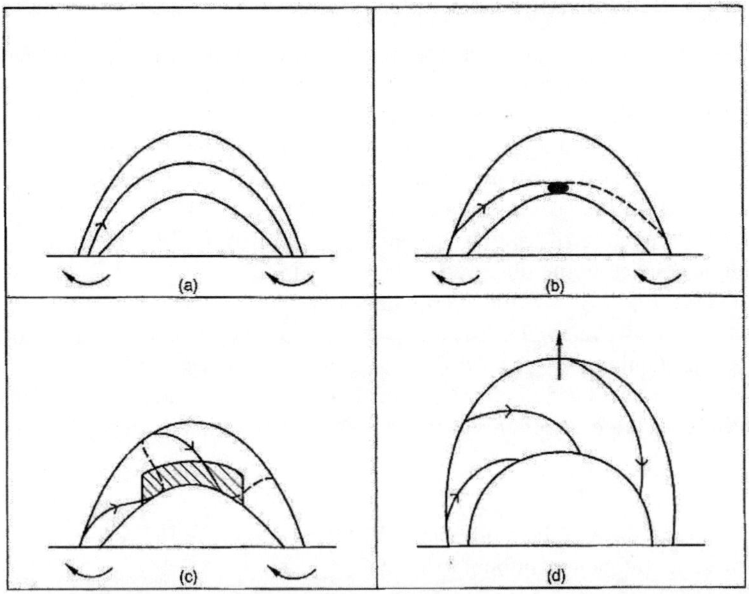


**Fig. 12.2** The formation of a prominence in a twisted-flux-tube model by Priest et al [1989].

tube that keeps the cross-field diffusion small. The filling and heating of the plasma in the coronal loops belong to important problems among coronal processes.

In addition to the large variety of temperatures, the spatial structures of the coronal loops are also variable. They can form long *arcades*, have an S-shaped appearance known as a *sigmoid*, create structures looking like a bow-tie, etc. The evolution toward the different shapes is related to the motion of the footpoints of the magnetic flux tubes forming the structures in the photosphere and to their expansion associated with the filling and heating of the plasma. When looking at coronal EUV/X-ray pictures, one cannot avoid the impression that the active corona is composed of a web of many loops at different temperatures. (For more details on coronal loops, see Aschwanden [2004]).

At some point of time in some location of the coronal magnetic web the conditions become conducive for a large-scale magnetic reconnection process and a flare erupts (Sect. 12.3). The flare may or may not be associated with an eruptive prominence, and it may or may not be associated with a CME. It is quite clear that there is still much to learn at this end of the space storm sequence.

## 12.2 Radio Storms on the Sun

During World War II amateur radio operators noticed radio noise (or hiss) that occurred only in daytime, and the first radar systems at meter wavelengths were occasionally jammed by radio interference apparently coming from the direction of the Sun. Soon after the war intense solar radio bursts were also detected.

The value of radio wave observations in solar research is based on two factors. First, the solar spectrum at these wavelengths is highly variable (Fig. 1.2) reflecting both high coronal temperatures and strong activity. Second, the Earth's atmosphere is transparent to electromagnetic waves that are longer than a few mm (i.e., frequencies below 100 GHz). The low-frequency cut-off, in turn, is determined by the ionospheric plasma frequency, which, depending on the peak plasma density, is 4–10 MHz. However, already below 20 MHz there is so much terrestrial interference that detailed observations from the ground are difficult, and space-borne observations are called for.

As is evident from Fig. 1.2, the energy density of solar radio waves is extremely small compared to visible wavelengths. To cope with small energy fluxes radioastronomers have introduced a particular unit, the *Jansky*, equal to $10^{-26}\,\mathrm{W\,Hz^{-1}\,m^{-2}}$. The Sun as seen from the Earth is a much brighter radio source than other astronomical sources and, consequently, in solar radio astronomy a four orders of magnitude larger unit, the *solar flux unit* $(SFU)$ is used

$$1\,SFU = 10^{-22}\,\mathrm{W\,Hz^{-1}\,m^{-2}}\,. \tag{12.1}$$

The thermal emission from the quiet Sun at 40 MHz is about 3 $SFU$, whereas typical radio bursts at the same frequency amount to $10^5\,SFU$.

**Train your brain**

Estimate the total power of solar radio waves hitting the surface of the Earth in the wavelength range 10 cm – 1 m.

### 12.2.1 Classification of radio emissions

There is a whole zoo of different types of solar radio emissions. They are named mostly according to their observed properties, which may be a nuisance to a student but practical for scientific study because the terminology is free from evolving physical interpretation. The most important emissions are known as Types I–IV (Fig. 12.3).

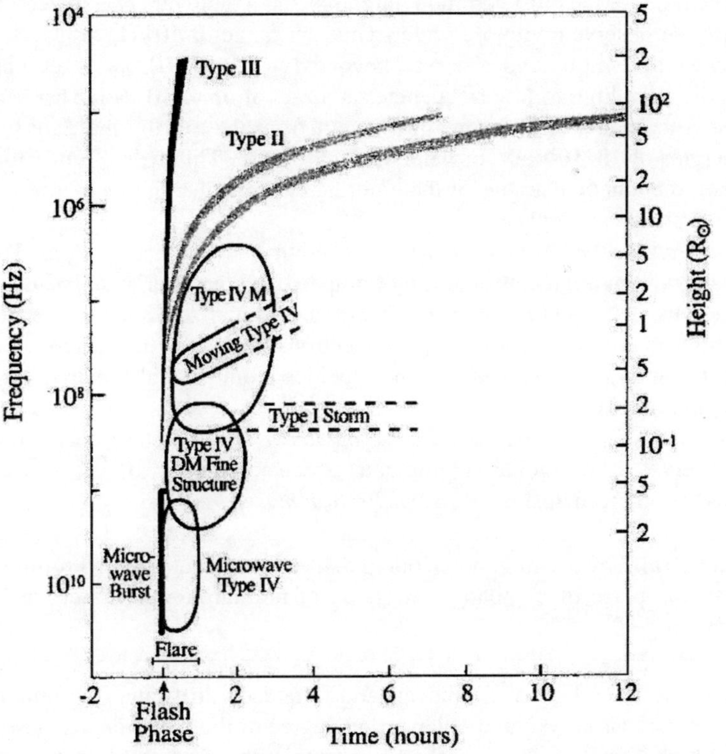


**Fig. 12.3** Schematic signatures of solar radio events in a dynamic spectrum. The timing is relative to the flash phase of a typical solar flare. On the vertical axis the frequency decreases upward to correspond to the increasing altitude of the emission region from the surface of the Sun. (From Lang [2000].)

*Type I bursts* are very short (< 1 s) emissions but they appear in large numbers to form irregular emission structures called *Type I noise storms*. These storms last from hours to days. The *brightness temperatures* (i.e., the temperature that the source would have if the emission would be thermal radiation from a hot gas) of Type I bursts are $10^7 - 10^9$ K. The bursts are believed to be generated by electrons accelerated to a few times their thermal energies by energy release in closed coronal loops. In the tenuous solar corona the particles emitting electromagnetic waves do not need to be in thermal equilibrium with the surrounding plasma, nor need the radiation mechanism be thermal.

*Type II bursts* are narrow-band emissions at frequencies 0.1–100 MHz. They drift slowly to lower frequencies at a rate of about 0.1–1 MHz per second suggesting an outward motion at the speed of about 1000 km $s^{-1}$ (see the discussion in Sect. 12.2.2 below), which has been attributed to outward propagating shock waves. The lowest frequencies (kilometer wavelengths) are emitted farther out from the Sun and often called *interplanetary Type II bursts*. Type II bursts can be associated with both flares and CMEs, but there is no one-to-one correspondence with either of them.

*Type III bursts* are the most common flare-associated radio bursts at meter wavelengths, but they can be observed within a wide frequency range of 10 kHz – 1 GHz, the lowest frequencies being emitted and observed beyond 1 *AU*. Type III bursts are characterized by a fast drift from high to low frequencies at a rate of up to 100 MHz per second. They are attributed to beams of electrons thrown out from the Sun by the flare process with kinetic energies of 10–100 keV, or velocities up to $0.5\,c$. This type of emission is clearly *non-thermal*, the kinetic energies of the emitting electrons being much above the thermal energy of the surrounding plasma.

*Type IV bursts* are the most common type of activity at meter wavelengths. The emission is broad-band continuum radiation lasting for up to one hour after an *impulsive flare* onset. The radiation from a Type IV burst is partly circularly polarized, and has been attributed to gyro synchrotron emission from energetic electrons trapped within magnetic clouds that travel into the interplanetary space with velocities from several hundreds of km $s^{-1}$ to about one thousand km $s^{-1}$.

As precursors of solar flares *microwave impulsive bursts* lasting only a few milliseconds are observed. Their radiation temperatures can reach up to $10^{15}$ K, which requires a coherent radiation mechanism, e.g., a *synchrotron maser*.

**Feed your brain** by finding out from the literature what a synchrotron maser is. What takes the place of population inversion of normal atomic masers or lasers in a synchrotron maser?

In addition to the intensity, frequency, and frequency drift rate, it is important to determine the polarization type and polarization degree of the radiation, as these depend on the emission mechanism and geometry. For example, the electron plasma emission is unpolarized whereas cyclotron and synchrotron emissions in the strong magnetic fields are circularly polarized.

### 12.2.2 Physical mechanisms for solar radio emissions

Radio emissions occur in perturbed plasma layers of the solar atmosphere and are mainly due to free electrons moving in the magnetized plasma. An important lesson from basic plasma physics is that a free-space electromagnetic wave can propagate only if its frequency is higher than the local (electron) plasma frequency

$$f_p = \frac{1}{2\pi}\sqrt{\frac{n_e e^2}{\varepsilon_0 m_e}}\,, \tag{12.2}$$

whose approximate numerical value is given by

$$f_p(\mathrm{Hz}) \approx 9\sqrt{n_e(\mathrm{m}^{-3})}\,. \tag{12.3}$$

In the chromosphere the electron density drops from $2.5\times10^{17}\,\mathrm{m}^{-3}$ to $10^{16}\,\mathrm{m}^{-3}$, which corresponds to the plasma frequency range 4.5–0.9 GHz, or wavelengths 6.7–33.3 cm. To calculate the plasma frequency in the corona there are several density models. For our purposes good enough is the *Baumbach-Allen formula*

$$n_e(\mathrm{m}^{-3}) = (0.036\,r^{-1.5} + 1.55\,r^{-6} + 2.99\,r^{-16}) \times 10^{14}\,, \tag{12.4}$$

where the distance is in solar radii (from the center of the Sun). At $2R_\odot$ the plasma frequency is about 14 MHz corresponding to the wavelength of 21 m. If we observe waves longer than 21 m, we know that they must have been emitted at least $2R_\odot$ from the center of the Sun (see Fig. 12.3).

In the interplanetary space, the solar wind velocity does not vary much as a function of distance. Thus the density decreases like $r^{-2}$ and we can scale the density to the distance of 1 $AU$. The scaled density is typically $3\text{–}10\times10^6\,\mathrm{m}^{-3}$, corresponding to plasma frequencies of 15–30 kHz. Thus, e.g., interplanetary Type III emissions at lower frequencies than these must be created *and observed* beyond the Earth's orbit.

The basic radiation mechanisms discussed in Chap. 9, thermal bremsstrahlung and cyclotron and gyro synchrotron radiation, are *incoherent* radiation processes where all electrons radiate independently. Thus the observed brightness temperatures are of the order of the kinetic temperature of the electrons (e.g., Type I bursts). However, electrons can also radiate *coherently* (cf. coherent scattering in Chap. 9). Coherent emissions can have brightness temperatures up to $10^{15}$ K.

The main coherent radiation mechanism in the corona and the solar wind is the *plasma emission* of which Type II and III emissions are examples. They have been interpreted to originate from warm beam–plasma instability at frequencies slightly above the local plasma frequency. When the source of the radiation moves outward from the Sun, the frequency of the emission decreases with the decreasing plasma frequency. Consequently, we can calculate the source velocity $v_{src}$ from the drift rate $df/dt$ of the frequency assuming that we have a reliable density model.

**Train your brain**

Derive an expression for the relationship between the frequency drift rate $df/dt$ and the source velocity $v_{src}$.

The mode driven by the beam–plasma instability is the Langmuir wave (Eq. 7.66). However, the Langmuir wave is an electrostatic mode and becomes rapidly Landau-damped outside the unstable plasma region. Thus it must somehow convert to a transverse electromagnetic wave. To illustrate the mode conversion let us first consider Type III because it has a simpler generation mechanism than Type II. In Type III emission the Langmuir wave is driven by an energetic electron beam accelerated by a solar flare. As the electrons move along the magnetic field line with a very high velocity, they move rapidly toward lower plasma density and the frequency drift rate is very high, up to 100 MHz per second. Ginzburg and Zhelezniakov suggested in 1958 that the mode conversion from the electrostatic Langmuir wave to a transverse electromagnetic wave was due to a nonlinear wave–wave interaction mechanism. The idea was later confirmed through in situ observations with the *Helios* spacecraft by Gurnett and Anderson [1976], who identified both the locally generated Langmuir wave and the escaping electromagnetic wave at the appropriate frequencies.

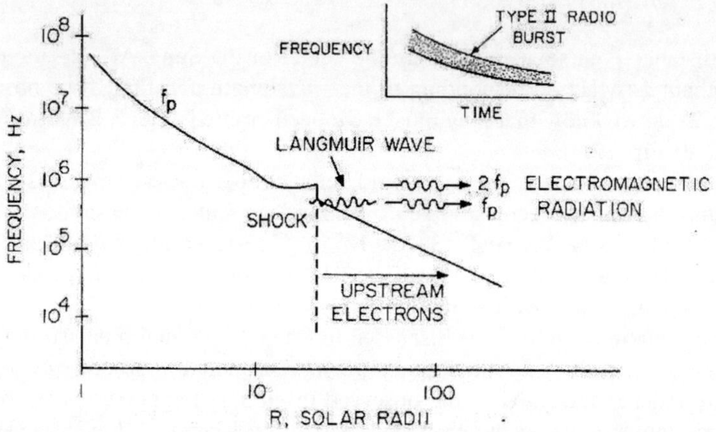


**Fig. 12.4** Energetic electrons streaming from the Sun excite Langmuir waves at the local plasma density. The electromagnetic radiation is produced at the plasma frequency $f_p$ and at $2f_p$ by mode conversion from the Langmuir waves. (From Gurnett [1995].)

The details of the mode conversion require a more thorough discussion of nonlinear wave–wave interactions than is possible in the present text. The underlying idea can, however, be illustrated in terms of elementary coupled oscillators and conservation laws. The Langmuir wave ($L$) can couple to a low-frequency ion–acoustic wave ($IAC$) and produce a transverse electromagnetic wave ($T$), a process that we formally describe as $L + IAC \to T$, under the condition that the frequencies and wave vectors fulfill the *matching relations*

$$\omega_L + \omega_{IAC} = \omega_T \tag{12.5}$$

$$\mathbf{k}_L + \mathbf{k}_{IAC} = \mathbf{k}_T \; . \tag{12.6}$$

These relations are actually expressions of conservation of energy and momentum in the mode coupling process (note the analogy to quantum mechanics by multiplying both equations by $\hbar$). Recall from Chap. 5 that $\omega_{IA} \ll \omega_L$ and $\mathbf{k}_L \ll \mathbf{k}_T$. Thus the transverse wave has a frequency somewhat above the plasma frequency $f_p$, and can thus escape into the directions of non-increasing plasma density. On the other hand the Langmuir wave and the ion–acoustic wave must have roughly the same wavelength with wave vectors pointing in opposite directions. The detailed investigation to find out how efficient the process is to transfer energy to the electromagnetic mode requires complicated calculations involving the solution of the dispersion equations for appropriate plasma parameters under realistic plasma conditions in the corona and the solar wind.

The emission at $2f_p$ is believed to result from another wave–wave coupling process involving two Langmuir waves (or wave "quanta"): $L + L' \rightarrow T$. The energy conservation requires that $\omega_T \simeq 2\,\omega_p$ and the momentum conservation that the Langmuir waves must propagate to opposite directions. Thus, while often called a "harmonic" of the fundamental plasma emission, it actually arises from a different physical process.

Type II bursts are associated with shock waves propagating outward through the solar corona and the interplanetary space. As the shock waves propagate much more slowly than the flare-accelerated electrons, the frequency drift rates are also slower. Close to the Sun the shock waves can be driven both by the flares and CMEs but the flare-associated shocks do not propagate to the interplanetary space. Again the primary emission is the Langmuir wave, which is converted to the transverse electromagnetic wave by a similar process as in the case of Type III bursts. What is different here, is the origin of the Langmuir waves that are now due to electrons accelerated upstream of the shock as discussed in Chap. 11. (For further description of these wave emissions, see Gurnett [1995] and references therein.)

## 12.3 Solar Flares

A solar flare is a huge magnetic energy release process on the Sun. The total power of a flare is about $10^{20} - 10^{22}$ W and the total energy release may be up to $10^{25}$ J within about 10 min. This is a considerable amount of energy, corresponding to 300 million years' energy production of a 1000-MW power plant. While large, this is not unreasonable when compared to the energy stored in coronal loops. The total magnetic energy of an arcade with a radius of 20 000 km, length 100 000 km, and shear angle 45°, is about $6 \times 10^{25}$ J, which is enough for a large flare.

### 12.3.1 Observational characteristics of solar flares

Only very large flares, such as the Carrington flare, can be seen in white light from the Earth. More characteristic for flares are certain line emissions (e.g., H$\alpha$) and the consequences of the rapid magnetic energy release in form of particle acceleration. In a flare electrons are typically accelerated to energies of 10–100 keV, sometimes up to 10 MeV, and the highest energy nuclei reach to hundreds of MeV. These particles emit electromagnetic radiation throughout the spectrum from radio waves to X- and $\gamma$-rays. Figure 12.5 shows the development of a typical flare as seen at various wavelengths.

Characterization of a flare sequence is somewhat non-trivial because the different signatures evolve in different ways. The chromospheric eruption is easiest to observe using the H$\alpha$ line, in which a "flash" of the flare lasting a few minutes is seen. In the early part of the flash $\gamma$-rays, hard X-rays, EUV radiation, and microwaves indicate an *impulsive phase* of the flare. This is followed by the *main phase* (or *decay phase*), which lasts from 30 min

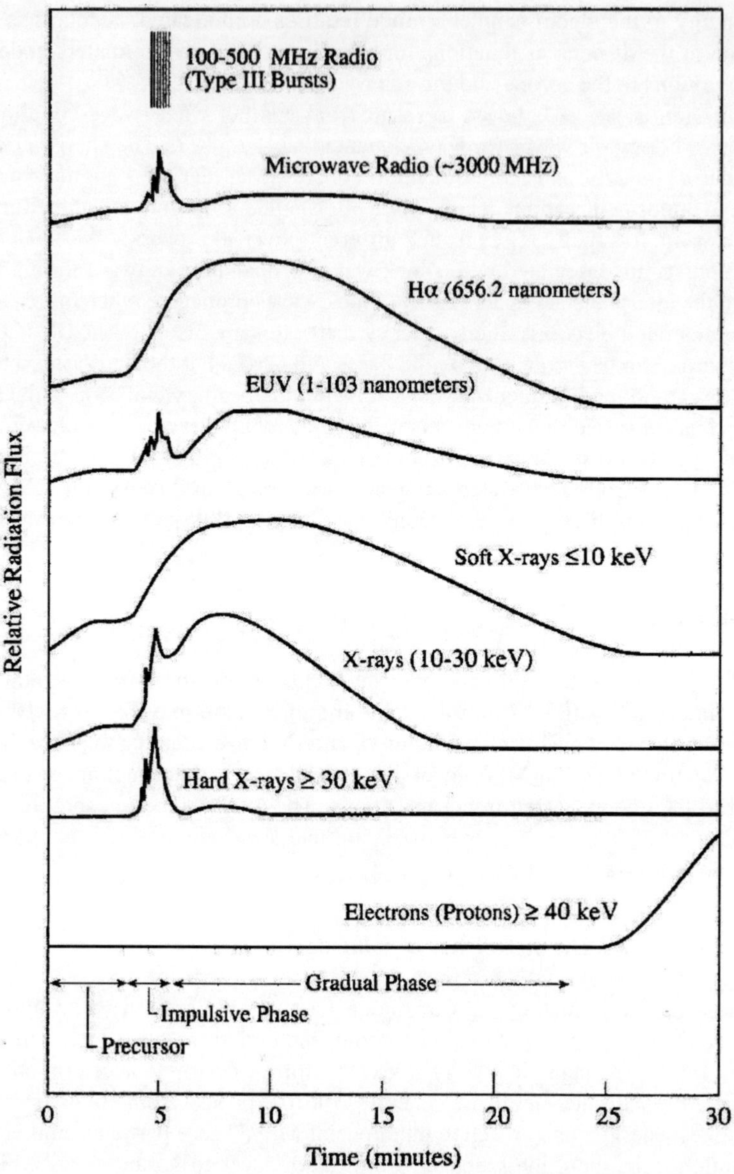


**Fig. 12.5** Various observational indicators of a flare from radio waves to hard X-rays. The flare-accelerated energetic particles arrive to 1 *AU* 20–30 min after the electromagnetic signals. (From Lang [2000], adapted from Kane [1974].)

to 1 h. Just before the flash there is a brief precursor characterized by thermal radiation corresponding to a temperature up to $10^7$ K.

There are no reliable methods of predicting flares in advance. Because the flare accelerated particles are of concern for spaceflight, the development of the active regions is monitored continuously and information of possible flare activity is spread throughout the world. However, the quality of these predictions compared to the statistical appearance of flares (so-called *skill*) is still modest.

### *Flares and radio waves*

The most important flare emissions at radio wavelengths are Type II and III bursts. Type III emission drifts rapidly down in frequency, which is interpreted to indicate a fast motion ($10^7$–$10^8$ m s$^{-1}$) of electrons in the 10–100-keV energy range. This is consistent with the energy required to produce the hard X-rays observed simultaneously. Type II emission appears somewhat later and indicates a lower velocity of $10^6$ m s$^{-1}$. The emission is interpreted as being emitted by a shock wave generated by the flare, or by an associated CME that propagates outward through the corona.

### *X-ray flares*

The solar X-rays are absorbed completely by the Earth's atmosphere and can thus be observed only in space. The first observation of flare X-rays was made by Peterson and Winckler [1959] using a high-altitude balloon in 1958. They observed radiation in the energy range 200–500 keV, lasting less than a minute, coincident with a solar radio burst and an H$\alpha$ flare.

Today solar X-rays are monitored regularly and their intensity as measured by geostationary satellites is readily available on the internet from the NOAA website (`http://www.swpc.noaa.gov/`). The intensity is indexed as A, B, C, M, and X as given in Chap. 1, Table 1.3.

There are two main components in the solar X-ray spectrum: *soft X-rays* between 1 and 10 keV, which are mainly due to thermal radiation of hot electrons, and *hard X-rays* in the range 10–100 keV, originating from non-thermal radiation of electrons accelerated to velocities of a sizable fraction of the speed of light. The energy range is not the only difference between soft and hard X-rays, also the spectra are different. The soft X-ray spectrum has an exponential shape, whereas the non-thermal hard X-ray spectrum has a power law spectrum at large energies. In both cases the dominant radiation mechanism is *bremsstrahlung* of electrons moving among ions (mostly protons) of the ambient gas (Chap. 9). As a 1-keV photon has a wavelength of 1.24 nm, the X-ray spectra are in the wavelength range 0.01–1 nm.

The soft X-ray flux builds up gradually and peaks a few minutes after the impulsive emission. The *Skylab* observations in 1973 produced first clear pictures of soft X-ray loops of a million-degree coronal gas. These loops are associated with dynamical magnetic field structures during the flare activity.

The temperatures of flaring soft X-ray loops are about ten times hotter than the quiescent non-flaring coronal loops. The temperature up to several times $10^7$ K is enough to strip almost all electrons from the iron atoms. Consequently the X-ray spectra also contain line emissions from the inner shells of multiple times ionized elements, e.g., at 0.1778 nm (Fe XXVI), 0.185 nm (Fe XXV), 0.3177 nm (Ca XIX), 0.5039 nm (S XV), 0.917 nm (Mg XI), 1.346 nm (Ne IX). The plasma densities of the loops have been estimated from the density-sensitive lines to be $10^{17}$–$10^{18}$ electrons per $m^3$.

The hard X-rays often have a double source which is nearly co-located with the magnetic footpoints of the soft X-ray loops and H$\alpha$ emission in the chromosphere. The two spots flash within 10 s of each other. This indicates that the hard X-rays are produced in the low corona and dense chromosphere by non-thermal electrons injected down along the legs of the coronal loop. This is further supported by the similar time profile of radio waves at centimeter wavelengths also produced by the non-thermal electrons.

### *Flares and γ-rays*

$\gamma$-rays have energies above 100 keV. Nuclear interactions of flare-accelerated protons and helium nuclei (energies 1–100 MeV) with nuclei in the dense solar atmosphere below the acceleration site produce $\gamma$-rays at energies between 0.4 and 7.1 MeV. Furthermore, protons with energies above 300 MeV interact with hydrogen in the solar atmosphere and produce mesons. The decay of neutral mesons produces a broad $\gamma$-ray peak around 70 MeV, whereas the decay of charged mesons leads to bremsstrahlung with a continuum $\gamma$-emission extending to several MeV. Also neutrons with energies above 1 GeV are produced through nuclear interactions associated with the flare process.

There are two particularly strong $\gamma$-ray lines in the solar spectrum: 511 keV and 2.223 MeV. The former is due to electron-positron annihilation

$$e^+ + e^- \rightarrow \gamma + \gamma \, (511\,\mathrm{keV\,each})\,, \tag{12.7}$$

where the positrons originate from decay of radioactive nuclei. The latter is a stronger line and results from the capture of a neutron by a proton

$$n + p \rightarrow d + \gamma \, (2.223\,\mathrm{MeV})\,. \tag{12.8}$$

Two important *spallation* reactions lead to $\gamma$-ray emission through a transition from an excited state to the ground state of one of the spallation products

$$\begin{aligned} p + {}^{16}O &\rightarrow {}^{12}C^* + \alpha + p \\ {}^{12}C^* &\rightarrow {}^{12}C + \gamma \, (4.438\,\mathrm{MeV}) \end{aligned} \tag{12.9}$$

and

$$\begin{aligned} p + {}^{20}Ne &\rightarrow {}^{16}O^* + \alpha + p \\ {}^{16}O^* &\rightarrow {}^{16}O + \gamma \, (6.129\,\mathrm{MeV})\,. \end{aligned} \tag{12.10}$$

In the latter reaction $\gamma$-lines at 6.917 and 7.117 MeV are also prominent.

The heated flaring loops are at a temperature corresponding to that of the center of the Sun. Although the corona is much more tenuous, there is a sufficient amount of fusion reaction to lead to observable $\gamma$ emissions, e.g.,

$$\begin{aligned} {}^4\mathrm{He} + \alpha &\rightarrow {}^7\mathrm{Be}^* + \mathrm{n} \\ {}^7\mathrm{Be}^* &\rightarrow {}^7\mathrm{Be} + \gamma\,(0.431\,\mathrm{MeV}) \end{aligned} \tag{12.11}$$

and

$$\begin{aligned} {}^4\mathrm{He} + \alpha &\rightarrow {}^7\mathrm{Li}^* + \mathrm{n} \\ {}^7\mathrm{Li}^* &\rightarrow {}^7\mathrm{Li} + \gamma\,(0.478\,\mathrm{MeV})\,. \end{aligned} \tag{12.12}$$

Furthermore, there are several emission lines of nuclei excited by protons, e.g., $^{14}$N (5.105 MeV), $^{20}$Ne (1.634 MeV), $^{24}$Mg (1.369 MeV), $^{28}$Si (1.779 MeV), and $^{56}$Fe (0.847 and 1.238 MeV).

### 12.3.2 Physics of solar flares

A large number of models describing the flare process have been discussed in the literature. Since the *Yohkoh* mission in the 1990s the improved observations have put tighter constraints to the models, but there is still much room for theorizing and speculations. Any proposed flare mechanism must explain the release of magnetic energy arising from an active region. The energy release cannot be simple diffusion because the magnetic diffusion times ($\tau = l^2/\eta$) are very long, of the order of hundreds of years for scale sizes of 10 000 km. However, by compressing the gradient scale length to 1 km or below the diffusion times are reduced to minutes. Thus the formation and stability of thin current sheets belong to central issues in flare research.

People familiar with reconnection in the terrestrial magnetosphere may think that the case for reconnection would be easier in the solar atmosphere, but, in fact, it may actually be the opposite. The flux tube structure in the solar atmosphere is much more complicated than the configuration of the magnetosphere, and there are hardly any such relatively simple structures as the magnetotail current sheet. Another important feature is the *line-tying*, i.e., tying the feet of the flux tubes to the photosphere at both ends. On the one hand, line-tying is a strongly stabilizing feature, but on the other, it facilitates large-scale motion of the flux tubes following the plasma motion in the photosphere. This leads to a complicated web of current sheets and magnetic null-points (or at least nearly null), strongly limiting the applicability of two-dimensional reconnection models discussed in Chap. 8.

Possibilities for observational characterization of solar reconnection are different from magnetospheric observations. The high-resolution images of X- and $\gamma$-ray emissions from several spacecraft during the early 21st century have made the reconnection almost visible. These observations together with radio emissions give us important information on the particles energized in the process and also about the temperature evolution. On the other hand, we can never obtain such detailed in situ plasma and field information at and near the reconnection regions as we have from the Earth's magnetosphere.

There are some features that make the comparison of flares and magnetospheric substorms (to be discussed in Chap. 13) meaningful. In both cases the magnetic Reynolds number is large, so the magnetic field evolution is convection-dominated. While the plasma is much denser in the solar atmosphere than in the magnetosphere, also the magnetic field is stronger. Consequently, the Alfvén speeds in both environments are of the same order of magnitude. This is important because $v_A$ determines the outflow velocity from the reconnection region (Chap. 8). Neither are the scale sizes too different in these two regimes. Typical lengths of prominences or coronal loop flux tubes are of the order of $10^5$–$10^6$ km, i.e., 16–160 $R_E$, which is not far from length of the magnetotail involved in the substorm process (Chap. 13). The time-scale of the process is proportional to the size divided by $v_A$ and, consequently, a typical flare has somewhat shorter time span than a typical substorm. Of course, the released energy in a flare is some ten orders of magnitude larger because there is much more magnetic energy stored in the system. Furthermore, the flare scale sizes are much more variable than is possible within the magnetosphere.

The present understanding is that in the flare process magnetic energy is released by *explosive reconnection* above the top of a coronal loop. The heated loop radiates soft X-rays whereas the hard X-rays and $\gamma$-rays and most of the intense radio waves are due to the more energetic non-thermal particles. As there are several possible configurations leading to thin current sheets to facilitate reconnection in the corona, several models have been proposed to explain the observed structures.

Figure 12.6 illustrates Shibata's attempt at a "unified" flare model [Shibata et al, 1995; Shibata, 1999]. This model includes an upward release of a magnetic structure, known as a *plasmoid*, which suggest that the release of a CME is an essential part of the process. In Shibata's model the flux rope footpoints in the photosphere are at some distance from the underlying soft X-ray loop. The twisted flux rope acts like a piston that stretches the field below. This enforces the plasma flow toward the current sheet leading to explosive reconnection. In this sense the formation and ejection of the plasmoid enhances the reconnection rate. The reconnection then both heats the plasma and accelerates electrons. Electrons accelerated toward the Sun are finally decelerated closer to the surface and emit hard X-rays.

Figure 12.7 is another representation of the same scenario in the case where the reconnection takes place between a prominence (filament) above and a long arcade below the reconnection region. Such a configuration is easy to imagine as a CME with a core of prominence matter as is often seen in coronagraph images (e.g., Fig. 12.8 in the next section). The H$\alpha$ emission emerges mostly from two ribbons in the chromosphere along the footpoints of the coronal arcade.

If we rotate Fig. 12.6 90 degrees, it is somewhat analogous to a plasmoid or flux rope release from the terrestrial magnetotail in a substorm process (Fig. 13.5 in the next chapter) although the line-tying into the photosphere may be much more rigid than the connection of the core field of a terrestrial flux rope to the ionosphere. Also the footpoints have much more freedom to move in the photosphere than is possible in the terrestrial ionosphere. While the plasmoid formation in the magnetosphere is usually seen as a consequence of near-Earth reconnection, in Shibata's model it drives the reconnection process. This driving must be powered externally. One can imagine that this is a result of *magnetic buoyancy* of the flux rope in the core of the plasmoid.

While this is a suggestive picture, it is only one of many proposed scenarios. In fact, the solar corona exhibits such a richness of magnetic structures and can support so many

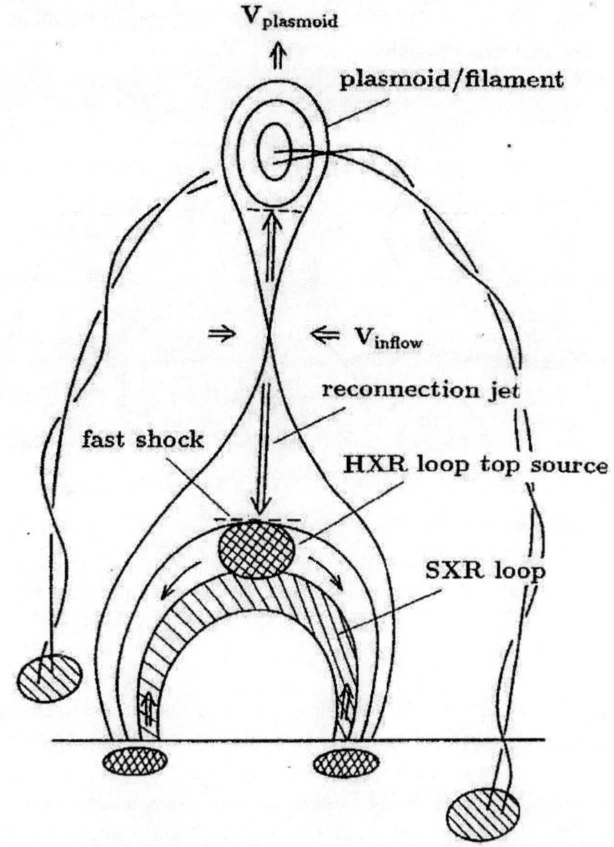


**Fig. 12.6** Shibata's plasmoid-driven reconnection model for flares. Hard X-rays can be produced both at the top of the loop and near its footpoints (the cross-hatched areas). Also $\gamma$-rays are produced at low altitudes where very energetic protons hit the nuclei as discussed above. (From Shibata et al [1995].)

different instabilities [see, e.g., Aschwanden, 2004], that it may not be really fruitful to look for *the* flare model. Instead the magnetic energy release can, most likely, find several structurally different ways of taking place. Nor is there a one-to-one association between CMEs and flares. During some CMEs no large flares are observed; on the other hand, flaring is a much more common phenomenon in a great variety of sizes. Recall that, in addition to the energy scale of $10^{21}$–$10^{25}$ J, we discussed briefly in Sect. 1.1.6 microflares (of the order of $10^{19}$ J) and nanoflares (of the order of $10^{16}$ J) as possible mechanisms to heat the corona. How similar to or different from these large-scale flares they are, is not known. From the viewpoint of space storms in the Earth's environment the large flares associated with CMEs are of the highest interest.

Of course, flares cannot be completely described by the MHD flow theory alone. In the simple quasi-static current sheet models the reconnection appears to be very much driven by the external boundary conditions, i.e., by the plasma flow toward the current

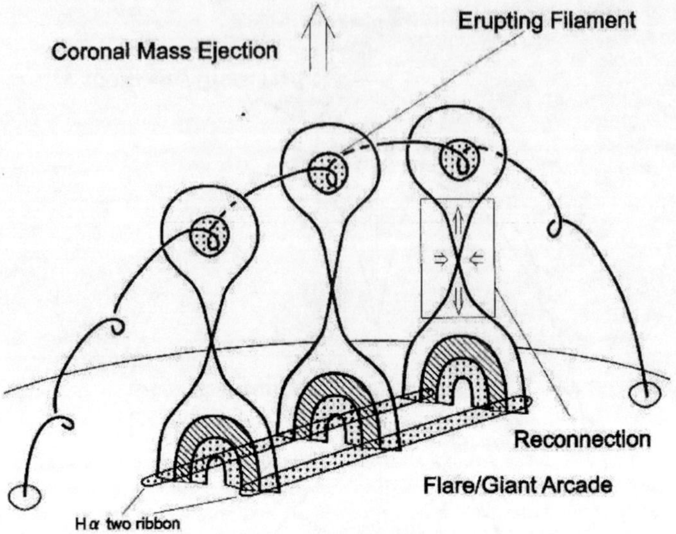


**Fig. 12.7** Schematic picture showing essentially the Shibata flare–CME model in case of an arcade below the eruptive prominence. The shaded domain below the reconnection line is the source of soft X-ray radiation, the hatched region emits EUV and the dotted regions $H_\alpha$. (From Shiota et al [2005].)

sheet. But what finally determines when the reconnection can set in, is the microscopic physics of the reconnecting region. In the case of more complicated geometries this issue may be even more important than in the effectively two-dimensional examples discussed in Chap. 8. Consequently, the interplay between microscopic and macroscopic physics needs to be understood much better than is the case today. Furthermore, the beautiful pictures of coronal activity need to be interpreted with care. What we see there *is not* the magnetic field but emissions from hot plasma being frozen-in to the magnetic field. The actual current sheets probably are much thinner than the visual plasma sheets, as we know from the in situ observations of the Earth's magnetotail current sheet.

## 12.4 Coronal Mass Ejections

Coronal mass ejections are large plasma and magnetic clouds leaving the Sun. The terminology needs to be interpreted with some care. Most of the *matter* in the mass ejections originates from the lower solar atmosphere and thus is not "coronal mass". The word "coronal" refers to the *observation* of the mass ejections in the corona. Typical CME masses are in the range $5 \times 10^{12} - 5 \times 10^{13}$ kg and angular sizes are in the range 40–50°.

It is of some interest to note that the kinetic energy leaving the Sun with a CME is of the same order of magnitude as the flare energy, of the order of $10^{24} - 10^{25}$ J. Thus from the total energy viewpoint the flares and CMEs look rather similar. However, most of the flare energy is released as electromagnetic radiation and radiated to a wide angle. The reasons, why fast CMEs are much more effective as drivers of magnetospheric activity than the flares, are the mass flux and strong magnetic field carried with a high speed to the vicinity of the Earth. The fast magnetic flux transport leads to a large motion-induced electric field to become imposed on the magnetosphere.

### 12.4.1 CMEs near the Sun

Although they are huge, CMEs are difficult to observe. They remained undiscovered until the early 1970s when they were found with white-light coronagraphs onboard *OSO* 7, the first one on December 14, 1971 [Tousey, 1973], and *Skylab*. A *coronagraph* produces an artificial occultation of the Sun, which allows the faint scattered sunlight from the CME to be observed around the occulting disk. Coronagraphs are also used at terrestrial observatories, but there is always lot of straylight in the atmosphere that makes detailed observations more difficult than with space-borne instruments.

A CME itself does not radiate. The observed light is produced by Thomson scattering (Chap. 9) of the solar photons from the electrons in the cloud. CME coronagraphs are typically designed to use white light because most of the solar photons are in the visible range. The white-light brightness of the scattered emission varies in proportion to the electron density of the CME but not to the temperature. Thus the brightness can be used to estimate the density structure of the ejected plasma cloud.

Figure 12.8 is a prototypical picture of a CME observed with the LASCO instrument on *SOHO*. The bright structure in the core of the cloud is interpreted to be matter from the eruptive prominence, whereas the surrounding structure is shocked plasma driven by the fast expansion of the magnetic flux rope around the core just as Fig. 12.7 suggests.

CMEs are quite common. According to the LASCO CME data base at the CDAW data service of the NASA Goddard Space Flight Center [Gopalswamy et al, 2009] the whole-Sun occurrence-rate averaged over Carrington rotation periods during solar cycle 23 was slightly less than one event per day at solar minimum and 4–6 events per day around the maximum years 2000–2003. The latter number is about a factor of two larger than estimates from previous maxima. This is more likely to be due to the improved sensitivity of LASCO to weak CMEs as compared to earlier observations than actually increased CME activity. Similar to many other space storm related phenomena, different selection criteria at the limit of weak events give different statistical results. For example, the automated LASCO CME catalog presented by Robbrecht et al [2009] contains about 2 events per day at minimum and 8–10 events at maximum. The automatic procedure used to compile this catalog includes a large number of bright but spatially narrow radially outward moving structures that traditionally have not been identified as CMEs.

**Feed your brain** by familiarizing yourself with the LASCO CME lists at `http://cdaw.gsfc.nasa.gov/CME_list/` [Gopalswamy et al, 2009] and `http://sidc.be/cactus/`
With these data bases you can easily initiate your own studies of various properties of CMEs.

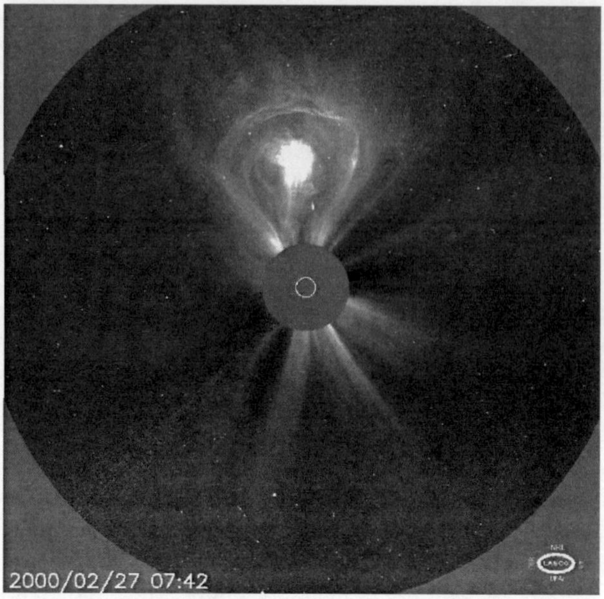


**Fig. 12.8** White-light observation of a CME by the LASCO C2 coronagraph. The occulting disk hides the Sun, which is indicated by the little circle in the middle of the disk. (Courtesy of ESA.)

The CME rate increases with the increasing sunspot activity during the solar cycle, but their latitudinal evolution is different. The first sunspots after the solar minimum appear at mid-latitudes (about 30–40°) and the sunspot belt moves toward the equator, becoming more latitudinally narrower during the cycle as indicated by the butterfly diagram (Fig. 1.5). Conversely, the minimum-time CMEs originate close to the equator and the source region widens toward the maximum. Furthermore, the structure of the CMEs becomes more complicated with increasing activity reflecting the more complicated magnetic structure of the active Sun.

The CMEs may actually play an important role in the total magnetic flux budget of the Sun, as they carry away excessive magnetic flux and helicity produced by the solar dynamo. When the differential rotation creates toroidal field, the field accumulates at low latitudes in the regions of closed magnetic loops. While the persistent solar wind carries magnetic flux only from the regions of open field lines, the ejection of large magnetic clouds contribute to getting rid of excessive closed flux.

That the CMEs originate from the closed field line regions determines their magnetic topology. However, as the magnetic structure has to be torn off from the Sun, the field must open locally. *Yohkoh* soft X-ray images have produced numerous examples of the escaping cloud, after which a soft X-ray arcade remains for several hours like a wound on the Sun at the place from which the CME was ejected. This phenomenon is called a *gradual flare* and it is associated with the restructuring of the magnetic field after the major ejection.

Already the early observations established that CMEs are more often associated with eruptive prominences than with impulsive flares. The exact relationship between the flares and CMEs is unknown but only some 40% of CMEs have an associated flare close to the site of the ejection. However, a flare may take place before, simultaneously with, or after the lift-off of the CME. On the other hand, for some 70% of ejections a disappearance of a dark filament or eruptive prominence has been identified. The prominence material can often be identified in the coronagraph images but only very rarely in direct plasma observations in the solar wind close to the Earth. Thus the interaction between CMEs and the ambient solar wind belongs to the central problems in solar wind physics.

### 12.4.2 Propagation time to 1 *AU*

When a CME leaves the Sun, its speed at $5\,R_\odot$ varies from less than $200\,\mathrm{km\,s^{-1}}$ to more than $2000\,\mathrm{km\,s^{-1}}$. At 1 *AU* the speed only seldom is larger than $750\,\mathrm{km\,s^{-1}}$ and most likely never smaller than the minimum solar wind speed of about $280\ \mathrm{km\,s^{-1}}$. The originally slow CMEs are accelerated toward the ambient solar wind speed, whereas the very high-speed CMEs are decelerated. The energy carried by a CME is of the order of $10^{24}$ J, which is comparable to large flares. The released energy is mostly in the kinetic energy of the plasma cloud with a small fraction in high-energy particles.

The determination of the CME speed from a coronal observation is a non-trivial task, particularly in the most interesting cases when the CME is heading toward the Earth, when only an expanding faint halo is seen in the coronagraph images. These events are known as *halo CMEs*. The determination of the speed in the corona is critical to ascertain that an *interplanetary CME (ICME)* observed in situ at 1 *AU* can be associated with the right CME appearing in the corona. For practical space storm forecasting an early estimate of the travel time is evidently of utmost importance.

There have been quite a few attempts to estimate the interplanetary propagation speeds and transit times to the Earth (cf., Schwenn et al [2005] and the extensive list of references therein). Based on a small number of first LASCO observations just after the solar minimum 1996–97 Brueckner et al [1998] noted that most ICMEs arrive in about 80 h, which is sometimes referred to as *Brueckner's 80-hour rule*. This seems to work pretty well near solar minima, whereas the fastest ICMEs around solar maxima arrive much more quickly. Thus something more accurate is needed.

In whatever direction a CME is released, a coronagraph gives only a projection of its motion at the plane of the sky. Estimation of the propagation speed of the front of a CME, including halo CMEs ejected toward (or away from) the Earth, is always subject to the projection effect. Schwenn et al [2005] noted that the only parameter that can be measured uniquely for any CME is the *lateral expansion speed* $V_{exp}$. By inspecting a large number of CMEs on the limb of the Sun they were able to establish a fairly good relationship (linear correlation coefficient 0.86) between $V_{exp}$ and the radial speed $V_r$ as

$$V_r = 0.88 \times V_{exp}\,. \tag{12.13}$$

The formula seems to be equally good for both fast and slow CMEs in the field-of-view of the LASCO coronagraph.

This does not yet give the transit time to the Earth accurately enough because the interaction between the ejecta and the ambient solar wind is poorly known. By comparing

CMEs with measured $V_{exp}$ to ICMEs at 1 $AU$ when one-to-one association could safely be demonstrated, Schwenn et al [2005] found that the transit time follows the law

$$T_{tr}\,(\mathrm{h}) = 203 - 20.77 \times \ln V_{exp}\,(\mathrm{km\,s^{-1}})\,. \tag{12.14}$$

The scattering in the data was, however, quite significant. Standard deviation was 14 h and the 95% certainty margin was defined to be two standard deviations, i.e., a little more than $\pm 1$ day from the transit time predicted by (12.14). The measured transit times were within 30–105 h.

Consequently, the determination of the transit times based on LASCO observations leaves a rather large uncertainty to the transit time predictions. In addition to this Schwenn et al [2005] warned that 15% of front side halo events, of which half were full halos, i.e., halos encircling the entire coronagraph occulting plate, were not observed near the Earth, which would lead to false alarms. On the other hand 20% of ICME events, e.g., an ICME-driven shock, were not preceded by a halo, not even even a partial one, which in this study was defined to encompass at least 120° of the occulting plate. Magnetospheric storms driven by these events would not have been warned for, based on these criteria.

Siscoe and Schwenn [2006] compared the empirical propagation time estimates to various physics-based numerical propagation schemes in use at that time. The best models had marginally smaller errors in the arrival time predictions, about 12 h instead of 14 h. However, what is more serious from a space weather forecaster's viewpoint is that these models led to 50% false alarm rates and missed about 25% of shocks that that actually hit the magnetosphere. Thus significant progress in shock propagation modeling and simulations is required before they can replace the actual ICME observations in situ upstream of the Earth's magnetosphere.

### 12.4.3 Magnetic structure of ICMEs

At the distance of 1 $AU$ from the Sun the ICMEs are huge structures with sizes of a significant fraction of 1 $AU$. With a single or even with two spacecraft it is practically impossible to determine the global structure an ICME (Fig. 12.9). Sometimes an ICME looks like an almost ideal force-free flux rope, sometimes the magnetic structure is quite unclear [e.g., Cane and Richardson, 2003]. Also the plasma parameters vary largely, and the cool dense prominence material often visible in coronagraph images (e.g., Fig. 12.8) can only seldom be recognized in in situ observations.

The magnetic structure of an ICME is particularly important for its efficiency to drive magnetospheric storms (Chap. 13). As determined from observations of counter streaming electrons along the magnetic field lines, an ICME may reach 1 $AU$ with both ends of the magnetic fields still tied to the Sun. A more typical topology seems to be such that only one end is tied to the Sun, and structures completely detached from the Sun are also observed.

At least one-third of all ICMEs can be represented as magnetic flux ropes with organized magnetic field rotation, almost 100% at solar minimum and about 15% near maximum [Cane and Richardson, 2003]. However, it has been suggested that when leaving the Sun all CMEs would be flux ropes [Krall, 2007] and the failure to recognize that in the ICMEs at 1 $AU$ might be due to observations made too far from the center of the flux rope.

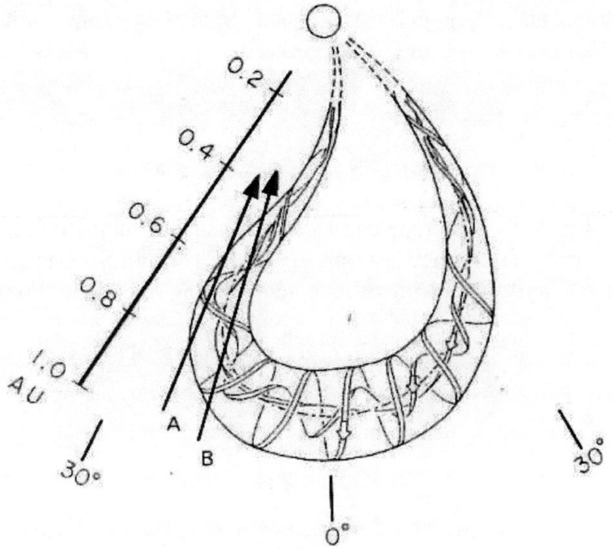


**Fig. 12.9** ICME as a flux rope. The arrows A and B illustrate the projection of two different spacecraft tracks when the ICME passes by. It is challenging to reconstruct the global structure from in situ spacecraft observations even in the case of an ideal force-free flux rope. (Adapted from Marubashi [1997].)

Considerable effort has been paid to reconstruct the magnetic structure from spacecraft observations with variable tools reaching from global MHD simulations to local analysis based on simplifying assumptions on flux rope geometries (for a review, see Riley et al [2004]).

The traditional method of studying the flux rope structure of the magnetic clouds [e.g., Lepping et al, 1990; Bothmer and Schwenn, 1998; Huttunen et al, 2005] is based on fitting in situ spacecraft observations to the solution to the cylindrically symmetric constant-$\alpha$ force-free configuration introduced by Lundquist [1950]

$$
\begin{aligned}
B_R &= 0 \\
B_A &= B_0 J_0(\alpha r) \\
B_T &= H B_0 J_1(\alpha r) \, .
\end{aligned}
\tag{12.15}
$$

Here $B_R$, $B_A$ and $B_T$ are the radial, axial and tangential components of the magnetic field, $B_0$ is the maximum magnetic field, $r$ is the radial distance from the flux rope axis, $\alpha$ is a constant related to the size of the flux rope, $J_0$ and $J_1$ are Bessel functions and $H = \pm 1$ defines the sense of the magnetic helicity.

The linear force-free fitting applying minimum variance techniques gives good results only when the spacecraft passes close to the flux rope axis. There is also no guarantee how good an approximation the force-free configuration is and several non-force-free models have been developed to study the flux ropes without this assumption. For example Hidalgo et al [2002] described the magnetic field as a sum of toroidal ($B_\phi$, corresponding to $B_A$ in

(12.15)) and poloidal ($B_\psi$, $B_T$ in (12.15)) magnetic field components, which are expressed in terms of poloidal ($J_\psi$) and torodial ($J_\phi$) currents

$$B_\phi = \frac{1}{2}\mu_0 J_\psi r$$
$$B_\psi = \mu_0 J_\phi (R - r)\,, \tag{12.16}$$

where $R$ is the radius of the flux rope and $r$ is the distance from the flux rope axis.

Another non-force-free method was presented by Hu and Sonnerup [2002] who described the magnetohydrostatic pressure balance using the Grad–Shafranov equation (6.82)

$$\frac{\partial^2 A}{\partial x^2} + \frac{\partial^2 A}{\partial y^2} = -\mu_0 \frac{d}{dA}\left(P(A) + \frac{B_z^2(A)}{2\mu_0}\right)\,, \tag{12.17}$$

where $\mathbf{A} = A(x,y)\mathbf{e}_z$ is the vector potential, the magnetic field is

$$\mathbf{B} = \left[\frac{\partial A}{\partial y}, -\frac{\partial A}{\partial x}, B_z(A)\right] \tag{12.18}$$

and the left-hand side of the equation is the axial current density $-\mu_0 J_z(A)$.

Riley et al [2004] performed a blind test to compare various methods, including force-free models, Hidalgo's model and the Grad–Shafranov approach, for a given $2\frac{1}{2}$-dimensional flux rope structure calculated with an MHD simulation. The "data" given to independent analysis teams was created by letting a hypothetical spacecraft cross the same structure at two different locations, one closer to the center of the rope, the other closer to the flank. That it actually was the same structure, was not told to the teams in advance. Fitting of the data to different flux rope models gave quite different results. However, Riley et al [2004] concluded that "it is *how* the technique is applied, rather than *which* technique is applied" that impacts the results of the fit most significantly. At the time of writing this book the methods of determining the magnetic structure of ICMEs are still in their infancy.

## 12.5 CMEs, Flares and Particle Acceleration

Both flares and CMEs accelerate charged particles to high energies. Typical flare-accelerated protons have energies of the order of 10 MeV, but can reach 1 GeV. Electrons are less energetic, typically 100 keV, but may in rare events reach 100 MeV. Particles are accelerated in all directions, some of them give rise to X- and $\gamma$-rays when they interact with solar plasma, some produce radio waves in the strong magnetic field structures, and some escape from the Sun and become observable in the heliosphere. In this section we discuss particle acceleration from the terrestrial point of view.

Note that the fluxes of energetic particles are much less than the flux of the ambient solar wind. A typical flare causes at 1 *AU* a flux of $10^7$ particles $\mathrm{m}^{-2}\,\mathrm{s}^{-1}$, whereas typical solar wind flux is $5\times10^{12}$ particles $\mathrm{m}^{-2}\,\mathrm{s}^{-1}$. The galactic cosmic ray fluxes at 1 *AU* are even smaller, about $6\times10^{2}$ particles $\mathrm{m}^{-2}\,\mathrm{s}^{-1}$.

When considering the propagation of energetic particles from the Sun to the heliosphere, the Parker spiral has to be understood properly. The bulk solar wind plasma expands almost radially out from the Sun and *causes* the spiraling magnetic field because the magnetic field lines are tied to the surface of the rotating Sun. However, this is an integral picture of all particles in the energy range of the bulk solar wind. Each individual particle, of any energy, is bound to the Larmor motion around the bent magnetic field lines with an individual speed along the magnetic field. Thus high-energy particles reaching the Earth in a short time mostly originate from the western half of the solar surface, which is the most typical footpoint of a solar wind field line intersecting the Earth's orbit.

The *solar energetic particle events (SEPs)* are divided into two main categories: *impulsive* and *gradual*. The impulsive events are much more common than the gradual events. Table 12.1 lists the main observational characteristics of these two classes.

**Table 12.1** Properties of impulsive and gradual solar energetic particle events (according to Lang [2000]).

| | impulsive | gradual |
|---|---|---|
| Particles: | electron-rich | proton-rich |
| $^3$He/$^4$He | $\approx 1$ | $\approx 0.0005$ |
| Fe/O | $\approx 1$ | $\approx 0.1$ |
| H/He | $\approx 10$ | $\approx 100$ |
| Duration of X-ray flare | impulsive (minutes) (hard & soft X-rays) | gradual (hours) (soft X-rays only) |
| Duration of particle event | hours | days |
| Radio bursts | Types III and IV | Types II and IV |
| Coronagraph observations | typically nothing | CME in 96% of cases |
| Solar wind observations | energetic particles | very energ. particles |
| Longitudinal extent | $< 30°$ | $\approx 180°$ |
| Events/year (solar max.) | $\approx 1000$ | $\approx 100$ |

Particles in impulsive events are thought to be accelerated close to the Sun both by the rapid energy release in the impulsive phase of a flare and by strong wave activity associated with the process. The very high abundance of $^3$He in impulsive events is a curious fact because the fraction of $^3$He nuclei of all helium in the solar atmosphere is about $5 \times 10^{-4}$. This indicates that, whatever the acceleration mechanisms are, at least one of them must be very efficient in accelerating this particular species. The same acceleration mechanism may accelerate also $^4$He, as the time profiles of both species are similar. This points to gyro resonant wave–particle interaction with waves having frequencies below the proton gyro frequency ($f_{cp}$), as the gyro frequencies of $^3$He$^{++}$ and $^4$He$^{++}$ are 2/3 $f_{cp}$ and 1/2 $f_{cp}$, respectively. In their high-frequency domain the Alfvén waves propagate as electromagnetic ion cyclotron waves at frequencies below $f_{cp}$ and are potential candidates for acceleration. Note that modern instruments have also found $^3$He/$^4$He-ratios of the order of 0.01–0.05 also in several gradual events, which supports the case for a similar mechanism in at least some gradual events.

The very strong association of gradual solar energetic particle events with CMEs suggests that in gradual events particles are most likely accelerated by the shock wave that the ICME drives in the solar wind plasma. This also explains the much longer duration of the gradual events than the impulsive flare-associated events. One might, furthermore, expect that the long duration could explain the acceleration to higher energies. However, it is unlikely that shock acceleration could lift the ambient solar wind temperature so much. Thus *pre-acceleration* in the corona is an obvious requirement for particles reaching energies of several tens or hundreds of MeV.

Again the common jargon may mislead an uninitiated reader. One must not mix up the *gradual flares* and *gradual particle events*. The gradual flares are post-CME phenomena in the corona and they are not responsible for particle acceleration in the gradual particle events.

The details of SEP acceleration belong to the many unsolved questions in solar physics. From particle observations we know that very efficient acceleration must take place and it is quite clear that the energy must have magnetic origin. In fact, only the very strongly stressed and sheared magnetic field structures have enough energy to explain the rapid acceleration. But the road from these conditions to a satisfactory physical explanation of acceleration in the great variety of explosive phenomena is long and winding. For example, a direct acceleration in a reconnecting current sheet cannot explain the high energies, as the outflow speed is limited to the order of the Alfvén speed in the inflowing plasma. It is more likely that *stochastic acceleration* by strongly fluctuating fields, including shocks, ion cyclotron waves, turbulence, etc., provides the main routes to particle energization. Several of these mechanisms may need to be combined before a particle is lifted from the quasi-thermal background to the observed energy level.

# 13. Magnetospheric Storms and Substorms

Strongly perturbed conditions in the magnetosphere form a class of phenomena that we call *magnetospheric storms*. Historically, they have been observed as strong perturbations of the Earth's magnetic field, and thus they are often referred to as *magnetic storms*. Storms are truly global and can be seen in magnetic recordings all over the Earth. During a magnetic storm a number of more localized *substorms* may take place. Despite their suggestive name substorms should not be seen just as building blocks of storms. A storm is more than the sum of substorms. On the other hand, isolated substorms, i.e., non-storm-time substorms, are actually much more common than substorms occurring during strong magnetospheric perturbations.

The main drivers of the magnetospheric storms are the interplanetary coronal mass ejections (ICMEs, Chap. 12) and the fast solar wind coming from the coronal holes. As the ICMEs are intermittent events, they give rise to *non-recurrent storms*. Another major class of storms are *recurrent storms* driven by high-speed solar wind and reappearing after about 27 days, when the same coronal hole is facing the Earth. The strongest global magnetic perturbations take place during ICME-driven stroms, whereas the high-speed wind is more strongly related to enhanced radiation belt electron fluxes and substorm activity.

## 13.1 What are Magnetic Storms and Substorms?

Depending on which signatures we are considering magnetospheric storms can look quite different from one event to another. A colleague of the author once remarked, "if you have seen one storm, you have seen one storm". Due to this variability and the actually relatively small number of storm events during, say, one solar cycle the results of very popular "statistical" studies must be interpreted with great care. In this section we review some of the major characteristics that most storms and substorms have.

H.E.J. Koskinen, *Physics of Space Storms: From the Solar Surface to the Earth*,
Springer Praxis Books, DOI 10.1007/978-3-642-00319-6_13,

### 13.1.1 Storm basics

A magnetic storm is a period of strongly disturbed magnetic field in the magnetosphere and on the ground. The storm conditions can last from several hours to a few days, and sometimes a new storm commences before the magnetosphere has fully recovered from the previous perturbation. There is no unique lower threshold for the magnetic perturbation above which it should be called a storm. In Sect. 1.5.1 we adopted the convention that low-latitude magnetic perturbation yielding the minimum *Dst* of –50 nT represents the threshold between weak and moderate storms. Identification of weak storms from observational data is often ambiguous, which in practice leads to somewhat different statistical results in studies that start from, say –30, –40, or –50 nT. In this book we call storms with *Dst* from –50 to –100 nT *moderate*, from –100 to –200 nT *intense*, and those with $Dst < -200$ nT *big*. There is no commonly agreed terminology here and storm classifications based on other indices, e.g. $Kp$, can place a given storm into a different category.

Figure 13.1 illustrates an intense storm as seen in magnetograms from four low-latitude stations. If the storm is driven by an ICME with a shock, as in this particular case, the storm begins with a rapid positive deviation of the magnetic north component ($H$), here at about 02 UT. This *storm sudden commencement* (SSC) is a signature of an ICME shock hitting the Earth's magnetopause. As the ICME pushes the magnetopause closer to the Earth, the Chapman–Ferraro current must increase to shield the enhanced geomagnetic flux density

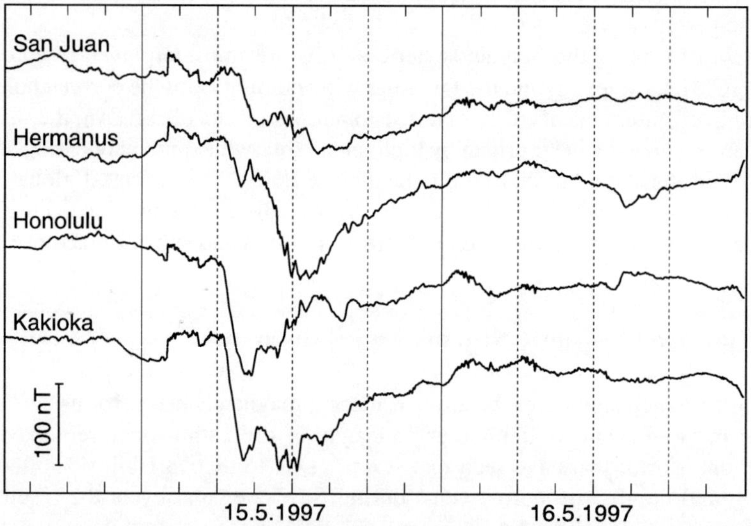


**Fig. 13.1** The horizontal component ($H$) of the magnetic field measured at four low-latitude stations during a magnetic storm on May 15, 1997. The sudden commencement took place on May 15 at about 02 UT, which is indicated by a sudden positive jump of the $H$ component at all stations. The main phase of the storm started after 06 UT as indicated by the strong negative deviation of the $H$ component. The solid vertical lines indicate the change of UT day and the horizontal tick-marks are given for each 3 hours. (Figure by courtesy of L. Häkkinen)

from the solar wind. Because the direction of the dayside $\mathbf{J}_{CF}$ is eastward, i.e., opposite to the ring current, the enhanced current, which is at the same time pushed closer to the Earth, causes a positive deviation in the $H$ component. This is the effect that is removed from the pressure corrected index $Dst^*$ using the formula (1.64). The SSC is a global phenomenon and the shaking of the magnetosphere is visible also in auroral region magnetometers and even in the nightside.

Magnetic storms can also be driven by low-speed ICMEs and by fast solar wind without a preceding shock. Thus there are storms without a sudden commencement signature. On the other hand, a shock wave hitting the magnetopause does not necessarily lead to a storm development. In such cases the positive deviation in the magnetograms is called a *sudden impulse* (SI).

After the SSC a period called the *initial phase* begins, which in Fig. 13.1 is best seen in the Honolulu magnetogram. The initial phase can have a very variable length depending on the structure of the solar wind driver (Sect. 13.4). If the IMF in the the sheath region between the shock and the ejecta is southward, the initial phase may be very short and the *main phase* commences as soon as the energy transfer into the ring current, and at the same time also to the tail current, has become strong enough. If the sheath IMF is northward, the main phase will not begin until a southward field of the ejecta enhances reconnection on the dayside magnetopause. If there is no southward IMF within the part of the ICME interacting with the magnetosphere, no regular global storm is expected to take place, unless the event is followed by fast and long-duration enough fast solar wind with southward component of IMF capable of driving a storm on its own. However, a pressure pulse also with northward IMF can shake the magnetosphere enough to trigger a substorm sequence.

The main phase is characterized by a rapid decrease of the $H$ component. This is due to strong enhancement of the westward ring current. The enhanced energy input from the solar wind leads both to energization of the current carriers and to increase of their number in the inner magnetosphere. Note that the ring current enhancement is typically not quite symmetric because a major fraction of the current carrying energetic ions are not necessarily on closed drift shells but pass the Earth in the evening sector and continue toward the dayside magnetopause. This is consistent with Fig. 13.1, in which the Honolulu and Kakioka magnetometers show steepest main phase development between 0630 and 0800 UT. Storms have also strong effects on the electron energy and content in the radiation belts. We will discuss the storm-time ring current and radiation belt dynamics more thoroughly in Chap. 14.

At some point in time the energy input from the solar wind ceases and the losses of energetic ring current carriers exceed their sources. Consequently, the $Dst$ index starts to return toward the background level. This phase is called the *recovery phase*. It is typically much slower than the main phase because the loss processes of the current carriers are much slower than the enhancement of the ring current. During the main phase and around the minimum $Dst$ several substorms may occur.

### 13.1.2 The concept of substorm

So what are the substorms? The evolution of the magnetic $H$ component during a storm can be quite irregular and, in particular in the auroral zone magnetograms, significant shorter-term activity can be seen. Birkeland [1908] called these activations *polar elementary storms*. The term substorm (more exactly, DP substorm, where DP stands for polar disturbances) was introduced by Akasofu and Chapman [1961]. A few years later Akasofu published the landmark paper [Akasofu, 1964] in which he used the term *auroral substorm* and described the process in terms of visual auroras using a large number of auroral all-sky images. While these polar elementary storms or auroral substorms are most pronounced within the night sector ionosphere, they involve large parts of the magnetosphere and the electrodynamic coupling between the ionosphere and the magnetosphere is an essential element in the process.

Similarly to storms the substorms can be divided into three distinct phases. The first one is called the *growth phase*, during which one or more quiet auroral arcs in the midnight sector slowly drift equatorward and the auroral electrojets enhance. Akasofu [1964] actually did not consider the growth phase as part of the substorm but defined the beginning of the substorm, the *substorm onset*, to take place when the most equatorward auroral arc suddenly brightens and the activation starts to expand to the west, to the east, and poleward. This *auroral breakup* signals the beginning of the substorm *expansion phase*. The growth phase was introduced later by McPherron et al [1973] to describe the phase when the magnetotail field becomes stretched until it breaks at the expansion phase onset. The growth phase may not necessarily lead to an auroral breakup and the expansion phase. During the growth phase also smaller localized activations called *pseudobreakups*, not leading to a full-fledged expansion phase evolution, may take place [e.g., Koskinen et al, 1993]. Sometimes the process stops at a pseudobreakup, but often the growth phase continues after the pseudobreakup to the expansion phase onset. The expansion phase lasts typically for a half-hour, after which the magnetosphere and ionosphere return to quiet conditions during the *recovery phase*. The whole substorm sequence takes typically 2–4 hours.

This basic sequence describes quite well typical isolated substorms. As we shall see later, the storm-time auroral activations are more complicated: new expansions may start before the previous ones have recovered and no growth phase is necessary, and also the observational characteristics both in space and on the ground may be different.

### 13.1.3 Observational signatures of substorms

Let us begin with the definition of the coordinate system that we use in the discussion of magnetospheric dynamics, the *Geocentric Solar Magnetospheric coordinates* (GSM). In this system $X$ points toward the Sun and the Earth's dipole axis is in the $XZ$-plane. Thus $Z$ points nearly northward and $Y$ is roughly opposite to the Earth's orbital motion around the Sun. As the Earth rotates, the $XZ$-plane flaps about the $X$-axis such that the dipole remains in that plane but can be tilted maximally 34° from the $Z$-direction (the sum of the angles of rotation and dipole axes).

### *Solar wind*

We start the discussion of substorm observations from the driver of the system, the solar wind. Fairfield and Cahill [1966] were the first to establish a clear statistical correlation between the direction of the IMF and auroral activity. This is consistent with the picture of reconnecting magnetosphere (Fig. 1.23). The southward IMF facilitates the dayside reconnection and, consequently, enhances the dynamo action over the nightside magnetopause, which in turn leads to enhanced convection. As we have seen in Chap. 8, the convective electric field gives a measure of how effective the reconnection process is. As the magnetospheric magnetic field on the dayside magnetopause points to the north, the rectified dawn-to-dusk pointing component of the interplanetary electric field $E_Y = VB_s$, where $V$ is the solar wind speed and $B_s$ is the southward component of IMF (zero if the IMF north-south component points toward the north), is an essential parameter. In fact, $VB_s$ is one of the most popular functions in studies on the dependence of the *Dst* index on the solar wind [Burton et al, 1975].

Another, a bit more sophisticated, widely used solar wind–magnetosphere coupling function is Akasofu's *epsilon parameter* [Perreault and Akasofu, 1978; Akasofu, 1981] given in SI units as

$$\varepsilon = \frac{4\pi}{\mu_0} VB^2 \sin^4\left(\frac{\theta}{2}\right) l_0^2 , \tag{13.1}$$

where $l_0$ is an empirically determined scale length, set to $7\,R_E$, $B$ the magnitude of the IMF, $V$ the solar wind speed and $\theta$ the "clock angle" between $Z$-axis and the projection of IMF onto the $YZ$-plane in the GSM coordinates. Thus the gate function $\sin^4(\theta/2)$ is zero, when the IMF projection in the $YZ$-plane points directly to the north ($B_Y = 0$), and one, when it points directly to the south.

We will discuss the physics behind the epsilon parameter in Sect. 13.6 but note already here that it is given in units of power (W), scaled empirically to estimated energy transfer to the inner magnetosphere, and its dependence on the clock angle allows for weak energy transfer also during northward IMF. During northward IMF epsilon is typically less than $10^{10}$ W. When the IMF turns toward the south, epsilon increases rapidly. As a rule of thumb, when epsilon exceeds $10^{11}$ W, a substorm is expected to break up. During a strong substorm epsilon can exceed $10^{12}$ W, and during intense storms $10^{13}$ W. While these are rough estimates only and the actual energy input is an integral over finite time, we can conclude that the magnitude and direction, in particular the southward and northward turnings, of the IMF and the solar wind speed belong to the critical parameters of the substorm process.

### *Ionosphere*

Let us then jump to the low-altitude end of the system, the ionosphere. During the substorm growth phase the energy input from the solar wind to the magnetosphere increases, which enhances the magnetospheric convection (Chap. 1). Mapped along the magnetic field lines into the ionosphere the motion is across the polar cap from the dayside to the nightside and returns back to the dayside through the evening and morning sector auroral zones. Recall that in the ionospheric E-layer the ion motion is constrained by collisions with neutrals, whereas the electrons follow the convection pattern yielding the electrojet currents directed opposite to the convective flow. During the growth phase the electrojet currents are enhanced and the *AE* index (Sect. 1.5.1) rises, although slowly. Note that, instead of the electrojets, it is also possible to measure the convection across the polar cap using a magnetometer close to the magnetic pole. This is actually done to calculate the so-called *PC* index, which has a good correlation with the *AE* index [Troshichev et al, 1988].

Another ionospheric signature of the growth phase is the enhancement of energetic (>30 keV) electron precipitation from the magnetosphere. The ionization caused by these electrons attenuates interstellar radio noise at 30 MHz, which can be observed from the ground using a special radio receiver called *riometer*. When stopped in the atmosphere, the electrons emit bremsstrahlung, which is possible to measure with high-altitude balloon-borne X-ray detectors.

At the time of the substorm onset (and auroral breakup) a strong westward current, *substorm electrojet*, appears in the ionosphere around the magnetic midnight. This leads to a rapid negative drop of the *AL* index (and enhanced *AE*). The *AE* network is quite sparse and thus the index does not give an accurate timing for the onset, unless one of the magnetometers happens to be located just below the newly established substorm electrojet. There are also more global methods to time the onset. One is to measure magnetic micropulsations with periods of 40–150 s, known as Pi2 pulsations (i stands for "irregular" and 2 indicates the frequency range). The pulsations are a response of the global magnetohydrodynamic system to the establishment of a new field-aligned current loop, the *substorm current wedge* (SCW), that connects the substorm electrojet to the current sheet in the magnetotail (Fig. 13.2). These ionospheric signatures of the onset take place within 1–2 min of each other, which is comparable to the period of Pi2 pulsations and thus to their temporal resolution. Accurate timing is, however, one of the most critical issues in substorm research. The onset signatures in different parts of the magnetosphere and ionosphere have different time constants and the observational tools have different time resolutions, which often complicate the analysis.

The expansion phase is the visually most impressive and physically most strongly disturbed part of the substorm process. The auroral activity expands from the original breakup close to the local magnetic midnight to the east, to the west and poleward (Fig. 13.3). The western edge of the expansion, associated with the upward current part of the SCW, is known as the *westward traveling surge* (WTS). There has, however, been some debate whether the surge really propagates in the ionosphere, or whether the propagation is merely an illusion arising from the formation of consecutive new fronts ahead of the previous ones as long as the expansion takes place.

During the expansion phase the ionospheric currents and electric fields behave in much more complicated ways than the sketch in Fig. 13.2 would suggest. The upward current has filamentary small-scale structure and it does not need to be in full balance with the current flowing down at the eastern edge of the wedge. Some of the upward current filaments appear to have a closure to local downward filaments in the near vicinity of the WTS [e.g., Marklund et al, 1998]. Note that the optical auroras also display fine-structuring that most likely is caused by local low-altitude processes and does not necessarily have any direct connection to the processes in the outer parts of magnetosphere. The total ionospheric

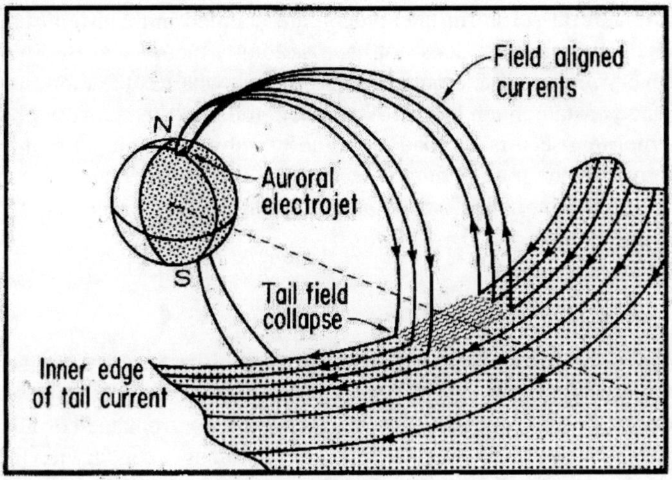


**Fig. 13.2** The establishment of the substorm current wedge according to McPherron et al [1973]. Note that the picture is highly idealized. The coupling of the FAC to the tail current is not as sharp as the figure suggests but takes place within a much larger volume. Recall from Chap. 6 that, in the steady state, current flows field-aligned only in the region where $\nabla P$ is negligible.

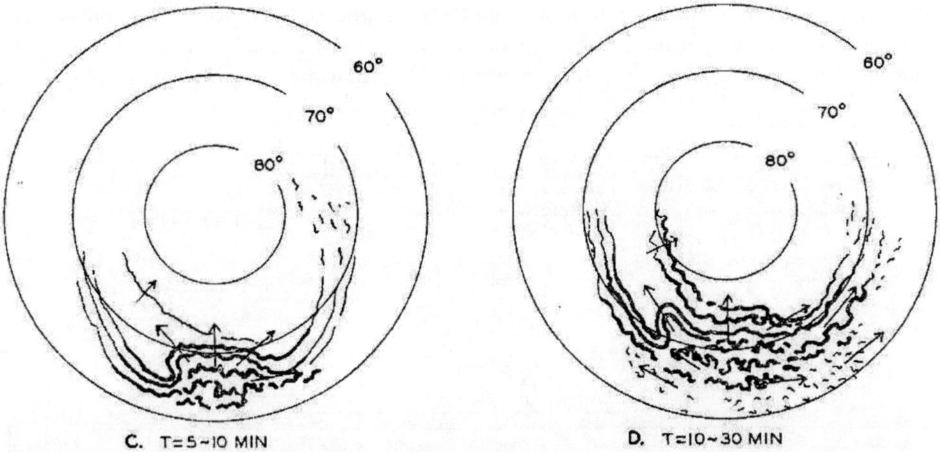


**Fig. 13.3** A part of the original illustration by Akasofu [1964] of the expansion of the auroral substorm in the ionosphere. Times are relative to $T = 0$ that refers to the onset of the expansion phase.

current in the horizontal part of the SCW is of the order of 1 MA, but a part of the observed current can be due to the enhanced ionospheric electric field, i.e., the enhanced convection, and does not need to be deviated from the tail current.

At some moment the substorm has exhausted the energy stored during the growth phase. When the direct energy input from the solar wind ceases, the recovery phase commences. Note that while the term is the same, the substorm recovery is different from the recovery of a storm. The auroral substorm phenomena are coupled more strongly to processes in the mid-tail, where the system does not have as long a memory as the ring current. The substorm recovery in the tail is a rather rapid process, whereas the more inert ionosphere remains in an active state much longer. At the beginning of the recovery phase the oval in the night and morning sectors is broad in latitude exhibiting bright auroral forms. Most of the activity moves to the post-midnight sector, and toward the end of the recovery large eastward traveling forms, called *omega bands* according to their spatial appearance, can sometimes be recognized.

### *Geostationary orbit magnetosphere*

It is a fortunate coincidence that the *geostationary orbit* at $6.6 R_E$ from the center of the Earth is a favorable region for substorm observations in the magnetosphere. Thus numerous geostationary satellites, having their main objectives from meteorology to military applications, have carried space physics instrumentation as secondary payloads.

Figure 13.4 is a classic illustration of such a dual-use of a military satellite equipped with energetic particle detectors for investigation of the substorm process. Before the substorm onset the magnetic field became increasingly stretched ($\theta_B$ in the figure increased toward 90°). In the same time the positive anisotropy parameter indicated a cigar-shaped electron distribution function with $T_\parallel > T_\perp$. These are signatures of the growth phase. The energetic electron fluxes first decreased and then disappeared just prior to the onset. This did not mean that the particles had disappeared from the plasma sheet. Instead the thinning plasma sheet moved away from the spacecraft. Note that the spacecraft was not quite at

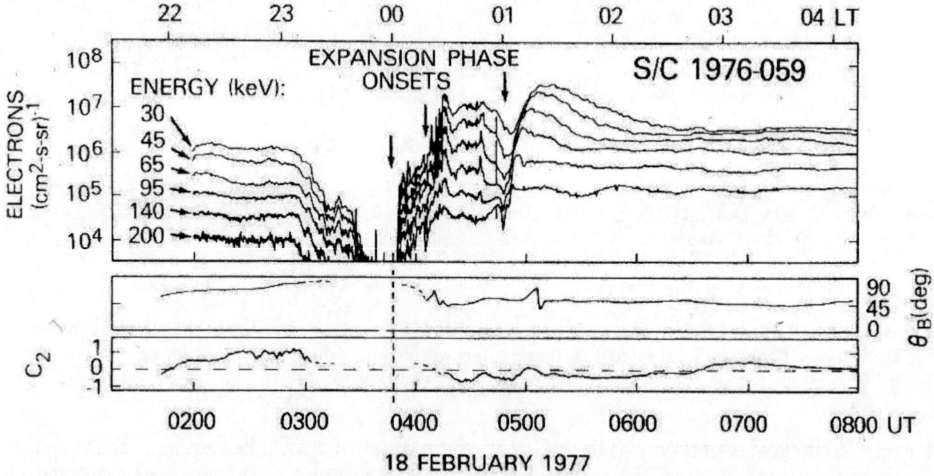


**Fig. 13.4** Observations of energetic (> 30 keV) electrons on the geostationary orbit. The top panel shows differential electron fluxes in 6 energy channels from >30 keV to >200 keV. The second panel gives the direction of the magnetic field that in this case was determined from the direction of the atmospheric loss cone ($\theta_B$ is the angle from the GSM Z-direction). The curve in the bottom panel indicates the anisotropy of the electron distribution. (Adapted from Baker and McPherron [1990].)

the center of the sheet even before or after the event as indicated by $\theta_B \approx 45^\circ$ during these periods.

The stretching of the magnetic field configuration also explains the previously mentioned enhanced energetic electron precipitation causing the attenuation of cosmic radio noise and X-ray emissions. As the field is stretched, its curvature radius approaches the gyro radii of the electrons. The electron motion across the current sheet becomes chaotic and some of the electrons jump into the atmospheric loss cone in the velocity space as discussed at the end of Sect. 3.4.

Soon after the expansion phase onset the spacecraft observed an almost *dispersionless electron injection* with particles at all energies arriving simultaneously. Thus the particles had not experienced much energy-dependent eastward gradient and curvature drifts (Chap. 3) and were likely injected from farther in the tail in the time sector where the spacecraft was located.

The two later injections indicated in Fig. 13.4 were dispersive, i.e., electrons with highest energies arrived before those with lower energies. At these times the satellite had moved toward the east with the rotation of the Earth, and the energetic electrons reaching the satellite had gradient and curvature drifted some distance eastward before reaching the satellite instrument. If the observing spacecraft had been in the evening sector, i.e., west of the injection longitude, similar behavior would be seen in energetic ion data. These injections are strongly correlated with expansion phase onset signatures in the ionosphere. The injected particles reach the geostationary orbit from the tail, but the distance to the origin of the injection is difficult to determine. Consequently, we have one more substorm onset signature whose exact timing is difficult.

Another important observation within the vicinity of the geostationary orbit is a rapid *dipolarization* of the magnetic field after the onset, i.e, the stretched magnetic field configuration returns toward a more dipole-like configuration. This is evidence of the relaxation of the magnetic tension accumulated during the growth phase. Dipolarization is generally thought to be associated with the establishment of the substorm current wedge through the midnight ionosphere.

#### ***Magnetotail***

The thinning of the tail plasma and current sheets during the substorm growth phase can also be seen farther out. The most popular view of the substorm process is that the near-Earth particle injections and the dipolarization as well as the establishment of the substorm current wedge are consequences of magnetic reconnection taking place somewhere at the distance of 8–30 $R_E$ from the Earth. This view is known as the *near-Earth neutral line (NENL) model* (for a review, see Baker et al [1996]). The word "near" refers to the creation of a new neutral line much closer to the Earth than the distant X-line somewhere beyond 100 $R_E$ in the steady-state Dungey picture.

Once the near-Earth reconnection starts to evolve the NENL cuts the magnetic connection between the near-Earth and far-tail plasmas with outflow jets toward the Earth and into the downwind direction ($T = 0$ in Fig. 13.5. Tailward of the NENL the large-scale magnetic structure known as a *plasmoid* looks in a two dimensional cut like an island bounded by the distant X-line). The reconnection outflow pushes the plasmoid to the downwind di-

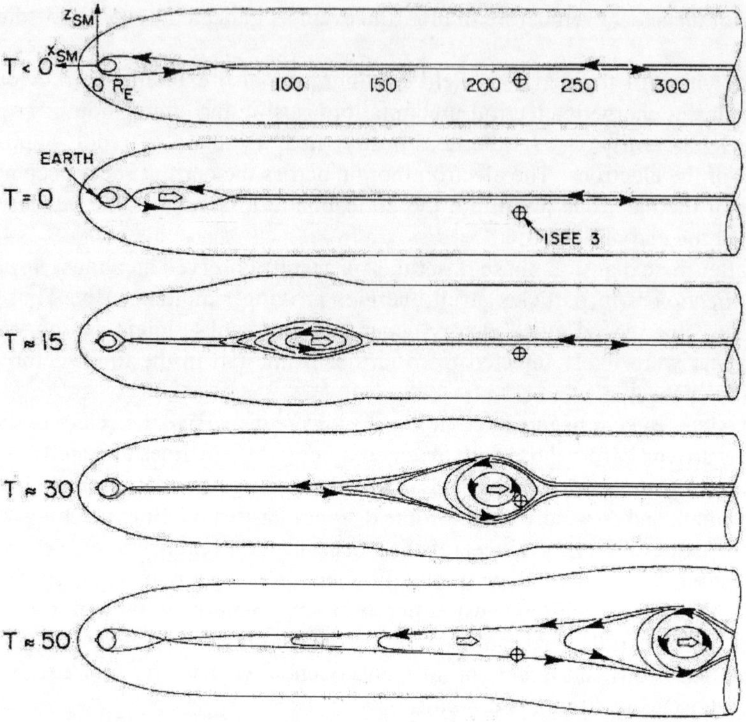


**Fig. 13.5** Formation of the near-Earth neutral line and the tailward motion of the associated plasmoid according to Hones et al [1984].

rection and eventually the plasmoid becomes detached from the magnetosphere . In reality plasmoids hardly are as symmetric as the two-dimensional cartoon of Fig. 13.5 suggests, but rather look like large flux ropes as discussed in the next section.

For a long time the empirical evidence for reconnection and NENL formation was indirect and based on observations of theoretically inferred consequences, such as outflow jets, or magnetic field perturbations known as *traveling compression regions* (TCR) in the far-tail being interpreted as bypassing plasmoids. However, in some fortunate cases spacecraft have passed through the ion diffusion region in such a way that it has been possible to infer even the Hall fields (Fig. 8.5) from the data. The first reports came from single-spacecraft observations by *Geotail* [Nagai et al, 2001] and *Wind* [Øieroset et al, 2001]. Using the four-spacecraft *Cluster* observations Runov et al [2003] were able to determine also the gradient scale size of the Hall-structure to be about 1500–2500 km, which according to the basic reconnection models should be of the order of the local ion inertial length $c/\omega_{pi}$.

While the Hall-type quadrupolar magnetic field structure has been recognized in some cases, the actual structure of the region where the tail is cut off may be more complicated and also different from one substorm to another. For example, Lui et al [2007] studied *Cluster* tail observations during a substorm expansion at the distance of about 19 $R_E$ downtail. They estimated the terms of the generalized Ohm's law written in the form

(8.40), where the resistive term $\eta \mathbf{J}$ was replaced by anomalous resistivity arising from the electromagnetic and plasma fluctuations as

$$\eta \mathbf{J} = -\frac{1}{n}\left[\langle \delta \mathbf{E} \delta n \rangle + \langle \delta (n\mathbf{V}_e) \times \delta \mathbf{B} \rangle\right] . \tag{13.2}$$

Within the uncertainties of the procedures for estimating the terms on the right hand side of this expression from *Cluster* observations the anomalous resistivity term was found to be the largest, the next was the Hall term, then electron pressure, and smallest was the inertial term. Furthermore, the electron diffusion region, where the anomalous resistivity was dominating, was much larger than suggested by the Hall MHD model in which the electron diffusion region is very small compared to the diversion of the ion flow by the Hall effect. Lui et al [2007] thus concluded that their observations were not consistent with X-line formation in the MHD theory but, instead, turbulence was the cause of the breakdown of the frozen-in condition.

In this context it is also worthwhile recalling from the discussion of the Hall effect in Chap. 8 that the 3D simulation results by Pritchett and Coroniti [2004] suggest that the Hall fields are strongly reduced, if there is strong enough guide field $B_y$. On the other hand, in an event study based on *Cluster* observations by Eastwood et al [2007] clear Hall structures in both the electric and magnetic fields were found around the reconnection site. In that event a strong guide field was observed within a small flux rope forming earthward of the diffusion region, but there was no significant guide field outside of the flux rope.

The signatures discussed in this section describe quite well an isolated substorm of a moderate size. However, not all of them are observed in all substorms and their relationships can be quite complicated. For example, there can be several expansion phase onsets following each other without clear growth phase signatures in between; there may not be a one-to-one correspondence between plasmoid release and near-tail dipolarization or enhancement of the substorm electrojet; and, as said earlier, storm-time substorms are even more intricate.

## 13.2 Physics of Substorm Onset

The NENL model is a comprehensive attempt to organize a wide variety of observations rather than a detailed description of the physical processes associated with the substorm process. Over the years several competing models have also been proposed. Most of the models focus on the onset mechanism, the tail current disruption, and early expansion, whereas less consideration has been given to the relaxation processes during the recovery phase. In this book we do not want to dwell on the history of the, sometimes quite heated, debate between the different models, nor on the large amount of observational details that have been brought to support the different views. As a starting point for an interested reader we suggest two comprehensive reviews that appeared in the same issue of the *Journal of Geophysical Research* in 1996 by Baker et al [1996] on the NENL model and by Lui [1996] on the *current disruption* (CD) model. The main difference between these approaches today concerns whether the reconnection at the NENL is the primary process, as discussed above, or is the near-tail current disruption an independent process that may or may not lead to the mid-tail X-line formation. These two views have more recently been termed as "outside–in" and "inside–out" views. There are several other approaches, in particular, the attempts to explain the onset as a result of a global instability in the current system coupling the magnetosphere and ionosphere together, termed *magnetosphere–ionosphere coupling* (MIC) model [e.g., Kan, 1993]. While the MIC model has not gained as much support as the NENL and CD models, the communication between the magnetosphere and ionosphere is an important issue in any approach to the substorm process.

We can think the initiation of the substorm expansion as a global instability of the magnetospheric configuration, or rather a complex of several instabilities from the microscopic to the macroscopic level. The auroral breakup, the establishment of the substorm current wedge and the release of a plasmoid are macroscopic phenomena, whereas the details of breaking the magnetic connection between the plasmoid and the inner magnetosphere, tail current disruption, or wave–particle interactions in the auroral acceleration region are essentially microphysical phenomena, of which we still, after more than 40 years of intense studies, can say remarkably little that is definitive.

The main physical building blocks, whose causal relationships need to be understood, are the large-scale flows in the magnetosphere and the ionosphere (the focus of the NENL model), the local instability mechanisms (the focus of the CD model), and the dynamics of the current systems between the magnetosphere and ionosphere (the focus of the MIC model). These elements may be causally dependent on each other or develop independently in different parts of the magnetosphere–ionosphere system. While they may be independent at some phase of the substorm sequence, they can become coupled to each other at a later stage.

### 13.2.1 The outside–in view

The NENL model is based on the dynamics of large-scale MHD flows where the tail reconnection plays the key role. The model invokes in a natural way the plasma sheet thinning during the growth phase, the earthward and tailward flows from the reconnection site after the expansion onset, and the plasmoid release. The modern versions of the model are more sophisticated than the original two-dimensional cartoons (e.g., Fig. 13.5). This is important, for example, for the description of the current diversion from the tail current sheet into the ionosphere as the substorm current wedge.

In the NENL model large-scale reconnection is assumed to start when the current sheet thinning has reached an unspecified threshold determined by local plasma parameters. Thus the onset of reconnection is an example of a global instability driven by external input, i.e., piling of magnetic flux to the tail lobes. However, the reconnection cannot start until the local parameters become conducive to the process. Traditionally the local (microscopic) instability process has been assumed to be the collisionless tearing instability, but as discussed in Chap. 8 the microphysics of collisionless tearing is a complicated issue. One should again be careful not to hook up on terminology of poorly understood phenomena. "Tearing" may be an attractive descriptive term, but the underlying physics is another matter.

An early problem with the NENL model was that it, at least implicitly, assumed that the substorm onset takes place when the tailward part of the plasma sheet becomes magnetically disconnected from the Earth. In that case the NENL would map to the boundary of open and closed field lines in the ionosphere, i.e., on the poleward edge of the auroral oval. As Akasofu [1964] has already noted, the auroral breakup usually takes place on the most equatorward auroral arc, poleward of which there often is a wide band of precipitating plasma sheet particles. Magnetic mapping of the breakup signatures and the location of the substorm electrojet, even assuming a considerable stretching of the near-tail magnetic configuration during the growth phase, suggested that both the source of the particle precipitation and the SCW were on the closed field lines not far beyond the geostationary distance. This is much closer to the Earth than the statistically determined location of the X-line at or beyond $-20R_E$ based on observations of reconnection outflows.

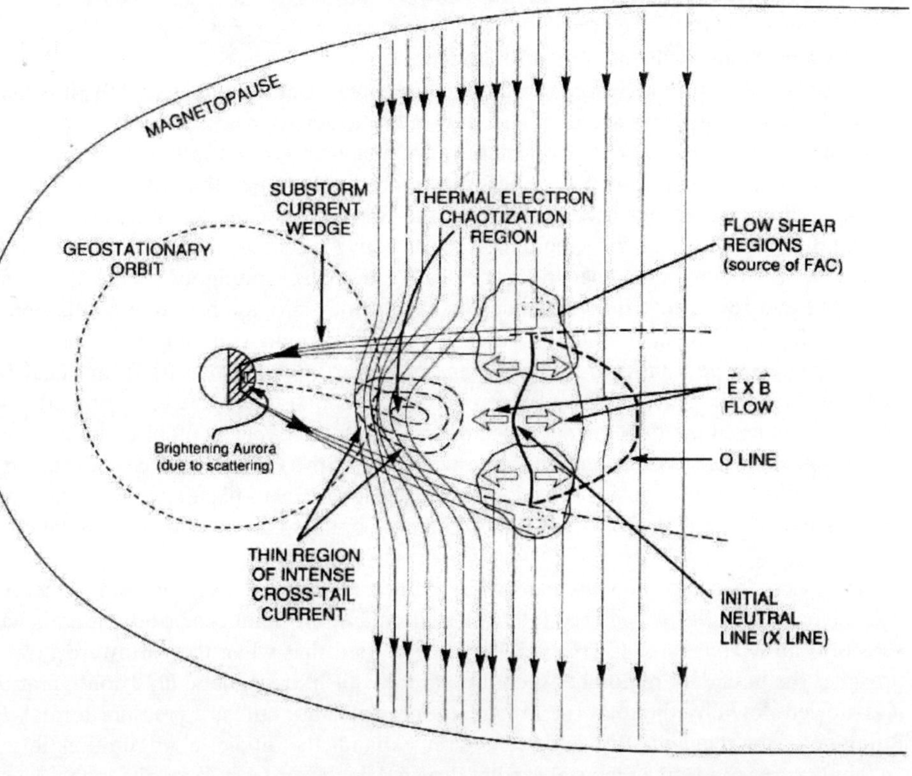


**Fig. 13.6** A schematic of how the SCW generation is described in the modern NENL picture according to Baker et al [1993]. The FAC rises from shear flows at the dawn and dusk flanks of the reconnection region. In addition to the SCW this picture addresses the brightening of the breakup auroral arc. In the domain labeled "thermal electron chaotization region" the current sheet is so thin that the adiabatic invariance of the magnetic moment of thermal electrons is broken. A fraction of these particles is scattered to the atmospheric loss cone leading to enhanced precipitation along the equatorward edge of the auroral oval.

As discussed by Baker et al [1996], by the mid-1990s overwhelming observational evidence from space-borne and ground-based observations indicated that the reconnection process must start within closed magnetic field lines and eat itself up to the open–closed field line boundary during the expansion phase within, say 30 min, consistent with the poleward expansion of the auroral bulge in the ionosphere. Figure 13.6 shows how both the generation of the SCW and the brightening of the equatormost auroral arc at the substorm onset can be reconciled with the NENL model. The SCW is explained to rise from the flow shears due to the fast outflows from the reconnection line, whereas the auroral brightening is due to enhanced particle scattering into the atmospheric loss cone through the chaotization of particle motion in the thin current sheet earthward of the reconnection line.

In the original NENL view when the cross-tail current in the vicinity of the X-line is cut off, the current has to find another way to maintain current continuity, which would give rise to the SCW. One can expect that the current must take the route of least resistance, but it is not evident that such a route would be through the resistive ionosphere. This is actually a problem pertinent to the CD model as well.

To explain the current closure through the ionosphere, at least heuristically, it is useful to recall that the "field-alignedness", i.e., how close to zero $\mathbf{J} \times \mathbf{B}$ is, depends in the steady state on the pressure gradient. By whichever mechanism current is fed from the current sheet toward a region of lower $\beta$, it becomes more field-aligned the farther it is from the high-$\beta$ region. We do not need to think of the formation of the SCW as just a passive current diversion. Instead, we should look for an active mechanism of generating the FAC, of course still maintaining that the large scale current is continuous, i.e., $\nabla \cdot \mathbf{J} = 0$. In Chapter 6 we found that time-dependent vorticity may provide a source and a sink for the FAC from and to the magnetosphere. In the NENL model suitable flow shears exist at the dawnward and duskward edges of the earthward-flowing plasma from the X-line. These shears are in the right sense to feed the SCW. At first the X-line has a limited dawn-to-dusk extent and the flow channel is narrow. During the expansion phase the earthward flow channel widens corresponding to the widening of the ionospheric part of the current wedge. In this view the current diversion to the ionosphere takes place somewhat earthward of the actual X-line (Fig. 13.6) and the brightening breakup arc is evidently on the closed field lines.

Another scenario for how the earthward outflow from the NENL can lead to the SCW was advocated by Shiokawa et al [1998] based on a multi-point case study of an isolated substorm on March 1, 1985 (Fig. 13.7). They argued that when the earthward flow approaches the boundary region between dipolar and tail-like magnetic field configurations, it is slowed down by the total (i.e., magnetic plus plasma) tailward pressure force $\nabla P_{tot}$. Piling up of the magnetic flux at this interface expands the dipolar configuration outward and thus corresponds to an inertial current directed dawnward, i.e., opposite to the tail current. In order to maintain total current continuity FACs in the sense of the SCW must be created. This can be called current disruption, but here it is a consequence of the NENL farther out. Shiokawa et al [1998] note that this scenario can explain the establishing of the SCW, but later in the expansion phase something else must sustain the current, as the initial fast flow lasts only about 10 min.

**Train your brain**

Show that both Fig. 13.6 and Fig. 13.7 are consistent with the current directions of the SCW, i.e., the source of the FAC flowing into the ionosphere is in the dawnward part and the sink of FAC on the duskward part of the wedge. Show further that the pattern is the same for both hemispheres.

The mechanisms suggested in Figs. 13.6 and 13.7 appear to create the FAC at somewhat different distances in the tail. Which one of them is closer to reality, or whether both of them can have a role in the substorm process, has not yet been firmly established. The braking mechanism may be important at the early phase of the SCW development whereas the shear flow can sustain the current system throughout the expansion phase.

Whatever the mechanism of current diversion is, building up the current loop through the ionosphere takes a finite time. The earthward flow creates a dawn-to-dusk directed polarization electric field $\mathbf{E} = -\mathbf{V} \times \mathbf{B}$. In ideal MHD the magnetic field lines are equipotentials and can be mapped to the ionosphere as discussed in Chap. 1. In reality this mapping is not established immediately but the electric field must be propagated along the magnetic field lines as an Alfvén wave, which is the fastest mode of propagating information of large-scale changes in the tail magnetic field configuration to the ionosphere. The polarization electric field in front of the wave sustains a current that is closed behind the wave by FACs in the same sense as the SCW. In the ionosphere, the electric field initiates an equatorward plasma flow. However, the inertia of the ionospheric plasma retards the flow

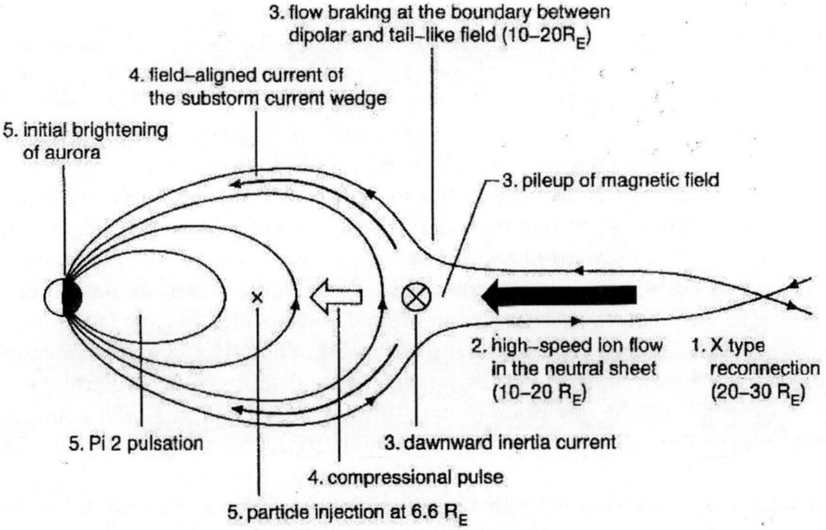


**Fig. 13.7** Schematic view of how the braking of fast earthward flow at the interface between the dipolar and tail-like field leads to the SCW through the dawnward inertial current. The numbering indicates the temporal sequence of different phenomena. (From Shiokawa et al [1998].)

and most of the electric field is reflected back to the magnetosphere. In the plasma sheet the wave is again bounced back to the ionosphere, and so on. At each bounce the total amount of FAC and ionospheric electric field grows until the tail flow and the ionospheric flow correspond to each other.

Thus the magnetosphere–ionosphere coupling is a natural and important part of the NENL model. The difference from the MIC model, to which we return in the context of the inside–out picture, is that in the NENL model the expansion phase onset is driven by the near-tail reconnection and the magnetosphere–ionosphere current system responds to it, whereas in the MIC model the onset is a consequence of the enhanced coupling.

Another important issue to understand is the plasmoid release. The early 2D picture where the entire tail is cut in a symmetric fashion is easy to criticize. The reconnection starts locally, potentially at several sites independently of each other associated with initiation of bursty bulk flows (Chap. 1) within closed field lines. Thereafter the system slowly evolves toward the large-scale plasmoid release. Furthermore, any finite background magnetic field component in the $Y$-direction superposed on a closed loop of magnetic field lines leads to a helical flux rope structure. The local $B_Y$ can be due to the stress imposed by the IMF $Y$ component on the magnetosphere or associated with the bending of the field-lines due to the local structure of the expanding reconnection region.

Assume that the flux rope begins to develop on closed field lines and $B_Y > 0$, i.e., pointing from dawn to dusk. Now the field is wound such that a field line originating from the southern hemisphere in the dawn sector connects through the flux rope to the northern hemisphere in the dusk. Such a flux rope can be called "right-handed", and for $B_Y < 0$ the helical structure is the opposite, "left-handed". Thus the ionospheric footpoints may have a considerable shift in the local time and during the process the ionospheric connection may be lost at different times in different sectors, which should cause a considerable azimuthal rotation of the flux rope axis. In fact, the first analysis of flux rope observations in the magnetotail [Sibeck et al, 1984] had already come to the conclusion that the flux rope axis can be nearly parallel to the tail axis, a result that the 2D space cartoonists seem to have often forgotten.

Slavin et al [2003] investigated *Geotail* observations of flux-ropes during the satellite's nightside season between November 1998 and April 1999 within $X$ distances from about $-14\,R_E$ to about $-30\,R_E$. They identified 73 flux ropes of which 35 were moving earthward and 38 tailward. The earthward-moving structures were associated with bursty bulk flows and thus named BBF flux ropes, whereas the tailward structures were named plasmoid flux ropes. The BBF flux ropes had smaller mean diameters ($1.4\,R_E$) and larger mean core fields (20 nT) than the plasmoid flux ropes ($4.4\,R_E$ and 14 nT). The azimuthal orientation of the flux ropes varied strongly, emphasizing the uneven detachment from Earth. Only 60% of the flux ropes could be described as cylindrical force-free ($\mathbf{J} \times \mathbf{B} = 0$) configuration. Note that while a force-free field tends to form a flux rope structure, the converse does not need to be true. Thus not finding a force-free configuration does not invalidate the MHD reconnection-based description of flux rope formation as sometimes has been claimed. Slavin et al [2003] interpreted the relatively large observation frequency of the BBF flux ropes as a signature of multiple X-line reconnection going on simultaneously at separated locations.

These features can also be found in numerical MHD simulations. Farr et al [2008] simulated a substorm event on August 11, 2002, which had been well-observed by several spacecraft and thus the simulation results could be benchmarked with actual data. The simulation produced initially two spatially separated plasmoid flux ropes, which coalesced after about 20 min. The process started within closed field lines and continued there for about 30 min, after which a complicated web of intertwined open and closed field lines emerged. After about 5 more minutes the flux rope finally consisted of open field lines predominantly detached from the Earth. The disconnection took place earlier in the dawn sector and the flux rope rotated towards an almost tail-axis-aligned orientation so that the duskward edge of the rope was much closer to the Earth when the structure was finally detached. Both the statistical results by Slavin et al [2003] and the event simulation study by Farr et al [2008] indicate that spacecraft observations of an individual substorm in the tail can look very different depending on at what time in what part of the evolving flux rope they have been made.

The NENL model is a flexible framework within which it is possible to embed new theoretical and observational elements along the way. Perhaps the most critical issue today is whether or not the outside–in concept is possible to confirm beyond reasonable doubt. Of course, it is quite possible, perhaps even likely, that there are substorms progressing from the outside in, while in other events the sequence may be the reverse.

### 13.2.2 The inside–out view

The inside–out view is advocated by the supporters of the CD model and also by many auroral researchers using ground-based instrumentation who are concerned by the fact that the auroral breakup takes place on the equatorwardmost auroral arc that is problematic to map along the magnetic field lines far into the magnetotail. According to the CD model the substorm process is initiated in the inner magnetosphere, not far from the geostationary orbit, and the mid-tail reconnection is a possible consequence of this process. The basic idea is that some current sheet instability in the near-Earth tail inhibits the current flow and forces its closure through the ionosphere as the SCW. The current disruption means also disappearance of cross-tail current and dipolarization of the stretched magnetic field structure.

An acceptable substorm model, even if limited to the onset and the early expansion phase, must also explain the observations of plasmoid release and mid-tail reconnection. In the CD model the disruption process is assumed to launch a rarefaction wave tailward. This tailward propagating rarefaction both induces earthward flow behind the rarefaction front and leads to the thinning of the current sheet to an unstable level and finally lets the reconnection go off (Fig. 13.8).

As discussed by Lui [1996] there are quite a few intriguing observations in the near-Earth tail that call for physical explanation and are not immediately related to the NENL model. These include the sometimes very turbulent magnetic field behavior close to the interface of dipolar and tail-like field lines [Takahashi et al, 1987] (Fig. 13.9) and very strong inductive electric fields associated with the rapid field line dipolarization [Aggson et al, 1983].

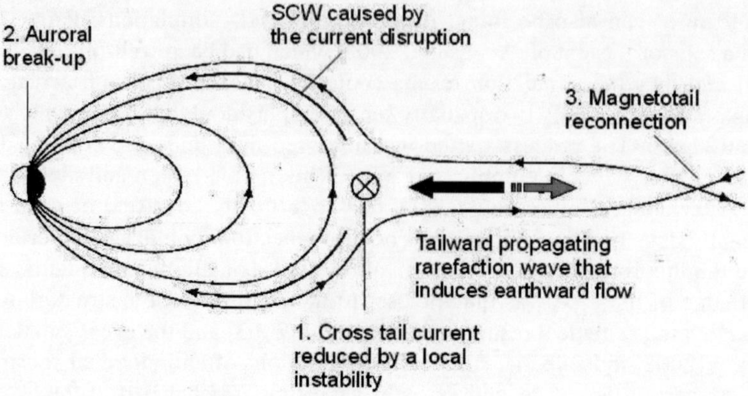


**Fig. 13.8** Figure 13.7 adapted to conform with the inside–out view of substorm expansion. The numbers indicate the expected temporal sequence.

Also the MIC model belongs to the inside–out family. In the MIC model the Alfvén waves bouncing between the ionosphere and magnetosphere during the growth phase lead to a feedback instability [cf., Lysak, 1991]. At the time when the system goes unstable, the SCW grows rapidly and reduces the cross-tail current leading to the dipolarization of the near-Earth field. In this picture no local instability process in the magnetospheric end of the current circuit is needed. The exchange of Alfvén waves during the growth phase is observable as magnetic pulsations because the strengthening of the magnetospheric convection must be communicated to the ionosphere. What is critical to the MIC model is whether it drives the substorm expansion, or not.

Another macroscopic instability proposed to explain the setting up the SCW is the ballooning instability [Roux, 1985]. As discussed in Chap. 7 (Fig. 7.4) a perturbation in the interface region between dipolar and tail-like field lines leads to an unstable undulation of the magnetic field configuration. The instability causes a build-up of space charge regions of alternating sign that are neutralized by electrons moving along the magnetic field lines, i.e., FACs. These elementary FAC loops coalesce and finally create the SCW. In the MHD picture this can be interpreted as enhanced Alfvén wave activity leading to the SCW formation.

Considerable effort has been given to finding out local microscopic current sheet instabilities that could be responsible for the current disruption spontaneously once the current sheet has reached an unstable state. Physically this is actually not so different from the search for the mechanism for setting up the reconnection process on closed field lines further out in the tail, although the proponents of the CD model sometimes wish to stress the difference of their approach from the microscopic studies of the reconnection process.

In the microscopic picture the free energy for the investigated instabilities is in the different velocity distribution functions of the particle species near the current sheet. For the current disruption, in particular, the different streaming of electrons and ions across the magnetic field is important. Some of the candidate instabilities are the ion Weibel instability (IWI), modified two-stream instability (MTSI), and the lower hybrid drift instability

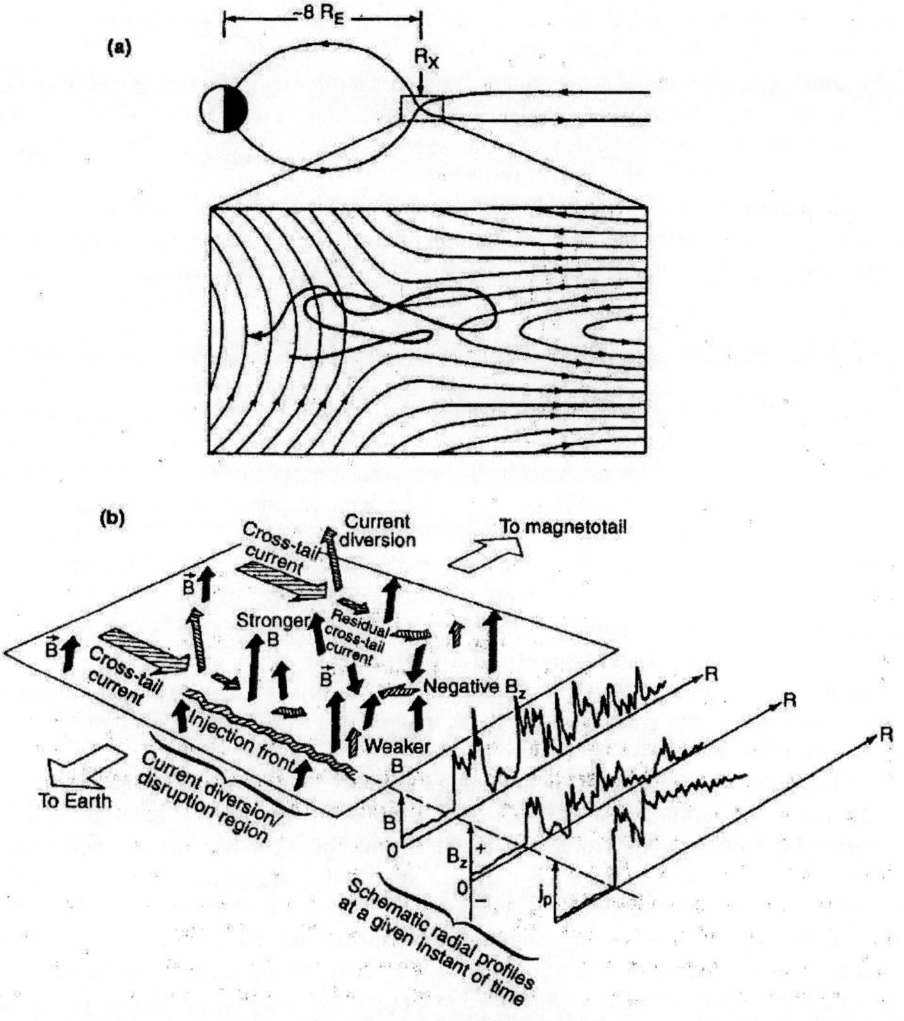


**Fig. 13.9** Illustration of turbulent magnetic field observations by *AMPTE/CCE* spacecraft around the interface region between dipolar and tail-like fields. The upper figure illustrates the interpretation by Takahashi et al [1987] according to which a neutral line was sloshing back and forth around the spacecraft. The lower figure describes the CD-model interpretation by Lui et al [1988], according to which the magnetic field fluctuations are associated with the current sheet disruption instability. (The figure is from Lui [1996].)

(LHDI) (Chap. 7). The unstable waves are electromagnetic modes on the whistler–lower hybrid surface of Fig. 5.4. The waves driven by IWI and MTSI can be called oblique whistler modes at frequencies above the ion gyro frequency, whereas the lower hybrid drift wave is an almost electrostatic mode propagating nearly perpendicular to the magnetic field. These modes are different from the tearing mode in that their wave vectors are perpendicular to the tail axis, whereas the tearing wave vector is approximately in the direction of the tail. So the physics of the instabilities is different.

In addition to the issue of what comes first and what follows, the supporters of the CD model tend to criticize the NENL model because of its foundations in the MHD description of magnetospheric plasma physics. Of course, MHD has its physical limitations, many of which are discussed in various parts of this book. In particular, we must not forget that the frozen-in picture is not always valid, neither is reconnection a magic wand that solves all problems in space plasma physics. On the other hand, when MHD picture is valid, it also is very powerful. Thus an open mind is needed on both sides of the debate.

### 13.2.3 External triggering of substorm expansion

Both the NENL and CD models explain the substorm expansion onset as a result of an internal instability in the magnetotail that goes off spontaneously once the appropriate conditions are obtained. However, the triggering of the onset can also be a consequence of an external perturbation, e.g., a change in the solar wind magnetic field or pressure that is strong enough to cause a sufficient change in the magnetotail.

Both a rapid reduction of the southward component (sometimes called a northward turning) of the IMF direction or a reduction of IMF $Y$-component have been found to often take place just before the expansion phase onset. In both cases the solar wind electric field is reduced. Lyons [1995] suggested a substorm theory in which this reduction is propagated into the magnetotail. In this theory the reduction of the magnetospheric electric field weakens the earthward convection in the magnetotail from the enhanced growth phase level. This reduces the cross-tail current and leads to the creation of the SCW.

While the theory has remained in the shadow of the NENL and CD models, the question of external triggering has also received considerable attention in studies of solar wind–magnetosphere interactions more recently. It is quite possible that the magnetosphere–ionosphere system is driven toward a marginal large-scale stability after a sufficiently long growth phase. Under such conditions even a weak external push might drive the system unstable. But how weak is strong enough, and what role does the preconditioning have in determining the threshold for the triggering to be effective? Evidently, the criterion, of how small solar wind perturbations are included in studies searching for possible triggers, may lead to different conclusions on how large fraction of substorms are triggered. As we do not yet know very well the transmission of the triggering signal from the solar wind to the nightside magnetosphere, finding potential triggers in the solar wind does not necessarily mean that they also act as triggers for the substorms. Correlation does not always imply a causal relationship.

### 13.2.4 Timing of substorm onset

The major disagreement between the supporters of the "outside–in" and "inside–out" views is in the timing between observations of different substorm-associated phenomena and, consequently, their causal relationships.

For example, Liou et al [2002] investigated dipolarization events observed at geostationary orbit together with auroral breakup observations using UV images from the *Polar* spacecraft. They required that the dipolarization had to take place within a longitude range of 2 hours in magnetic local time from the auroral breakup in the ionosphere in order to minimize the propagation delays in their analysis. They identified 32 clear dipolarization events suitable for the analysis.

The result of this study was that the auroral expansion preceded the geostationary dipolarization on the average by $1.7 \pm 2.7$ min (note that there were also negative time lags). Liou et al [2002] argued that this delay was mostly a propagation time effect within the auroral bulge latitudes and longitudes. They estimated that the location of the initiation of substorm process was at about $X = -8.3\,R_E$. Based on this they argued that the observations were inconsistent with the assumption that the substorm process would be initiated at the reconnection X-line, because the X-line has been statistically found considerably further down the tail.

In addition to the temporal resolution and the cadence of critical observations a serious problem has been the lack of comprehensive enough data sets of simultaneous observations of the mid-tail reconnection effects, the near-tail dipolarization events and ionospheric signatures. Thus the scientists performing very careful data analysis focusing on some part(s) of the issue have had to rely on statistical results on data not directly available simultaneously [e.g., Liou et al, 2002].

To address this problem NASA launched the five-satellite constellation *Time History of Events and Macroscale Interactions during Substorms (THEMIS)* in February 2007 [Angelopoulos, 2008]. The orbital periods of the individual spacecraft were selected so that they, during the northern winter, become recurrently aligned along a common nightside meridian above the North American continent, where an extensive ground-based network of optical and magnetic instruments for ionospheric observations was established to support the mission. One of the first contributions of the *THEMIS* mission was to provide a very detailed temporal sequence of a substorm that took place on 26 February 2008 [Angelopoulos et al, 2008]. At the time of the substorm onset the five *THEMIS* satellites were nearly aligned along a common meridian in the tail at $X$ distances from $-5.5\,R_E$ to $-21.5\,R_E$ and less than $1\,R_E$ from the nominal current sheet center.

This particular substorm was not particularly strong. The auroral electrojet index calculated using specific magnetometer stations for the *THEMIS* mission reached about 200 nT at its maximum. The magnetometers showed the appearance of the substorm electrojet at 0454:00 UT. Before this Pi2 pulsations had already been observed to start at an auroral station at 0452:00 UT and at a mid-latitude station at 0453:05 UT. The auroral intensification was observed for the first time at 0451:39 UT at 67.9° geomagnetic latitude and expanded poleward of 68.2° at 0452:21 UT, which is the time that Angelopoulos et al [2008] interpreted as the time of the substorm expansion onset.

The outermost *THEMIS* spacecraft (P1, $X = -21.5\,R_E$) observed at 0450:28 UT a tailward flow that was interpreted as the reconnection outflow, whereas the second farthest spacecraft (P2, $X = -17.2\,R_E$) observed an earthward flow at 0450:38 UT. From the observations of the flow speeds the reconnection onset was inferred to have taken place between the spacecraft at about $X = -20\,R_E$ at 0450:03 UT. This was 96 s before the auroral intensification, 117 s before the high-latitude Pi2 onset and 138 s before the auroral expansion onset.

The third *THEMIS* spacecraft (P3, $X = -10.9 R_E$) observed the earthward flow onset at 0452:27 UT and and a transient magnetic field dipolarization at the same time as the mid-latitude Pi2 pulsations were observed on the ground (0453:05 UT). The more permanent dipolarization was not observed until 0454:50 UT.

This study together with a somewhat more detailed analysis of a substorm on 16 February by Gabrielse et al [2009] represent good evidence that the tail reconnection can precede both the auroral intensification and the dipolarization in the near-Earth magnetosphere. While these are strong cases for the NENL model, also the opposite inside–out view has been supported by analysis of other substorm events observed by the very same satellite constellation.

Lui et al [2008] analyzed a series of three activations that took place on 29 January 2008. The *THEMIS* spacecraft were aligned close to the $X_{GSM}$ axis from $-8$ to $-30 R_E$. Furthermore, data from the geostationary *GOES* 11 and 12 satellites, which nicely bracketed the meridian of *THEMIS*, and from the ground-based all-sky camera network were used.

The first auroral activation took place at 0714 UT and was interpreted as a small substorm. The minimum *AL* was only about –120 nT and the activation lasted about 25 min. The activation was evidently confined into the closed field lines and never progressed to the open-close field line boundary. This activation might be better to describe as a pseudo-breakup than a full-fledged substorm onset.

The second activation at 0742 UT started a rather complicated sequence of a multiple-onset substorm. Lui et al [2008] interpreted the observed data to be inconsistent with the outside–in view until the activation at 0833 UT, which if analyzed alone would give support to the NENL initiation of the substorm process. This sequence of events was certainly more complicated than the 26 February event analyzed by Angelopoulos et al [2008]. The different activations did not produce well-defined signatures at all spacecraft and the timing of different elements could not be made as rigorously as in the 26 February case. However, it is clear that these data are difficult to interpret in terms of the NENL model or outside–in picture.

Thus to give the definitive answer to the inside–out versus outside–in question with the *THEMIS* constellation has not turned out to be as straightforward as some may have hoped. Instead, it looks like the magnetosphere would be capable of organizing the substorm energy release in many different ways. Furthermore, while the multipoint approach of *THEMIS* is very useful, each individual spacecraft makes observations only in a tiny little spot in the magnetotail. During substorm activity the tail bends, the spatial distribution of BBFs most likely is inhomogeneous, the role of the guide field is an issue, the plasmoid release can be very asymmetric, and, of course, there may be various localized instabilities that sometimes lead to current disruption, sometimes to reconnection, but may often just settle down without leading to observable large-scale signatures. Consequently, it may finally turn out to be hopeless to squeeze the substorm process into any unifying picture of the type of Figs. 13.7 or 13.8, very much analogous to the solar corona where the flares may take place in several different ways.

**Feed your brain**

Read carefully the articles Angelopoulos et al [2008] and Lui et al [2008] and think how *you* would reconcile two such opposite conclusions. When thinking about the problem, pay attention to the propagation speeds of the onset signatures between the different observations and consider your conclusions in terms of Alfvén velocity in the magnetosphere.

## 13.3 Storm-Time Activity

From the second half of the 1990s an increasing amount of interest has been paid to problems in the relationship between storms and substorms. While we can identify clear connections between substorm elements and fundamental plasma physics, such as reconnection, current sheet instabilities, Alfvén wave propagation, sources and sinks of field-aligned currents, etc., the situation with storm-time phenomena is more complicated, in particular if we want to make some sense of the great variety of activations taking place during magnetospheric storms. In fact, by the time of writing this book there have been remarkably few studies on storm-time activations that would in a comprehensive way address all relevant issues from the solar wind driver properties to phenomena in far- and mid-tail, and in the ionosphere.

Common to all storm-time processes is that they take place during strong solar wind driving and the ring current is at an enhanced level. Because we cannot deal with all observational details here, we group the strongly driven activity into three main categories: substorm-like activations, *sawtooth events*, and *steady magnetospheric convection.* Substorm-like activations share many of the observational characteristics of prototypical isolated substorms discussed above. The sawtooth events form a particular class of quasi-periodic global oscillations that are typically identified through particle injections observed by geostationary satellites. Steady magnetospheric convection is not characteristically a storm phenomenon because it is not associated with particularly strong ring current.

### 13.3.1 Steady magnetospheric convection

Sustained southward IMF driving of the magnetosphere leads in most cases to a sequence of substorm activations, but sometimes the system finds a quasi-steady state during which the convection stays at high level and the auroral electrojets remain strong and steady for a long time, from several hours to a half day or so. This state of the magnetosphere has been named as *convection bay* [Pytte et al, 1978] or steady magnetospheric convection (SMC) (for a review, see Sergeev et al [1996]).

There are theoretical arguments claiming that steady convection should be impossible and continuous IMF driving should always lead to substorm expansions in order to resolve the so-called *pressure balance inconsistency* [Erickson and Wolf, 1980; Erickson, 1992]. In simple terms the pressure balance inconsistency arises from two reasonable assumptions, namely that the magnetic field and the plasma are frozen-in to each other and that the convection on closed field lines is adiabatic. The latter assumption can be expressed as

$$\frac{d}{dt} P \mathcal{V}^{\gamma} = 0 \,, \tag{13.3}$$

where $\gamma = 5/3$ for 3D adiabatic transport, $P$ is pressure and $\mathcal{V}$ the *flux tube volume*

$$\mathcal{V} = \int \frac{ds}{B} \,. \tag{13.4}$$

Here the integral is taken along the magnetic field line from the ionosphere on one hemisphere to the ionosphere on the other hemisphere. Due to the $1/B$ dependence $\mathcal{V}$ is very large for flux tubes crossing the current sheet in mid-tail but becomes much smaller when the flux tube is convected to the near-Earth region. In order to conserve $P\mathcal{V}^{\gamma}$ the pressure must grow to an unreasonable level unless the system is relaxed episodically, e.g., by substorms.

**Feed your brain**

$P\mathcal{V}^{\gamma}$ is sometimes called the *entropy function* and (13.3) is interpreted as conservation of entropy. Figure out the connection to the definition of entropy given by (5.3) and explain when $P\mathcal{V}^{\gamma}$ is a valid measure of entropy.

However, as discussed by Sergeev et al [1996], there are, although relatively rarely, SMC events, during which there is no pressure release in form of substorms up to about 10 hours. During these events the auroral oval is very wide indicating that the amount of closed magnetic flux through the plasma sheet is particularly large. Based on low-altitude signatures of precipitating particles of solar wind origin Sergeev et al [1996] argued that the distant neutral line during SMC events lies somewhere at the distance of 50–100 $R_E$. According to ionospheric observations the corresponding open–closed field line boundary is located at 70–72° corrected geomagnetic latitude.

**Feed your brain**

Find from the literature the definitions of magnetic local time (MLT) and corrected geomagnetic latitude (CGL). Under what conditions there is no unique way to determine CGL?

To find out the near-Earth tail configuration, observations of the boundaries of isotropic precipitation of ions and electrons are useful. As discussed in Sect. 3.4, the motion of charged particles becomes chaotic when the field line curvature radius ($R_C$) becomes comparable to the gyro radii of the particles ($r_L$). The particles are effectively scattered and fill the atmospheric loss cone when

$$\frac{R_C}{r_L} \lesssim 8 \,. \tag{13.5}$$

Using this criterion observations of precipitating isotropic 30-keV ions during SMC events indicate that an equatorial field of 5 nT, i.e., quite a thin current sheet, reaches as close as 6–8 $R_E$ geocentric distance. This is not too far from the isotropic boundary of 30-keV electrons, corresponding in turn to a 40-nT field. Thus there must be strong gradient in the magnetic field at these distances and the field goes over from dipolar to tail-like within a relatively narrow region in the $X$-direction. This conclusion is consistent with direct magnetic field observations. Thus the near-tail configuration is rather similar to a substorm growth phase.

On the other hand, farther out in the mid-tail region relatively large equatorial fields of $B_z > 6$ nT have been observed during SMC events. Thus the mid-tail configuration resembles more the recovery phase than the growth phase. This "hybrid state" of tail magnetic configuration was modeled by Sergeev et al [1994] using the Tsyganenko 89 magnetic field model [Tsyganenko, 1989] modified to fit to actual in situ magnetic field observations during the SMC event on November 24, 1981 (Fig. 13.10).

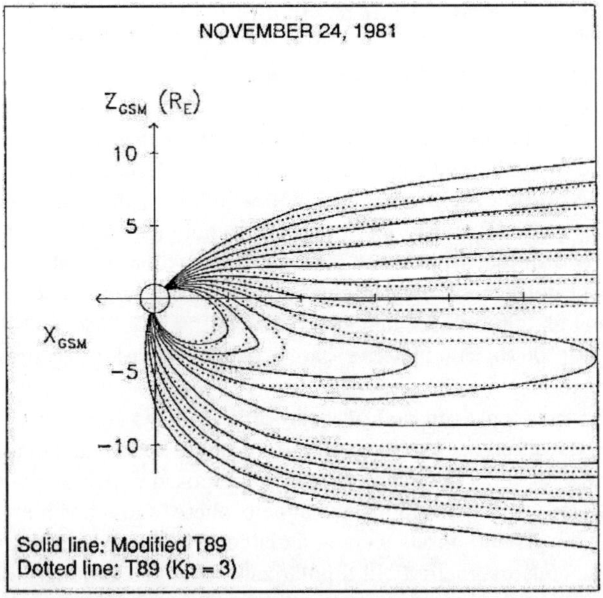


**Fig. 13.10** Model of the hybrid state magnetic field configuration during the SMC event on November 24, 1981. (From Sergeev et al [1994].)

According to this modeling exercise there is a magnetic field minimum at the distance of about 12 $R_E$. This allows steady convection to reach this distance without too strong compression. Closer to the Earth the strong magnetic gradient causes rapid azimuthal drift. Sergeev et al [1994] argued that this helps to remove the particles to the dayside before excessive pressure can build up. Another factor limiting the pressure build-up is the deviation of a part of the plasma sheet flow toward the flanks of the magnetosphere already at the distance of about $20 R_E$ [Sergeev and Lennartsson, 1988].

In fact a magnetic field configuration with a mid-tail magnetic field minimum capable of sustaining SMC was independently found as a result of 2D MHD calculation by Hau et al [1989]. They, however, questioned the realism of their solution expecting it to be tearing-unstable. While this magnetic field configuration may be able to avoid the pressure balance inconsistency, it remains unclear how such a peculiar magnetic field structure can be sustained over the long SMC period.

In order to avoid misunderstanding we wish to stress that the steadiness of the convection refers to the average large-scale convection. Also during SMC periods the actual plasma flow consists of intermittent bursty bulk flows (BBF). Furthermore, small auroral activations do take place during SMC events, mostly near the poleward boundary of the auroral oval. They resemble substorm onsets but their effects on, e.g., the *AL* index remain weak and the large-scale convection continues in a quasi-steady manner.

### 13.3.2 Substorm-like activations and sawtooth Events

Sawtooth events are examples of magnetospheric dynamics during strong external driving, but they are completely different from steady magnetospheric convection. Sawtooth events are large quasi-periodic oscillations of energetic particle fluxes and the magnetic field observed at geostationary orbit (for a detailed data description, see Henderson et al [2006]). The period of the oscillations is 2–4 h, which is in the same range as the recurrence rate of quasi-periodic substorms. In fact, it is not clear if the sawtooth oscillations are a phenomenon different from other recurrent storm-time activations, of which many exhibit several, but not necessarily all, characteristics of prototypical isolated substorms [e.g., Henderson et al, 2006; Pulkkinen et al, 2007b].

Pulkkinen et al [2007b] conducted a statistical analysis of 150 storm-time activation events during 10 storms with the peak $Dst < -75$ nT in 2004. The activations were identified as rapid enhancements (onsets) of westward electrojet signatures (about 200 nT or more) in ground-based magnetometer observations in the Scandinavian and Canadian sectors. The onset was taken as the epoch time in superposed epoch analysis of various observables in the solar wind, geostationary magnetosphere, and ionosphere.

About 48% of the ground onsets were associated with energetic particle injections and 67% with simultaneous magnetic field dipolarizations at geostationary orbit. 45% of the events were preceded by decreasing inclination of the geostationary distance magnetic field, which can be interpreted as intensification of the cross-tail current sheet resembling the substorm growth phase. However, the growth phase signatures were weak and they could not be recognized in the *AL* index, which may not be so surprising, as these activations took place during moderate, intense, and big storms. The electron fluxes and the magnetic field inclination recovered within 1.5–2 h after the onset, and the periodicity of the activations was 2–3 h. Regardless of possible selection effects (e.g., the phase of the solar cycle, a small total number of storms), it is fair to say that about half of the activations identified in the ionospheric current systems show clear substorm onset and recovery signatures at geostationary orbit.

In order to compare the substorm-like activations with sawtooth events Pulkkinen et al [2007b] made a similar analysis of an independent data set of 138 individual sawtooth oscillations during 1999–2002. In this case the zero epoch time was the proton injection at geostationary orbit. The magnetic field inclination was found to behave quite similarly to that during substorm-like activations. The sawtooth events were associated with slightly weaker auroral activity (in terms of *AL*), but the ring current was quite similar in both cases. Also the periodicity of 2–3 h was the same as in substorm-like activations.

When comparing the solar wind properties during sawtooth events and other storm-time activations Pulkkinen et al [2007b] found that the solar wind electric field is comparable in both cases $E_Y \sim 3$–$4\,\mathrm{mV\,m^{-1}}$ as is also $B_Z \lesssim -5\,\mathrm{nT}$. However, when the solar wind velocity is considered separately, the sawtooth events tend to take place during, on the average, slower solar wind ($<500\,\mathrm{km\,s^{-1}}$) than other storm-time activations (on the average $\sim 600\,\mathrm{km\,s^{-1}}$). Both of these are faster than typical solar wind velocities driving steady magnetospheric convection.

Concerning the possible triggering of storm-time activity by sudden changes in solar wind parameters Pulkkinen et al [2007b] found that of all storm-time activations 30% had signatures in solar wind parameters that were interpreted as potential triggers, whereas the fraction for sawtooth events was only 20%. However, at the geostationary orbit the potentially triggered storm-time activations show clear injection features only in the midnight sector and do not have associated dipolarizations. Thus if the triggering is due to a pressure pulse or IMF turning, its magnetic consequences do not reach the inner magnetosphere. The process releasing the injection takes place farther out and only the injected particles reach the geostationary orbit. For the sawtooth events the situation was similar but the injections were stronger, which may have been a selection effect of the sawtooth events based on the injections.

When the level of fluctuations in the IMF was measured in terms of either $\delta B = (\sigma_{B_X}^2 + \sigma_{B_Y}^2 + \sigma_{B_Z}^2)^{-1/2}$ or of $\delta B / \langle B \rangle$, where $\sigma$'s are the standard deviations and $\langle B \rangle$ is the average magnetic field, both triggered storm-time activations and sawtooth events showed markedly stronger fluctuations than events without identified triggers. During periods of stronger solar wind fluctuations there are more potential triggers.

As a conclusion, the differences between recurrent substorm-like storm-time activations and sawtooth events are not so significant that they should be considered to belong to different classes of magnetospheric activity. This view was also expressed by Henderson et al [2006] and several other investigators. However, what determines the 2–3-hour recurrence time of magnetospheric activity, remains unclear. This recurrence issue may, in fact, be a key to an improved understanding of the whole solar wind–magnetosphere system.

The reader may now wonder what are the remaining about 50% of stormtime activations that do not exhibit clear substorm characteristics. There have not been too many investigations to address this question. Pulkkinen et al [2004] analyzed 6 storms in the range from intense to very large during 2000–2001. One of the storms (21–22 October, 2001) exhibited evident sawtooth oscillations, although the authors did not yet use this, at that time emerging, terminology. During the other 5 storms 3 classes of events were identified: There were typical substorm-like events of poleward electrojet expansion associated with geostationary injection. The second class was called "non-substorm" events, in which the electrojet expansion was equatorward and there were no geostationary injections. The third class consisted of triggered events that lacked clear geostationary injections. Note that about 50% of the substorm-like events also had potential IMF triggers.

As a possible scenario to account for at least some of the non-substorm events Pulkkinen et al [2004] suggested that the tail current sheet was much thicker during these than during the substorm-like events. A fast flow channel intrudes to the inner magnetosphere, possibly as a large BBF, but it does not necessarily reach the geostationary distance. The ionospheric current is associated with the earthward edge of the flow channel with downward current originating from the dawn flank and upward current joining the dusk flank of the channel to a westward Hall current in the ionosphere. The same scenario on the coupling of BBFs to the ionosphere was discussed by Kauristie et al [2000], although not in the storm context.

## 13.4 ICME–Storm Relationships

In this section we turn to the drivers of magnetospheric storms. Based on observations of 10 storms in 1978–1979 Gonzalez and Tsurutani [1987] presented a useful rule-of-thumb on when to expect a strong storm to occur. According to them the IMF must have a long-duration (more than 3 h), large negative ($<-10$ nT) southward component associated with duskward electric field $E = VB_s > 5\,\mathrm{mV\,m^{-1}}$. As fast ICMEs and their shocked sheath regions can enhance both $V$ and $B_s$ and expose the magnetosphere to these conditions for several hours, they are by far the most efficient drivers of the strongest magnetospheric activity. The response of the magnetosphere is complicated and depends on the detailed structure of the driver.

### 13.4.1 Geoeffectivity of an ICME

The speeds of the ICMEs, their shocks, and the post-shock streams at 1 *AU* can easily be larger than twice the background solar wind speed. The IMF in the sheath region between the shock and the ejecta is strongly compressed and if the IMF ahead of a fast ICME has a southward component, the sheath region can drive a strong storm even if the ICME itself passes by the magnetosphere not actually hitting it. The southward IMF component may be further amplified by draping the magnetic field around the ICME [Gosling and McComas, 1987], which can lead to a southward IMF component even in cases where the pre-existing IMF is slightly northward.

Depending on the background solar wind conditions and on the magnetic structure of the ICME a large number of different storm evolutions can take place. In the following we define a magnetic storm as *sheath-associated* if 85% of the *Dst* minimum occurs while the dayside magnetosphere is embedded in the ICME sheath region. For a *magnetic cloud-associated storm* we require that during a magnetic cloud passing the magnetosphere *Dst* reaches the intense storm level of $-100$ nT. There are storms that do not fall into either of these categories because not all ejecta exhibit the magnetic cloud structure.

The different storm sequences are most illustrative to consider for cases where the ICME has a well-defined flux rope configuration, which is the case for at least 30% of all ICMEs. If the inclination of the flux-rope from the ecliptic plane is small, the north-south magnetic structure is bipolar and the magnetic cloud can arrive with northward (NS) or southward (SN) magnetic field ahead, which give different temporal storm evolutions. For example, if a southward sheath field is followed by an NS-type cloud with sufficiently strong and long-lasting northward IMF, a double-peaked *Dst*-storm or even two separate storms may follow. Double- or multiple-peaked storms may also take place when several ICMEs from the same active region on the Sun are heading toward the Earth.

A flux rope can also have a large inclination with respect to the ecliptic. In such cases the IMF can have a unipolar, either northward (N) or southward (S), orientation throughout the passage of the flux rope. In the northward case the ICME will most likely pass the Earth with only minor perturbations, whereas the southward case may lead to a particularly strong and long-lasting storm because the Earth may remain within the southward pointing flux rope much longer than in a bipolar case.

Figure 13.11 shows the results of an analysis of 73 magnetic cloud events identified in *Wind* and *ACE* observations during solar cycle 23 [Huttunen et al, 2005]. Unipolar southward clouds always caused at least a medium-size storm ($Dst < -50\,\text{nT}$), whereas in northward cases only sheath regions caused storms. Note that about one-third of the

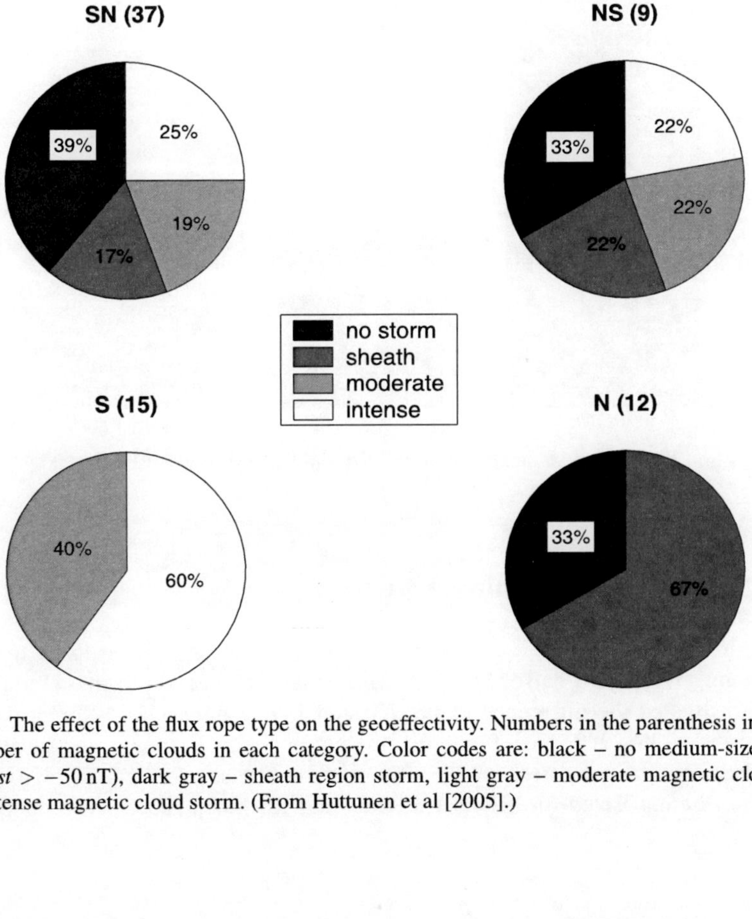


**Fig. 13.11** The effect of the flux rope type on the geoeffectivity. Numbers in the parenthesis indicate the total number of magnetic clouds in each category. Color codes are: black – no medium-size or larger storms ($Dst > -50\,\text{nT}$), dark gray – sheath region storm, light gray – moderate magnetic cloud storm, white – intense magnetic cloud storm. (From Huttunen et al [2005].)

bipolar, either NS or SN, clouds did not lead to a medium-size or larger storm, which evidently is a nuisance for space weather forecasters.

The importance of the sheath regions as efficient storm drivers was demonstrated by Tsurutani et al [1988], but their significance was not fully appreciated before the more extensive analyses of in situ observations from solar cycle 23. Huttunen and Koskinen [2004] showed that during the ascending phase of the cycle (1997–2002) 45% of 53 intense ($Dst < -100$ nT) storms were caused by a sheath region. When the threshold was changed to $Dst < -150$ nT, already 60% of the remaining storms were sheath-driven (Fig. 13.12). The number of events in that study was too small to make statistical conclusions, but the importance of sheath regions as storm drivers was clear. A more complete catalog of intense storms during 1996–2005, consistent with these results, was compiled later by Zhang et al [2007].

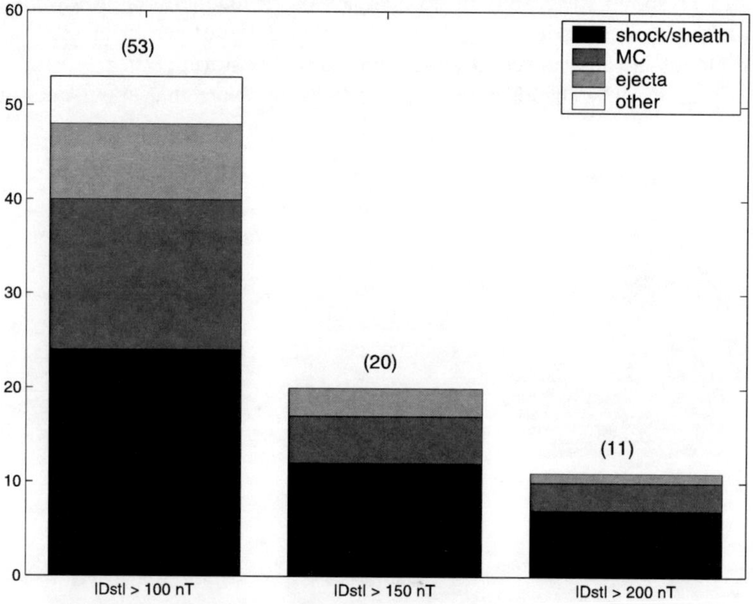


**Fig. 13.12** Drivers of intense *Dst*-storms. (From Huttunen and Koskinen [2004].)

### 13.4.2 Different response to different drivers

Selecting just one activity index to represent the magnetosphere's response to the solar wind driving gives a too narrow perspective to magnetospheric storms. Huttunen et al [2002] investigated the difference of the $Kp$ and $Dst$ responses to different solar wind drivers during 1996–1999. They found that the fast post-shock streams and sheath regions had a relatively stronger effect on $Kp$, whereas the effects of ejecta favored $Dst$. This tendency was emphasized further by Huttunen and Koskinen [2004] who compared the evolution of several magnetic indices (*Dst*, *SYM-H*, *ASY-H*, *AE*, and *Kp*) during magnetic cloud and sheath region storms. The difference in the response of magnetic indices is most most evident when comparing sheath regions and magnetic clouds, excluding ejecta without a well-organized magnetic structure, because solar wind dynamic pressure and the magnetic field configuration are most different under these two types of storm drivers.

Figure 13.13 shows the maximum *Kp* and minimum *Dst* indices of all intense storms ($Kp_{max} \geq 7-$ or $Dst_{min} < -100\,\text{nT}$) during 1997–2003 that were possible to associate uniquely with a sheath region or with a magnetic cloud. From these data it is evident that most of the large *Kp* storms were sheath storms as were all large-*Kp* – smaller-*Dst* events; whereas large-*Dst* – smaller-*Kp* events were mostly associated with magnetic clouds.

A proximate explanation for this behavior is that *Kp* is more sensitive to auroral zone current systems than *Dst*. This is supported by the investigation of the storm response in the *SYM–H* and auroral electrojet indices by Huttunen and Koskinen [2004], who illustrated using four sample events that the high *Kp*-activity really was due to strongly enhanced auroral activity and not just an artifact produced by the procedure to derive the *Kp* index.

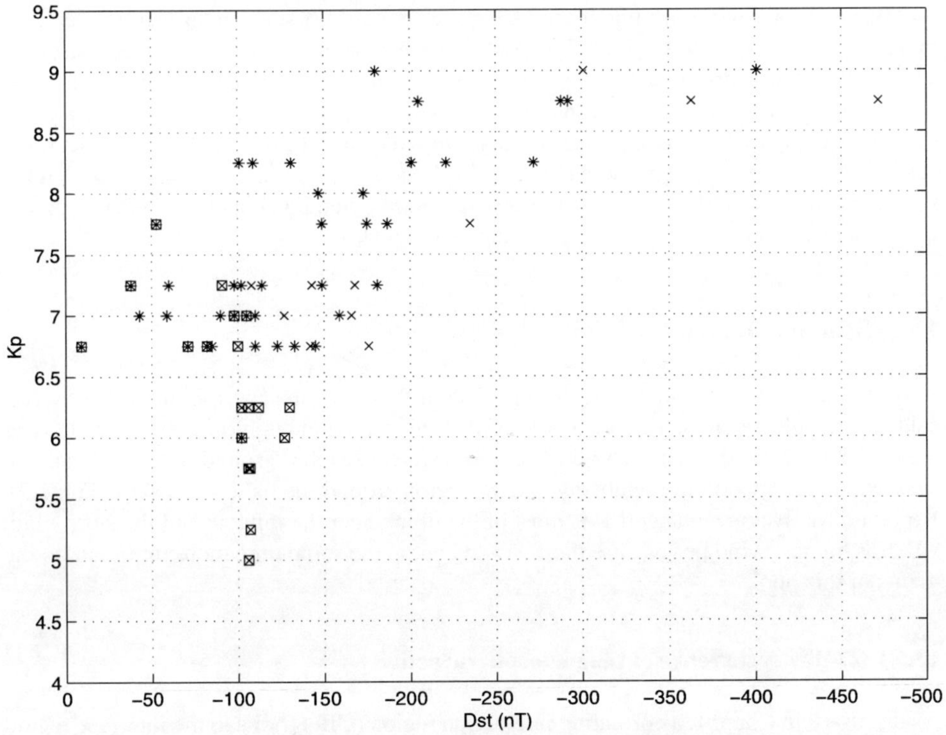


**Fig. 13.13** *Kp* and *Dst* indices for all intense storms ($Kp_{max} \geq 7-$ or $Dst_{min} < -100\,\text{nT}$) during 1997–2003 that were possible to associate clearly with a sheath region (asterisks) or with a magnetic cloud (crosses). The rectangles identify storms that did not fulfill the additional condition: $Kp \geq 6-$ over three 3-h intervals.

While the ultimate explanation for the sheath regions favoring high-latitude activity is unknown, a possible scenario is the following. The irregularities in the sheath region cause perturbations in the low-latitude boundary layer of the magnetosphere. These enhance the Region 1 current system, which couples to the auroral current systems in the ionosphere. Consequently the auroral activity is enhanced and it shows up more strongly in $Kp$ than in $Dst$. On the other hand, the smooth rotation of the magnetic cloud field does not cause the same effect on the high-latitude current systems but strengthens the large-scale magnetospheric convection, resulting in a relatively stronger ring current build-up and enhanced *Dst* effect.

This is, of course, not an either–or question because we are dealing with relatively large storms where both ring current and auroral current systems are activated. The large-scale convection also enhances the auroral currents. In fact, *Dst* storms without significant high-latitude activity seem to be very exceptional (if there are any), but there are examples of $Kp$ storms with very weak *Dst* response (Fig. 13.13).

Another viewpoint on this question was given by Pulkkinen et al [2007a]. They conducted a superposed epoch analysis of 14 sheath storms and 14 cloud storms, including solar wind data, activity indices and geostationary particle observations. Their conclusion was that the solar wind driving at the beginning of a sheath storm is harder and leads to more stretched magnetic field configuration in the near-tail region. This leads to stronger auroral substorm activity and to more particle injections into the geostationary distance, whereas a smaller amount of ions ends up on closed trajectories to form the symmetric ring current as compared to cloud-driven storms. We will return to this issue when discussing storms in the inner magnetosphere in Chap. 14. Pulkkinen et al [2007a] also noted that the strong fluctuations in the sheath region provide more potential substorm triggers than the much more smoothly rotating IMF of a magnetic cloud.

## 13.5 Storms Driven by Fast Solar Wind

Another mechanism for imposing enough $E_Y$ on the magnetosphere is provided by fast solar wind with a southward component of IMF for a long enough time. In the context of the 3D heliosphere, the fast solar wind can be considered as the "ground state", whereas the slow wind is limited to the relatively narrow region around the ballerina's skirt (Fig. 1.9). However, we live on a planet that is most of the time under the influence of the slow wind, whereas the fast wind periods are more limited but at the same time more important to the theme of this book.

### 13.5.1 27-day recurrence of magnetospheric activity

As discussed in Chap. 1, a corotating interaction region (CIR) forms in the interface region where fast solar wind is overtaking slow wind, and gradually steepens to a shock typically somewhere beyond $1\,AU$. If the $Z$-component of the IMF in the fast wind is southward, the passing of a CIR is a signal of enhanced magnetospheric activity to follow. Consequently, the fast wind-driven storms are often called CIR-driven storms [e.g., Borovsky and Denton,

2006]. This terminology is a little misleading because the storm driver in most cases is the fast wind following the CIR, not the CIR itself. However, as the interaction region is relatively wide, the solar wind already during the passage of the CIR itself often is sufficiently strong driver for a storm in the magnetosphere. We prefer to call these storms CIR-related storms, but remind the reader again that instead of hanging on to terminology we should focus on physics.

The appearance of CIR-related storms have a strong 27-day periodicity. This periodicity in the geomagnetic activity was already known long before the solar wind was found [Chree and Stagg, 1927]. Bartels [1932] interpreted it to be due to magnetically active regions on the solar surface, which he called *M-regions*. Today we know that more or less the opposite is true. The fast wind originates from the coronal holes, which are actually the magnetically most quiet regions on the Sun, whereas the intermittent ICME-related storms originate from more active regions. The large coronal holes are also the most stable regions of the corona and extend to low solar latitudes during the declining phase of the solar activity cycle. The stability is the cause of the recurrence. The same hole returns toward the direction of the Earth with the 27.3-day synodic period of the solar rotation. Thus the fast solar wind episodes have a 27-day periodicity and the recurrent storms take place preferentially during the declining phase of the solar cycle.

There are also non-recurrent high-speed episodes throughout the solar cycle, but they often are of shorter duration and lead to weaker storms or to one or several substorms.

### 13.5.2 Differences from ICME-driven storms

Borovsky and Denton [2006] listed 21 differences between ICME-driven and CIR-related storms. In this section we discuss briefly some of these but, for more details, refer to that study and references therein.

At the start of a CIR-related storm there usually is no SSC. The reason is that while the CIR is characterized by a pressure enhancement, this enhancement has not yet grown to a shock and thus the effect of the fast wind upon the magnetopause grows much more smoothly than in the case of a fast ICME. Of course, as not all ICMEs are fast, there are also ICME-driven storms without the SSC signature.

The CIR-related storms typically follow a period of slow solar wind lasting a few days. The CIRs are usually associated with an IMF transit from an away sector to a toward sector, or vice versa. Thus if the IMF in the fast wind region has a southward component, as required for the storm to occur, it most likely had a northward component in the leading region of slow wind. Thus before a CIR-related storm there often has been a long period of very weak solar wind driving, during which the magnetosphere has had time to reach a particularly calm state.

This "calm before the storm" is a time when so called *cold dense plasma sheet* (CDPS) is known to form. This may, at first, sound contradictory because there is no dayside reconnection to let solar wind plasma to flow into the magnetosphere. However, a practically closed magnetosphere is also less leaky than a strongly reconnecting open magnetosphere. As noted by Lavraud et al [2006], there is ample evidence that both high-latitude reconnection beyond the cusps and viscous interaction, e.g., in form of Kelvin–Helmholtz instability (Chap. 7), can contribute to the filling of the plasma sheet with low-energy plasma. As the magnetosheath is denser than the magnetosphere, any diffusive process across the boundary should have a net inward flux of plasma.

Calm before the storm can also take place before CME-related storms, and the CDPS can form before them as well, but less frequently than is the case with CIR-related storms. Lavraud et al [2006] analyzed the consequences of the CDPS for the storm development. They found that the models commonly used to calculate *Dst*, e.g., (1.64):

$$Dst^* = Dst - b\sqrt{P_{dyn}} + c$$

tend to underestimate the ring current after long quiescence by 10–20% during the early part of the storm main phase. Lavraud et al [2006] suggested that the CDPS provides an enhanced seed population for ring current carriers when pushed closer to the Earth by the enhanced convection. During the storm the cold plasma is removed and the remaining storm-time plasma sheet is hot, actually hotter during CIR-related storms than during ICME-related storms. In both cases the plasma sheet density is a factor of 2 or more larger than the average non-storm plasma sheet density with ICME-related storms being more "superdense" than CIR-related storms. In the CIR cases the solar wind supply is large only during the brief passage of the CIR itself, whereas the fast solar wind density is relatively small.

$E_Y$ in a fast solar wind stream does not become as large as can be the case within a strongly compressed ICME sheath plasma or in a strong magnetic cloud. Consequently, *Dst* remains smaller, but because the fast stream with southward IMF may last much longer, the CIR-related storms are of longer duration. Thus, the cumulative effects of the fast wind-driven storms may become more severe than the effects of ICME storms with larger peak *Dst*. We will discuss the processes leading to large relativistic electron fluxes in the radiation belts more deeply in Chap. 14, but already note here that the CIR-related storms are known to produce much larger relativistic electron fluxes in the inner magnetosphere than ICME storms. These, together with the hotter plasma sheet temperature, lead to enhanced risks for spacecraft charging problems. On the other hand, as the CIRs are related to neither flares nor CMEs, there is no direct association with solar energetic particle events. Thus there is no enhanced risk of single-event upsets due to high solar proton fluxes. Also the most severe GIC events have been associated with ICMEs but not with CIR-related storms.

As will be discussed in Chap. 14, the large-scale magnetic ULF oscillations in the Pc5 range (150–600 s) are an important candidate for the radiation belt electron acceleration. The ULF power is high during both fast wind and fast ICMEs, which is consistent with the observations that the ULF wave amplitude is proportional to the solar wind speed. Due to the longer duration of fast flows the ULF oscillations have a longer duration during CIR-related storms.

Finally, both most dramatic auroral displays and global sawtooth oscillations are characteristically ICME-related phenomena. However, this does not mean weaker auroral activity during fast solar wind episodes. On the contrary, Tanskanen et al [2005] found in an analysis extending over the 11-year period 1993–2003 that substorms occurring during years of large occurrence of high-speed solar wind were 32% more intense (in terms of the *AE* index) and transferred twice as much energy to the polar ionosphere as substorms during years when the Earth was less exposed to fast solar wind flow.

Physical explanations for these differences are, for the time being, mostly speculative and underline our lack of understanding of the solar wind-driven magnetosphere. For example, Borovsky and Denton [2006] associated the appearance of global sawtooth oscillations during magnetic cloud events with the fact that in those cases, due to the strong IMF, the magnetosonic Mach number ($M_{ms} \sim 1$–$3$) and the magnetosheath plasma beta ($\beta < 1$) are much smaller than normally. The low $M_{ms}$ reduces the plasma compression ratio at the bow shock from the maximal 4 toward slightly over 1 (cf. Chap. 6). It is fair to say that the role of the magnetosheath, the different plasma behavior behind the quasi-parallel and quasi-perpendicular sectors of the bow shock, and the turbulent flow that actually interacts with the magnetopause belong to the most critical unknowns in the solar wind–magnetosphere interaction at the time of writing of this book.

## 13.6 Energy Budgets of Storms and Substorms

Magnetospheric storms and substorms can be seen as energy transfer processes in which the solar wind is the primary energy source and the magnetosphere is an engine that reprocesses the incoming energy and distributes it to the different domains of the system where it is dissipated by various mechanisms. Both energy input and output are difficult to measure because the entire system as well as its different parts are large and in continuous interaction with their surroundings.

Furthermore, it is useful to keep in mind that electromagnetic energy cannot be localized. An illustrative example of this is a simple capacitor consisting of two circular plates. When the capacitor is being charged, the charge to the capacitor comes along the wires connected to the plates. But if you calculate the Poynting vector during the charging, it points radially into the capacitor from the direction perpendicular to the electric field being created between the plates.

### 13.6.1 Energy supply

Let us start with order of magnitude estimates of how much energy is available in the solar wind for magnetospheric activity. Assume, for simplicity, that the solar wind at 1 *AU* consists of protons with the density of 5 $\mathrm{cm^{-3}}$, that its velocity is 400 $\mathrm{km\,s^{-1}}$ and the interplanetary magnetic field 10 nT. Assume further, that the magnetospheric obstacle has a radius of 15 $R_E$. Now the solar wind kinetic power flux density is about $5 \times 10^{-4}\,\mathrm{W\,m^{-2}}$ and the total power over the obstacle about $1.4 \times 10^{13}$ W, i.e., 14 TW. Assuming that $\mathbf{B} \perp \mathbf{V}$, the electromagnetic power flux density is about $3 \times 10^{-5}\,\mathrm{W\,m^{-2}}$, yielding the corresponding power of $0.8 \times 10^{12}$ W. Note that

$$\frac{\text{kinetic energy flux}}{\text{electromagnetic energy flux}} \sim \frac{V \rho V^2}{V B^2/\mu_0} = \frac{V^2}{v_A^2} = M_A^2 \,. \tag{13.6}$$

As the solar wind at $1\,AU$ is practically always super-Alfvénic, the solar wind kinetic energy flux is larger than the electromagnetic energy flux, typically a few tens of times larger.

These numbers can be compared with the estimated power needed for the maintenance of the magnetosphere, which was estimated to be $1.2 \times 10^{12}$ W by Siscoe and Cummings [1969]. As noted by Koskinen and Tanskanen [2002] this may be a slight underestimate, but for our discussion the right order of magnitude is sufficient.

The velocity of a fast ICME and its magnetic field intensity can be several times larger than in the example above. For example, on November 20, 2003, the north–south magnetic field component of an ICME reached $-53$ nT. Thus the total kinetic power over the magnetospheric obstacle can easily exceed $3 \times 10^{15}$ W and the electromagnetic power be of the order of $5 \times 10^{13}$ W. As we will see below, these numbers are much larger than the estimates for the energy dissipation in the strongly disturbed magnetosphere. There is sufficient amount of energy available for magnetospheric activity, but the question is how does the magnetospheric engine process this energy.

From simplified cartoons of reconnecting magnetosphere one may get the impression that reconnection simply transports solar wind magnetic energy to the magnetosphere. According to the numbers above this would imply that, except in cases of very fast solar wind with strongly southward IMF, practically all magnetic energy incident on the magnetopause would need to be transported, and even then the input would be marginal. However, as discussed in Chap. 8, the solar wind flow supplies kinetic energy to a dynamo on the magnetopause, which is responsible for the increase of magnetic energy inside the magnetosphere. The solar wind kinetic energy flux is always more than sufficient to maintain the magnetosphere and fully capable of powering the magnetospheric processes, energy requirements of which will be discussed next.

### 13.6.2 Ring current energy

When Perreault and Akasofu [1978] introduced the epsilon parameter (13.1) to quantitatively describe the energy input into the inner magnetosphere and ionosphere, they estimated the ring current to be the largest energy sink, considerably larger than the polar ionosphere. Subsequent studies turned this picture around (see, e.g., the reviews by Stern [1984] and Weiss et al [1992]). The overestimation of the ring current energy may have been partly due to the focus of Perreault and Akasofu [1978] on magnetic storms but equally well to the, at that time, underestimated power of ionospheric Joule heating and electron precipitation.

Without dwelling on the details of the ring current dynamics, to be discussed in Chap. 14, we can take the traditional approach and estimate the ring current energy from the *Dst* index. Assuming that the current is carried by particles trapped in the dipole field, the magnetic deviation observed $\triangle\mathbf{B}$ on ground is associated with the energy of the current carriers $W_{RC}$ through the *Dessler–Parker–Sckopke (DPS) relationship* [Dessler and Parker, 1959; Sckopke, 1966]

$$\frac{\triangle\mathbf{B}}{B_0} = -\frac{2}{3}\frac{W_{RC}}{W_{dip}}\mathbf{e}_z\,, \tag{13.7}$$

where $B_0$ is the dipole field strength on the equatorial surface of the Earth, $W_{dip}$ is the total energy of the dipole field above the Earth's surface, and $\mathbf{e}_z$ indicates the horizontal (north) direction at the equator. Denoting the deviation of the north component by $\triangle H$ we can write this as

$$\triangle H = -\frac{\mu_0}{2\pi}\frac{W_{RC}}{B_0 R_E^3} . \tag{13.8}$$

> **Train your brain**
>
> Derive the DPS relation (13.8) assuming a single ion species drifting in the dipole field in the equatorial plane. If you feel like a more challenging exercise, perform the calculations in Sckopke [1966].

Because $\triangle H$ essentially gives $Dst$, we have in (13.8) the zeroth order relationship between the ring current energy $W_{RC}$ and $Dst$, the first correction to which is the pressure-corrected index $Dst^*$ (1.64). The numerical relationship between $W_{RC}$ and $Dst^*$ is according to (13.8)

$$W_{RC}(\mathrm{J}) \approx -4\times 10^{13}\, Dst^*(\mathrm{nT}) . \tag{13.9}$$

Let us denote the energy injection rate into the ring current by $P_{RC}$ and the loss rate by $L_{RC}$ with a time constant $\tau$ and write $L_{RC} = Dst^*/\tau$. Thus the rate of change of the ring current energy is given by

$$\frac{\partial W_{RC}}{\partial t} = P_{RC} - L_{RC} . \tag{13.10}$$

Now we get for the energy injection rate, i.e., the power into the ring current

$$P_{RC}(\mathrm{W}) \approx -4\times 10^{13}\left(\frac{\partial Dst^*(\mathrm{nT})}{\partial t} + \frac{Dst^*(\mathrm{nT})}{\tau}\right) . \tag{13.11}$$

While this equation is simple, Akasofu [1981] had already warned about its uncritical use. The real ring current dynamics is more complicated than this. The decay time is not a constant but varies during the progress of the actual loss mechanisms. The decay also takes place during the main phase when it is hidden behind the rapid increase of the current. For example, Lu et al [1998] used different time constants $\tau = 4$–$20\,\mathrm{h}$ for different levels of $Dst$. Note that $\tau = 20\,\mathrm{h}$ is between the ring current $H^+$ and $O^+$ lifetimes as will be discussed in the context of Fig. 14.3.

An even more serious problem is the simple fact that the magnetometers used to determine the ring current are also sensitive to a variety of other currents. The pressure correction (1.64) is the easiest to take into account. As already discussed in Chap. 1, the estimates of the tail current contribution vary from 25% to 50% and the ground induced currents due to large $\partial\mathbf{B}/\partial t$ during the main phase may contribute up to 25% of the observed $Dst$. Thus (13.11) most likely overestimates the power into the ring current by a factor of about 2. This is one of the reasons why the relative role of the ring current was overemphasized in early energy budget studies.

The most straightforward way of determining the energy of the ring current would be to measure the energy of the current carriers. While all particles throughout the vast ring current region cannot ever be observed, this method has been useful in some fortunate cases when suitably equipped spacecraft, e.g., *AMPTE/CCE* and *CRRES*, have been traversing the most important $L$-shells in the magnetosphere. Turner et al [2001] used the ion composition instrument data onboard the *Polar* satellite together with a ring current model and concluded that the "real" ring current energy actually was about 50% of that given by the DPS relation. For example, during a storm in May 1998 with a peak $Dst \approx -250\,\mathrm{nT}$ on May 4 the peak ring current energy was found to be about $4 \times 10^{15}$ J, i.e., 40% of the prediction of (13.9).

### 13.6.3 Ionospheric dissipation

The energy dissipation into the ionosphere takes place mostly through two main mechanisms: Joule heating and auroral electron precipitation. As both of these are enhanced during substorms, the data base for statistical studies is much larger than is the case with the energetics of the ring current, which is mostly a storm-time phenomenon. Also here we need to turn to activity index-based proxies if we want to statistically determine the global energy input into the ionosphere. However, there are several means of benchmarking the proxies, including magnetic and radar observations of the ionospheric electrodynamics, multi-wavelength auroral imaging and spectroscopy from both space and ground, and direct observations of particle precipitation, electric field and field-aligned currents above the auroral zone,

The Joule heating occurs when field-aligned currents close through the resistive ionosphere. The Pedersen current associated with this current loop is in the same direction as the ionospheric electric field and thus energy dissipation is given by

$$\int \mathbf{J} \cdot \mathbf{E}\, d^3r = \int \sigma_P E^2\, d^3r\,. \tag{13.12}$$

The Pedersen current associated with the FACs is hard to determine from ground-based magnetic effects, in the case of a homogeneous ionosphere even impossible [Fukushima, 1976]. However, the ratio between Hall and Pedersen conductances (i.e., height-integrated conductivities) is typically about 2. Thus, the ground-based measurements of Hall currents give us a fairly good picture of the Pedersen currents as well, and we can calibrate, e.g., the *AE* indices as proxies for the Joule heating [e.g., Ahn et al, 1983] using a simple formula

$$P_J(\mathrm{W}) = C \times 10^8 AL(\mathrm{nT})\,. \tag{13.13}$$

Different studies have resulted in slightly different factors of the proportionality $C$. It is in the range 2–5 with $C = 3$ as a good rule-of-thumb value for statistical studies (e.g., Lu et al [1998] and references therein). Thus a 500-nT substorm dissipates energy at the rate of 150 GW *per hemisphere*. While the energy dissipation may be not be equal on both hemispheres, considering all other uncertainties multiplication of the power given by (13.13) by 2 is a reasonable assumption.

The energy carried by the precipitating electrons can be estimated using direct particle measurements by polar-orbiting spacecraft. A commonly used formula derived by Spiro et al [1982] is

$$P_A(\mathrm{W}) = (1.75AE(\mathrm{nT}) + 160) \times 10^8 \,, \tag{13.14}$$

which for a 500-nT substorm yields the power of about 100 GW per hemisphere. According to (13.13) with $C \approx 3$ and (13.14) the Joule heating power is somewhat larger than the power precipitated by the electrons, except during very weak electrojet activity ($AE \lesssim 30\,\mathrm{nT}$). The reason for the 16-GW offset in (13.14) is the ever-present soft electron precipitation from the central plasma sheet.

The worst weakness of the standard $AE$ index in storm studies is its limited latitudinal coverage. Fortunately, there are several more widely distributed magnetometers in the northern polar region that can be utilized in case studies. An example of a method using global magnetometer observations supplemented with particle precipitation, electric field, ionosonde and radar data is the so-called *assimilative mapping of ionospheric electrodynamics* (AMIE) technique (for practical examples, see e.g., Knipp et al [1998] and Lu et al [1998]). For statistical studies the collection of large amounts of data is impractical. A useful compromise is to use data from the local networks with longitudinally more limited but latitudinally sufficient extent, e.g., the IMAGE chain in the Scandinavian sector or the CARISMA network in Canada.

Tanskanen et al [2002] used the IMAGE chain to investigate the Joule heating during all 352 substorms in 1997 that took place during the time interval 1600–2000 UT. In this time sector the westward electrojet index $IL$ derived from the IMAGE observations is a good surrogate for $AL$ rising from the SCW-related electrojet with the advantage of sufficient latitudinal coverage [Kauristie et al, 1996]. The conversion factor in (13.13) was set to $C = 3$. Tanskanen et al [2002] integrated the dissipation power from the beginning of the substorm growth phase to the end of the recovery phase. The events were, furthermore, divided into storm-time substorms ($Dst \leq -40\,\mathrm{nT}$, 60 events) and isolated substorms (292 events). Multiplying the obtained median values of energy deposition in the northern hemisphere by a factor of 2, the median Joule heating over isolated substorms was $0.6 \times 10^{15}$ J and over the storm-time substorms $2 \times 10^{15}$ J. As the substorms and solar flares share many similar physical features, it may be of interest to note that a typical solar flare energy is of the order of $10^{10}$ larger than a typical substorm energy.

There usually are several substorms during a storm and the electrojet activity remains large even between the substorms. Thus the ionospheric Joule heating over a long storm period can exceed $10^{17}$ J. For example, Knipp et al [1998] examined an 8-day CIR-related storm period in November 1993 using a large amount of satellite and ground-based data. Using the AMIE technique they found the total Joule heating to have been $13.7 \times 10^{16}$ J. It was 60% of the total dissipation and about $4 \times 10^{16}$ J of it took place during the first 24 h after the storm onset.

Storms driven by large ICMEs are of shorter duration and thus the total dissipation may remain smaller, but at the time of maximum driving the energy numbers can be much larger. A recent study by Rosenqvist et al [2006] of the "Halloween storm" focused on the 3-h period 1900–2200 on 30 October 2003. The proxies used in this study indicated that $0.9 \times 10^{16}$ J was deposited into the ring current and $1.2 \times 10^{16}$ J dissipated as Joule heating. However, the AMIE technique produced an even larger Joule heating of $4.5 \times 10^{16}$ J. Thus the Joule heating in 3 h was more than during the entire most intense day of the November 1993 storm.

Adding to these numbers the electron precipitation, it is clear that the ionosphere is the main energy sink in the inner magnetosphere, not only for isolated substorms but also for strong storms. All other physical processes in the ionosphere and on the auroral field lines contribute much less to the energy budget. These include ion precipitation, acceleration of ionospheric ions away from the ionosphere, auroral kilometric radiation, etc. The largest of these is the power required to sustain the ion outflow, which is of the order of $10^{10}$ W.

**Feed your brain**

Find from the literature relevant data on the ions escaping from the ionosphere (number flux, typical energy) to substantiate the power estimate $10^{10}$ W.

### 13.6.4 Energy consumption farther in the magnetosphere

The epsilon parameter was scaled only to the energy output into the inner magnetosphere and thus does not represent the total energy input through the magnetopause. We have already noted that sustaining the magnetosphere requires power of the order of $10^{12}$ W, which over a 2-h substorm period means an energy of some $10^{16}$ J, and over a one-day storm $10^{17}$ J, which actually corresponds to substantial storm-time dissipation in the ionosphere.

Most of the tail energy dissipation takes place in the plasma sheet and a significant fraction of it is delivered to particles being injected into the ring current and precipitated in the ionosphere, but not all. When Akasofu introduced the epsilon parameter, the role of plasmoids in substorm dynamics was not yet known. According to the statistical analysis by Ieda et al [1998] based on *Geotail* observations of 824 plasmoids the average energy carried by individual plasmoids in the mid-tail region was $0.16 \times 10^{15}$ J. There are, on average, 1.8 plasmoids per substorm; thus the plasmoids can be estimated to carry about $0.3 \times 10^{15}$ J during a typical substorm. In addition the post-plasmoid plasma sheet outflow was estimated to carry twice as much energy as the plasmoid itself. Summing up all these, Ieda et al [1998] concluded that the fast tailward flow amounts to $10^{15}$ J per substorm, i.e. it is of the same order of magnitude as the ionospheric Joule heating.

### 13.6.5 Energy transfer across the magnetopause

Even after the estimates discussed above, we still have a rather vague picture of total energy transfer through the magnetopause to the magnetosphere. The main mechanisms are thought to be processes leading to anomalous resistivity on the magnetopause and the magnetopause dynamo driven by the opening of the dayside magnetopause as discussed in Chap. 8. Of these the latter is generally thought to be the dominant process responsible for some 90% of the energy transfer.

In the MHD picture the magnetic stress at the magnetopause extracts the flow energy in the magnetosheath and deposits it as magnetic energy inside the magnetopause. We can use elementary electrodynamics to describe this process quantitatively. The components of the magnetic part of Maxwell's stress tensor $\mathcal{T}$ (2.67) are

$$T_{ij} = \frac{1}{\mu_0}\left(B_i B_j - \frac{1}{2}\delta_{ij}B^2\right). \tag{13.15}$$

The divergence of $\mathcal{T}$ integrated over a volume $\mathcal{V}$ gives us the force in the volume, which can be written as surface integral using Gauss's law

$$\mathbf{F} = \int_{\mathcal{V}} \nabla \cdot \mathcal{T}\, d^3r = \oint_A \mathcal{T} \cdot d\mathbf{a}\,. \tag{13.16}$$

The force density $\nabla \cdot \mathcal{T}$ is, of course, $\mathbf{J} \times \mathbf{B}$. Denoting the magnetosheath flow velocity by $\mathbf{V}$ the force $\mathbf{F}$ performs work with power

$$P = \mathbf{V} \cdot \mathbf{F} = \oint_A \mathbf{V} \cdot \mathcal{T} \cdot d\mathbf{a}\,, \tag{13.17}$$

which is in the component form

$$P = \oint_A \sum_{ij} V_i T_{ij}\, da_j = \oint_A \frac{B_n B_t}{\mu_0} V_t\, da_n\,, \tag{13.18}$$

where the subscripts $t$ and $n$ denote the tangential and normal components. The integrand of the surface integral is the normal component of the Poynting vector $S_n$. Thus both the opening of the magnetopause giving a finite $B_n$ and the magnetosheath flow $V_t$ are essential to produce energy flux through the magnetopause.

If we follow the Poynting vector from the upstream solar wind to the magnetosphere, the bow shock, being a fast shock compressing the magnetic field, already begins to convert solar wind kinetic energy to electromagnetic energy. Thus $\nabla \cdot \mathbf{S} > 0$ at the bow shock. Closer to the magnetopause the Poynting vector field lines begin to bend toward the magnetopause in the plane of the IMF, whereas they are deviated around the magnetopause in the plane perpendicular to the IMF [Papadopoulos et al, 1999; Palmroth et al, 2003]. This is in itself a trivial consequence of the definition of the Poynting vector. What is more interesting is that only in the case of open magnetopause ($B_n \neq 0$) the Poynting vector has component through the magnetopause and electromagnetic energy can flow into the magnetosphere through the high-latitude magnetopause (Fig. 13.14) as predicted by the classical Dungey picture of reconnecting magnetosphere.

Palmroth et al [2003] calculated the energy flow through the magnetopause directly from an MHD simulation of a magnetospheric storm on 6–7 April, 2000. Correct determination of the magnetopause from an MHD simulation output is a critical but non-trivial task. Palmroth et al [2003] found that identification of solar wind streamlines encompassing the magnetosphere suited best for their simulation data. Thereafter they divided the

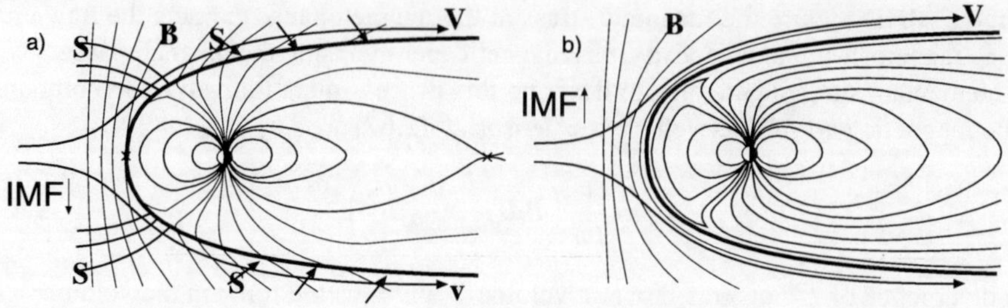


**Fig. 13.14** A schematic view of the Poynting vector on the magnetopause in the plane of the IMF for purely southward (a) and purely northward (b) IMF orientation. In the southward case **S** has a component through the high-latitude magnetopause, whereas in the northward case $\mathbf{S} \cdot \mathbf{n} \approx 0$ on the magnetopause. (Figure by courtesy of M. Palmroth.)

surface into quadrangular surface elements and calculated the energy flux $dE_q$ across each element as

$$dE_q = \mathbf{K} \cdot d\mathbf{a} \,, \tag{13.19}$$

where $d\mathbf{a}$ is the vector surface element (positive outward from the magnetopause) and **K** the total energy flux (W m$^{-2}$). Noting the total energy density by $u$ and the pressure by $P$ the energy flux can be written in the form (Chap. 6)

$$\mathbf{K} = \left( u + P - \frac{B^2}{2\mu_0} \right) \mathbf{V} + \frac{1}{\mu_0} \mathbf{E} \times \mathbf{B} \,. \tag{13.20}$$

This is evaluated from the simulation data at every surface element and the total power over any domain of interest can be calculated by simple integration.

The results presented by Palmroth et al [2003] illustrate that the energy influx is in the regions where the Poynting vector has a component across the magnetopause. When the energy influx was integrated over the magnetopause surface to the tailward distance of $X = -30 R_E$ followed surprisingly well the epsilon parameter during the storm main phase but was about 4 times larger. This supports the view that the epsilon parameter underestimates the energy input but otherwise is a parameter consistent with the MHD picture, at least during the period of southward-pointing IMF.

However, it is important to understand that while Fig. 13.14 indicates electromagnetic energy flux penetration through the magnetopause, it is not the upstream solar wind electromagnetic energy flux nor the electromagnetic energy flux enhanced by the bow shock that turns out as the *enhanced* magnetic energy in the tail lobes. Instead the energy transfer mechanism is the dynamo driven by the magnetosheath flow that works against the open magnetic flux producing power according to (13.18) to enhance the magnetopause current system. This in turn means enhanced tail lobe magnetic flux, and thus enhanced magnetic energy density.

This all means that the magnetopause acts as a Poynting vector source $\nabla \cdot \mathbf{S} > 0$ when moving inward, which cannot be seen by simply following the Poynting vector field lines in an MHD simulation [e.g., Papadopoulos et al, 1999], but one has to actually calculate

$\nabla \cdot \mathbf{S}$. However, calculating $\nabla \cdot \mathbf{S}$ from the simulation output is very sensitive to small inaccuracies in the determination of the magnetopause or, in case of reconnection, to numerical effects within the diffusion region. In order to avoid these problems Laitinen et al [2006] introduced the concept of *energy conversion surface density*. It is a line integral of $\nabla \cdot \mathbf{S}$ across the boundary surface being investigated

$$\sigma_{Ec} = -\int_{l_1}^{l_2} \nabla \cdot \mathbf{S} \, dl \,. \tag{13.21}$$

The integration limits must be chosen to enclose the entire energy conversion region to ensure that $\sigma_{Ec}$ really gives the correct energy conversion, i.e., the conversion *due to the divergence of the Poynting vector*. The units of $\sigma_{Ec}$ are readily W m$^{-2}$ and it can be applied to both magnetic energy annihilation (reconnection) and magnetic energy creation (dynamo). Note that the sign in (13.21) has been selected to give a positive number for the reconnection power, i.e., annihilation of the magnetic energy.

To actually measure the energy flux across the magnetopause boundary is very challenging because the magnetopause is always moving with respect to a spacecraft and it is necessary to separate the temporal and spatial changes in physical parameters. Rosenqvist et al [2006] utilized the four-satellite constellation *Cluster* to determine both the $\mathbf{J} \times \mathbf{B}$ force and the velocity field to calculate the energy flux locally during the very strong Halloween storm in October 2003. They found the local energy flux rate of 0.25 mW m$^{-2}$, which they integrated over a simple model magnetopause. Depending on the distance to the neutral line this resulted in total power of 17–40$\times 10^{12}$ W, of which the lower end was argued to be more plausible.

A curious fact is that the epsilon parameter calculated from the upstream parameters was at the time of the *Cluster* observations $37 \times 10^{12}$ W. This is a factor of 8 larger than the epsilon parameter during the April 2000 storm investigated by Palmroth et al [2003]. On the other hand the energy input derived from the MHD simulation was of the same order of magnitude as the *Cluster*-derived input rate above. These rather different relations between the epsilon parameter and other ways of determining the energy input may well result from uncertainties in the various methods of calculating the energy input, but they may also indicate a saturation of the system. After some threshold the magnetosphere may not be able to acquire more energy from the solar wind even if the solar wind driver parameters continue to increase. We conclude this chapter with a brief discussion of the saturation, which has become quite a popular topic during the first decade of the 21st century.

## 13.7 Superstorms and Polar Cap Potential Saturation

The saturation of the solar wind energy input into the magnetosphere at very strong driving is most clearly seen in the response of the polar cap potential to the increased solar wind electric field $\mathbf{E} = -\mathbf{V} \times \mathbf{B}$. Indications of such a saturation were found already some time ago, for example, from the spacecraft data discussed by Reiff et al [1981]. However, the quantification and, in particular, the theoretical explanation of the saturation has turned out to be very difficult (for a review, see, Shepherd [2007]).

### 13.7.1 Quantification of the saturation

The main methods of estimating the polar cap potential are either to integrate the $\mathbf{V} \times \mathbf{B}$ electric field along the orbit of a polar cap traversing spacecraft, utilizing direct electric field or plasma drift observations, or to determine it from ground-based ionospheric radar observations of the plasma flow. Unfortunately, these methods are subject to significant observational uncertainties because they provide spatially incomplete maps of the potential distribution and finding the maximum and minimum potentials require fitting the data to models. The observations can be supplemented by other ionospheric data, e.g., as is done in the AMIE modeling discussed in Sect. 13.6.3, but also in this process the error bars remain large. Another problem is that the saturation takes place only during particularly strong solar wind driving, sometimes called *superstorms*, which makes the number of useful events small. During strong storms the polar cap expands moving the auroral oval equatorward of the view of polar cap monitoring radars, which further increases the uncertainty in the determination of the actual polar cap potential. Consequently, even the existence of the saturation was questioned for a long time.

The first determinations of the saturated polar cap potential underestimated the actual potential. The estimates for the associated solar wind electric field, at which the saturation becomes observable, varied from 0.5 to $10\,\mathrm{mV\,m^{-1}}$. After the extensive study by Hairston et al [2005], including several storm events from 1998–2002 and the superstorms of October and November 2003, the best estimates for the maximum potential are about $200 \pm 65$ kV, within a wide range of solar wind electric fields from below $10\,\mathrm{mV\,m^{-1}}$ to about $40\,\mathrm{mV\,m^{-1}}$. In fact, Hairston et al [2005] claim that "it is unlikely that we will ever observe potential drop much (if any) in excess of 260 kV."

While there is no longer much doubt that the saturation, or at least a nonlinear response, of the polar cap potential is a real phenomenon during very strong solar wind driving, no generally accepted physical explanation has emerged. Shepherd [2007] and Borovsky et al [2009] list several theories, or models, attempting to explain the potential saturation. Concerning the great scatter in the data and the rarity of the events, it is no surprise that there is little observational evidence to distinguish between the different approaches. We will discuss two of the established theories and a recent proposal, which all highlight, from somewhat different angles, the critical physical questions that need to be understood in this context.

### 13.7.2 Hill–Siscoe formulation

One of the most popular approaches to the saturation problem can be named as the *Hill–Siscoe formulation* [Hill et al, 1976; Siscoe et al, 2002a,b]. Its basic idea is that the Region 1 current system closing the ionospheric current associated with the polar cap potential to the magnetopause does not produce a magnetic field that would exceed a significant fraction of the Earth's dipole field near the reconnection site. This is a reasonable argument because these magnetic fields are oppositely directed and tend to cancel each other. Once the maximum available current is reached, the maximum potential is determined by the height-integrated Pedersen current across the polar cap.

Let $\Phi_{PC}$ be the polar cap potential, $\Phi_M$ the potential over the dayside reconnection line, which in the classical MHD reconnection picture without saturation is the magnetospheric convection potential, and $\Phi_S$ the "saturation potential". In the Hill–Siscoe model these potentials are related to each other as

$$\Phi_{PC} = \frac{\Phi_M \Phi_S}{\Phi_M + \Phi_S} . \tag{13.22}$$

The argumentation behind this formula is the following. If $\Phi_M \ll \Phi_S$, the polar cap potential corresponds to the magnetospheric convection potential $\Phi_{PC} \approx \Phi_M$, i.e., there is no saturation. At the other limit $\Phi_M \gg \Phi_S$ we get $\Phi_{PC} \approx \Phi_S$, i.e., if the driver of the magnetospheric convection tries to impose a larger potential than $\Phi_S$ on the polar cap, the polar cap potential saturates. If $\Phi_M = \Phi_S$, half of the saturation level is reached.

The total current in the polar cap driven by the potential is $\Sigma_P \Phi_{PC}$, where $\Sigma_P$ is the height-integrated Pedersen conductivity, assumed to be uniform, for simplicity. This is related to the Region 1 field-aligned current driven by the magnetopause generator as

$$I_1 = \xi \, \Sigma_P \Phi_{PC} , \tag{13.23}$$

where $\xi$ is a numerical factor depending on the actual geometry of the current systems. Similarly

$$I_S = \xi \, \Sigma_P \Phi_S . \tag{13.24}$$

With these (13.22) can be rewritten as

$$\Phi_{PC} = \Phi_M - \frac{\Phi_M}{I_S} I_1 . \tag{13.25}$$

If we think in terms of equivalent current circuits, $\Phi_M / I_S$ represents the effective internal resistance of the current generator. Its role is to regulate the amount of $\Phi_M$ that is imposed over the polar cap. At some level of driving the generator becomes current limited, i.e, it cannot deliver an increasing amount of current to the magnetosphere even if the solar wind driver, represented here by the solar wind electric field, increases. Thus the amount of magnetic flux in the tail lobe saturates.

Next, the expressions for $\Phi_M$ and $I_S$ need to be found. It is a somewhat nontrivial task and requires assumptions about the reconnection process and the Region 1 current system. After some calculations Siscoe et al [2002a] find the expression that we call the Hill–Siscoe formulation of the polar cap potential

$$\Phi_{PC} = \frac{57.6 E_{SW} P_{SW}^{1/3} F(\theta)}{P_{SW}^{1/2} + 0.0124 \xi \, \Sigma_P E_{SW} F(\theta)} , \tag{13.26}$$

where $E_{SW}$ is in $\mathrm{mV\,m^{-1}}$, $P_{SW}$ is the solar wind dynamic pressure in nPa, $F(\theta)$ is the IMF clock angle dependence (e.g., $F(\theta) = \sin^2(\theta/2)$), $\xi$ has a value between 3 and 4. Note that we have omitted the scaling factor between the present dipole field and an arbitrary dipole field included in Siscoe et al [2002a].

**Feed your brain**

With the help of Siscoe et al [2002a] find out how (13.26) is obtained.

There have been several attempts to explain the physics behind the internal resistance of the current generator, but there is lack of good enough experimental data to judge between different approaches. Borovsky et al [2009] criticize the current limited model based on their global MHD simulations of the polar cap potential saturation. However, the validity of the simulation models in such extreme conditions as is the case here is of some concern. Generally, our understanding of the physics of the boundary layer dynamo driving the Region 1 current system is yet insufficient.

### 13.7.3 The Alfvén wing approach

Another, from the fundamental plasma physics viewpoint, rather interesting idea is related to one of the problems with the global MHD models for the solar wind – magnetosphere interaction during very strong driving. The models usually assume, implicitly or explicitly, super-Alfvénic flows, which can be motivated by the typical Alfvén Mach numbers $M_A$ of the order of 8 or larger. Furthermore, the boundary conditions for MHD simulations are much easier in the case of super-Alfvénic solar wind. However, the solar wind electric field $E_Y = VB_z$ of, say, $20\,\mathrm{mV\,m^{-1}}$ can result, e.g., from $V = 1000\,\mathrm{km\,s^{-1}}$ and $B_z = -20\,\mathrm{nT}$. Usually these numbers are obtained within magnetic clouds, where the plasma density is very small, making the Alfvén speed large. Thus while $1000\,\mathrm{km\,s^{-1}}$ is a high velocity, the Alfvén Mach number $M_A$ may approach unity, or be even smaller, making the solar wind interaction with the magnetopause temporarily *sub-Alfvénic*.

We can look at the low $M_A$ case from the reverse viewpoint. If you place a magnetic obstacle with a quasi-dipolar field and an open polar cap into a sub-Alfvénic flow, the polar cap flux tubes do not fold to form the tail lobes of a prototypical magnetosphere. Instead they become tilted toward the downwind direction forming so-called *Alfvén wings*, one above (north) and one below (south) the obstacle. The tilting depends on $M_A$ and for large enough $M_A$ we obtain the familiar magnetospheric configuration with tail lobes separated by the cross-tail current sheet.

The formation of Alfvén wings is actually independent of the magnetic field of the obstacle. In fact, the concept was applied already long time ago to the sub-Alfvénic motion of the non-magnetic Io in the Jovian low-density but strong magnetosphere [Neubauer, 1980]. An example of sub-Alfvénic magnetized object with Alfvén wings in the Jovian system is Ganymede.

Ridley [2007] investigated with MHD simulations the formation of Alfvén wings in the Earth's magnetosphere during conditions of low $M_A$ and showed how the Alfvén wings evolved from the tail lobes when $M_A$ is decreased from 8 to 0.7. What makes this idea interesting is that while the solar wind electric field $E_{SW}$ increases, the electric field within the Alfvén wing does not change much. In other words the Alfvén wing shields the polar cap from $E_{SW}$. Ridley [2007] found that as long as $M_A > 1$ the simulated polar cap potential rose until it quickly saturated, and slightly decreased when $M_A$ decreased below 1. That the saturation depends on $M_A$ may partly explain the large variation in the observed threshold $E_{SW}$ for the saturation, because

$$M_A = \frac{V}{v_A} = \frac{V}{B}\sqrt{\mu_0 \rho_m} \tag{13.27}$$

in addition to $V$ and $B$ also depends on the solar wind density.

According to the Io-study by Neubauer [1980] the shielding depends only on the conductivity of the shielded body, in our case $\Sigma_P$ in the polar cap ionosphere, and on the *Alfvén conductivity*

$$\Sigma_A = \frac{1}{\mu_0 v_A} \tag{13.28}$$

in the solar wind. Actually it might be a little more appropriate to discuss this in terms of the *Alfvén wave impedance* $\Sigma_A^{-1}$, since it makes the idea of the shielding more transparent. Namely, the transmission of the perturbed potential from the solar wind to the polar cap ionosphere takes place as an Alfvén wave. If the wave impedance in the solar wind is larger than the impedance in the ionosphere $\Sigma_P^{-1}$, the wave is partially reflected, which limits the potential in the ionosphere.

Kivelson and Ridley [2008] modified the Io-specific analysis of Neubauer [1980] to the Earth's polar cap potential problem and found the expression

$$\Phi_{PC} = \frac{2 E_{SW} F(\theta)\, d\, \Sigma_A}{\Sigma_P + \Sigma_A} \, . \tag{13.29}$$

Here $E_{SW} F(\theta)$ is the same expression for the reconnection electric field as in (13.26) and $d$ is the distance across the unperturbed solar wind that contains field lines that reconnect at the dayside magnetosphere.

While (13.29) is formally not very different from (13.26), the underlying physical argumentation is. The saturation is a natural product of information transfer in MHD plasma and there is no need to refer to artificial current circuits. According to Borovsky et al [2009] this model was the only one that did not contradict their simulation results among the nine models they considered.

### 13.7.4 Magnetosheath force balance

It should always be kept in mind that it is not the upstream solar wind plasma flow with its electric field but the magnetosheath plasma that interacts with the magnetopause. While an argumentation based on forces in the magnetosheath has already been introduced by Siscoe et al [2002b] considering the roles of the current systems and solar wind ram pressure in the Hill–Siscoe formulation, we discuss it separately because Lopez et al [2010] have recently made the argument more explicit and related it to the magnetosheath plasma parameters during strong solar wind driving.

The magnetosheath force balance argument is based on the change in the relative magnitude of the pressure force ($\nabla P$) and magnetic force ($\mathbf{J} \times \mathbf{B}$) when the upstream Alfvén Mach number and, as a consequence, the magnetosheath plasma beta decrease. During typical high $M_A$ conditions the magnetosheath flow is controlled by the pressure force and the length of the reconnection line is independent of the changes in the solar wind parameters. Thus a larger solar wind electric field leads to a larger potential over the reconnection line. However, when the IMF becomes large enough, the magnetosheath plasma beta just behind the shock becomes less than 1 and the magnetic force begins to dominate over the pressure force causing increasingly efficient deviation of the plasma flow around the magnetopause. Consequently, the fraction of the solar wind electric field that is imposed on the reconnection line decreases. In other words, the length of the reconnection line mapped along the magnetic field lines back to the upstream solar wind becomes shorter and the geoeffective length of the reconnection line is reduced in the solar wind. Correspondingly, the potential over the reconnection line does no more grow linearly with the solar wind $\mathbf{V} \times \mathbf{B}$ electric field.

The magnetosheath force balance argument as stated by Lopez et al [2010] is different from the Hill–Siscoe model, as it is independent of the Region 1 current dynamics. However, these two approaches certainly are closely related to each other. An interesting property of the Region 1 current closure pointed out by Siscoe et al [2002b] and consistent with the magnetosheath force balance argument is that near the saturation level the magnetopause Chapman–Ferraro current can no longer close the entire Region 1 current. Instead the MHD simulations indicate that part of the current closes to the current at the bow shock. For southward IMF the direction of the current in the magnetosheath is outward from the magnetopause in the duskward side of the subsolar direction and inward in the dawnward side, consistent with the Region 1 system. It should be no surprise that there is a current system at the bow shock, as its role is to compress the magnetic field. However, this current system has received surprisingly little attention in the solar wind–magnetosphere interaction investigations.

# 14. Storms in the Inner Magnetosphere

The inner magnetosphere containing the plasmasphere, ring current and radiation belts is a key domain of magnetospheric storms from both physical and practical viewpoints. The growth, decay and asymmetries of the ring current are considered as main indicators of the large-scale storm evolution and the electron belts respond strongly to the storm evolution leading to the most hazardous conditions for spacecraft in orbit. The outer electron belt reaches beyond the geostationary orbit, which contains the largest number of satellites. The growing importance of global navigation satellite systems also brings the even more hazardous conditions deeper inside the electron belt ($L \simeq 4$) into focus.

A particular feature of the physics of the inner magnetosphere is the overlapping plasma populations of widely different and variable temperatures, densities and particle contents: the cold but dense plasmasphere, the highly variable ring current carried by ions with energies up to about 200 keV, and the ion and electron radiation belts where electron energies reach relativistic levels. These populations do not only overlap spatially, but they also affect each other through wave–particle interactions. For example, the plasma gradients at the plasmapause provide free energy for wave modes that interact with both ring current and radiation belt particles. Furthermore, the energetic ions interact with the cold tenuous neutral hydrogen atom exosphere through charge exchange collisions, which leads to enhanced loss cone in the ion population. The enhanced anisotropy further amplifies the waves that are also capable of interacting with relativistic electrons, and so on.

These properties of the inner magnetosphere pose significant challenges to observations, theories and modeling. The scientific instruments need to cover a wide range of physical parameters and operate in a very hostile radiation environment. Theoretical treatment of multicomponent, non-Maxwellian, inhomogeneous plasmas is nontrivial indeed and requires use of computationally demanding numerical methods. For example, inclusion of the ring current and its coupling to the ionosphere in MHD simulation schemes has turned out to be a tough challenge. The particle tracing and diffusion models must, so far, rely on prescribed background electric and magnetic field models and the treatment of transport, acceleration and loss of energetic particles due to different mechanisms from wave–particle interactions to charge exchange collisions requires ingenious solutions to calculate the applicable diffusion coefficients.

H.E.J. Koskinen, *Physics of Space Storms: From the Solar Surface to the Earth*,
Springer Praxis Books, DOI 10.1007/978-3-642-00319-6_14,

## 14.1 Dynamics of the Ring Current

The ring current is the key element of storm evolution in the magnetosphere. The main ring current carriers are protons in the energy range 10–200 keV, which during storm periods are supplemented by significant amounts of oxygen ions originating from the ionosphere.

Because substorms and other activations are frequent during the storm main phase, the original idea was that the substorm particle injections would be the main source of energetic plasma of the ring current (for an early review, see Akasofu [1966]). However, it has turned out to be difficult to find a straightforward relationship between individual injections and the enhancement of the ring current [see, e.g., Kamide, 1992]. Consequently, the question was turned around and quite some effort was put on trying to understand the ring current evolution based on the enhanced large-scale convection alone. But this approach does not address properly the question of how the relatively low-temperature plasma sheet ions gain energies to the order of 100 keV. Thus transient perturbations addressed to substorm-like activations are again called for [e.g., Fok et al, 1999; Ganushkina et al, 2005].

### 14.1.1 Asymmetric structure of the ring current

Before going to the physical mechanisms of ring current growth and decay, let us consider the large-scale interpretation of the magnetic deviations measured on the ground.

The *Dst* index and its high-time resolution variant *SYM-H* are constructed to describe the average westward ring current using ground-based observations of magnetic deviation at low latitudes.[1] As discussed in Chaps. 1 and 13, *Dst* and *SYM-H* are strongly contaminated by other magnetospheric current systems and by induction effects during the rapid main phase evolution. A direct observation of 10–200-keV ions on trapped orbits would be a preferable method to determine the real ring current. Unfortunately, such observations are only seldom available. Observations of energetic neutral atoms to be discussed shortly may, in the future, give quantitative information of the ring current, but so far their role has been to provide supplementary information with rather large uncertainties.

During magnetic storms the asymmetry of the ring current becomes an essential part of the story. The asymmetry can be estimated from the differences between the recordings at low-latitude magnetometer stations on different longitudes and indexed by the *Asym* or *ASY-H* indices. The reason for asymmetry is usually described in terms of *partial ring current*, but again the terminology is not unique. There is a persistent partial ring current corresponding to the closure of the ever-present Region 2 FAC system. The Region 2 FAC enters the ring current region in the dawn sector and closes across midnight to the FAC toward the ionosphere in the dusk sector. This partial ring current strengthens with the strengthening FACs during magnetospheric activity, but in ground-based magnetic observations this evolution is overshadowed by simultaneous strengthening of the cross-tail current.

---

[1] While the acronym *SYM-H* suggests that it is the symmetric part of the ring current, it actually is a weighted average over the longitudes of the observed magnetic deflections at low-latitude stations similar to *Dst*.

A more direct storm-time asymmetry rises, however, from ions drifting on open trajectories around the dusk toward the dayside magnetopause. Also this large-scale current must be continuous and any divergence of the perpendicular current be compensated by a divergence of a FAC. However, as the current is carried by fresh ions approaching from the plasma sheet and departing toward the magnetopause, the closure current does not need to connect to the ionosphere at low $L$ shells. In fact, during enhanced convection there is an upward FAC emerging from the Harang discontinuity near the magnetic midnight (Fig. 1.28) on field lines that connect to the pre-midnight sector plasma sheet beyond $10 R_E$ [Koskinen and Pulkkinen, 1995]. On the dayside there is a downward current to the poleward edge of the polar cusp region slightly past noon (Fig. 1.28). Whether these ionospheric source and sink regions are connected to the partial ring current in the dusk sector or not, is not fully clear.

The different magnetospheric response to magnetic clouds and ICME sheath regions discussed in Sect. 13.4 is evident also in the ring current asymmetry. Huttunen et al [2006] found that during intense sheath-driven storms the asymmetric component (*ASY-H*) dominates the symmetric component (*SYM-H*) while during most cloud-driven storms the situation is the opposite. Furthermore, throughout the main phase of a sheath storm the asymmetry is highly variable with variations related to auroral activity.

### 14.1.2 Sources of the enhanced ring current

The ring current never disappears because there always are charged particles in the near-Earth space drifting around the Earth. For our topic it is important to understand the processes that lead to the growth and decay of the storm-time ring current.

Both the ionosphere and the solar wind are sources of magnetospheric plasma and thus of ring current as well. The two main ring-current-carrying ion populations are energetic $H^+$ and $O^+$ ions. While singly charged oxygen must be of ionospheric origin, the protons may come from both sources. Table 14.1 summarizing conclusions based on the *AMPTE/CCE* and *CRRES* satellite observations has been adapted from the review by Daglis et al [1999]. Note that the data are based on a relatively small number of storm-time observations and thus there are considerable uncertainties in the numbers and individual storms can show large deviations from these values.

Furthermore, the ion composition does not vary due to the variability of the source alone but also due to the very different lifetimes of the different ion species in the ring current. As discussed in Sect. 14.1.4 below, the main mechanism causing the decay of the current, the charge exchange with the neutral hydrogen in the geocorona, has very different consequences for the evolution of the $H^+$ and $O^+$ contents at different energies.

The ion energies in the ionosphere and the solar wind are much smaller than the energies of the main carriers of the ring current. While the solar wind seed population already is in the keV-range, the ionospheric plasma has to be accelerated all the way from a few eV to the ring current energies. On the other hand, the current carriers do not arrive directly at the ring current but are first transported to the plasma sheet where they undergo quite significant acceleration before the injection into the ring current.

**Table 14.1** Relative abundances of different ion species and total ion energy densities at $L = 5$ in the ring current during quiet times and under different levels of storm activity based on *AMPTE/CCE* and observations [Daglis et al, 1999]. Note that in experimental space physics it is customary to give the particle energy density in units of keV cm$^{-3}$.

| Source and species | | Quiet times | Small & medium storms | Intense storms |
|---|---|---|---|---|
| Solar wind $H^+$ | (%) | $\geq 60$ | $\sim 50$ | $\leq 20$ |
| Solar wind $He^{++}$ | (%) | $\sim 2$ | $\leq 5$ | $\geq 10$ |
| Ionospheric $H^+$ | (%) | $\geq 30$ | $\sim 20$ | $\leq 10$ |
| Ionospheric $O^+$ | (%) | $\leq 5$ | $\sim 30$ | $\geq 60$ |
| Solar wind total | (%) | $\sim 65$ | $\sim 50$ | $\sim 30$ |
| Ionosphere total | (%) | $\sim 35$ | $\sim 50$ | $\sim 70$ |
| Total energy density | (keV cm$^{-3}$) | $\sim 10$ | $\geq 50$ | $\geq 100$ |

The acceleration and heating of the outflowing ionospheric plasma is likely to take place in several steps (see, e.g., Chap. 2 of Hultqvist et al [1999]). The strongest outflow occurs along the field lines connected to the auroral oval. Some heating of plasma by fluctuating electric fields already takes place in the ionosphere. The more energy the ions gain, the more efficiently the magnetic mirror force pushes them up. Further acceleration is provided by the same electric potential structures that accelerate auroral electrons downward in the range of 1–10 keV. Due to the strong FACs and particles moving up and down and drifting across the magnetic field lines the regions above the auroras host a large variety of plasma waves, including electrostatic ion cyclotron waves, lower hybrid waves, ion–Bernstein waves, whistler mode waves, etc., many of which can contribute to the energization of the ionospheric plasma to the keV-range, i.e., to the same level as the plasma coming from the solar wind.

In the magnetotail current sheet electromagnetic energy is transferred to particles ($\mathbf{J} \cdot \mathbf{E} > 0$), which is the main reason why the power of some $10^{12}$ W is needed to maintain the magnetotail, as discussed in Sect. 13.6. While transient processes like reconnection can be very effective particle accelerators, the energy transfer also takes place during quiescence. Ions crossing the current sheet with a finite but small normal magnetic field component ($B_n$) are transported for a short while in the direction of the electric field and thus gain energy [see, e.g., Lyons and Speiser, 1982]. This is essentially a diffusion process in pitch angle and energy, which is due to the loss of exact guiding center motion, i.e., breaking of the first adiabatic invariant and chaotization particle motion [Chen and Palmadesso, 1986; Büchner and Zelenyi, 1989].

Figure 14.1 illustrates how a low-energy ion entering the nightside magnetosphere from the high-latitude mantle is transported first to the distant tail and from there earthward with the large-scale convection. The closer to the Earth it comes, the more frequently it crosses the current sheet. Numerical test-particle simulation results by Ashour-Abdalla et al [1993] together with an analytic estimate based on the analysis by Lyons and Speiser [1982] for

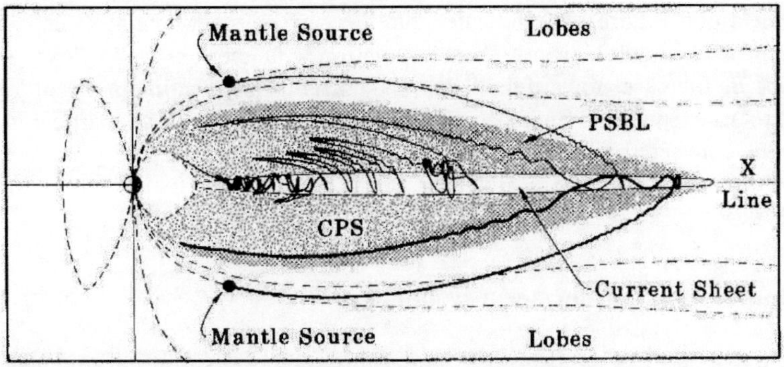


**Fig. 14.1** Schematic picture of transport of a solar wind particle from two different source locations in the mantle to the inner plasma sheet. (From Ashour-Abdalla et al [1993].)

the maximum energy gained in such a model

$$W_1(x) = \frac{m_i}{2}\left[\left(v_x + \frac{2E_y}{B_n(x)}\right)^2 + v_y^2 + v_z^2\right] \tag{14.1}$$

are shown in Fig. 14.2. Note that sometimes the particle gains energy, sometimes it loses energy when crossing the current sheet. This leads to a complicated spatial structuring of the distribution functions (for more details, see Ashour-Abdalla et al [1993]).

Because the stretching is strongest in the distant tail, the particle motion is most chaotic there and, consequently, the acceleration is most efficient for particles entering the plasma

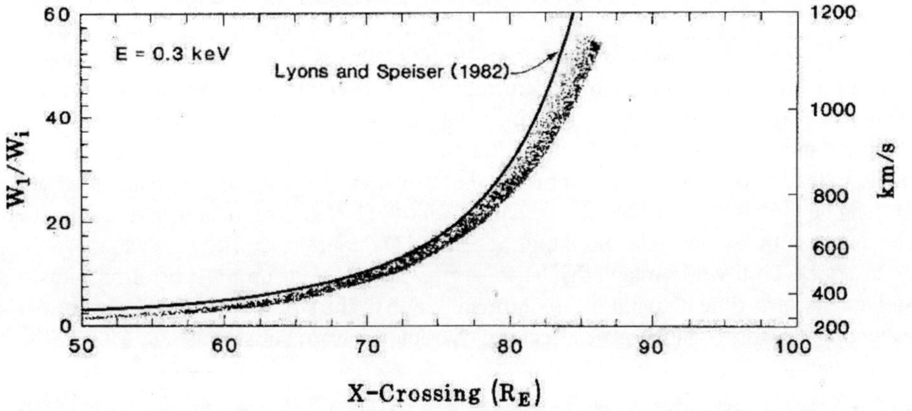


**Fig. 14.2** Results of test-particle simulation by Ashour-Abdalla et al [1993]. The scale on the right gives the energy gain, that is proportional to the convection velocity $V = E_y/B_n$ given on the right. The horizontal scale illustrates the distance from the Earth to the point where the first current sheet crossing has taken place.

sheet furthest out. Figure 14.2 indicates that the current sheet heating is capable of producing ions of several keV from a sheet population with energies well below 1 keV.

Because $B_n$ increases toward the Earth, the current sheet heating is less efficient in the near-Earth space. In the absence of inductive electric fields or wave–particle interactions the particles convected adiabatically to the inner magnetosphere still gain energy by the drift betatron mechanism (3.51). However, this is not sufficient to account for ion energies above 100 keV.

### 14.1.3 Role of substorms in storm evolution

It has been difficult to establish direct connection between storm development and substorm expansions, at least if we focus on magnetic indices such as *Dst* and *AL*. On the other hand, if we look at the storms from the particle viewpoint, it is difficult to neglect the role of substorms. After all, substorm onsets are associated with energetic particle injections into the inner magnetosphere (Chap. 13), a fraction of which are bound to end up in the ring current. Furthermore, substorm activity both heats up the plasma sheet and enhances ionospheric ion outflows.

Baumjohann et al [1996] showed that the average ion temperature in the near-Earth plasma sheet (radial distances 10–19 $R_E$) is significantly higher around storm-time substorm onsets (about 7 keV) than around onsets of isolated substorms (about 3 keV). In both cases the substorm onset was found to lead, on the average, to heating of the ions by about 2 keV. Thus a sequence of substorm expansion phase onsets during a storm main phase increases the temperature of the ring current seed population and most likely contributes to the enhancement of the current. As this mechanism also takes place during isolated substorms, they also can contribute to the source population for a storm-time ring current to be enhanced at some later time.

Enhanced $O^+$ ion ouflows are observed during substorm growth and expansion phases in the ionosphere by ground-based radars and directly by satellites traversing the auroral field-lines. Thus the observed high storm-time $O^+$ fluxes (Table 14.1) are not surprising. Substorm dipolarizations contribute further to ion acceleration through strong transient inductive electric fields, whose role in reaching 100-keV energies may be critical [e.g., Pellinen and Heikkila, 1984; Ganushkina et al, 2005]. The inductive electric fields are likely to lead to preferential acceleration of $O^+$ over $H^+$ because all adiabatic invariants of $O^+$ can be violated while the magnetic moment of $H^+$ may remain conserved. This idea is consistent with test-particle simulation results of Delcourt et al [1990].

In conclusion, while magnetospheric storms are not built up by substorms, substorms and other storm-time activations are intimately tied to the storm evolution. Substorms contribute to the storms and storms affect the characteristics of substorms.

### 14.1.4 Loss of ring current through charge exchange collisions

The actual level of ring current is determined by a balance between the current carrier sources and losses. The loss of current takes place all the time, but is overshadowed by the injection of new current carriers during the storm main phase. The energy relationship

(13.11) illustrates this balance. The term $Dst^*/\tau$ is always negative describing the average energy loss with time constant $\tau$. The slope $\partial Dst^*/\partial t$ is negative during the main phase describing the increase of ring current energy. During the recovery phase, the derivative is smaller and positive. Note that neither of these terms is directly associated with a single physical source or loss process. Furthermore, $\tau$ is not a constant because different loss processes evolve differently during the progress of a storm.

The main loss processes for the ions are *charge exchange* and Coulomb collisions, as well as interaction with electromagnetic ion cyclotron waves. Of these the most important is the charge exchange between the ring current ions and the neutral hydrogen atoms of the *geocorona* that is an extension of the Earth's exosphere [e.g., Chamberlain, 1963].

Charge exchange collisions are ubiquitous in the solar system. The most typical charge exchange reaction is a collision between an ion and a neutral atom, in which the ion acquires an electron from the atom. After the process the charge state of the ion is reduced by one and the neutral particle becomes positively charged

$$\mathrm{X}^{n+} + \mathrm{Y} \rightarrow \mathrm{X}^{(n-1)+} + \mathrm{Y}^{+} . \tag{14.2}$$

There are more complicated reactions, including charging of dust particles, but they are not essential to our topic.

An example of charge exchange is the interaction between solar wind ions and interstellar hydrogen flowing through the heliosphere. Although the collision frequency is very small, the volume is huge and the process has an observable effect. Measuring the solar Lyman-$\alpha$ radiation scattered from the interstellar hydrogen atoms it is possible to map the large-scale structure of the solar wind. The more there is solar wind in some direction, the larger the fraction of interstellar hydrogen atoms that become ionized and no longer scatter the solar Lyman-$\alpha$ photons. Most of the incident ions are protons, which are converted to neutral hydrogen. However, their speeds are the same as the solar wind speed, i.e., of the order of 20 times greater than the speed of the incoming interstellar matter (about 25 km s$^{-1}$) and thus the Lyman-$\alpha$ radiation scattered by the newly-born neutrals is Doppler-shifted and the two hydrogen populations can be distinguished from each other. Such an instrument, called SWAN, is actually a part of the payload of *SOHO*. Other extraterrestrial examples are the interaction of the solar wind with the upper atmospheres of non-magnetized planets, Venus and Mars, as well as with the comets.

Dessler and Parker [1959] suggested that charge exchange between ring current ions and the cool hydrogen geocorona would be an efficient mechanism for the ring current decay. At ring current altitudes the collisionless exosphere consists almost purely of hydrogen atoms and the main charge exchange processes are

$$\begin{aligned}
\mathrm{H}^{+} + \mathrm{H} &\rightarrow \mathrm{H} + \mathrm{H}^{+} \\
\mathrm{O}^{+} + \mathrm{H} &\rightarrow \mathrm{O} + \mathrm{H}^{+} \\
\mathrm{He}^{++} + \mathrm{H} &\rightarrow \mathrm{He}^{+} + \mathrm{H}^{+} \\
\mathrm{He}^{+} + \mathrm{H} &\rightarrow \mathrm{He} + \mathrm{H}^{+} .
\end{aligned} \tag{14.3}$$

Note that ions mirroring at low altitudes also undergo charge exchange with oxygen atoms, which must be included in detailed model calculations.

The temperature of the geocorona is about 1000 K (0.1 eV). Thus when charge exchange with a ring current ion of tens or hundreds of keV takes place, the emerging particles are a very slow ion and an *energetic neutral atom* (ENA). Charge exchange does not directly decrease the number of current carriers, but transports the charge from efficient current carriers to very inefficient ones, as the current carried by a drifting particle is directly proportional to its energy (Chap. 3).

The efficiency of charge exchange as a loss mechanism depends on the lifetimes of the current carriers. These in turn depend on the density profile of the geocorona and are different for different ion species at different energies. Furthermore, the $L$ shells and pitch angles of the incident ions need to be considered because ions mirroring at different altitudes encounter different exospheric densities. The atomic hydrogen density is almost spherically symmetric and drops from more than 1000 atoms cm$^{-3}$ at $2R_E$ (geocentric) to less than 50 at the geostationary distance $6.6R_E$ [Rairden et al, 1986]. For equatorial particles ($\alpha = 90°$) the lifetime can be given as

$$\tau_e = \frac{1}{n(r)\sigma_{che} v}, \tag{14.4}$$

where $n(r)$ is the neutral hydrogen density in the equatorial plane, $\sigma_{che}$ is the energy- and mass-dependent charge exchange cross-section and $v$ is the ion velocity. Ions mirroring at off-equatorial latitudes are lost more quickly. The lifetimes can be estimated using the formula

$$\tau_m = \tau_e \cos^{\delta} \lambda_m, \tag{14.5}$$

where $\lambda_m$ is the mirror latitude and $\delta$ has been found to be in the range 3–4 at ring current altitudes. Thus charge exchange leads to an anisotropic ion pitch angle distribution that is strongly peaked at 90°, in particular at low $L$ shells where the process is strongest due to the largest neutral density.

Unfortunately, the collisional cross-sections cannot be determined theoretically and also experimental determination of the charge exchange cross-sections $\sigma_{che}$ is difficult because the exosphere is a much better vacuum than can be created in laboratories. Figure 14.3 illustrates predicted ion lifetimes together with lifetimes inferred from *Explorer* 45 observations during a geomagnetic storm in February 1972. The very different energy-dependence of $H^+$ and $O^+$ is evident. High-energy $O^+$ disappears by charge exchange much faster than high-energy $H^+$. At larger $L$ shells the lifetimes are longer, at smaller shells shorter. For more details see Smith et al [1981].

From these considerations it is clear that the inclusion of charge exchange collisions in numerical ring current models is a challenging task. However, it has been done quite successfully during a long process since the mid-1990s, for example in the RAM code described in Sect. 10.5. There is a consensus that charge exchange is the dominant ring current loss process; but the wave–particle interactions are also of great importance.

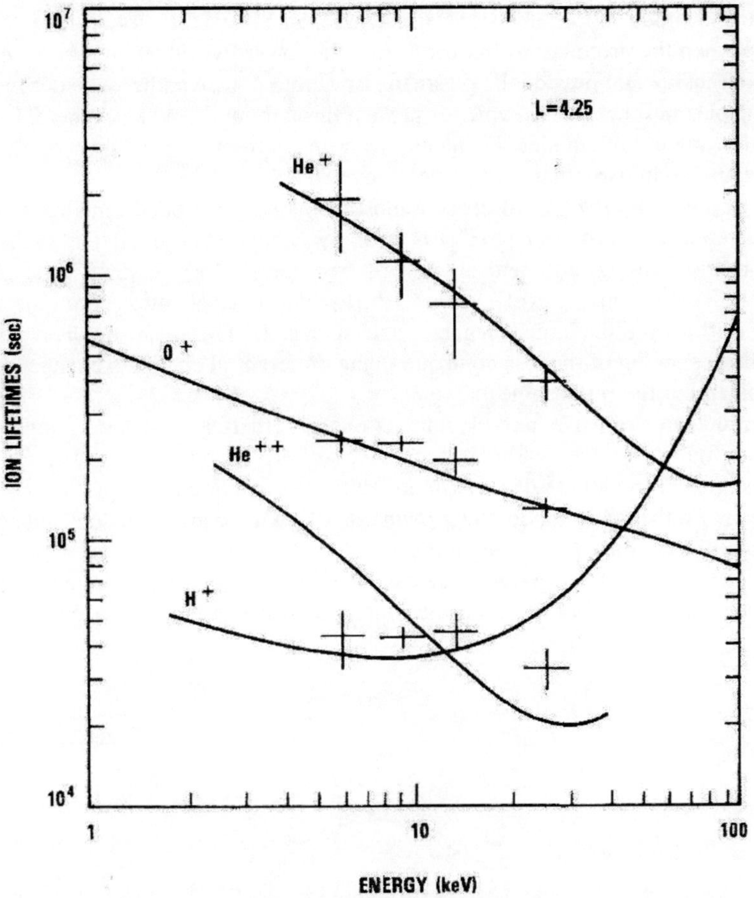


**Fig. 14.3** Theoretically predicted (solid lines) and observationally inferred (crosses) lifetimes of ions in the energy range 1–100 keV at $L = 4.25$ during a geomagnetic storm in February 1972. The mirror latitude $\lambda_m = 14°$ has been assumed. (From Smith et al [1981].)

### 14.1.5 Pitch angle scattering by wave–particle interactions

Ring current is also lost by direct removal of current carriers to the atmospheric loss cone by pitch angle scattering. Part of this scattering takes place through Coulomb collisions, which are most efficient at lower energies (<10 keV). However, charge exchange and Coulomb collisions jointly do not remove enough ions with energies over a few tens of keV and above 100 keV they lead to too flat pitch angle distributions, i.e., smaller loss cones than observed [Fok et al, 1996].

These problems point to the role of wave–particle interactions, which can result in very efficient pitch angle scattering. Waves on the whistler mode surface (see Fig. 5.4) and electromagnetic ion cyclotron (EMIC) waves are capable of interacting with ring current ions. Whistler mode chorus emissions or magnetosonic waves at very oblique propagation angles between the proton gyro frequency and the lower hybrid frequency have been observed both inside and outside the plasmapause (Fig. 14.4). Another whistler mode emission is the plasmaspheric hiss, which is present throughout the plasmasphere. For oblique propagation angles both modes can interact with ring current ions and provide an important loss-mechanism, in particular, at energies above 80 keV.

During quiet times EMIC waves occur mostly beyond $L = 7$ but during magnetic storms the temperature anisotropy ($T_{\perp} > T_{\|}$) is strongly enhanced at smaller $L$ shells near the plasmapause providing sufficient amount of free energy for the EMIC waves to grow (Fig. 14.4). The anisotropy evolves due to drift-betatron acceleration when the ions adiabatically drift toward the larger magnetic field (Chap. 3). The anisotropy is further amplified by the deepening of the loss cone due to the preferential charge exchange loss of ions with small equatorial pitch angles.

Proper inclusion of wave–particle interactions in numerical ring current models is even more challenging than the charge exchange losses because both the growth and decay of the waves must be modeled self-consistently with the evolution of the particle populations. For example, in the RAM model the growth rate of EMIC waves is calculated solving the hot plasma dispersion equation simultaneously with the kinetic equation (10.43). From the growth rate the wave amplitudes are calculated using an empirical relation. The effect of the wave–particle interactions on the ions is thereafter treated as a diffusion process

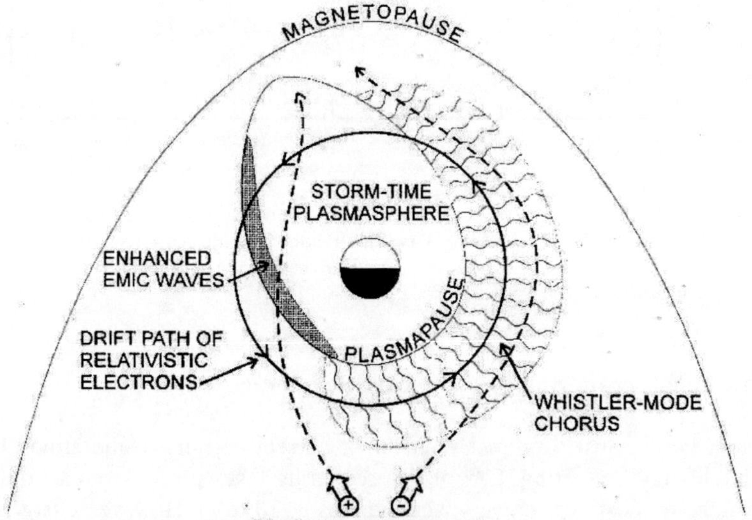


**Fig. 14.4** Schematic picture of the storm-time inner magnetospheric waves and key particle populations. During storms the afternoon bulge of the plasmasphere, introduced in Chap. 1, is shifted towards noon by the strong convection, which affects the location of EMIC wave generation region. EMIC waves can interact both ring current ions and relativistic electrons. (From Summers et al [1998].)

where the diffusion coefficients are determined using the calculated wave amplitudes (see Jordanova et al [2006] and references therein).

### 14.1.6 ENA imaging of the ring current

Imaging of energetic neutral atoms (ENAs) produced by charge exchange collisions was introduced in the 1990s as a new observational tool for studies of the inner magnetosphere. While the ions are confined to helical paths around the magnetic field lines, a newly-born ENA moves along a straight line in the direction of its momentum at the time of its formation. Thus, an appropriately designed instrument looking to the ring current could in principle form an image of the ring current. The ENA flux at the detector is a line-of-sight integral over the ion intensity and the neutral gas number density in the exosphere and upper atmosphere. Methods have been devised to deconvolve the ion intensity distribution from the ENA images [e.g., DeMajistre et al, 2004; Perez et al, 2004]. Most useful information on the ring current evolution can likely be inferred from the ENA data when they are analyzed in combination with data assimilation methods and good numerical ring current models [e.g., Nakano et al, 2008].

The first space-borne ENA observations were made using ion instruments. The counts caused by ENAs penetrating to the detector can to some extent be distinguished from ion counts because the ENAs arrive from the direction of the Earth and their fluxes do not reflect the changing relative directions of the magnetic field and the Earth when the spacecraft moves. Evidently, such observations based on the doctrine "one man's noise is other man's data" do not tell much more than indicate the ENA production in the ring current region. However, the relative changes in the ENA fluxes obtained this way from the ion instruments onboard *Geotail* and *Polar* spacecraft (see, e.g., Daglis et al [1999] and references therein) have actually provided useful information on the dynamics of the ring current.

The first dedicated ENA instrument was PIPPI[2] onboard the Swedish low-altitude microsatellite *Astrid* in 1995 [Barabash et al, 1997]. Its further evolution versions have been sent to make ENA observations around Mars and Venus onboard ESA's *Mars Express* and *Venus Express* spacecraft. While PIPPI demonstrated that it is also possible to make ENA observations from a low-altitude vantage point [Brandt et al, 2001], an ENA instrument well above the ring current region gives a more global view. The *IMAGE* satellite of NASA launched in 2000 was designed for global magnetospheric imaging utilizing several different techniques, including ENA imagery [Burch, 2000].

As an example of results relevant to the understanding of magnetic storms in the inner magnetosphere Brandt et al [2002] studied *IMAGE* ENA observations from several storm main phases in 2000 and 2001. They found that the peak ion concentration in the energy range 27–198 keV was near midnight or slightly skewed toward the post-midnight region. This skewing was strongest during large positive IMF $B_y$. This suggests that under such conditions the near-Earth electric field is strong enough for the $\mathbf{E}\times\mathbf{B}$ drift to overcome the

[2] PIPPI is an acronym for Prelude in Planetary Particle Imaging. It was carefully selected to honor Pippi Långstrump who was a child heroine created by the Swedish writer Astrid Lindgren.

magnetic drift in the sector east of the midnight meridian until the ions are transported to relatively low $L \simeq 4$–5.

An important goal of the ENA imagery is to convert the ENA spectra to represent the ring current more directly than the *Dst* or *SYM-H* indices. Ohtani et al [2005] investigated ENA data of the *IMAGE* satellite together with the *SYM-H* index and magnetic observations from geostationary orbit in order to assess the role of substorm dipolarizations in the storm evolution. The substorm onset can be expected to have a two-fold effect on the storm progress as indicated by the *SYM-H* index. The original picture was that the substorms would inject current carrying particles to the ring current and thus enhance the current and lead to the (negative) enhancement of *SYM-H*. On the other hand, the substorm dipolarization leads to a substantial weakening of the cross-tail current in the near-Earth region, which has an opposite effect on *SYM-H*. In fact, it is the latter that Ohtani et al [2005] found in their analysis. But this is not the whole story. The ENA emission was found to enhance, suggesting an increase of the "true" ring current, which was overcompensated by the decrease of the tail current contribution to the ground-based observations. While the interpretation of ENA data is still far from complete, ENA imaging is the only tool to image the ring current ion distribution from which the plasma pressure can be derived. It therefore has a promise to become a diagnostic tool for investigating the 3D pressure-driven current system that makes up the Region 2 currents and the partial ring current that close through the sub-auroral ionosphere.

**Feed your brain**

Read carefully the articles by Ohtani et al [2005] and Nakano et al [2008]. Pay particular attention to the difficulties in interpreting the ENA intensity as ring current intensity pointed out in section 4 of Ohtani et al [2005].

## 14.2 Storm-Time Radiation Belts

In principle the trapped part of the ring current could be considered as a low-energy tail of the radiation belts. However, it is convenient to distinguish the ring current and the radiation belts from each other because their sources, composition, and spatial structures are different, as are their roles in storm processes. Furthermore, the theoretical and model studies of radiation belts require relativistic formulation, in particular for the electrons. We start this discussion from the radiation belt ions.

### 14.2.1 Sources of radiation belt ions

As discussed in Chap. 1 (Fig. 1.20) the energetic particle content of the inner radiation belt ($L \simeq 1.5$–3) is dominated by protons in the energy range 0.1–40 MeV. While the spectrum of trapped ions at energies larger than 100 keV appears to turn quite smoothly from the ring current carriers to radiation belt protons, it is important to understand that the histories of the ions are different.

As discussed in the previous section, most protons up to about 100–200 keV are believed to originate from the much lower-energy ionosphere and solar wind, and accelerated by internal magnetospheric processes. Substorm onset-related inductive electric fields may be able to energize particles to $\sim$1 MeV, depending on the actual X–O-line geometry and $\partial \mathbf{B}/\partial t$, which define the integral $\int \mathbf{E} \cdot d\mathbf{s}$ along the path of the particle being accelerated (see Pellinen and Heikkila [1984] and references therein). However, the number of such particles ending up in closed drift paths in the inner radiation belt is most likely very small. Therefore other mechanisms are needed to produce ions up to tens of MeV.

The solar flares and CMEs (Chap. 12) produce large fluxes of solar energetic particles (SEP), of which most are shielded beyond $L \approx 4$ by the geomagnetic field. The innermost magnetosphere is a relatively steady magnetic bottle, which is equally difficult to get into as to escape from. SEPs arriving at the region with pitch angles within the atmospheric loss cone are lost, whereas most ions are deflected by the magnetic field. At geostationary orbit particle spectra are rather consistent with the solar wind source. There the field is also more variable than closer to the Earth, which allows a small fraction of the ions to experience sufficient amount of pitch angle scattering by inductive electric fields or wave–particle interactions that can move them to trapped orbits. Thereafter they can be transported radially inward through diffusion determined by wave–particle interactions and due to spatial/temporal inhomogeneities of the electromagnetic field along their orbits.

Consequently, the solar storms have a two-fold role in the radiation belt ion dynamics. They provide intermittent source populations and the solar wind perturbations drive perturbations in the magnetosphere that are necessary for trapping the particles. The response of the inner magnetosphere does not need to be immediate. The populations persist in the geostationary region for a few days after a major SEP event and thus provide a long-lasting source for inward transport. On the other hand the energetic solar particles arrive at the Earth much faster than the associated ICME and are already waiting for major perturbation leading to enhanced trapping and transport during the commencing magnetospheric storm.

The mechanism for introducing the highest energy protons that can be confined within the inner radiation belt is called *CRAND (cosmic ray albedo neutron decay)*. The cosmic ray bombardment of the atmosphere produces neutrons that move in all directions. Although the average neutron lifetime is about 14 min 38 s, during which a multi-MeV neutron escapes far from the Earth, a small fraction of them decay to protons while still in the magnetosphere. At energies below 30–50 MeV the proton spectra are too intense and variable to be explained by the CRAND mechanism. Note also that CRAND is too inefficient by far to account for the observed electron fluxes discussed below.

### 14.2.2 Losses of radiation belt ions

The charge exchange cross-sections decrease rapidly for energies above 100 keV. Thus charge exchange can remove radiation belt ions only after some other mechanisms, e.g., Coulomb collisions have first decreased their energies. In fact, the main effect of Coulomb collisions on the inner belt ions is to transfer them in the phase space toward lower energies when they are finally lost through charge exchange. This is an important loss mechanism for protons at energies $>10$ MeV.

Wave–particle interactions lead the ion loss through pitch angle scattering them into the atmospheric loss cone. The most important wave modes are EMIC waves and plasmaspheric hiss. Because the inner belt region is an excellent magnetic bottle, the lifetimes of radiation belt ions are long and their loss is much less important to the storm dynamics than the loss of ring current ions.

### 14.2.3 Transport and acceleration of electrons

Some of the most challenging theoretical and most important practical questions in magnetospheric dynamics are the source and loss processes of relativistic radiation belt electrons. The appearance of the electrons is strongly correlated with fast solar wind speed, but the physical mechanisms for the particle transport and acceleration are surprisingly poorly understood (see, e.g., Baker and Kanekal [2008]). The practical side of the problem is that some of the most fatal space weather related spacecraft anomalies have taken place during large relativistic electron fluxes. Consequently, relativistic electrons inside the magnetosphere have earned the nickname "killer electrons".

Energetic electrons populate both the inner and outer belts (Fig. 1.20). Between the belts there normally is a *slot region*, where energetic electron fluxes are very low, except after some major storms, when even the slot region may become filled by electrons. The weakly relativistic electrons respond to weaker storms, whereas the enhancements of ultrarelativistic electrons are related with the strongest storms. However, there is no one-to-one correspondence between relativistic electrons and geomagnetic storms. Reeves et al [2003] investigated the fluxes of 1.8–3.5-MeV electrons measured by geostationary spacecraft and of 1.2–2.4-MeV electrons observed at a polar orbit of the *Polar* satellite during 276 moderate to strong geomagnetic storms in 1989–2000. Of these storms only 53% increased the relativistic electron fluxes by more than a factor of 2, whereas 19% actually decreased the fluxes by the same amount. This trend was found to be independent of the $L$ shell and the storm strength determined by the $Dst$ minimum. Furthermore, no correlation was found between the pre-storm and post-storm fluxes.

The most dramatic increases of relativistic electrons, including the inner belt and the slot region follow strong ICME events, but it is noteworthy that even relatively modest geomagnetic storms can lead to rapid enhancements of the electron belts during periods of high solar wind speeds. As discussed by, e.g., Baker and Kanekal [2008], whenever the solar wind speed substantially exceeds 500 km s$^{-1}$, the relativistic electron population at $L \sim 2.5$–$6.0$ is enhanced, whereas if the solar wind speed is below 500 km s$^{-1}$, the electron fluxes remain small. In fact, the periods of recurrent high solar wind speed streams from large coronal holes during the declining phase of the solar cycle produce much larger average fluxes of relativistic electrons than the ICME-dominated storms around solar maxima. It is quite likely that different mechanisms may lead to electron acceleration under different driver conditions, or at least their relative importance is different.

But where do the electrons that are accelerated come from? The ionospheric electrons have temperatures of about 1 eV and the solar wind electrons about 10 eV. There are also high-energy electrons in the solar wind, but their phase space density is not sufficient to account for large fluxes within the magnetosphere without substantial acceleration inside the magnetosphere [Li et al, 1997]. The electron temperature in the plasma sheet is somewhat smaller than the ion temperature, about 1 keV. However, as discussed in Chap. 13 the substorm activations inject hot electrons into the vicinity of the geostationary orbit. These electrons provide a sufficient seed population of about 10–300 keV, which resides in the inner magnetosphere long enough to be accelerated to MeV-energies by, e.g., strong shock waves or wave–particle interactions [Baker et al, 1998]. That electron acceleration really takes place in the inner magnetosphere was demonstrated by the phase space density distribution analysis based on observations at several different locations during two storm periods by Chen et al [2006]. However, the authors did not exclude some contribution from an external source, i.e., that part of the acceleration to very high energies would also take place in the plasma sheet before the particles are injected to the inner magnetosphere.

A strong ICME-driven shock hitting the magnetosphere may lead to immediate acceleration of electrons from relatively low energies of 0.1–1.0 MeV to more than 10–20 MeV, as was the case of the largest storm during the *CRRES* observations on March 24, 1991. The storm led to a formation of a new relativistic radiation belt within the slot region, whose remains were detectable several years later. The storm sudden commencement launched a very strong electromagnetic pulse inside the magnetosphere. The peak-to-peak electric field of the bipolar pulse was about 80 mV $m^{-1}$ and the unipolar magnetic field reached about 140 nT lasting 120 s at the location of *CRRES* ($L = 2.5$, magnetic local time 0300). It was likely that the pulse was even stronger in the dayside magnetosphere. The pulse was modeled by Li et al [1993], who were able to demonstrate that such a pulse could energize electrons up to 50 MeV in less than 100 s (Fig. 14.5).

While it is plausible that strong transient inductive electric fields play a decisive role during these shock-driven "superstorms", they cannot explain the more typical evolution of relativistic electron fluxes that *decrease* during the storm main phase and *increase* during several hours in the recovery phase finally reaching a higher level than before the storm. Thus the relativistic electron fluxes often behave opposite to the ring current evolution. Obviously the loss mechanisms, e.g., loss to the dayside magnetopause during strong main phase compression of the magnetopause or to the atmosphere due to enhanced pitch angle scattering into the loss cone, may become stronger than any available source is able to supply new electrons. The imbalance between the source and loss mechanisms is consistent with the findings by Reeves et al [2003] cited above.

The enhancement of relativistic electron fluxes during the storm recovery phase may well have the same origin as the creation of high electron fluxes during fast solar wind-driven storms because the post-ICME solar wind velocity may remain high for an extended period. For the time being the physical mechanisms to lead to strong electron acceleration during fast solar wind remain to be explained. For this purpose various wave modes, in particular whistler mode chorus waves and ultra low-frequency (ULF) waves, have been invoked. These two wave modes act on electrons quite differently.

The classical theory of electron belt formation is based on inward radial diffusion due to some large-scale low-frequency electromagnetic fluctuations. It was developed during the

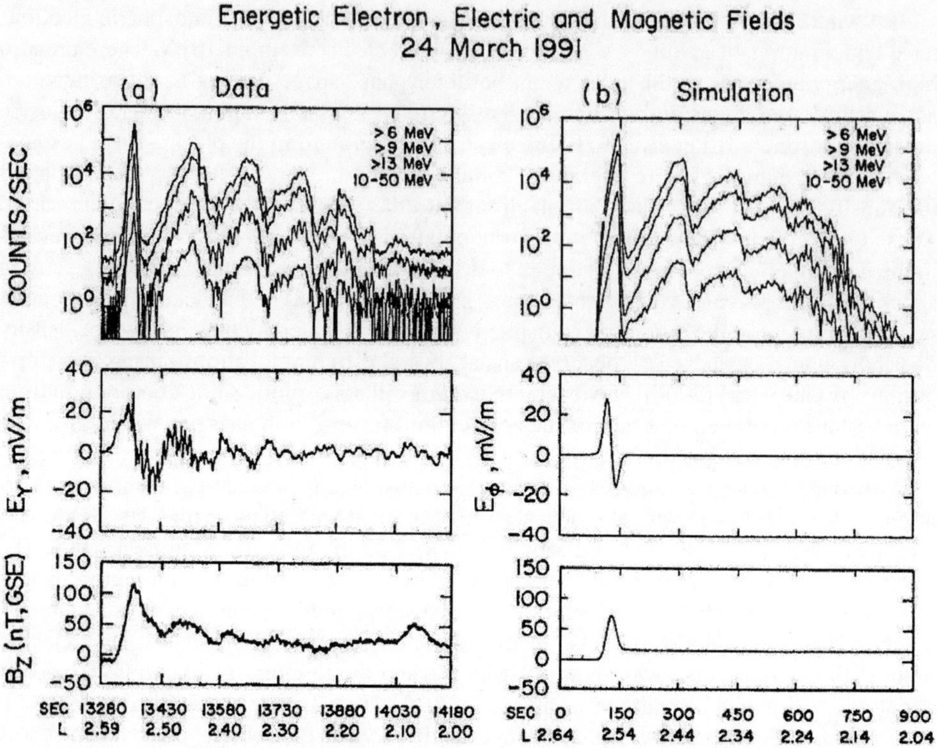


**Fig. 14.5** Electron, electric field and magnetic field data from the *CRRES* satellite at the time following the SSC on March 24, 1991 (left) and simulated electron counts with a model electromagnetic pulse corresponding to the electric and magnetic field observations. Note that only one component of the electric field magnitude is given and thus the amplitude is smaller than the total electric field referred to in the text. (From Li et al [1993].)

1960s (see, e.g., the textbook by Schulz and Lanzerotti [1974]). The fluctuations conserve the first and second adiabatic invariants but break the third invariant, which in radiation belt calculations usually is $L$ or $L^*$ (Chap. 10). Thus the diffusion equation can be written as

$$\frac{\partial f}{\partial t} = L^2 \frac{\partial}{\partial L}\left(\frac{D_{LL}}{L^2}\frac{\partial f}{\partial L}\right) - \Lambda f + S, \tag{14.6}$$

where some average electromagnetic fluctuations determine the radial diffusion coefficient $D_{LL}$, and $\Lambda f$ and $S$ describe the losses and sources of the particles. When the seed population is transported adiabatically toward larger magnetic field, the particles gain energy due to the conservation of $\mu = W_{\perp}/B$.

The challenge is to determine $D_{LL}$. It requires understanding of both the nature of the fluctuations and reasonably good background electric and magnetic field models. In practice one has to make quite a few assumptions and approximations. Already a slightly distorted dipole field geometry together with standard convection electric field models lead to complicated calculations and in order to verify the results of the calculations it is necessary to look for empirical estimates for the diffusion coefficients using direct satellite observations (e.g., Elkington et al [2003] and references therein).

For radial diffusion to be efficient the fluctuations should be global and take place in the time scales of the electron drift period, of the order of $10^3$ s. Thus global large-amplitude ULF waves in the Pc4–Pc5 frequency band (0.5–15 mHz) are natural candidates for inward diffusion. Ground-based observations of ULF waves in association with high relativistic electron fluxes during geomagnetic storms [Rostoker et al, 1998] further support this idea.

Elkington et al [2003] came to the conclusion that the Pc5 waves are capable of adiabatically accelerating electrons from about 100 keV to MeV energies and transporting them inward in the magnetosphere. They considered equatorial electrons ($\alpha = 90°$) in a magnetic field of the form

$$B(r,\phi) = \frac{B_0 R_E^3}{r^3} + b_1(1 + b_2 \cos\phi) \,, \tag{14.7}$$

where the first term is the dipole field in the equatorial plane and the coefficients $b_1$ and $b_2$ describe the distortion of the magnetic field and thus depend on the applied field model. The drift contours (2D drift shells) are determined by the constant magnetic field strength, i.e., the $L$ parameter is replaced by

$$\mathscr{L} = \left( \frac{R_E^3}{r^3} + \frac{b_1 b_2}{B_0} \cos\phi \right)^{-1/3} \,. \tag{14.8}$$

**Train your brain**

Calculate the relationship between $\mathscr{L}$ (14.8) and $L^*$ (10.18) and show that for $b_1 \ll B_0$, $\mathscr{L} \approx L^*$ within the range of $\mathscr{L}$ relevant to radiation belt studies.

The electric field of the ULF waves was written by Elkington et al [2003] as

$$\begin{aligned} \mathbf{E}(r,\phi,t) = \mathbf{E}_0(r,\phi) &+ \sum_{m=0}^{\infty} \delta E_{rm} \sin(m\phi \pm \omega t + \xi_{rm})\, \mathbf{e}_r \\ &+ \sum_{m=0}^{\infty} \delta E_{\phi m} \sin(m\phi \pm \omega t + \xi_{\phi m})\, \mathbf{e}_\phi \,. \end{aligned} \tag{14.9}$$

Here $\mathbf{E}_0(r,\phi)$ is the time-independent convection electric field. $\delta E_{rm}$ are the amplitudes of the *toroidal* modes and $\delta E_{\phi m}$ of the *poloidal* modes, and $\xi_{rm}$ and $\xi_{\phi m}$ represent their phase lags.

According to Eq. (3.50) the adiabatic acceleration is given by

$$\frac{dW}{dt} = q\mathbf{E}\cdot\mathbf{v}_d + \mu \frac{\partial B}{\partial t} \,. \tag{14.10}$$

The magnetic perturbation of the toroidal mode $\delta B_\phi$ and the dominant magnetic field component of the poloidal mode $\delta B_r$ both have a node at the equator and the compressional component $\delta B_\parallel$ of the poloidal mode is so small that the pure betatron acceleration $\mu \partial B/\partial t$ can be neglected and the energization is due to the drift-betatron term only.

Figure 14.6 illustrates how a drift resonant ($\omega = \omega_d$) electron is accelerated by the toroidal $m = 2$ mode in a distorted dipole. When the electron has a maximal outward radial velocity, it sees an inward electric field, whereas when it has a maximal inward radial velocity, it sees an outward electric field. Thus the electron gains energy twice during one drift period. For arbitrary $m$ the resonance condition is

$$\omega - (m \pm 1)\,\omega_d = 0\,. \tag{14.11}$$

The asymmetric compression of the inner magnetospheric magnetic field is an essential factor in the process. The increasing distortion increases $v_r$ on the dawn and dusk sectors

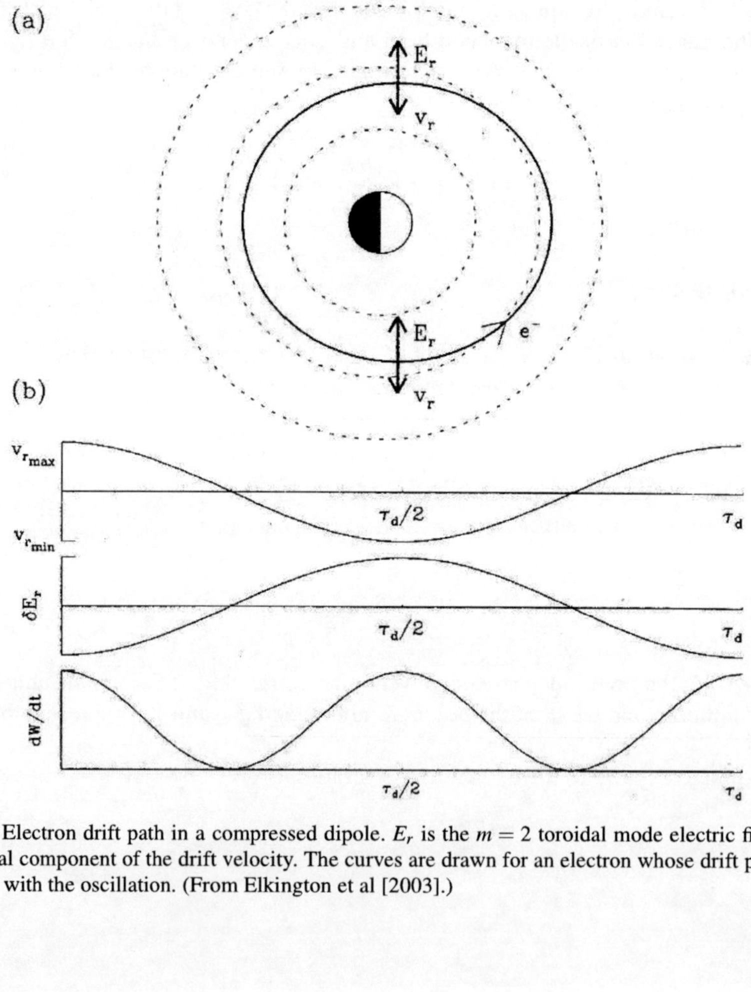


**Fig. 14.6** Electron drift path in a compressed dipole. $E_r$ is the $m = 2$ toroidal mode electric field and $v_r$ is the radial component of the drift velocity. The curves are drawn for an electron whose drift period is in resonance with the oscillation. (From Elkington et al [2003].)

and thus increases the energy gain which is proportional to $E_r v_r$. Also the convection electric field contributes to the acceleration, allowing particles with energies below the resonant energy to be accelerated.

The poloidal mode can also lead to efficient acceleration. In this case a resonant electron that encounters an electric field opposing the drift motion on the nightside, and is accelerated, encounters an electric field with the same direction as its velocity on the dayside, and is thus decelerated. In a distorted dipole the deceleration is weaker than the acceleration, leading to net energization over the course of a drift orbit. In contrast to the toroidal mode, a static convection electric field imposed on a single frequency poloidal mode will cause electrons to lose energy. However, if the poloidal modes are distributed over a range of frequencies or there is an additional non-static convection electric field acting on an electron, the dominant component of the electron's drift velocity in the azimuthal direction will permit more efficient acceleration than interaction with purely toroidal modes of the same amplitude.

The numerical calculations with a continuum of frequencies by Elkington et al [2003] show that the resonant mechanism can lead to very efficient radial diffusion. However, the analysis was limited to equatorial particles and the results are model-dependent. More theoretical work and, in particular, more comprehensive observations of both electrons and ULF waves are needed before we can be sure how much of the acceleration and transport finally is attributed to this mechanism.

The acceleration by radial diffusion alone may not be efficient enough to establish the observed high relativistic electron fluxes in the radiation belts. It has been suggested that interaction with whistler mode chorus waves in the kHz frequency range can increase the electron flux in the inner magnetosphere by more than three orders of magnitude within one or two days [Summers et al, 1998]. The process is quite different from the ULF interaction because it takes place through the gyro resonance between the waves and the electrons in regions where the waves have finite amplitude (Fig. 14.4). Thus the mechanism already breaks the first adiabatic invariance.

The right-hand polarized whistler mode chorus waves are driven by an unstable anisotropic $\sim$10 keV electron population [Kennel and Petschek, 1966]. They interact with a small fraction of more energetic electrons through the Doppler-shifted gyro resonance, which must now be written relativistically

$$\omega - k_\parallel v_\parallel = \frac{n\omega_{ce}}{\gamma}, \tag{14.12}$$

where $\gamma$ is the Lorentz factor and $\omega_{ce}$ the non-relativistic gyrofrequency calculated for the electron rest mass. Of these variables $\omega$, $v_\parallel$ and, $\omega_{ce}$ are straightforward to measure but $k_\parallel$ must, in practice, be determined by solving the relevant dispersion equation, which in turn depends on plasma density and ion composition. This is actually how the low-energy background plasmasphere is coupled to the evolution of very high-energy electrons.

The whistler mode interaction is efficient in regions where $\omega_{pe}/\omega_{ce}$ is small ($\lesssim$4). This usually is not the case at low $L$ shells, but, e.g., during the Halloween storm in the autumn 2003, the high-density plasmasphere was confined inside $L = 2$ on October 31, and remained inside $L = 2.5$ in the pre-noon sector (06–12 MLT) until November 4. Horne et al

[2005] analyzed this event using relativistic electron data from the *SAMPEX* satellite, *Kp* and *Dst* indices, ground-based ULF observations and kHz-range wave observations from the *Cluster* spacecraft. They argued that the radial diffusion due to the ULF waves could not explain the strong increase of 2–6-MeV electron fluxes between *L* shells from 2 to 3 at the late phase of the storm from November 1 onward. Instead, the Fokker–Planck calculations by Horne et al [2005] based on diffusion rates calculated for chorus wave amplitudes measured by *Cluster* at somewhat higher *L* shell ($L = 4.3$) suggest that the gyro resonant interaction really was sufficient to explain the establishment of exceptionally high electron fluxes in the exceptional location, i.e., the slot region, during an exceptionally strong storm period.

### 14.2.4 Electron losses

Several physical mechanisms contribute to the loss of radiation belt electrons, which takes place mainly through pitch angle scattering into the atmospheric loss cone. Figure 14.7 illustrates the most important of these mechanisms [Abel and Thorne, 1998].

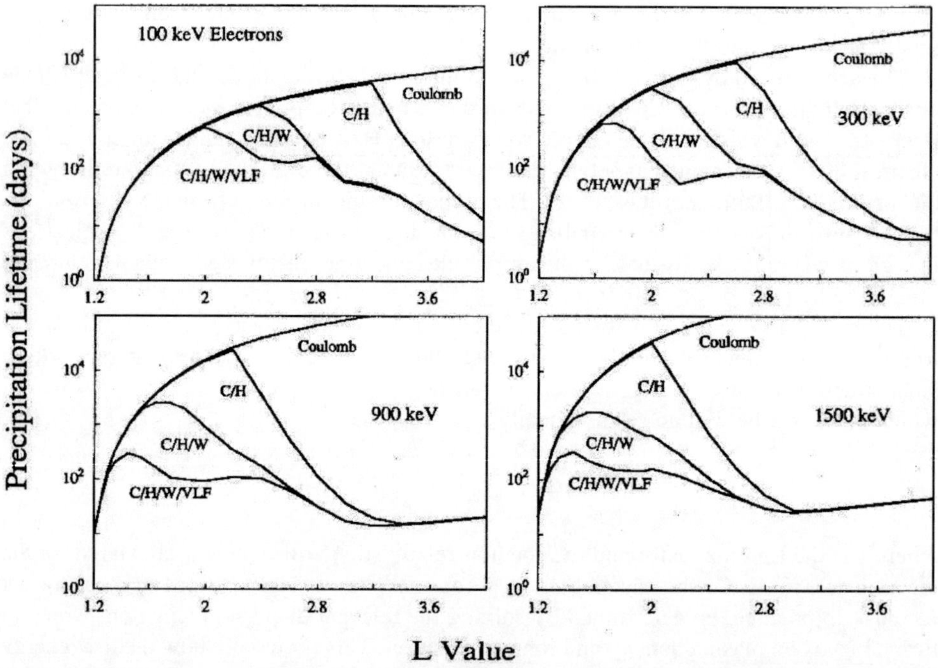


**Fig. 14.7** Theoretical results of electron lifetimes as a function of *L* shell for radiation belt electrons from weakly to strongly relativistic energies. The uppermost curve in each pane is calculated for Coulomb collision only. The indicate the lifetimes when the plasmaspheric hiss (C/H), lightning-induced whistlers (C/H/W), and man-made VLF emissions (C/H/W/VLF) are included. (From Abel and Thorne [1998].)

The uppermost curve in each pane of Fig. 14.7 gives the electron lifetimes if Coulomb collisions alone are considered. The 100-keV electron lifetime exceeds 1 year beyond $L = 1.8$ and is about 30 years at $L = 5$. Thus it is clear that the Coulomb collisions are inefficient in removing electrons from the radiation belts. The symbol C/H in Fig. 14.7 indicates the lifetimes when the whistler mode plasmaspheric hiss is included. This leads to a significant reduction of the lifetimes, to the order of a few tens of days, around and beyond the plasmapause. Adding lightning-induced whistlers (symbol W) and man-made VLF signals reduces the lifetimes at low $L$ shells as well.

Writing (14.12) in the form

$$\omega - k_{\parallel} v_{\parallel} = \pm \frac{|\omega_{ce}|}{\gamma} , \qquad (14.13)$$

where $n$ has been limited to $\pm 1$, it becomes evident that electrons can be in resonance both with right-hand ($R$; $+$) and left-hand ($L$; $-$) polarized waves. Modes considered in Fig. 14.7, as well as the chorus waves discussed above, are all $R$ mode waves in the frequency range $\omega_{ci} < \omega < \omega_{ce}$. As discussed in Chaps. 4 and 5 the EMIC waves are left-hand polarized waves with $\omega < \omega_{ci}$, which in multi-ion plasmas appear in the frequency bands below the gyro frequency of each ion species. The gyro resonance of EMIC waves with electrons requires that the Lorentz factors $\gamma$ of the electrons are large enough. According to the calculations by Summers and Thorne [2003] the minimum resonant energies can under suitable conditions reach below 1 MeV but this requires that the ratio between electron gyro and plasma frequencies be very small ($\omega_{ce}^2/\omega_{pe}^2 \sim 10^{-3}$).

Based on wave and particle analysis of *CRRES* observations Meredith et al [2003] concluded that conditions could be suitable for strong diffusion of electrons at energies from less than 2 MeV upward during about 1% of the electron drift periods around the Earth. While this may seem quite restrictive, it actually is in favor of the explanation since it keeps the diffusion time scale in the range from hours to one day. If the interaction were to take place within a much wider region, the electrons would disappear too quickly compared to the observations.

Also the comprehensive model computations by Jordanova et al [2008] of the intense storm on 21 October 2001 came to similar conclusions. The model was the up-to-date version of the RAM code (Eq. 10.43). It contained all major loss processes and was coupled with a dynamic plasmasphere model, including 77% $H^+$, 20% $He^+$ and 3% $O^+$. The EMIC wave amplitudes were calculated self-consistently with evolving plasma populations. The calculations were performed considering separately EMIC scattering only, all processes except EMIC waves, and all scattering processes including EMIC waves. The conclusion was that scattering by EMIC waves enhances the loss relativistic (>1 MeV) electrons and can cause significant electron precipitation during the storm main phases.

In conclusion there are several wave–particle interaction mechanisms that can explain both the decrease of relativistic electron fluxes during the main phase and subsequent increase of electron fluxes during the recovery phase. Thus the previously cited result [Reeves et al, 2003] that some storms decrease electron fluxes whereas others increase them, no longer looks so surprising. However, it is still unclear why any individual storm behaves as it does.

# 15. Space Storms in the Atmosphere and on the Ground

Space storm effects reach all the way through the atmosphere to the ground. In the Sun–Earth system the storms take place in the weather time scales, lasting maximally a few days from the release of the CME to the time when the ICME has passed the Earth. On the other hand, the effects of solar proton events can last in the middle atmosphere months and thus become coupled to the atmospheric climate cycles. This is further amplified through the solar cycle variability both in the total solar irradiance and at EUV wavelengths. To have any intelligent discussion of the climate issues would require extensive treatment of middle atmospheric ion and neutral chemistry and atmospheric circulation patterns, and lead us too far from the topic of this book. Therefore we limit our discussion to a short description of some immediate signatures of space storms in the middle atmosphere.

The physical manifestations of space storms on the ground are, in turn, short-term phenomena caused by electromagnetic induction due to rapid changes in the magnetospheric and ionospheric current systems. They are most interesting for their effects on technological systems. While the technological consequences are also beyond the context of the present book, some of the basic principles of geomagnetically induced currents belong to the fundamental knowledge base on physics of space storms.

## 15.1 Coupling to the Neutral Atmosphere

In Sect. 1.4 our discussion of the neutral atmosphere did not reach below the bottom of the thermosphere, i.e., the upper boundary of the *mesosphere* known as the *mesopause*. The mesosphere is located at altitudes of about 50–85 km. Below the mesosphere is the *stratosphere* reaching down to altitudes of about 8 km at high geographic latitudes, and to 12–15 km at low latitudes. The stratosphere and mesosphere form the region that is called here the *middle atmosphere*, although again the terminology is not unique. Finally, below the stratosphere is the *troposphere*, where the familiar atmospheric weather phenomena take place.

The boundary between the troposphere and stratosphere is called the *tropopause*. At the tropopause the atmospheric temperature is at its minimum of about $-60^\circ$ C. Water vapor cannot lift through the tropopause to the cold dry stratosphere, where it would dissociate by

H.E.J. Koskinen, *Physics of Space Storms: From the Solar Surface to the Earth*,
Springer Praxis Books, DOI 10.1007/978-3-642-00319-6_15,

solar irradiation. This would lead to rapid hydrogen escape, dehydration of the atmosphere, and finally disappearance of the oceans.

In the stratosphere the temperature slowly increases with altitude. Most of the atmospheric ozone ($O_3$), which protects us from the solar UV radiation lies in the altitude range 15–40 km. The boundary between the stratosphere and mesosphere is called the *stratopause*. Above the stratopause the temperature starts again to decrease with altitude up to the mesopause. In the thermosphere the altitude-dependence of the temperature turns once more as the gas becomes more tenuous and the degree of ionization due to the solar EUV irradiance begins to increase, to which we turn next.

### 15.1.1 Heating of the thermosphere

We have already discussed the ionosphere (Sect. 1.4) and the increase of Joule heating during storms and substorms (Chap. 13). The ionospheric E-layer, where the horizontal ionospheric currents flow, is a weakly ionized domain within the thermosphere and the Joule heating is essentially frictional heating of thermospheric neutral atoms and molecules. Also the energy deposited by electron precipitation shows as heat in the thermosphere, whereas the energy density of the auroral light emissions is very small.

Heating of the thermosphere leads to increased scale height (Eq. 1.61) and thus expansion of the neutral atmosphere. This increases the atmospheric drag of low-altitude spacecraft, whose orbital lifetimes are strongly affected by the solar activity. In fact the neutral density at 500 km altitude may increase by a factor of 10 during a strong storm. This must also be taken carefully into account in order to safely de-orbit large structures that do not burn in the atmosphere. For example, in March 2001 the space weather was "better" than expected during the solar maximum epoch. This prolonged the natural orbital decay of the *MIR* station for several days. The increased drag also has a positive effect by cleaning off a fraction of low-altitude space debris.

During solar maximum years the EUV radiation at wavelengths below 100 nm also increases by a factor of three from solar minimum. This energy is stopped through ionization at higher altitudes, in particular around the F-region maximum. Thus the effects are two-fold: increased thermospheric temperature and enhanced ionization.

### 15.1.2 Solar proton events and the middle atmosphere

The strongest *solar proton events* last a few days (Table 12.1) and produce high energy particles that precipitate into the upper atmosphere, mostly at high latitudes in the auroral regions and the polar caps. As discussed in Chap. 3 the penetration of cosmic rays through the geomagnetic field is proportional to the latitude as $\cos^4 \lambda$. At auroral latitudes particles with rigidities of 1 GV or higher can reach the atmosphere. Also a fraction of solar energetic particles at lower energies entering the magnetosphere are scattered into the atmospheric loss cone and precipitate to the middle atmosphere. Precipitating 3-MeV protons deposit their energy and are stopped around 80 km in the mesosphere, whereas protons in excess of 30 MeV reach down to the stratosphere. Protons at these energies are abundant in solar particle events.

Protons penetrating to the mesosphere cause ionization and dissociation of atmospheric gases leading to enhancements of *odd-hydrogen* ($HO_x$: H, OH, $HO_2$) and *odd-nitrogen* ($NO_x$: N, NO, $NO_2$) constituents. $NO_x$ can also be produced in the lower thermosphere by precipitation of electrons with energies in the 100-keV range and transported down to the mesosphere. Also this population is enhanced during space storms and it is not easy to distinguish between $NO_x$ populations of different origin. Other constituents of importance to ozone chemistry, being enhanced by solar proton events, include $HNO_3$, $N_2O_5$, $ClONO_2$, HOCl, and ClO.

$HO_x$ and $NO_x$ molecules produced by solar proton events lead to both short- and long-term catalytic ozone destruction in the lower mesosphere and stratosphere. During large proton events $NO_x$ abundances of a factor of ten larger than normally have been observed and included model calculations (Jackman et al [2008] and references therein). The lifetimes of $HO_x$ molecules are not longer than hours, and thus they induce only short-period, but still large, ozone depletions above 50 km. $NO_x$ lifetimes are longer and there is a large difference between the summer and winter hemispheres. In the sunlit mesosphere photodissociation

$$
\begin{aligned}
&\mathrm{NO} + h\nu \rightarrow \mathrm{N} + \mathrm{O} \qquad \text{followed by}\\
&\mathrm{N} + \mathrm{NO} \rightarrow \mathrm{N_2} + \mathrm{O}
\end{aligned}
\tag{15.1}
$$

reduces significantly the number of odd-hydrogen constituents and their transport to the stratosphere. In the dark hemisphere $NO_x$ enhancements can survive months and end up being transported to the middle and lower stratosphere. Consequently, they have similarly long-term effects on ozone dynamics.

For a more detailed discussion we refer to the extensive modeling study by Jackman et al [2008] of strong solar proton events in 1963–2005. They used the *Whole Atmosphere Community Climate Model* (WACCM3) and compared the results with various data sets available from the different times. WACCM3 includes interactive dynamics, radiation and chemistry. Its dynamics is not limited to the neutral atmosphere because it is coupled the *Thermosphere–Ionosphere–Mesosphere–Electrodynamics General Circulation Model* (TIME-GCM). Another module keeps track of ozone and other trace gases. Many of the major features of the solar proton effects in the middle atmosphere can presently be modeled fairly well. However, discrepancies between observed and modeled enhancements of $HNO_3$, $N_2O_5$, $ClONO_2$, and ClO still are considerable. According to Jackman et al [2008] the underlying causes for these discrepancies are difficult to identify in the global WACCM3 studies. Instead, more detailed studies with simpler models are needed to understand the fundamental physical and chemical processes behind the complex global system.

## 15.2 Coupling to the Surface of the Earth

The basic idea behind the *geomagnetically induced current* (GIC) phenomenon is simple. A time-variable ionospheric current gives rise to a finite $\partial \mathbf{B}/\partial t$, which propagates to the ground. According to Faraday's law there is an associated rotational electric field $\nabla \times \mathbf{E}$, the *geoelectric field*, at the same location. The geoelectric field causes a voltage between two points $a$ and $b$ in any conductor embedded in the field

$$\varphi_{ab} = \int_a^b \mathbf{E} \cdot d\mathbf{l}\,, \tag{15.2}$$

which in turn drives a current, the GIC, in the conductor.

There are thus two different parts in the GIC problem: the *geophysical problem* to determine the geoelectric field and the *engineering problem* to calculate the current in the conductor system one is interested in. As the electric field is not a potential field, the integral (15.2) depends on the path along which it is calculated. The configuration and also the topology of the conductor system, e.g., the electric power transmission network, are important to the engineering problem because they determine where in the network the current actually flows. Because we are not dealing with the technological effects of space storms in this book, we limit our discussion to the geophysical problem.

To find out the geoelectric field requires knowledge not only of the ionospheric and magnetospheric currents but also of the conductivity structure in the ground. The Earth is a conductor and the changing primary currents in space induce secondary currents in the ground. In fact, the Earth is such a good conductor that the contributions from the primary and secondary currents to the horizontal electric field on the surface have almost the same magnitude but the opposite sign. Thus the practical numerical computations of the electric field need to be accurate enough and be based on a good enough model for the conductivity structure of the ground.

To keep the discussion simple consider the primary field as a plane wave with frequency $\omega$ propagating vertically downward (taken as the positive $z$-direction) and assume the conductivity structure of the ground to be homogeneous. Denote the horizontal component of the electric field by $E_y$ and the horizontal component of the magnetic field perpendicular to $E_y$ by $B_x$. Assuming $\mu = \mu_0$ and that the ground is a good conductor in the sense $\sigma \gg \varepsilon\omega$, the equation for the wave impedance (4.27) gives us the electric field in terms of the magnetic field as

$$E_y(\omega) = -\sqrt{\frac{\omega}{\mu_0 \sigma}} \exp\left(\frac{-i\pi}{4}\right) B_x(\omega)\,. \tag{15.3}$$

Note that in (4.27) the magnetic field was assumed to be in the $y$-direction, thus here $E_x$ is replaced by $-E_y$.

In practice we are interested in the time series $E(t)$ as a function of the observed magnetic field. This can be calculated taking the inverse Fourier transform of (15.3)

$$E_y(t) = \frac{-1}{\sqrt{\pi\mu_0\sigma}} \int_0^{\infty} \frac{B_x'(t-u)}{\sqrt{u}}\, du\,, \tag{15.4}$$

where the prime denotes the time derivative $d/dt$. Thus the past values of $dB_x/dt$ affect the geoelectric field but their weight decreases with increasing time interval $u$.

This calculation can be generalized to inhomogeneous conductivities and general three-dimensional current systems. The practical computations are tedious and time-consuming. Pirjola and Viljanen [1998] showed that the so-called *complex image method* (CIM) is a very accurate approximation when the geoelectric field is caused by an auroral (substorm) electrojet with a finite length connected to the magnetosphere through vertical currents at its ends, which clearly is the case of most interest for our purposes. The CIM provides analytical solutions that are much faster to compute than applying the exact approach.

**Feed your brain**

Read carefully the article Pirjola and Viljanen [1998] and perform the calculations indicated therein. Note that their sign convention in the plane wave approximation is opposite to ours, thus be careful, e.g., in the derivation of (15.3).

For an introduction to the solution of the engineering problem in electric transmission networks, see Chapter 10 written by R. Pirjola in Bothmer and Daglis [2007]. The treatment of GIC effects on pipelines buried in the ground requires somewhat different methods, e.g., the distributed source transmission line (DSTL) theory [Pulkkinen et al, 2001].

Figure 15.1 shows an example of measured GIC in the natural gas pipeline at Mäntsälä in southern Finland at the beginning of the so-called Halloween storm on October 29, 2003. At the measurement site the pipeline is nearly east–west-aligned. Consequently, $-dX/dt$ measured at the nearby magnetometer station in Nurmijärvi correlates well with the GIC along the pipe.

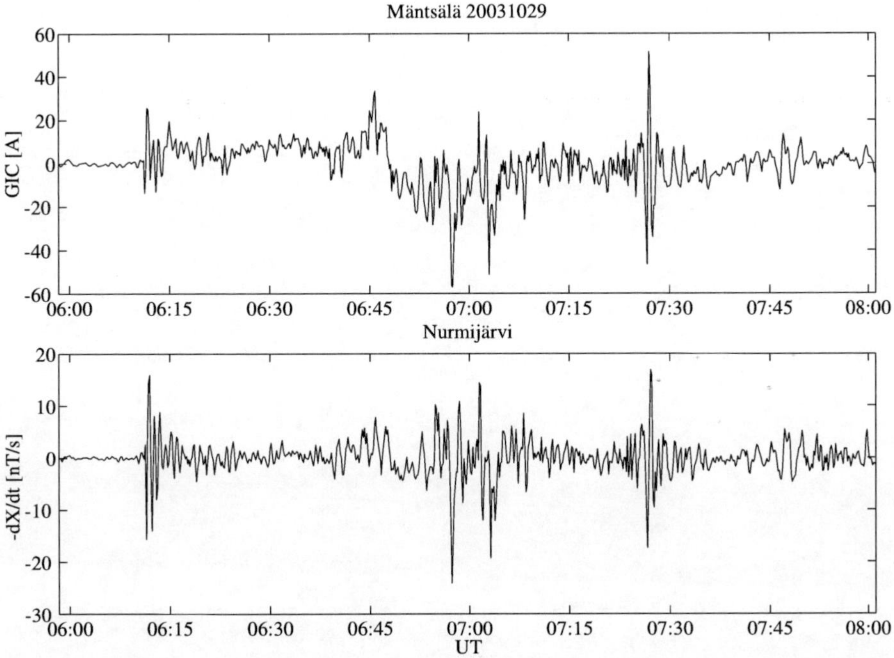


**Fig. 15.1** Upper panel: Geomagnetically induced current flowing along the Finnish natural gas pipeline at the Mäntsälä compressor station of the natural gas company Gasum Oy on 29 October 2003. The peak value (–57 A) is the largest during the measurement period since November 1998. Lower panel: Negative of the time derivative of the northward magnetic field at the nearby Nurmijärvi Geophysical Observatory of the Finnish Meteorological Institute. (Figure by courtesy of A. Viljanen.)

# References


Abel B. and Thorne R.M. (1998) Electron scattering loss in Earth's inner magnetosphere 1. Dominant physical processes. *J. Geophys. Res.*, 103:2385–2396.

Aggson T.L., Heppner J.P., and Maynard N.C. (1983) Observations of large magnetospheric electric fields during the onset phase of a substorm. *J. Geophys. Res.*, 88:3981–3990.

Ahmed S.N., SNO Collaboration (2004) Measurement of the total active $^8$B solar neutrino flux at the Sudbury Neutrino Observatory with enhanced neutral current sensitivity. *Phys. Rev. Lett.*, 92(18):181,301, 1–6.

Ahn B.H., Akasofu S.I., and Kamide Y. (1983) The Joule heat production rate and the particle energy injection rate as a function of the geomagnetic indices AE and AL. *J. Geophys. Res.*, 88:6275–6287.

Akasofu S.I. (1964) The development of the auroral substorm. *Planet. Space Sci.*, 12:273–282.

Akasofu S.I. (1966) Electrodynamics of the magnetosphere: Geomagnetic storms. *Space Sci. Rev.*, 6:21–143.

Akasofu S.I. (1981) Energy coupling between the solar wind and the magnetosphere. *Space Sci. Rev.*, 28:121–190.

Akasofu S.I. and Chapman S. (1961) The ring current, geomagnetic disturbance, and the Van Allen radiation belts. *J. Geophys. Res.*, 66:1321–1350.

Alexeev I.I., Belenkaya E.S., Feldstein Y.I., and Grafe A. (1996) Magnetic storms and magnetotail currents. *J. Geophys. Res.*, 101:7737–7747.

André M. (1985) Dispersion surfaces. *J. Plasma Phys.*, 33:1–19.

Angelopoulos V. (2008) The THEMIS mission. *Space Sci. Rev.*, 141:5–34, DOI 10.1007/s.11214-008-9336-1.

Angelopoulos V., Baumjohann W., Kennel C.F., Coroniti F.V., Kivelson M.G., Pella R., Walker R.J., Lühr H., and Paschmann G. (1992) Bursty bulk flows in the inner central plasma sheet. *J. Geophys. Res.*, 97:4027–4039.

Angelopoulos V., McFadden J.P., Larson D., Carlson C.W., Mende S.B., Frey H., Phan T., Sibeck D.G., Glassmeier K.H., Auster U., Donovan E., Mann I.R., Rae I.J., Russell C.T., Runov A., Zhou X.Z., and Kepko L. (2008) Tail reconnection triggering substorm onset. *Science*, 321, DOI 10.1126/science.1160495.

Anzer U. and Priest E. (1985) Remarks on the magnetic support of quiescent prominences. *Solar Phys.*, 95:263–268.

Aschwanden M.J. (2004) *Physics of the Solar Corona: An Introduction*. Springer, Berlin, Germany.

Ashour-Abdalla M., Berchem J.P., Büchner J., and Zelenyi L.M. (1993) Shaping of the magnetotail from the mantle: Global and local structuring. *J. Geophys. Res.*, 98:5651–5676.

Axford W.I. and Hines C.O. (1961) A unifying theory of high-latitude geophysical phenomena and geomagnetic storms. *Can. J. Phys.*, 39:1433–1464.

Baker D.N. and Kanekal S.G. (2008) Solar cycle changes, geomagnetic variations, and energetic particle properties in the inner magnetosphere. *J. Atm. Sol. Terr. Phys.*, 70:195–206.



H.E.J. Koskinen, *Physics of Space Storms: From the Solar Surface to the Earth*,
Springer Praxis Books, DOI 10.1007/978-3-642-00319-6,

Baker D.N. and McPherron R.L. (1990) Extreme energetic particle decreases near geostationary orbit: A manifestation of current diversion within the inner plasma sheet. *J. Geophys. Res.*, 95:6591–6599.

Baker D.N., Pulkkinen T.I., McPherron R.L., Crave J.D., Frank L.A., Elphinstone R.D., Murphree J.S., Fennell J.F., Lopez R.E., and Nagai T. (1993) CDAW 9 analysis of magnetospheric events on May 3, 1986: Event C. *J. Geophys. Res.*, 98:3815–3834.

Baker D.N., Pulkkinen T.I., Angelopoulos V., Baumjohann W., and McPherron R.L. (1996) Neutral line model of substorms: Past results and present view. *J. Geophys. Res.*, 101:12,975–13,010.

Baker D.N., Pulkkinen T.I., Li X., Kanekal S.G., Blake J.B., Selesnick R.S., Henderson M.G., Reeves G.D., Spence H.E., and Rostoker G. (1998) Coronal mass ejections, magnetic clouds, and relativistic magnetospheric electron events: ISTP. *J. Geophys. Res.*, 103:17,279–17,291.

Baker D.N., Klecker B., Schwartz S.J., Schwenn R., and von Steiger R. (eds) (2007) *Solar Dynamics and its Effects on the Heliosphere and Earth, Space Sciences Series of ISSI*, Vol. 22. Springer, Dordrecht, Holland.

Bale S.D., Balikhin M.A., Horbury T.S., Krasnoselskikh V.V., Kucharek H., Mobius E., Walker BA S N, Burgess D., Lembege B., Lucek E.A., Scholer M., Schwartz S.J., and Thomsen M.F. (2005) Quasi-perpendicular shock structure and processes. *Space Sci. Rev.* 118:161–203.

Barabash S., Brandt P.C., Norberg O., Lundin R., Roelof E.C., Chase C.J., Mauk B.H., and Koskinen H. (1997) Energetic neutral atom imaging by the *Astrid* microsatellite. *Adv. Space Res.*, 20:1055–1060.

Bartels J. (1932) Terrestrial-magnetic activity and its relations to solar phenomena. *Terr. Mag. Atmos. Elec.*, 37:1–52.

Baumjohann W. and Treumann R.A. (1996) *Basic Space Plasma Physics*. Imperial College Press, London, U.K..

Baumjohann W., Kamide Y., and Nakamura R. (1996) Substorms, storms, and the near-earth tail. *J. Geomagn. Geoelectr.*, 48:177–185.

Bellan P.M. (2006) *Fundamentals of Plasma Physics*. Cambridge University Press, Cambridge, U.K..

Benz A.O. (2002) *Plasma Astrophysics: Kinetic Processes in Solar and Stellar Coronae*. Kluwer, Dordrecht.

Bernstein I.B. (1958) Waves in a plasma in a magnetic field. *Phys. Rev.*, 109:10–21.

Biermann L. (1951) Kometenschweife und solare Korpuscular-strahlung. *Z. Astrophys.*, 29:274–286.

Biermann L. (1957) Solar corpuscular radiation and the interplanetary gas. *Observatory*, 109:109–110.

Birkeland K. (1908) *The Norwegian Aurora Polaris Expedition 1902–1903*, Volume I. A.W. Brøggers Printing Office, Christiania, Norway.

Birn J. and Priest E.R. (eds) (2007) *Reconnection of Magnetic Fields*. Cambridge University Press, Cambridge, U.K..

Birn J., Drake J.F., Shay M.A., Rogers B.N., Denton R.E., Hesse M., Kuznetsova M., Ma Z.W., Bhattacharjee A., Otto A., and Pritchett P.L. (2001) Geospace environmental modeling (GEM) magnetic reconnection challenge. *J. Geophys. Res.*, 106:3715–3719.

Borovsky J.E. and Denton M.H. (2006) Differences between CME-driven storms and CIR-driven storms. *J. Geophys. Res.*, 111, DOI 10.1029/2005JA011447.

Borovsky J.E., Lavraud B., and Kuznetsova M.M. (2009) Polar cap potential saturation, dayside reconnection and changes to the magnetosphere. *J. Geophys. Res.*, 114, DOI 10.1029/2005JA014058.

Bothmer V. and Daglis I.A. (eds) (2007) *Space Weather, Physics and Effects*. Springer, Praxis Publishing, Chichester, UK.

Bothmer V. and Schwenn R. (1998) The structure and origin of magnetic clouds in the solar wind. *Ann. Geophys.*, 16:1–24.

Boyd T.J.M. and Sanderson J.J. (1969) *Plasma Dynamics*. Barnes and Noble, New York, NY.

Boyd T.J.M. and Sanderson J.J. (2003) *The Physics of Plasmas*. Cambridge University Press, Cambridge, U.K..

Brandt P.C., Barabash S., Roelof E.C., and Chase C.J. (2001) Energetic neutral atom imaging at low altitudes from the Swedish microsatellite *Astrid*: Extraction of the equatorial ion distribution. *J. Geophys. Res.*, 106:25,731–25,744.

Brandt P.C., Ohtani S., Mitchell D.G., Fok M.C., Roelof E.C., and Demajstre R. (2002) Global ENA observations of the storm mainphase ring current: Implications for skewed electric field in the inner magnetosphere. *Geophys. Res. Lett.*, 29, DOI 10.1029/2002GL015160.

Brueckner G.E., Delaboudinière J.P., Howard R.A., Paswaters S.E., St Cyr O.C., Schwenn R., Lamy P., Simnett G.M., Thompson B., and Wang D. (1998) Geomagnetic storms caused by coronal mass ejections (CMEs): March 1996 through June 1997. *Geophys. Res. Lett.*, 25:3019–3022.

Büchner J. and Zelenyi L.M. (1987) Chaotization of the electron motion as the cause of and internal magnetotail instability and substorm onset. *J. Geophys. Res.*, 92:13,456–13,466.

Büchner J. and Zelenyi L.M. (1989) Regular and chaotic charged particle motion in magnetotaillike field reversals. *J. Geophys. Res.*, 94:11,821–11,842.

Budden K.G. (1985) *The propagation of radio waves: The theory of radio waves of low power in the ionosphere and magnetosphere*. Cambridge University Press, Cambridge, U.K..

Burch J.L. (2000) IMAGE mission overview. *Space Sci. Rev.*, 91:1–14.

Burgess D., Lucek E.A., and Scholer M. (2005) Quasi-parallel shock structure and processes. *Space Sci. Rev.*, 118:205–222.

Burton R.K., McPherron R.L., and Russell C.T. (1975) An empirical relationship between interplanetary conditions and *Dst. J. Geophys. Res.*, 80:4204–4214.

Cane H.V. and Richardson I.G. (2003) Interplanetary coronal mass ejections in the near-Earth solar wind during 1996–2002. *J. Geophys. Res.*, 108, DOI 10.1029/2002JA009817.

Carrington R.C. (1859) Description of a singular appearance seen in the Sun on September 1, 1859. *Mon. Not. R. Astron. Soc.* XX:13.

Cassak P.A. and Shay M.A. (2007) Scaling of asymmetric magnetic reconnection: General theory and collisional simulations. *Phys. Plasmas* 14, DOI 10.1063/1.2795630.

Chamberlain J.W. (1963) Planetary coronae and atmospheric evaporation. *Planet. Space Sci.*, 11:901–960.

Chapman S. (1957) Notes on the solar corona and the terrestrial ionosphere. *Smithsonian Contribution to Astrophysics*, 2:1–12.

Chapman S. and Ferraro V.C.A. (1931) A new theory of magnetic storms. *Terr. Magn. Atmos. Electr.*, 36:77–97.

Chappell C.R. (1972) Recent satellite measurements of the morphology and dynamics of the plasmasphere. *Rev. Geophys. Space Phys.*, 10:951–972.

Chappell C.R., Moore T.E., and Waite J.H. Jr. (1987) The ionosphere as a fully adequate source of plasma for the Earth's magnetosphere. *J. Geophys. Res.*, 92:5896–5910.

Chen J. and Palmadesso P.J. (1986) Chaos and nonlinear dynamics of single-particle orbits in a magnetotail-like magnetic field. *J. Geophys. Res.*, 91:1499–1508.

Chen Y., Friedel R.H.W., and Reeves G.D. (2006) Phase space density distributions of energetic electrons in the outer radiation belt during two Geospace Environment Modeling Inner Magnetosphere/Storms selected storms. *J. Geophys. Res.* 111, DOI 10.1029/2006JA011703.

Chew G.F., Goldberger M.L., and Low F.E. (1956) The Boltzmann equation and the one-fluid hydromagnetic equations in the absence of particle collisions. *Proc. Roy. Soc., Ser. A*, 236(1024):112–118.

Chree C. and Stagg J.M. (1927) Recurrence phenomena in terrestrial magnetism. *Phil. Trans. R. Soc., Ser. A*, 227:21–62.

Cliver E.W. and Svalgaard L. (2004) The 1859 solar-terrestrial disturbance and the current limits of extreme space weather activity. *Solar Phys.*, 224:407–422.

Coppi B., Laval G., and Pellat R. (1966) Dynamics of the geomagnetic tail. *Phys. Rev. Lett.*, 16:1207–1210.

Coroniti F.V. (1980) On the tearing mode in quasi-neutral sheets. *J. Geophys. Res.*, 85:6719–6728.

Crooker N., Joselyn J.A., and Feynman J. (eds) (1997) *Coronal Mass Ejections, Geophysical Monograph*, Vol. 99. American Geophysical Union, Washington, DC, USA.

Daglis I.A. (ed) (2001) *Space Storms and Space Weather Hazards, NATO Science Series II: Mathematics, Physics and Chemistry*, Vol. 38. Kluwer, Dordrecht.

Daglis I.A., Thorne R.M., Baumjohann W., and Orsini S. (1999) The terrestrial ring current: Origin, formation, and decay. *Rev. Geophys.*, 37:407–438.

Davidson W.F. (1940) The magnetic storm of March 24, 1940 – Effects in the power system. *Edison Electric Inst. Bull.* pp. 365–366 and 374.

Delcourt D.C., Sauvaud J.A., and Pedersen A. (1990) Dynamics of single-particle orbits during substorm expansion phase. *J. Geophys. Res.*, 95:20,853–20,865.

DeMajistre R., Roelof E.C., Brandt P.C., and Mitchell D.G. (2004) Retrieval of global mangetospheric ion distributions from high-energy neutral atom measurements made by the IMAGE/HENA instrument. *J. Geophys. Res.*, 109, DOI 10.1029/2003JA010322.

Dessler A.J. and Parker E.N. (1959) Hydromagnetic theory of geomagnetic storms. *J. Geophys. Res.*, 64:2239–2252.

Drury L.O. (1983) An introduction to the theory of diffusive shock acceleration of energetic particles in tenuous plasmas. *Rep. Prog. Phys.*, 46:973–1027.

Dungey J.W. (1953) Conditions for the occurrence of electrical discharges in astrophysical systems. *Phil. Mag.*, 44:725–738.

Dungey J.W. (1961) Interplanetary magnetic field and auroral zones. *Phys. Rev. Lett.*, 6:47–48.

Eastwood J.P., Phan T.D., Mozer F.S., Shay M.A., Fujimoto M., Retinò A., Hesse M., Balogh A., Lucek E.A., and Dandouras I. (2007) Multi-point observations of the Hall electromagnetic field and secondary island formation during magnetic reconnection. *J. Geophys. Res.*, 112, DOI 10.1029/2006JA012158.

Elkington S.R., Hudson M.K., and Chan A.A. (2003) Resonant acceleration and diffusion of outer zone electrons in an asymmetric geomagnetic field. *J. Geophys. Res.*, 108, DOI 10.1029/2001JA009202.

Erickson G.M. (1992) A quasi-static magnetospheric convection mdoel in two dimensions. *J. Geophys. Res.*, 97:6505–6522.

Erickson G.M. and Wolf R.A. (1980) Is steady convection possible in the Earth's magnetotail? *Geophys. Res. Lett.*, 7:897–900.

Fairfield D. and Cahill L. Jr. (1966) Transition region magnetic field and polar magnetic disturbances. *J. Geophys. Res.*, 71:155–169.

Farr N.L., Baker D.N., and Wiltberger M. (2008) Complexities of a 3-D plasmoid flux rope as shown by an MHD simulation. *J. Geophys. Res.*, 113, DOI 10.1029/2008JA013328.

Fok M.C., Moore T.E., and Greenspan M.E. (1996) Ring current development during storm main phase. *J. Geophys. Res.*, 101:15,311–15,322.

Fok M.C., Moore T.E., and Delcourt D.C. (1999) Modeling of inner plasma sheet and ring current during substorms. *J. Geophys. Res.*, 104:14,557–14,569.

Fukushima N. (1976) Generalized theorem for no ground magnetic effect of vertical currents connected with Pedersen currents in the uniform-conductivity ionosphere. *Rep. Ionos. Space Res. Japan*, 22:35–40.

Gabrielse C., Angelopoulos V., Runov A., Frey H.U., McFadden J., Larson D.E., Glassmeier K.H., Mende S., Russell C.T., Apatenkov S., Murphy K.R., amd Rae I.J. (2009) Timing and localization of near-Earth tail and ionospheric signatures during a substorm onset. *J. Geophys. Res.*, 114, DOI 10.1029/2008JA013583.

Galeev A.A. and Zelenyj L.M. (1976) Tearing instability in plasma configurations. *Sov. Phys. JETP*, 43:1113.

Galeev A.A., Kuznetsova M.M., and Zelenyi L.M. (1986) Magnetopause stability threshold for patchy reconnection. *Space Sci. Rev.*, 44:1–41.

Ganushkina N.Y., Pulkkinen T.I., and Fritz T. (2005) Role of substorm-associated impulsive electric fields in the ring current development during storms. *Ann. Geophys.*, 23:579–591.

Ginzburg V.L. (1959) Radio astronomy and the origin of cosmic rays. In: Bracewell R.N. (ed) *Paris Symposium of Radio Astronomy, IAU Symposium no. 9*, Stanford University Press, Stanford, CA, pp. 589–594.

Giovanelli R.G. (1946) A theory of chromospheric flares. *Nature*, 158:81–82.

Gold T. (1959) Motions in the magnetosphere of the Earth. *J. Geophys. Res.*, 64:1219–1224.

Goldston R.J. and Rutherford P.H. (1995) *Introduction to Plasma Physics*. IOP Physics Publishing, Ltd., Bristol, U.K..

Gonzalez W.D. and Tsurutani B.T. (1987) Criteria of interplanetary parameters causing intense magnetic storms ($D_{st} < -100n_T$). *Planet. Space Sci.*, 35:1101–1109.

Gopalswamy N., Yashiro S., Michalek G., Stenborg G., Vourlidas A., Freeland S., and Howard R. (2009) The SOHO/LASCO CME catalog. *Earth Moon Planet*, 104:295–313.

Gordon W.E. (1958) Incoherent scattering of radio waves by free electrons with applications to space exploration by radar. In: *Proc. I.R.E.*, Vol. 46, pp. 1824–1829.

Gosling J.T. (1993) The solar flare myth. *J. Geophys. Res.*, 98:18,937–18,949.

Gosling J.T. and McComas D.J. (1987) Field line draping about fast coronal mass ejecta: a source of strong out-of-the-ecliptic interplanetary magnetic fields. *Geophys. Res. Lett.*, 14:335–358.

Gurnett D.A. (1995) Heliospheric radio emissions. *Space Sci. Rev.*, 72:243–254.

Gurnett D.A. and Anderson R.R. (1976) Electron plasma oscillations associated with type III radio bursts. *Science* 194:1159–1162.

Gurnett D.A. and Bhattacharjee A. (2004) *Introduction to Plasma Physics. With Space and Laboratory Applications*. Cambridge University Press, Cambridge, U.K..

Hairston M.R., Drake K.A., and Skoug R. (2005) Saturation of the ionospheric polar cap potential during the October-November 2003 superstorms. *J. Geophys. Res.*, 110, DOI 10.1029/2004JA010864.

Häkkinen L.V.T., Pulkkinen T.I., Nevanlinna H., Pirjola R.J., and Tanskanen E.I. (2002) Effects of induced currents on Dst and on magnetic variations at midlatitude stations. *J. Geophys. Res.*, 107, DOI 10.1029/2001JA900130.

Hasegawa H., Fujimoto M., Phan T.D., Rème H., Balogh A., Dunlop M.W., Hashimoto C., and TanDokoro R. (2004) Transport of solar wind into Earth's magnetosphere through Kelvin–Helmholtz vortices. *Nature*, 430:755–757.

Hau L.N., Wolf R.A., Voigt G.H., and Wu C.C. (1989) Steady state magnetic field configurations for the earth's magnetotail. *J. Geophys. Res.*, 94:1303–1316.

Henderson M.G., Reeves G.D., Skoug R., Thomsen M.F., Denton M.H., Mende S.B., Immel T.J., Brandt P.C., and Singer H.J. (2006) Magnetospheric and auroral activity during the 18 April 2002 sawtooth event. *J. Geophys. Res.*, 111, DOI 10.1029/2005JA011111.

Hidalgo M.A., Cid C., Viñas A.F., and Sequeiros J. (2002) A non-force-free approach to the topology of magnetic clouds in the solar wind. *J. Geophys. Res.*, 106, DOI 10.1029/2001JA900100.

Hill T.W., Dessler A.J., and Wolf R.A. (1976) Mercury and Mars: The role of ionospheric conductivity in the acceleration of magnetospheric particles. *Geophys. Res. Lett.*, 3:429–432.

Hodgson R. (1859) On a curious appearance seen in the Sun, 1859. Mon Not R Astron Soc XX:15.

de Hoffmann F. and Teller E. (1950) Magneto-hydrodynamic shocks. *Phys. Rev.*, 80:692–703.

Hones E.W. Jr., Baker D.N., Bame S.J., Feldman W.C., Gosling J.T., McComas D.J., Zwickl R.D., Slavin J.A., Smith E.J., and Tsurutani B.T. (1984) Structure of the magnetotail at 220 $R_E$ and its response to geomagnetic activity. *Geophys. Res. Lett.*, 11:5–7.

Horne R.B., Thorne R.M., Shprits Y.Y., Meredith N.P., Glauert S.A., Smith A.J., Kanekal S.G., Baker D.N., Engebretson M.J., Posch J.L., Spasojevic M., Inan U.S., Pickett J.S., and Decreau P.M.E. (2005) Wave acceleration of electrons in the Van Allen radiation belts. *Nature*, 437:227–230.

Hu Q., and Sonnerup B.U.O. (2002) Reconstruction of magnetic clouds in the solar wind: Orientation and configurations. *J. Geophys. Res.*, 107, DOI 10.1029/2001JA000293.

Hultqvist B., Øieroset M., Paschmann G., and Treumann R. (eds) (1999) *Magnetospheric Plasma Sources and Losses, Space Sciences Series of ISSI*, Vol. 6. Kluwer Academic Publishers, Dordrecht, Holland.

Huttunen K.E.J. abd Koskinen H.E.J. (2004) Importance of post-shock streams and sheath regions as drivers of intense magnetospheric storms and high-latitude activity. *Ann. Geophys.*, 22:1729–1738.

Huttunen K.E.J., Koskinen H.E.J., and Schwenn R. (2002) Variability of magnetospheric storms driven by different solar wind perturbations. *J. Geophys. Res.*, 107, DOI 10.1029/2001JA900171.

Huttunen K.E.J., Schwenn R., Bothmer V., and Koskinen H.E.J. (2005) Properties and geoeffectiveness of magnetic clouds in the rising, maximum and early declining phases of solar cycle 23. *Ann. Geophys.*, 23:625–641.

Huttunen K.E.J., Koskinen H.E.J., Karinen A., and Mursula K. (2006) Asymmetric development of magnetospheric storms during magnetic clouds and sheath regions. *Geophys. Res. Lett.*, 33, DOI 10.1029/2005GL024894.

Ieda A., Machida S., Mukai T., Saito Y., Yamamoto T., Nishida A., Terasawa T., and Kokubun S. (1998) Statistical analysis of the plasmoid evolution with Geotail observations. *J. Geophys. Res.*, 103:4453–8851.

Iijima T. and Potemra T.A. (1976) Field-aligned currents in the dayside cusp observed by Triad. *J. Geophys. Res.*, 81:5971–5979.

Jackman C.H., Marsh D.R., Vitt F.M., Garcia R.R., Fleming E.L., Labow G.J., Randall C.E., López-Puertas M., Funke B., von Clarmann T., and Stiller G.P. (2008) Short- and medium-term atmospheric constituent effects of very large solar proton events. *Atmos. Chem. Phys.*, 8:765–785.

Jackson J.D. (1999) *Classical Electrodynamics*, 3rd edn. John Wiley & Sons, New York, NY.

Janhunen P. and Olsson A. (1998) The current-voltage relationship revisited: exact and approximate formulas wih almost general validity for hot magnetospheric electrons for bi-Maxwellian and kappa distributions. *Ann. Geophys.*, 16:292–297.

Johnson C.Y. (1969) Ion and neutral composition of the ionosphere. Annals of the IQSY 5:197–213.

Jordanova V.K., Miyoshi Y.S., Zaharia S., Thomsen M.F., Reeves G.D., Evans D.S., Mouikis C.G., and Fennell J.F. (2006) Kinetic simulations of ring current evolution during the Geospace Environment Modeling challenge events. *J. Geophys. Res.*, 111, DOI 10.1029/2006JA011644.

Jordanova V.K., Albert J., and Miyoshi Y. (2008) Relativistic electron precipitation by EMIC waves from self-consistent global simulations. *J. Geophys. Res.*, 113, DOI 10.1029/2008JA013239.

Kamide Y. (1992) Is substorm occurrence a necessary condition for a magnetic storm? *J. Geomagn. Geoelectr.*, 44:109–117.

Kan J. (1993) A global magnetosphere-ionosphere coupling model of substorms. *J. Geophys. Res.*, 98:17,263–17,272.

Kane S.R. (1974) Impulsive (flash) phase of solar flares: Hard X-ray, microwave, EUV and optical observations. In: Newkirk G Jr. (ed) *Coronal Disturbances*, D. Reidel, Boston, MA, Proceedings of IAU symposium no. 57, pp. 105–141.

Kauristie K., Pulkkinen T.I., Pellinen R.J., and Opgenoorth H.J. (1996) What can we tell about the AE-index from a single meridional magnetometer chain? *Ann. Geophys.*, 14:1177–1185.

Kauristie K., Sergeev V.A., Kubyshkina M., Pulkkinen T., Angelopoulos V., Phan T., Lin R.P., and Slavin J.A. (2000) Ionospheric signatures of transient plasma sheet flows. *J. Geophys. Res.*, 105:10,677–10,690.

Kavanagh L.D. Jr., Freeman J.W. Jr., and Chen A.J. (1968) Plasma flow in the magnetosphere. *J. Geophys. Res.*, 73:5511–5519.

Kelley M.C. (1989) *The Earth's Ionosphere: Plasma physics and electrodynamics*. Academic Press, San Diego, CA.

Kennel C.F. and Engelmann F. (1966) Velocity space diffusion from weak plasma turbulence in a magnetic field. *Phys. Fluids*, 9:2377–2388.

Kennel C.F. and Petschek H.E. (1966) Limit on stably trapped particle fluxes. *J. Geophys. Res.*, 71:1–28.

Kindel J.M. and Kennel C.F. (1971) Topside current instabilities. *J. Geophys. Res.*, 76:3055–3078.

Kivelson M.G. and Ridley A.J. (2008) Saturation of the polar cap potential: Inference from Alfvén wing arguments. *J. Geophys. Res.*, 113, DOI 10.1029/2007JA012302.

Kivelson M.G. and Russell C.T. (1995) *Introduction to Space Physics*. Cambridge University Press, Cambridge, U.K..

Knight S. (1973) Parallel electric fields. *Planet. Space Sci.*, 21:741–750.

Knipp D.J., Emery B.A., Engebretson M., Li X., McAllister A.H., Mukai T., Kokubun S., Reeves G.D., Evans D., Obara T., Pi X., Rosenberg T., Weatherwax A., McHarg M.G., Chun F., Mosely K., Codrescu M., Lanzerotti L., Rich F.J., Sharber J., and Wilkinson P. (1998) An overview of the early November 1993 geomagnetic storm. *J. Geophys. Res.*, 103:26,197–26,220.

Kopp G., Lawrence G., and Rottman G. (2005) The Total Irradiance Monitor (TIM): Science results. *Solar Phys.*, 230:129–139.

Koskinen H.E.J. and Pulkkinen T.I. (1995) Midnight velocity shear zone and the concept of Harang discontinuity. *J. Geophys. Res.*, 100:9539–9547.

Koskinen H.E.J. and Tanskanen E.I. (2002) Magnetospheric energy budget and the epsilon parameter. *J. Geophys. Res.*, 107, DOI 10.1029/2002JA009283.

Koskinen H.E.J., Lopez R.E., Pellinen R.J., Pulkkinen T.I., Baker D.N., and Bösinger T. (1993) Pseudo-breakup and substorm growth phase in the ionosphere and magnetosphere. *J. Geophys. Res.*, 98:5801–5813.

Kosovichev A.G., Schou J., Scherrer P.H., Bogart R.S., Bush R.I., Hoeksema J.T., Aloise J., Bacon L., Burnette A., DeForest C., Giles P.M., Leibrand K., Nigam R., Rubin M., Scott K., Williams S.D., Basu S., Christensen-Dalsgaard J., Dappen W., Rhodes E.J., Duvall T.L., Howe R., Thompson M.J., Gough D.O., Sekii T., Toomre J., Tarbell T.D., Title A., Mathur D., Morrison M., Saba J.L.R., Wolfson C.J., Zayer I., and Milford PN (1997) Structure and rotation of the solar interior: Initial results from the MDI medium-l program. *Solar Phys.*, 170:43–61.

Krall J. (2007) Are all coronal mass ejections hollow flux ropes? *Astrophys. J.*, 657:559–566.

Krall N.A. and Trivelpiece A.W. (1973) *Principles of Plasma Physics*. McGraw-Hill, New York, NY.

Kulsrud R.M. (2005) *Plasma Physics for Astrophysics*. Princeton University Press, Princeton, NJ.

Kuznetsova M.M., Hesse M., Rastätter L., Taktakishvili A., Toth G., De Zeeuw D.L., Ridley A., and Gombosi T.I. (2007) Multiscale modeling of magnetospheric reconnection. *J. Geophys. Res.*, 112, DOI 10.1029/2007A012316.

Laitinen T.V., Janhunen P., Pulkkinen T.I., Palmroth M., and Koskinen H.E.J. (2006) On the characterization of magnetic reconnection in global MHD simulations. *Ann. Geophys.*, 24:3059–3069.

Landau L.D. (1946) On the vibrations of the electronic plasma. *J. Phys. (U.S.S.R.)*, 10:25–34.

Lang K.R. (2000) *The Sun from Space*. Springer, Berlin, Germany.

Langel R.A. and Estes R.H. (1985) Large-scale, near-field magnetic fields from external sources and the corresponding induced internal field. *J. Geophys. Res.*, 90:2487–2494.

Lavraud B., Thomsen M.F., Borovsky J.E., Denton M.H., and Pulkkinen T.I. (2006) Magnetosphere preconditioning under northward IMF: Evidence from the study of coronal mass ejection and corotating interaction region geoeffectiveness. *J. Geophys. Res.*, 111, DOI 10.1029/2005JA011566.

Lepping R.P., Jones J.A., and Burlaga L.F. (1990) Magnetic field structure of interplanetary magnetic clouds at 1 AU. *J. Geophys. Res.*, 95:11,957–11,965.

Li X., Roth I., Temerin M., Wygant J.R., Hudson M.K., and Blake J.B. (1993) Simulation of the prompt energization and transport of radiation belt particles during the March 24, 1991 SSC. *Geophys. Res. Lett.*, 20:2423–2426.

Li X., Baker D.N., Temerin M., Larson D., Lin R.P., Reeves G.D., Looper M., Kanekal S.G., and Mewaldt R.A. (1997) Are energetic electrons in the solar wind the source of the outer radiation belt? *Geophys. Res. Lett.*, 24:923–926.

Lilensten J., Belehaki A., Messerotti M., Vainio R., Waterman J., and Poedts S. (eds) (2008) *Developing the scientific basis for monitoring, modelling and predicting Space Weather*. Final Report of EU COST Action 724, Opoce, Brussels, Belgium.

Lindemann F.A. (1919) Note on the theory of magnetic storms. *Philos. Mag.*, 38:669.

Liou K., Meng C.I., Lui A.T.Y., Newell P.T., and Wing S. (2002) Magnetic dipolarization with substorm expansion onset. *J. Geophys. Res.*, 107, DOI 10.1029/2001JA000179.

Lites B.W., Kubo M., Socas-Navarro H., Berger T., Frank Z., Shine R., Tarbell T., Title A., Ichimoto K., Katsukawa Y., Tsuneta S., Suematsu Y., Shimizu T., and Nagata S. (2008) The horziontal magnetic flux of the quiet-Sun internetwork as observed with the Hinode spectro-polarimeter. *Astrophys. J.*, 672:1237–1253.

Liu W.W. (1997) Physics of the explosive growth phase: Ballooning instability revisited. *J. Geophys. Res.*, 102:4927–4931.

Lopez R.E., Bruntz R., Mitchell E.J., Wiltberger M., Lyon J., and Merkine V.G. (2010) The role of magnetosheath force balance in regulating the dayside reconnection potential. *J. Geophys. Res.*, 115, DOI 10.1029/2009JA014597.

Lu G., Baker D.N., McPherron R.L., Farrugia J., Lummerzheim D., Ruohoniemi J.M., Rich F.J., Evans D.S., Lepping R.P., Brittnacher M., Li X., Greenwald R., Sofko G., Villain J., Lester M., Thayer J., Moretto T., Milling D., Troshichev O., Zaitzev A., Odintzov V., Makarov G., and Hayashi K. (1998) Global energy deposition during the January 1997 magnetic cloud event. *J. Geophys. Res.*, 103:11,685–11,694.

Lui A.T.Y. (1996) Current disruption in the Earth's magnetosphere: Observations and models. *J. Geophys. Res.*, 101:13,067–13,088.

Lui A.T.Y., Lopez R.E., Krimigis S.M., Zanetti L.J., and Potemra T.A. (1988) A case study of magnetotail current sheet disruption and diversion. *Geophys. Res. Lett.*, 15:721–724.

Lui A.T.Y., Zheng Y., Rème H., Dunlop M.W., Gustafsson G., and Owen C.J. (2007) Breakdown of the frozen-in condition in the Earth's magnetotail. *J. Geophys. Res.*, 112, DOI 10.1029/2006JA012000.

Lui A.T.Y., Angelopoulos V., LeContel O., Frey H., Donovan E., Sibeck D.G., Liu W., Auster H.U., Larson D., Li X., Nosé M., and Fillingim M.O. (2008) Determination of the substorm initiation region from a major conjunction interval of THEMIS satellites. *J. Geophys. Res.*, 113, DOI 10.1029/2008JA013424.

Lundin R. and Evans D.S. (1985) Boundary layer plasmas as a source for high-latitude, early afternoon, auroral arcs. *Planet. Space Sci.*, 33:1389–1406.

Lundquist S. (1950) Magneto-hydrostatic fields. *Ark. Fys.*, 2:316–365.
Lyons L.R. (1995) A new theory for magnetospheric substorms. *J. Geophys. Res.*, 100:19,069–19,081.
Lyons L.R. and Speiser T.W. (1982) Evidence of current sheet acceleration in the geomagnetic tail. *J. Geophys. Res.*, 87:2276–2286.
Lyons L.R. and Williams D.J. (1984) *Quantitative Aspects of Magnetospheric Physics*. D. Reidel, Dordrecht, NL.
Lysak R.L. (1991) Feedback instability of the ionospheric resonant cavity. *J. Geophys. Res.*, 96:1553–1568.
Lysak R.L. and Lotko W. (1996) On the kinetic dispersion relation for shear Alfvén waves. *J. Geophys. Res.*, 101:5085–5094.
Malyshkin L.M. (2008) A model of Hall reconnection. *Phys. Rev. Lett.*, 101, DOI 10.1103/PhysRevLett.101.225001.
Marklund G.T., Karlsson T., Blomberg L.G., Lindqvist P.A., Fälthammar C.G., Johnson M.L., Murphree J.S., Andersson L., Eliasson L., Opgenoorth H.J., and Zanetti L.J. (1998) Observations of the electric field fine structure associated with the westward traveling surge and large-scale auroral spirals. *J. Geophys. Res.*, 103:4125–4144.
Marubashi K. (1997) Interplanetary magnetic flux ropes and solar filaments. In: Crooker N., Josely J.A., Feynman J. (eds) *Coronal Mass Ejections, American Geophysical Union*, Washington, DC, Geophysical Monograph, Vol. 99, pp. 147–156.
Mayaud P.N. (1980) *Derivation, Meaning, and Use of Geomagnetic Indices, Geophysical Monograph*, Vol. 22. American Geophysical Union, Washington, DC.
McPherron R.L., Russell C.T., and Aubry M.A. (1973) Satellite studies of magnetospheric substorms on August 15, 1968; 9. Phenomenological model for substorms. *J. Geophys. Res.*, 78:3131–3149.
Meredith N.P., Thorne R.M., Horne R.B., Summers D., Fraser B.J., and Anderson R.R. (2003) Statistical analysis of relativistic electron energies for cyclotron resonance with EMIC waves observed on CRRES. *J. Geophys. Res.*, 108, DOI 10.1029/2002JA009700.
Mozer F.S. and Retinò A. (2007) Quantitative estimates of magnetic field reconnection properties from electric and magnetic field measurements. *J. Geophys. Res.*, 112, DOI 10.1029/2007JA012406.
Nagai T., Shinohara I., Fujimoto M., Hoshino M., Saito Y., Machida S., and Mukai T. (2001) Geotail observations of the hall current system: Evidence of magnetic reconnection in the magnetotail. *J. Geophys. Res.*, 106:25,929–25,949.
Nakano S., Ueno G., Ebihara Y., Fok M.C., Ohtani S., Brandt P.C., Mitchell D.G., Keika K., and Higuchi T. (2008) A method for estimating the ring current structure and electric potential distribution using energetic neutral atom data assimilation. *J. Geophys. Res.*, 113, DOI 10.1029/2006JA011853.
Neubauer F.M. (1980) Nonlinear standing Alfvén wave current system at Io: Theory. *J. Geophys. Res.*, 85:1171–1178.
Northrop T.G. (1963) *The Adiabatic Motion of Charged Particles*. Interscience Publishers, John Wiley & Sons, New York.
Nygrén T. (1996) *Introduction to Incoherent Scatter Measurements*. Invers Publications, Sodankylä, Finland.
O'Brien T.P. and McPherron R.L. (2000) An empirical phase analysis of ring current dynamics: Solar wind control of injection and decay. *J. Geophys. Res.*, 105:7707–7719.
Ohtani S., Brandt P.C., Mitchell D.G., Singer H., Nosé M., Reeves G.D., and Mende S.B. (2005) Storm–substorm relationship: Variations of the hydrogen and oxygen neutral atom intensities during storm-time substorms. *J. Geophys. Res.*, 110, DOI 10.1029/2004JA010954.
Øieroset M., Phan T.D., Fujimoto M., Lin R.P., and Lepping R.P. (2001) In situ detection of collisionless reconnection in the Earth's magnetotail. *Nature* 412:414–417.
Palmroth M., Pulkkinen T.I., Janhunen P., and Wu C.C. (2003) Stormtime energy transfer in global MHD simulation. *J. Geophys. Res.*, 108, DOI 10.1029/2002JA009446.
Papadopoulos K., Goodrich C., Wiltberger M., Lopez R., and Lyon J.G. (1999) The physics of substorms as revealed by the ISTP. *Phys. Chem. Earth, Part C* 24:189–202.
Parker E.N. (1957) Sweet's mechanism for merging magnetic fields in conducting fluids. *J. Geophys. Res.*, 62:509–520.
Parker E.N. (1958) Dynamics of the interplanetary gas and magnetic fields. *Astrophys. J.*, 128:664–676.

Parks G.K. (2003) *Physics of Space Plasmas: An Introduction*, 2nd edn. Westview Press, Boulder, CO.

Pellinen R.J. and Heikkila W.J. (1984) Inductive electric fields in the magnetotail and their relation to auroral and substorm phenomena. *Space Sci. Rev.*, 37:1–61.

Perez J.D., Zhang X.X., Brandt P.C., Mitchell D.G., Jahn J.M., and Pollock C.J. (2004) Dynamics of ring current ions as obtained from IMAGE/HENA and IMAGE/MENA images. *J. Geophys. Res.*, 109, DOI 10.1029/2003JA010421.

Perreault P. and Akasofu S.I. (1978) A study of geomagnetic storms. *Geophys. J. R. Astron. Soc.*, 54:547–573.

Peterson L.E. and Winckler J.R. (1959) Gamma-ray bursts from a solar flare. *J. Geophys. Res.*, 64:697–707.

Petschek H.E. (1964) Magnetic field annihilation. In: Hess W.N. (ed) AAS/NASA *Symposium on the Physics of Solar Flares*, Washington, DC, NASA SP-50, pp. 425–439.

Phillips J.L., Bame S.J., and Barnes A. (1995) Ulysses solar wind plasma observations from pole to pole. *Geophys. Res. Lett.*, 22:3301–3304.

Pirjola R. and Viljanen A. (1998) Complex image method for calculating electric and magnetic fields produced by an auroral electrojet of finite length. *Ann. Geophys.*, 16:1434–1444.

Prescott G.B. (1860) *History, Theory, and Practice of the Electric Telegraph*. Ticknor and Fields, Boston, MA.

Priest E.R. and Forbes T.G. (1986) New models for fast steady-state magnetic reconnection. *J. Geophys. Res.*, 91:5579–5588.

Priest E.R., Hood A.W., and Anzer U. (1989) A twisted flux-tube model for solar prominences. I. General properties. *Astrophys. J.*, 334:1010–1025.

Pritchett P.L. (2008) Collisionless magnetic reconnection in an asymmetric current sheet. *J. Geophys. Res.*, 113, DOI 10.1029/2007JA012930.

Pritchett P.L. and Coroniti F.V. (2004) Three-dimensional collisionless magnetic reconnectionin the presence of a guide field. *J. Geophys. Res.*, 109, DOI 10.1029/2003JA009999.

Pulkkinen A., Pirjola R., Boteler D., Viljanen A., and Yegorov I. (2001) Modelling of space weather effects on pipelines. *J. Appl. Geophys.*, 48:233–256.

Pulkkinen P. and Tuominen I. (1998) Velocity structures from sunspot statistics in cycles 10 to 22. *Astron. Astrophys.*, 332:748–754.

Pulkkinen T.I., Koskinen H.E.J., Kauristie K., Palmroth M., Reeves G.D., Donovan E., Singer H.J., Slavin J.A., Russell C.T., and Yumoto Y. (2004) Storm-substorm coupling: Signatures of stormtime substorms. In: Zelenyi L.M., Geller M.A., Allen JH (eds) *Auroral Phenomena and Solar-Terrestrial Relations. Proceedings of the Conference in Memory of Yuri Galperin, Boulder*, CO, CAWSES Handbook-1, pp. 309–316.

Pulkkinen T.I., Partamies N., Huttunen K.E.J., Reeves G.D., and Koskinen H.E.J. (2007a) Differences in geomagnetic storms driven by magnetic clouds and ICME sheath regions. *Geophys. Res. Lett.*, 34, DOI 10.1029/2006GL027775.

Pulkkinen T.I., Partamies N., McPherron R.L., Henderson M., Reeves G.D., Thomsen M.F., and Singer H.J. (2007b) Comparative statistical analysis of storm time activations and sawtooth events. *J. Geophys. Res.*, 112, DOI 10.1029/2006JA012024.

Pytte T., McPherron R.L., Hones E.W. Jr., and West H.I. Jr. (1978) Multiple satellite studies of magnetospheric substorms: Distinction between polar magnetic substorms and convection-driven negative bays. *J. Geophys. Res.*, 83:663–679.

Rairden R.L., Frank L.A., and Craven J.D. (1986) Geocoronal imaging with Dynamics Explorer. *J. Geophys. Res.*, 91:13,613–13,630.

Reeves G.D., McAdams K.L., and Friedel R.H.W. (2003) Acceleration and loss of relativistic electrons during geomagnetic storms. *Geophys. Res. Lett.*, 30, DOI 10.1029/2002GL016513.

Reiff P.H., Spiro R.W., and Hill T.W. (1981) Dependence of polar cap potetial drop on interplanetary parameters. *J. Geophys. Res.*, 86:7639–7648.

Ridley A.J. (2007) Alfvén wings at Earth's magnetosphere under strong interplanetary magnetic fields. *Ann. Geophys.*, 25:533–542.

Riley P., Linker J.A., Lionello R., Mikic Z., Odstrcil D., Hidalgo M.A., Cid C., Hu Q., Lepping R.P., Lynch B.J., and Rees A. (2004) Fitting flux ropes to a global MHD solution: a comparison of techniques. *J. Atmos. Sol. Ter. Phys.*, 66:1321–1331.

Robbrecht E., Berghmans D., and Van der Linden R.A.M. (2009) Automated LASCO CME catalog for solar cycle 23: Are CMEs scale invariant? *Astrophys. J.*, 691:1222–1234.

Roederer J.G. (1970) *Dynamics of Geomagnetically Trapped Radiation*. Springer, Berlin, Germany.

Rosenqvist L., Buchert S., Opgenoorth H., Vaivads A., and Lu G. (2006) Magnetospheric energy budget during huge geomagnetic activity using cluster and ground-based data. *J. Geophys. Res.*, 111, DOI 10.1029/2006JA011608.

Rostoker G., Skone S., and Baker D.N. (1998) On the origin of relativistic electrons in the magnetosphere associated with some geomagnetic storms. *Geophys. Res. Lett.*, 25:3701–3704.

Roux A. (1985) Generation of field-aligned current structures at substorm onsets. In: Rolfe E., Battrick B. (eds) *Proceedings of ESA Workshop on Future Missions in Solar, Heliospheric and Space Plasma Physics*, Noordwijk, The Netherlands, ESA Spec. Publ., vol SP-235, p 151.

Runov A., Nakamura R., Baumjohann W., Treumann R., Zhang T.L., Volwerk M., Vörös Z., Balogh A., Glassmeier K.H., and Klecker B. (2003) Current sheet structure near magnetic X-line observed by Cluster. *Geophys. Res. Lett.*, 30, DOI 10.1029/2002GL016730.

Scherer K., Fichtner H., Heber B., and Mall U. (eds) (2005) *Space Weather: The Physics Behind a Slogan (Lecture Notes in Physics)*. Springer, Berlin, Germany.

Schindler K. (1974) A theory of the substorm mechanism. *J. Geophys. Res.*, 79:2803–2810.

Schlichenmaier R. and Stix M. (1995) The phase of the radial mean field in the solar dynamo. *Astron. Astrophys.*, 302:264–270.

Schmidt G. (1979) *Physics of High Temperature Plasmas*, 2nd edn. Academic Press, New York, NY.

Schrijver C.J. (2001) The coronae of the Sun and solar-type stars. In: García López R.J., Rebolo R., Zapatero Osorio M.R. (eds) *The 11th Cool Stars, Stellar Systems and the Sun, Astronomical Society of the Pacific, San Francisco*, CA, ASP Conference Series, vol 223, pp. 131–144.

Schulz M. (1996) Canonical coordinates for radiation-belt modeling. In: Lemaire J.F., Heynderickx D., Baker D.N. (eds) *Radiation Belts: Models and Standards, American Geophysical Union*, Washington, DC, Geophysical Monograph, Vol. 97, pp. 153–160.

Schulz M. and Lanzerotti L.J. (1974) Particle Diffusion in the Radiation Belts, Physics and Chemistry in Space, Vol. 7. Springer, New York, NY.

Schwenn R., dal Lago A., Huttunen E., and Gonzalez W.D. (2005) The association of coronal mass ejections with their effects near the earth. *Ann. Geophys.*, 23:1033–1059.

Sckopke N. (1966) A general relation between the energy of trapped particles and the disturbance field near the Earth. *J. Geophys. Res.*, 71:3125–3130.

Sergeev V.A. and Lennartsson W. (1988) Plasma sheet at $X \sim -20R_E$ during steady magnetospheric convection. *Planet. Space Sci.*, 36:353–370.

Sergeev V.A., Pulkkinen T.I., Pellinen R.J., and Tsyganenko N.A. (1994) Hybrid state of the tail magnetic configuration during steady convection events. *J. Geophys. Res.*, 99:23,571–23,582.

Sergeev V.A., Pellinen R.J., and Pulkkinen T.I. (1996) Steady magnetospheric convection: A review of recent results. *Space Sci. Rev.*, 75:551–604.

Shelley E.G., Johnson R.G., and Sharp R.D. (1972) Satellite observations of heavy ions during a geomagnetic storm. *J. Geophys. Res.*, 77:6104–6110.

Shepherd S.G. (2007) Polar cap potential saturation: Observations, theory, and modeling. *J. Atm. Sol. Ter. Phys.*, 69:234–248.

Shibata K. (1999) Evidence of magnetic reconnection in solar flares and a unified model of flares. *Astrophys. Space. Sci.*, 264:129–144.

Shibata K., Masuda S., Shimojo M., Hara H., Yokoyama T., Tsuneta S., Kosugi T., and Ogawara Y. (1995) Hot-plasma ejections associated with compact-loop solar flares. *Astrophys. J.*, 451:L83–L85.

Shiokawa K., Baumjohann W., Haerendel G., Paschmann G., Fennell J.F., Friis-Christensen E., Lühr H., Reeves G.D., Russell C.T., Sutcliffe P.R., and Takahashi K. (1998) High-speed ion flow, substorm current wedge, and multiple Pi2 pulsations. *J. Geophys. Res.*, 103:4491–4507.

Shiota D., Isobe H., Chen P.F., Yamamoto T.T., Sakajiri T., and Shibata K. (2005) Self-consistent magnetohydrodynamic modeling of a coronal mass ejection, coronal dimming, and a giant cusp-shaped arcade formation. *Astrophys. J.*, 634:663–678.

Sibeck D.G., Siscoe G.L., Slavin J.A., Smith E.J., Bame S.J., and Scarf F.L. (1984) Magnetotail flux ropes. *Geophys. Res. Lett.*, 11:1090–1093.

Siscoe G.L., and Cummings W.D. (1969) On the cause of geomagnetic bays. *Planet. Space Sci.* 17:1795–1802.

Siscoe G. and Schwenn R. (2006) CME disturbance forecasting. *Space Sci. Rev.*, 123:453–470.

Siscoe G.L., Erickson G.M., Sonnerup B.U.O., Maynard N.C., Schoendorf J.A., Siebert K.D., Weimer D.R., White W.W., and Wilson G.R. (2002a) Hill model of transpolar potential saturation: Comparison with MHD simulations. *J. Geophys. Res.*, 107, DOI 10.1029/2001JA000109.

Siscoe G.L., Grooker N.U., and Siebert K.D. (2002b) Transpolar potential saturation: Roles of region 1 current system and solar wind ram pressure. *J. Geophys. Res.*, 107, DOI 10.1029/2001JA009176.

Slavin J.A., Lepping R.P., Gjerloev J., Fairfield D.H., Hesse M., Owen C.J., Moldwin M.B., Nagai T., Ieda A., and Mukai T. (2003) Geotail observations of magnetic flux ropes in the plasma sheet. *J. Geophys. Res.*, 108, DOI 10.1029/2002JA009557.

Smith P.H., Bewtra N.K., and Hoffman R.A. (1981) Inference of the ring current ion composition by means of charge exchange decay. *J. Geophys. Res.*, 86:3470–3480.

Song P., Singer H.J., and Siscoe G.L. (eds) (2001) *Space Weather, Geophysical Monograph*, vol 125. American Geophysical Union, Washington, DC, USA.

Sonnerup B.U.Ö. (1970) Magnetic-field reconnection in a highly conducting incompressible fluid. *J. Plasma Phys.*, 4:161–174.

Spiro R.W., Reiff P.H., and Maher L.J. Jr. (1982) Precipitating electron energy flux and auroral zone conductances – an empirical model. *J. Geophys. Res.*, 87:8215–8227.

Stern D.P. (1984) Energetics of the magnetosphere. *Space Sci. Rev.*, 39:193–213.

Stix M. (2002) *The Sun: An Introduction*, 2nd edn. Springer, Berlin, Germany.

Sturrock P.A. (1994) *Plasma Physics, An introduction to the theory of astrophysical, geophysical & laboratory plasmas*. Cambridge University Press, Cambridge, U.K..

Summers D. and Thorne R.M. (2003) Relativistic pitch-angle scattering by electromagnetic ion cyclotron waves during geomagnetic storms. *J. Geophys. Res.*, 108, DOI 10.1029/2002JA009489.

Summers D., Thorne R.M., and Xiao F. (1998) Relativistic theory of wave-particle resonant diffusion with application to electron acceleration in the magnetosphere. *J. Geophys. Res.*, 103:20,487–20,500.

Sweet P.A. (1958) The neutral point theory of solar flares. In: Lehnert B. (ed) *Electromagnetic Phenomena in Cosmical Physics*, Gambridge University Press, Gambridge, U.K., pp. 123–134.

Takahashi K., Zanetti L.J., Lopez R.E., McEntire R.W., Potemra T.A., and Yumoto K. (1987) Disruption of the magnetotail current sheet observed by AMPTE/CCE. *Geophys. Res. Lett.*, 14:1019–1022.

Tanskanen E.I., Viljanen A., Pulkkinen T.I., Pirjola R., Häkkinen L., Pulkkinen A., and Amm O. (2001) At substorm onset, 40% of the AL comes from underground. *J. Geophys. Res.*, 106:13,119–13,134.

Tanskanen E.I., Koskinen H.E.J., Pulkkinen T.I., Slavin J.A., and Ogilvie K. (2002) Dissipation to the joule heating: Isolated and stormtime substorms. *Adv. Space. Res.*, 30:2305–2311.

Tanskanen E.I., Slavin J.A., Tanskanen A.J., Viljanen A., Pulkkinen T.I., Koskinen H.E.J., Pulkkinen A., and Eastwood J. (2005) Magnetospheric substorms are strongly modulated by interplanetary high-speed streams. *J. Geophys. Res.*, 32, DOI 10.1029/2005GL023318.

Tousey R. (1973) The solar corona. *Space Research*, XIII:713–730.

Treumann R.A. and Baumjohann W. (1996) *Advanced Space Plasma Physics*. Imperial College Press, London, U.K..

Troshichev O.A., Andrezen V.G., Vennestrøm S., and Friis-Christensen E. (1988) Magnetic activity in the polar cap – A new index. *Planet. Space Sci.*, 36:1095–1102.

Tsurutani B.T., Gonzalez W.D., Tang F., Akasofu S.I., and Smith E.J. (1988) Origin of interplanetary southward magnetic fields responsible for major magnetic storms near solar maximum (1978–1979). *J. Geophys. Res.*, 93:8519–8531.

Tsurutani B.T., Gonzalez W.D., Kamide Y., and Arballo J.K. (eds) (1997) *Magnetic Storms, Geophysical Monograph*, Vol. 98. American Geophysical Union, Washington, DC, USA.

Tsyganenko N.A. (1989) Magnetospheric magnetic field model with warped tail current sheet. *Planet. Space Sci.*, 37:5–20.

Turner N.E., Baker D.N., Pulkkinen T.I., and McPherron R.L. (2000) Evaluation of the tail current contribution to *Dst*. *J. Geophys. Res.*, 105:5431–5439.

Turner N.E., Baker D.N., Pulkkinen T.I., Roeder J.L., Fennell J.F., and Jordanova V.K. (2001) Energy content in the storm time ring current. *J. Geophys. Res.*, 106:19,149–19,156.

Weiss L.A., Reiff P.H., Moses J.J., Heelis R.A., and Moore D.B. (1992) Energy dissipation in substorms. In: *Substorms 1*, Noordwijk, The Netherlands, ESA Spec. Publ., vol SP-335, pp. 309–317.

Yoon P.H. and Lui A.T.Y. (2006) Quasi-linear theory of anomalous resistivity. *J. Geophys. Res.* 111, DOI 10.1029/2005JA011482.

Zhang J., Richardson I.G., Webb D.F., Goplaswamy N., Huttunen E., Kasper J.C., Nitta N.V., Poomvises W., Thompson B.J., Wu C.C., Yasihro S., and Zhukov A.N. (2007) Solar and interplanetary sources of major geomagnetic storms ($D_{st} < -100n_T$) during 1996–2005. *J. Geophys. Res.*, 112, DOI 10.1029/2007JA012321.

# Index